Relativistic Effects in Atoms, Molecules, and Solids

NATO Advanced Science Institutes Series

A series of edited volumes comprising multifaceted studies of contemporary scientific issues by some of the best scientific minds in the world, assembled in cooperation with NATO Scientific Affairs Division.

This series is published by an international board of publishers in conjunction with NATO Scientific Affairs Division

A Life Sciences
B Physics
Plenum Publishing Corporation
New York and London

C Mathematical and Physical Sciences
D. Reidel Publishing Company
Dordrecht, Boston, and London

D Behavioral and Social Sciences
E Applied Sciences
Martinus Nijhoff Publishers
The Hague, Boston, and London

F Computer and Systems Sciences
G Ecological Sciences
Springer Verlag
Heidelberg, Berlin, and New York

Recent Volumes in Series B: Physics

Volume 84 —Physical Processes in Laser–Materials Interactions
edited by M. Bertolotti

Volume 85 —Fundamental Interactions: *Cargèse 1981*
edited by Maurice Lévy, Jean-Louis Basdevant, David Speiser, Jacques Weyers, Maurice Jacob, and Raymond Gastmans

Volume 86 —Surface Mobilities on Solid Materials
Fundamental Concepts and Applications
edited by Vu Thien Binh

Volume 87 —Relativistic Effects in Atoms, Molecules, and Solids
edited by G. L. Malli

Volume 88 —Collective Excitations in Solids
edited by Baldassare Di Bartolo

Volume 89a—Electrical Breakdown and Discharges in Gases: Fundamental Processes and Breakdown
edited by Erich E. Kunhardt and Lawrence H. Luesen

Volume 89b—Electrical Breakdown and Discharges in Gases: Macroscopic Processes and Discharges
edited by Erich E. Kunhardt and Lawrence H. Luessen

Relativistic Effects in Atoms, Molecules, and Solids

Edited by

G. L. Malli

Department of Chemistry and Theoretical Sciences Institute
Simon Fraser University
Burnaby, British Columbia, Canada

Plenum Press
New York and London
Published in cooperation with NATO Scientific Affairs Division

Proceedings of a NATO Advanced Study Institute on
Relativistic Effects in Atoms, Molecules, and Solids,
held August 10–21, 1981,
at the University of British Columbia, Vancouver, Canada

Library of Congress Cataloging in Publication Data

NATO Advanced Study Institute on Relativistic Effects in Atoms, Molecules, and Solids (1981: University of British Columbia)
Relativistic effects in atoms, molecules, and solids.

(NATO advanced science institutes series. Series B., Physics; v. 87)
Proceedings of a NATO Advanced Study Institute on Relativistic Effects in Atoms, Molecules, and Solids, held August 10–21, 1981, at the University of British Columbia.
Published in cooperation with NATO Scientific Affairs Division.
Bibliography: p.
Includes index.
1. Atomic structure—Congresses. 2. Molecular structure—Congresses. 3. Solid state physics—Congresses. 4. Relativity (Physics)—Congresses. I. Malli, G. L. II. North Atlantic Treaty Organization. Scientific Affairs Division. III. Title. IV. Series.
QC172.N37 1981 539′.1 82-16714
ISBN 0-306-41169-5

A Division of Plenum Publishing Corporation
233 Spring Street, New York, N.Y. 10013

Printed in the United States of America

"The general theory of quantum mechanics is now almost complete, the imperfections that still remain being in connection with the exact fitting in of the theory with relativity ideas. These give rise to difficulties only when high-speed particles are involved, and are therefore of no importance in the consideration of atomic and molecular structure and ordinary chemical reactions, in which it is, indeed, usually sufficiently accurate if one neglects relativity variation of mass with velocity and assumes only Coulomb forces between the various electrons and atomic nuclei. The underlying physical laws necessary for the mathematical theory of a large part of physics and the whole of chemistry are thus completely known, and the difficulty is only that the exact application of these laws leads to equations much too complicated to be soluble. It therefore becomes desirable that approximate practical methods of applying quantum mechanics should be developed, which can lead to an explanation of the main features of complex atomic systems without too much computation."

P.A.M. Dirac
Proceedings of the Royal Society
Volume 123A, p. 714, 1929

PREFACE

The NATO Advanced Study Institute (ASI) on "Relativistic Effects in Atoms, Molecules and Solids" cosponsored by Simon Fraser University (SFU) and Natural Sciences and Engineering Research Council of Canada (NSERC) was held at the University of British Columbia (UBC), Vancouver, Canada from August 10th until August 21st, 1981. A total of 77 lecturers and students with diverse backgrounds in Chemistry, Physics, Mathematics and various interdisciplinary subjects attended the ASI.

In the proposal submitted to NATO for financial support for this ASI, it was suggested that recent impressive experimental developments coupled with the availability of sophisticated computer technology for detailed investigation of the relativistic structure of atoms, molecules and solids would provide an excellent testing ground for the validity and accuracy of the theoretical treatment of the relativistic many-electron systems involving medium and heavy atoms. Such systems are also of interest to the current energy crisis because of their usage for photovoltaic devices, nuclear fuels (UF_6), fusion lasers (Xe^*_2), catalysts for solar energy conversion, etc.

In addition, a large number of the 'so-called' ab initio relativistic calculations on atoms and molecules are available in the literature; and various semi-rigorous computational schemes have been developed to incorporate the effect of relativity in fairly large systems including heavy transition metal clusters. Furthermore, various relativistic methods for band structure calculations have been developed for solids. A thorough and critical examination of the validity and limitations of the various approximate relativistic methods would be extremely desirable and highly informative for various workers in this field.

Thus, the main object of the proposed ASI was to discuss and review at an advanced level the recent developments in the relativistic many-electron problems in atomic, molecular and solid-state systems so as to stimulate and increase the awareness of the 'experimentalists' and 'theoreticians' for new possibilities in their domains of this rapidly expanding area of research activity. It should be remarked that a large number of subjects in which relativistic effects are significant (e.g. heavy-ion collisions) could not be included in the program of this ASI due to space-time limitations.

This book contains 'the contents' of the main invited lectures by experts in the various areas of research, as well as the abstracts (limited to one-page only due to space limitations) of the additional invited and contributed talks by active scientific workers in this very broad and deep area of vigorous scientific activity.

It was decided not to 'referee' the submitted manuscripts, and thus the whole responsibility regarding each account lies with the various authors themselves. Furthermore, it is expected that some 'repetition' would be the rule in such a collection, and therefore patience is requested of the reader in this matter. I would like the reader to share the informal style and atmosphere which prevailed at the ASI, and therefore, the lectures are printed here as submitted. Needless to say, these lectures stimulated much discussion amongst the participants and further discussions, comments and clarifications occurred during extra tutorials and panel discussions following these lectures. Unfortunately, these could not be included here but I hope that the various lectures presented at this Institute will be of as great an interest to the reader as they were to the participants in the Institute.

I sincerely thank Dr. Mario di Lullo, Executive Officer of the NATO Advanced Study Institutes Programme who helped me throughout various stages of this Institute. I am most grateful to the members of the Executive Committee of the Institute and to all the lecturers for their enthusiastic support and their valuable advice on various matters relating to the organization, etc. of this Institute. I would also like to express my sincere thanks to all the students because without their participation this ASI would not have been possible.

I am very grateful to Mrs. Rani Kaczmarek for her invaluable help and assistance beyond the call of her normal duties as the Conference Secretary throughout her association with the Institute.

Finally, I cannot fully express my appreciation for the inspiration and understanding given to me by my wife, Uma and my daughter, Sarada, during my directorship of the Advanced Study Institute.

G.L. Malli

March, 1982

CONTENTS

FOUNDATIONS OF THE RELATIVISTIC THEORY OF MANY-ELECTRON SYSTEMS

Joseph Sucher

Center for Theoretical Physics
Department of Physics and Astronomy
University of Maryland, College Park, Maryland

I. INTRODUCTION

A. Preliminary Remarks

Why is the study of the relativistic theory of many-electron bound systems interesting and useful? As regards utility, a minimal answer can be given by generalizing the response given long ago by a number theorist to a student who asked "Why is number theory useful?" His reply: "It is useful because one can get a Ph.D. with it!"

It is certainly true that today many such degrees are being acquired by work on problems for which the nonrelativistic theory of many-electron systems is inadequate, but which are not in the domain of high-energy physics. This strong interest has arisen in the past ten years or so and has diverse sources, including astrophysics, solar physics, and plasma diagnostics, where one encounters highly-ionized large-Z atoms, delicate problems in chemistry and solid-state physics, inner-shell excitations of heavy atoms, etc. We will hear about many such topics during the course of this Institute.

If I intepret "many" to mean simply "more than one" then I can state that my own interest in this subject began in the mid-

50's, with work on the simplest of all many-electron bound systems, the helium atom.[1,2] In view of the very accurate measurements of the ionization energy being then carried out by Hertzberg (appropriately enough, in Canada), it became of interest to calculate the ground state energy of He to α^3Ry accuracy, that is to one order beyond the α^2Ry level shift given by the expectation value of the nonrelativistic fine structure operator H_{fs}.

The formal starting point for this calculation was a four-dimensional Bethe-Salpeter-Schwinger type of equation for a two-body amplitude $\psi(x_1,x_2)$, generalized to include an interaction of the bound-state constituents with an external field, derived as an exact consequence of quantum electrodynamics (QED).[1] By an extension of techniques developed by Salpeter[3] one can derive from this equation an equation of the Schrödinger-Dirac type for the equal-times amplitude $\phi'(\vec{x}_1,\vec{x}_2;t) \equiv \psi(x_1,x_2)$ with $t_1=t_2=t$. For a stationary state of energy E, with $\phi'(\vec{x}_1,\vec{x}_2;t) = \phi(\vec{x}_1,\vec{x}_2)e^{-iEt}$, this equation assumes the form, when the effect of virtual electron-positron pairs and virtual photons are neglected,[4]

$$E\phi = H_+\phi \tag{1.1}$$

where

$$H_+ = H_{D;ext}(1) + H_{D;ext}(2) + L_{++}\frac{e^2}{r_{12}}L_{++} \tag{1.2}$$

with

$$H_{D;ext}(i) = \vec{\alpha}_i\cdot\vec{p}_i + \beta_i m + V_{ext}(i). \tag{1.3}$$

Here

$$L_{++} = L_+(1)L_+(2) \tag{1.4}$$

with $L_+(i)$ the projection operator onto the space spanned by the positive-energy eigenstates $u_n(i)$ of $H_{D;ext}(i)$:

$$L_+(i) = \sum_n u_n(i)\, u_n^\dagger(i) . \tag{1.5}$$

Eq. (1.1) was then used as the basis for the development of a practical perturbation theory which facilitated the calculation of the α^3Ry effects in question,[1,2] and, many years later, the calculation of the spin-dependent effects of order α^4Ry.[5] I will come back to eq. (1.2), with its unfamiliar projection operators, in a little while.

Let us now consider why, at this late date, it has become necessary to consider the foundations of the relativistic theory of many-electron systems.

B. Historical Perspective

In the extreme nonrelativistic limit, the energy levels of an atomic system are determined by the eigenvalues of

$$H_{nr}^{atom} \equiv \sum_{i=1}^{N} H_{nr}(i) + V_{ee}, \tag{1.6}$$

where

$$H_{nr}(i) = \vec{p}_i^{\,2}/2m + V_{ext}(i) \ . \tag{1.7}$$

Here $V_{ext}(i) = Ze^2/r_i$ for the case of a spin-less point nucleus, and V_{ee} is the operator representing the electron-electron electrostatic interaction:

$$V_{ee} = \sum_{i<j}^{N} e^2/r_{ij} \ . \tag{1.8}$$

Although the passage from the one-electron Schrödinger equation for a stationary state

$$H_{nr}(1)\psi(1) = W\psi(1) \tag{1.9}$$

to the many- electron equation

$$H_{nr}^{atom}\psi(1,2\ldots N) = W\psi(1,2\ldots N) \tag{1.10}$$

with H_{nr}^{atom} given by (1.6) seems obvious now, it should be remembered that in the early days things were not so clear cut and that the successful calculation of the ground state energy of the helium atom by Hylleraas, based on (1.10) with N=2, was of great importance in establishing the correctness of such ideas.

Let us now consider the relativistic version of (1.7), for the spin-1/2 electron. The sum of the electron mass and the nonrelativistic kinetic energy operator is first replaced by the free Dirac Hamiltonian

$$H_D(1) = \vec{\alpha}_1 \cdot \vec{p}_1 + \beta_1 m \, . \tag{1.11}$$

For an electron moving in an external electromagnetic field described by the four-potential $A^{\mu}_{ext}(x)$, the counterpart of (1.7) was taken by Dirac to be just that dictated by gauge invariance, viz.

$$H_{rel}(1) = H_{D;ext}(1) \equiv H_D(1) + V_{ext}(1) \tag{1.12}$$

where now $V_{ext}(1) = -e[A^{o}_{ext}(1) - \vec{\alpha} \cdot \vec{A}_{ext}(1)]$. As we all know, the one-electron external-field Dirac equation for a stationary state of energy E,

$$H_{rel}(1)\, \psi(1) = E\psi(1), \tag{1.13}$$

gives a result for the spectrum of the hydrogen atom which is in very good agreement with experiment. In addition, the calculation of radiative decay widths, proceeding from the relativistic matrix element for one-photon emission

$$M = \frac{-e}{\sqrt{2\omega}} \langle \psi_F | \vec{\alpha}_1 \cdot \vec{\varepsilon}^{\,*} e^{-i\vec{k}\cdot\vec{r}_1} | \psi_I \rangle \tag{1.14}$$

is also highly successful. Furthermore, the reduction of (1.13) to an equation for the large components $\psi^{+}(1)$, with

$$\psi^{\pm}(1) \equiv \frac{1 \pm \beta_1}{2} \psi(1), \tag{1.15}$$

yields, in the presence of an external magnetic field, the correct value of the electron magnetic moment, and also an effective Schrödinger-Pauli type Hamiltonian, $H_{SP}(1)$ of the form

$$H_{SP}(1) = H_{nr}(1) + H_{fs}(1), \tag{1.16}$$

where $H_{fs}(1)$ contains the usual fine-structure operators. Apart from the so-called "S-state correction," these operators had already been written down on the basis of semi-classical arguments before the invention of the Dirac equation.

Following one's nose, one would think that for an atom with more than one electron a suitable relativistic counterpart of (1.13) would be obtained by adding the relativistic one-body Hamiltonians $H_{rel}(i)$ for each of the electrons to get

$$H_{rel}^{(o)atom} = \sum_{i=1}^{N} H_{rel}(i) = \sum_{i=1}^{N} H_{D;ext}(i) \tag{1.17}$$

and then adding V_{ee}, given by (1.8), to this operator to get what I will call the "Dirac-Coulomb" Hamiltonian H_{DC},

$$H_{rel}^{atom} \stackrel{?}{=} H_{DC} \equiv H_{rel}^{(o)atom} + V_{ee} \; . \tag{1.18}$$

The analogue of (1.10) would then be

$$H_{DC}\chi = E\chi \tag{1.19}$$

where χ is a multi-Dirac spinor.

Eq. (1.19) is surely familiar. However, natural and time-honored as it may be, the approach to the study of relativistic effects in atoms which uses (1.19) rests on quicksand. As was pointed out many years ago by Brown and Ravenhall,[6] equations such as (1.19) simply have no normalizable eigenstates. To see this most simply, consider the case N=2 and switch off e^2/r_{12}. Then each bound state $\psi_{n_1 n_2}$ of $H_{rel}(1) + H_{rel}(2)$, constructed as a product $u_{n_1}(1)u_{n_2}(2)$ of normalizable eigenstates $u_{n_1}(1)$ and $u_{n_2}(2)$ of $H_{rel}(1)$ and $H_{rel}(2)$ respectively, is degenerate with a continuum of non-normalizable product eigenstates in which one electron is in a continuum positive-energy state and the other in a continuum negative-energy state. The turning on of e^2/r_{12} then causes $\psi_{n_1 n_2}$ to "dissolve into the continuum," in terminology borrowed from that used in discussions of autionization. This feature of eq. (1.19) might be termed "Brown-Ravenhall disease" or, in a less medical style, "continuum dissolution" (CD).

C. Implications

The fatal flaw of H_{DC} was stressed in a number of subsequent papers,[7,8,9] but these had little impact, if any, on the theoretical literature dealing with relativistic effects in many-electron atoms. The reasons for this are analyzed in some detail elsewhere.[10] The main points are the following: (i) Eq. (1.19) was used (for N=2) by Breit in his pioneering discussion of the effects arising from transverse-photon exchange between the electrons,[11] leading to the derivation of the Breit operator

$$B(i,j) = \frac{-e^2}{2r_{ij}} (\vec{\alpha}_i \cdot \vec{\alpha}_j + \vec{\alpha}_i \cdot \hat{r}_{ij}\ \vec{\alpha}_j \cdot \hat{r}_{ij}) . \qquad (1.20)$$

It was further used by him to find, via a "reduction to large components," the fine-structure operator H_{fs}, whose expectation value correctly gives the α^2Ry level shifts in atoms.[12] (ii) Eq. (1.19) has been used to derive relativistic analogues of the nonrelativistic Hartree-Fock (HF) equations, the so-called Dirac-Hartree-Fock (DHF) equations and variants thereof, which have been used with considerable success in the calculation of, e.g. X-ray transition energies, etc.[13] (iii) These appear to have been essentially the only applications of (1.19) made to date and it turns out that <u>both</u> of them fail to reveal the flaws of eq. (1.19). However, if one were to try to go beyond α^2Ry accuracy in level-shift calculations, or beyond a product-type ansatz for the wavefunction ψ which is supposed to satisfy (1.19), the difficulties inherent in the use of (1.19) would rapidly become apparent.[10]

With regard to calculations of fine structure, the neglect of the BR paper is to some extent understandable because Brown and Ravenhall concluded, using QED, that the final results of Breit for H_{fs} were correct. However, the situation is different with regard to calculations using the DHF equations. These are obtained by applying the variational principle

$$\frac{\delta}{\delta\tilde{\chi}} \langle\tilde{\chi}|H_{DC}|\tilde{\chi}\rangle = 0 \qquad (1.21)$$

with $\tilde{\chi} \equiv (N!)^{-1/2} \det (\tilde{\chi}(1)\ldots\tilde{\chi}_N(N))$, subject to the constraint $\langle\tilde{\chi}_i|\tilde{\chi}_j\rangle = \delta_{ij}$. The resulting equations do have normalizable solutions. However, since H_{DC} has no bound states, the physical meaning of such a $\tilde{\chi}$ is obscure. In particular $\tilde{\chi}$ is not an approximation to a bound state solution χ of (1.19), because such solutions don't exist!

Nevertheless, as mentioned above and as we will hear at this Institute, calculations carried out in the past decade which use (1.21) or elaborations thereof have been relatively successful in accounting for a variety of experimental measurements relating to the spectra of heavy atoms. It may be that because of this success the CD issue has continued to be neglected by practitioners of the art. However, it seems clear that this situation cannot continue indefinitely, for several reasons. First, if one asks how the calculation based on (1.21) may be improved, i.e. how one can go beyond a central field approximation (CFA) within the framework of H_{DC}, no simple answer is forthcoming. Second, since the wavefunctions $\tilde{\chi}$ are used to calculate further properties of atoms, such as transition rates, one needs to understand their physical meaning. Finally, even when the calculations based on (1.21) give good results one must ask why this is so. In order to be satisfied one must, at a minimum, tie the calculation to fundamental theory, by which I mean QED.

D. What Sayeth QED?

As we will see, QED leads in a quite straightforward way to equations which don't suffer from CD. In particular, one rather natural approach, not involving any four-dimensional machinery, leads to a configuration-space Hamiltonian which is the obvious generalization of (1.2) to more than two-electrons

$$H_+ = \sum_{i=1}^{N} H_{D;ext}(i) + L_+ V_{ee} L_+ , \qquad (1.22)$$

where V_{ee} is given by (1.8) and L_+ is the product positive-energy projection operator, defined by

$$L_+ = L_+(1) L(2) \ldots L_+(N) . \qquad (1.23)$$

It is this projection operator which leads to immunity from CD.

The central feature of QED which complicates the discussion of relativistic bound-state problems is that it involves not only the creation or annihilation of virtual photons but also of virtual electron-positron pairs. This means that even when photons are neglected, a full description of even the ground state of a so-called N-electron atom (or ion) with nuclear charge Ze, involves not only knowledge of a function f depending on N variables, but also a function $f_{1\text{-pair}}$ involving N+2 variables describing N+1 electrons and a positron moving in the nuclear field, a function $f_{2\text{-pair}}$ involving N+4 variables, etc. Fortunately, unless Z is very large, it is possible to treat the virtual-pair production and annihilation by perturbation theory. As will be seen, this is because the level shift contribution of such a process, when it involves just a single pair of electrons, is not large.

It may be worthwhile to note, *en passant*, that the pure two-body analogue of (1.19) with N=2, i.e. the equation describing two Dirac particles interacting via an attractive local potential, in the <u>absence</u> of an external field, viz.

$$[\vec{\alpha}_1\cdot\vec{p}_1 + \beta_1 m_1 + \vec{\alpha}_2\cdot\vec{p}_2 + \beta_2 m_2 + V(12)]\psi=E\psi\ , \qquad (1.24)$$

does <u>not</u> suffer from CD. The reason for this is that momentum conservation now forces a correlation between the possible (virtual) energies of the fermions. In particular, the quantity $E_1(\vec{p}_1)-E_2(\vec{p}_2)$ cannot assume a given value infinitely often as $\vec{p}_1$ and $\vec{p}_2$ are varied, because now $\vec{p}_1 + \vec{p}_2$ is fixed. This explains why although (1.24) was used in early studies of recoil effects in hydrogen and of fine-structure in hydrogen and positronium[14] no disasters were encountered thereby.

E. Outline

The main purpose of the remaining lectures is to discuss an approach to the relativistic theory of many-electron atoms that does not suffer from the above-mentioned difficulties and which can be used as the starting point for systematic calculations.[15] It is securely founded in QED, so that no guesswork is involved in its application to atomic physics problems.

By way of warming up, in Sec. II, I will consider an exactly soluble problem, the motion of a single Dirac electron in the presence of a static external electromagnetic field $A^o_{ext}(\vec{x})$, using however the language of second quantization. This will permit the easy introduction of a number of concepts, including that of a time-ordered Feynman diagram, which may not be familiar to everyone here. Armed with this knowledge, in Sec. III we will consider H_{QED}, the full Hamiltonian of QED, and see how a separation of H_{QED} into a "no-pair part" H^{np}_{QED} and a "pair part" H^{pair}_{QED}, with

$$H_{QED} = H^{np}_{QED} + H^{pair}_{QED}\ , \qquad (1.25)$$

leads to Hamiltonians such as (1.22). In Sec. IV I will derive relativistic HF-type equations associated with (1.22), and discuss their relationship with the usual DHF equations. In Sec. V, I will consider the effects of virtual pairs, which are not described by the use of H_+. In Sec. VI, I will briefly illustrate the application of this kind of approach to a particular physical problem, the $2\,{}^3S_1 \to 1\,{}^1S_0 + \gamma$ transition in He and He-like ions; this is an M1 transition which is forbidden in the nonrelativistic limit. A review of "nonrelativistic QED" is given in an Appendix, and used to discuss the concept of a relativistic effect in atomic physics. A concluding discussion is contained in Sec. VII.

II. WARM-UP EXERCISE: ONE ELECTRON IN AN EXTERNAL FIELD

A. c-number Theory

The Dirac equation for an electron moving in a static external electromagnetic field $A^\mu_{ext}(\vec{x})$ is

$$\frac{i\partial}{\partial t}\,\psi(x) = H_{D;ext}\,\psi(x) = (H_D + V_{ext})\,\psi(x) \qquad (2.1)$$

where $H_D = \vec{\alpha}\cdot\vec{p} + \beta m$ and $V_{ext} = -e[A^0_{ext}(\vec{x}) - \vec{\alpha}\cdot\vec{A}_{ext}(\vec{x})]$. For a stationary state, $\psi(x) = e^{-iEt}\psi(\vec{x})$ and (2.1) reduces to

$$E\psi(\vec{x}) = H_{D;ext}\,\psi(\vec{x})\,. \qquad (2.2)$$

The normalizable solutions $\psi_n(\vec{x})$ of (2.2) with $E_n > 0$ determine a discrete spectrum associated with the energy levels of a one-elecron atom. The amplitude for an electron to scatter in the external field from an initial planewave state $|\vec{k},\sigma\rangle$ to a final state $|\vec{k}',\sigma'\rangle$ is given by

$$T(\vec{k},\sigma';\vec{k},\sigma) = \langle\vec{k}',\sigma'|V_{ext}|\vec{k},\sigma;out\rangle = \langle\psi_{\vec{k}',\sigma'}|V_{ext}|\psi^{out}_{\vec{k},\sigma}\rangle\ . \qquad (2.3)$$

Here

$$\psi_{\vec{k}\sigma}(\vec{x}) = \langle x|\vec{k},\sigma\rangle = u_\sigma(\vec{k})e^{i\vec{k}\cdot\vec{x}}, \tag{2.4}$$

with $u_\sigma(\vec{k})$ a free Dirac spinor of momentum $\vec{k}$ and eigenvalue $\sigma = \pm 1$ for $\vec{\sigma}_D\cdot\hat{k}$, and $\psi^{out}_{\vec{k}\sigma}(\vec{x})$ is a continuum solution of (2.2) with $E \to E(\vec{k}) \equiv (\vec{k}^2+m^2)^{1/2}$, satisfying boundary conditions incorporated in the integral equation

$$\psi^{out}_{\vec{k},\sigma} = \psi_{\vec{k},\sigma} + (E-H_D+i\varepsilon)^{-1}\, V_{ext}\, \psi^{out}_{\vec{k},\sigma}\ . \tag{2.5}$$

The Born series for T generated by iteration of (2.5) is given by

$$T = T^{(1)} + T^{(2)} + \ldots \tag{2.6}$$

where

$$T^{(1)} = \langle\vec{k}',\sigma'|V_{ext}|\vec{k},\sigma\rangle, \tag{2.7}$$

$$T^{(2)} = \langle\vec{k}',\sigma'|V_{ext}\,\frac{1}{E(\vec{k})-H_D+i\varepsilon}\,V_{ext}|\vec{k},\sigma\rangle\ , \tag{2.8}$$

and so on.

Note that $T^{(2)}$ and all higher-order $T^{(n)}$ get contributions from negative energy intermediate states: In particular we may write, with $E(\vec{p}_{op}) \equiv (\vec{p}_{op}^{\,2}+m^2)^{1/2}$,

$$\frac{1}{E-H_D+i\varepsilon} = \frac{\Lambda_+}{E-E(\vec{p}_{op})+i\varepsilon} + \frac{\Lambda_-}{E+E(\vec{p}_{op})+i\varepsilon} \tag{2.9}$$

where

$$\Lambda_+ = \sum_{\tau=\pm 1}\int \frac{d\vec{q}}{(2\pi)^3}|\vec{q},\tau\rangle\langle\vec{q},\tau| \tag{2.10}$$

is the projection operator onto the space spanned by the positive-energy eigenfunction of H_D and $\Lambda_- = 1 - \Lambda_+$ is the negative-energy projection operator. Then (2.8) assumes the form

$$T^{(2)} = T^{(2)}_+ + T^{(2)}_- \tag{2.11}$$

where

$$T_{\pm}^{(2)} = \langle\vec{k}',\sigma'|V_{ext} \frac{\Lambda_{\pm}}{E \pm E(\vec{p}_{op})+i\varepsilon} V_{ext}|\vec{k},\vec{\sigma}\rangle \ . \tag{2.12}$$

However, according to the Dirac hole theory, the negative-energy sea is filled, so how can there be contributions from negative-energy states? The poor Hamiltonian in (2.2)has never heard of the Dirac hole theory, so it does what it has to in order to solve (2.2), subject to the boundary conditions imposed. Let's see how the language of second- quantization allows one to incorporate the ideas of Dirac hole theory in mathematical form and to remove these conceptual difficulties.

B. q-number theory

In quantum field theory, the c-number Dirac function $\psi(x)$ is replaced by a q-number $\psi_D(x)$ (quantum field) which still obeys (2.1), i.e.

$$i \frac{\partial}{\partial t} \psi_D(x) = H_{D;ext} \psi_D(x) \tag{2.13}$$

but also the equal-time anticommutation relations (ACR)

$$\{\psi_{D\alpha}(\vec{x},t),\psi^{\dagger}_{D\beta}(\vec{x}',t)\} = \delta_{\alpha\beta}\ \delta(\vec{x}-\vec{x}'),$$

$$\{\psi_{D\alpha}(\vec{x},t),\ \psi_{D\beta}(\vec{x}',t)\} = 0. \tag{2.14}$$

The "big" Hamiltonian H_{ext} which describes the time-evolution of $\psi_D(\vec{x},t)$ must be such that (2.13) is equivalent to the Heisenberg equation of motion

$$\frac{\partial\psi_D(\vec{x},t)}{\partial t} = i[H_{ext},\ \psi_D(\vec{x},t)] \tag{2.15}$$

This is satisfied if

$$H_{ext} = \int \psi^{\dagger}_D(\vec{x})\ H_{D;ext}\ \psi_D(\vec{x})d\vec{x} + \text{const.} \tag{2.16}$$

where

$$\psi_D(\vec{x}) \equiv \psi_D(\vec{x},o) \tag{2.17}$$

is the matter-field operator in a Schrödinger picture which coincides with the Heisenberg picture at t=0. In the S-picture, the evolution of the state vector $\Psi(t)$, lying in the physical Hilbert space H_{phys} in which ψ_D and H_{ext} operate is of course determined by

$$i \frac{\partial}{\partial t} \Psi = H_{ext} \Psi . \tag{2.18}$$

(a) <u>The smart thing</u>

To get back to the usual Dirac theory for a single electron, the smart thing to do is to follow Furry[16] and to expand $\psi_D(\vec{x})$ in the form

$$\psi_D(\vec{x}) = \sum_n A(n)\, u_n(\vec{x}) + \sum_m B^\dagger(m)\, v_m(\vec{x}) \tag{2.19}$$

where the $u_n(\vec{x})$ form an orthogonal basis for the space S_+ spanned by the set of all positive-eneregy eigenfunctions of $H_{D;ext}$ and the $v_m(\vec{x})$ serve the same purpose for the space S_- spanned by the negative-energy eigenfunctions, with

$$H_{D;ext}\, u_n(\vec{x}) = \varepsilon_+(n)\, u_n(\vec{x}) \tag{2.20a}$$

$$H_{D;ext}\, v_m(\vec{x}) = \varepsilon_-(m)\, v_m(\vec{x}) . \tag{2.20b}$$

The labels <u>n</u> and <u>m</u> lie in label spaces P_+ and P_- respectively, which we need not specify any further; the symbol Σ denotes a sum over the discrete states and an integral over the continuum. The condition (2.14) will be satisfied provided the operators A(n) and B(m) satisfy

$$\{A(n), A^\dagger(n')\} = \delta(n,n'), \quad \{B(n), {}^\dagger B\,(n')\} = \delta(m,m'), \tag{2.21}$$

with all other anti-commutators being zero. If we substitute

(2.19) into (2.16) and choose the constant in (2.16) to cancel the (divergent) sum $\Sigma\varepsilon_-(m)$, we get

$$H_{ext} = \sum_n A^\dagger(n)A(n)\,\varepsilon_+(n) - \sum_m B^\dagger(m)\,B(m)\,\varepsilon_-(m) \quad . \qquad (2.22)$$

It is easy to see that for any state Ψ in the physical Hilbert space

$$\langle\Psi|H_{ext}|\Psi\rangle \geqslant 0 \qquad (2.23)$$

and the equality holds only for a state Ψ_o with the property

$$A(n)\,\Psi_o = 0, \qquad B(m)\,\Psi_o = 0. \qquad (2.24)$$

We shall assume that one and only one such a state exists. Further, the operator

$$\vec{P}_{op} = \int d\vec{x}\;\psi_D^\dagger(\vec{x})\,\frac{\vec{\nabla}}{i}\,\psi_D(\vec{x})$$

is readily seen to have the property

$$\vec{\nabla}\,\psi_D(\vec{x}) = -i[\vec{P}_{op},\psi_D(\vec{x})]$$

and may therefore be identified with the operator which generates displacements in space. A short calculation shows that

$$\vec{P}_{op}\,\Psi_o = 0 \ . \qquad (2.25)$$

Thus Ψ_o is a state of both zero energy and zero three-momentum and we may identify it as the "vacuum":

$$\Psi_o = |Vac\rangle \ . \qquad (2.26)$$

It is easy to verify that the "number operators"

$$N_e \equiv \sum_n A^\dagger(n)\;A(n), \quad N_p = \sum_m B^\dagger(m)\;B(m) \qquad (2.27)$$

have the following properties:

(i) The charge operator Q_{op}, defined by

$$Q_{op} = -e \int d\vec{x}\ [\psi_D^\dagger(\vec{x})\ \psi_D(\vec{x}) - \langle\Psi_o|\psi_D^\dagger(\vec{x})\ \psi_D(\vec{x})|\Psi_o\rangle] \quad ,$$

has the form

$$Q_{op} = -eN_e + eN_p .$$

(ii) $[N_e, A^\dagger(n)] = N_e+1$, $[N_p, B^\dagger(m)] = N_p + 1$; the eigenvalues of N_e and N_p are all non-negative integers.

(iii) The operators N_e, N_p and H_{ext} commute with each other. It follows that a state Ψ such that

$$N_e\Psi = n_e\Psi, \quad N_p\Psi = n_p\Psi \tag{2.29}$$

may be interpreted as a state containing n_e "electrons" and n_p "positrons" and the $A^\dagger$ and $B^\dagger$ as creation operators for electrons and positrons, respectively

In particular, the general form of a state in the sector $n_e = 1$, $n_p = 0$, is

$$\Psi_{1-e} = \sum_{n'} f(n')\ A^\dagger(n')|Vac\rangle \ . \tag{2.30}$$

We may think of f(n') as the Fock-space wave function. If we ask what is the condition on f(n') so that

$$H_{ext}\ \Psi_{1-e} = E\Psi_{1-e} \ , \tag{2.31}$$

we find $\varepsilon_+(n')\ f(n') = E\ f(n')$. This implies that $E = \varepsilon_+(n)$ for some n. Hence, if we ignore degeneracies, $f(n') = C\delta(n',n)$. Thus, with C=1,

$$\Psi_{1-e} = A^\dagger(n)|Vac\rangle \tag{2.32}$$

and the Fock-space wavefunction of this one-electron state is

$$f_n(n') = \delta(n',n). \tag{2.33}$$

We can define a configuration-space equivalent to $f_n(n')$, via

$$\psi_n(\vec{x}) \equiv \Sigma_{n'} u_{n'}(\vec{x}) \; f_n(n') = u_n(\vec{x}) . \tag{2.34}$$

which of course satisfies

$$H_{D;ext} \; \psi_n(\vec{x}) = \varepsilon_+(n) \; \psi_n(\vec{x}) \quad . \tag{2.35}$$

Thus both the bound state and the scattering problem reduces to the one studied in subsection A, provided that the eigenstates of $H_{D;ext}$ are used for the expansion of $\psi_D(\vec{x})$.

(b) A different approach

A less clever but more instructive approach, which lets one see the conceptual difference between hole-theory and one-electron theory, is obtained by expanding $\psi_D(\vec{x})$ in plane-wave eigenfunctions of H_D. We define "free" creation operators $a^\dagger(k)$, $b^\dagger(\ell)$ and annihilation operators, $a(k)$, $b(\ell)$ via

$$\psi_D(\vec{x}) = \sum_k a(k) \; u_k(\vec{x}) + \sum_\ell b^\dagger(\ell) \; v_{-\ell}(\vec{x}), \tag{2.36}$$

where $k = (\vec{k},\sigma)$, $\ell = (\vec{\ell},\rho)$, and a vacuum state $\Phi_o = |vac\rangle$ by

$$a(k)\Phi_o = b(\ell)\Phi_o = 0 \qquad \text{(all } k \text{ and } \ell\text{)}. \tag{2.37}$$

The number operators defined by

$$N_- = \sum_k a^\dagger(k) \; a(k), \qquad N_+ = \sum_\ell b^\dagger(\ell) \; b(\ell) \tag{2.38}$$

commute with

$$H_o = \int \psi_D^\dagger (\vec{x}) \; H_D(\vec{x}) \; \psi_D(\vec{x}) + \text{const}' .$$

For a suitable choice of the constant we have

$$H_o = \sum_k E(\vec{k}) \; a^\dagger(k)a(k) + \sum_\ell E(\vec{\ell}) \; b^\dagger(\ell)b(\ell) \quad . \tag{2.39}$$

If we write

$$H_{ext} = H_o + H' \tag{2.40}$$

with

$$H' = \int d\vec{x}\ \psi_D^\dagger(\vec{x})\ V_{ext}\ \psi_D(\vec{x}), \tag{2.41}$$

the interaction term H' leads to scattering of an incoming electron, with amplitude

$$T(k',k) = T^{(1)}(k',k) + T^{(2)}(k',k) + \dots , \tag{2.42a}$$

where

$$T^{(1)}(k',k) = \langle a^\dagger(k')\Phi_o | H' | a^\dagger(k)\Phi_o\rangle , \tag{2.42b}$$

$$T^{(2)}(k',k) = \langle a^\dagger(k')\Phi_o | H' \frac{1}{E-H_o+i\varepsilon} H' | a^\dagger(k)\Phi_o\rangle , \tag{2.42c}$$

and so on.

We now want to see the connection of this point of view with the one discussed above, involving the expansion (2.19). The first thing to note is that the states $\Phi_o = |vac\rangle$ and $\Psi_o = |Vac\rangle$ are <u>not</u> the same: The "physical vacuum" Ψ_o is a complicated superposition of the "bare vacuum" Φ_o, one-pair states $\Phi_{1\text{-pair}}$, ...:

$$\Psi_o = C_o\Phi_o + C_1\ \Phi_{1\text{-pair}} + C_2\ \Phi_{2\text{-pair}} + \dots$$

with, e.g.

$$\Phi_{1\text{-pair}} = \sum_{k,\ell} f_{1\text{-pair}}(k,\ell)\ a^\dagger(\vec{k})\ b^\dagger(\vec{k})\ \Phi_o .$$

With $\Phi_{n\text{-pair}}$ normalized to unity, $\langle\Psi_o|\Psi_o\rangle = 1$ implies

$$1 = |C_0|^2 + |C_1|^2 + |C_2|^2 + \dots$$

so that $|C_n|^2$ can be interpreted as the probability for finding <u>n</u> pairs, if V_{ext} were suddenly switched off.

Now let's study the scattering problem within this context. The interaction part of H is

$$H' = H - H_o = \int d\vec{x} : \psi_D^\dagger(\vec{x}) V_{ext} \psi_D(\vec{x}) : + \text{ const"} \tag{2.43}$$

where the constant arises from the normal-ordering of the integrand. For a suitable choice of this constant, substitution of the expansion (2.36) into (2.43) then yields H' as a sum

$$H' = H_{sc}(e^-,e^-) + H_{sc}(e^+,e^+) + H_{cr}(e^-,e^+) + H_{ann}(e^-,e^+), \tag{2.43'}$$

where the various operators describing electron scattering, positron scattering, pair-creation, and pair-annihilation in the external field. It is convenient to introduce a graphical notation to describe these operators. With the assignment

k  $\leftrightarrow u_k$, ℓ 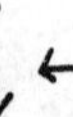 $\leftrightarrow v^\dagger_{-\ell}$, (2.44a)

k'  $\leftrightarrow u^\dagger_{k'}$, ℓ'  $\leftrightarrow v_{-\ell'}$, (2.44b)

we may write

$H_{sc}(e^-,e^-) = \Sigma$ 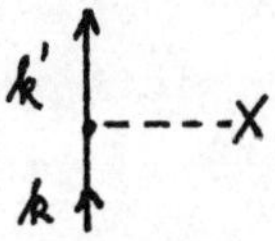

$a^\dagger(k')\, a(k)$ (2.45a)

$H_{sc}(e^+,e^+) = \Sigma$ 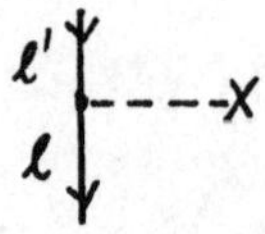

$b^\dagger(\ell')\, b(\ell)$ (2.45b)

and

$H_{cr}(e^-,e^+) = \Sigma$ 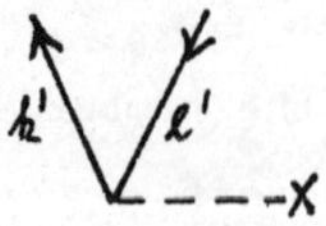

$a^\dagger(k')\, b^\dagger(\ell')$ (2.45c)

$H_{ann}(e^-,e^+) = \Sigma$ 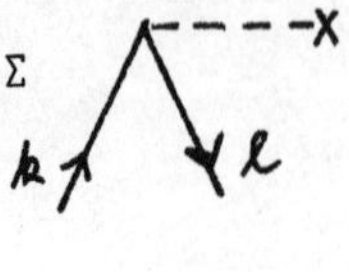

$b(\ell)\, a(k)$ (2.45d)

With $\tilde{V}$ the Fourier transform of V, the pictures in (2.45a-d) represent, respectively, the factors

$$\langle u_{k'} | V_{ext} | u_k \rangle = u^\dagger_{\sigma'}(\vec{k})\ \tilde{V}(\vec{k}'-\vec{k})\ u_\sigma(\vec{k}) \qquad (2.46a)$$

$$-\langle v_{-\ell} | V_{ext} | v_{-\ell'} \rangle = -v^\dagger_{-\rho}(-\vec{\ell})\tilde{V}(\vec{\ell}'-\vec{\ell})v_{-\rho'}(-\vec{\ell}') \qquad (2.46b)$$

$$\langle u_k | V_{ext} | v_{-\ell'} \rangle = u^\dagger_{\sigma'}(\vec{k}')\tilde{V}(\vec{k}'+\vec{\ell}')v_{-\rho}(-\vec{\ell}') \qquad (2.46c)$$

$$\langle v_{-\ell} | V_{ext} | u_k \rangle = v^\dagger_{-\rho}(-\vec{\ell})\tilde{V}(-\vec{k}-\vec{\ell})u_\sigma(\vec{k}). \qquad (2.46d)$$

Note the minus sign in (2.46b), which arises from the relation $:b(\ell)b^\dagger(\ell'): = -b^\dagger(\ell')b(\ell)$ and corresponds to the fact that the positron electric charge is the negative of the electron charge.

We can now study electron scattering in the external field as follows. We write,

$$H' = H'_{np} + H'_{pair} \qquad (2.47a)$$

where H'_{np} (read: "H-prime-no-pair") is defined by

$$H'_{np} = H_{sc}(e^-,e^-) + H_{sc}(e^+,e^+) \qquad (2.47b)$$

i.e., is the part of H' describing scattering only and

$$H'_{pair} = H_{cr}(e^-,e^+) + H_{ann}(e^-,e^+) \qquad (2.47c)$$

describes virtual-pair creation or destruction. Since H'_{np} commutes with N_- and N_+, we add it to H_o and define a "no-pair Hamiltonian"

$$H_{np} = H_o + H'_{np} \qquad (2.48)$$

which still conserves N_- and N_+, and treat H_{pair} by perturbation theory.

With H_{np} as the zero-order Hamiltonian, the eigenvalue problem in the $N_- = 1$, $N_+ = 0$ sector of Fock space reduces to a relatively simple configuration-space eigenvalue condition. In analogy with the procedure used in (a) above, we put

$$\Psi = \sum_{\vec{k},\sigma} f_\sigma(\vec{k})\ a_\sigma^\dagger(\vec{k})|vac\rangle \tag{2.49}$$

and require that

$$H_{np}\Psi = E\Psi\ . \tag{2.50}$$

Exercise Use the anticommutation relation

$$a_\sigma(\vec{k})\ a_{\sigma'}^\dagger(\vec{k}') + a_{\sigma'}^\dagger(\vec{k}')a_\sigma(\vec{k}) = \delta_{\sigma\sigma'}\ \delta(\vec{k}'-\vec{k})$$

to show that

$$H_o\Psi = \sum_k E(\vec{k})\ f_\sigma(\vec{k})\ a_\sigma^\dagger(\vec{k})|vac\rangle,$$

$$H'_{np}\Psi = \sum_{k',k} \langle u_{k'}|V_{ext}|u_k\rangle\ f_\sigma(\vec{k})\ a_\sigma^\dagger(\vec{k})|vac\rangle\ .$$

From the results of the exercise it follows that (2.50) can only be satisfied if

$$E(\vec{k})\ f_\sigma(\vec{k}) + \sum_{\vec{k}',\sigma'} \langle u_k|V_{ext}|u_{k'}\rangle\ f_{\sigma'}(\vec{k}') = Ef_\sigma(\vec{k})\ . \tag{2.51}$$

To convert (2.51) to a configuration-space spinor equation define

$$\tilde{\psi}(\vec{k}) = \sum_{\sigma=1}^{2} u_\sigma(\vec{k})\ f_\sigma(\vec{k})\ , \tag{2.52}$$

multiply (2.51) by $u_\sigma(\vec{k})$ and sum an σ. Since

$$\langle u_k|V_{ext}|u_{k'}\rangle = u_\sigma^\dagger(\vec{k})\ \langle\vec{k}|V_{ext}|\vec{k}'\rangle\ u_{\sigma'}(\vec{k}')\ ,$$

this gives

$$E(\vec{k})\ \tilde{\psi}(\vec{k}) + \Lambda_+(\vec{k})\ \sum_{\vec{k}'} \langle\vec{k}|V_{ext}|\vec{k}'\rangle\ \tilde{\psi}(\vec{k}') = E\tilde{\psi}(\vec{k}) \tag{2.53}$$

where

$$\Lambda_+(\vec{k}) \equiv \sum_\sigma u_\sigma(\vec{k})\ u_\sigma^\dagger(\vec{k}) \tag{2.54a}$$

is just the Casimir positive-energy projection operator in $\vec{k}$-space:

$$\Lambda_+(\vec{k}) = \frac{E(\vec{k}) + \vec{\alpha}\cdot\vec{k} + \beta m}{2E(\vec{k})} . \tag{2.54b}$$

Since from the definition (2.52) we have

$$\Lambda_+(\vec{k})\tilde{\psi}(\vec{k}) = \tilde{\psi}(\vec{k}) , \tag{2.55}$$

We may replace $E(\vec{k})$ by $\vec{\alpha}\cdot\vec{k} + \beta m$ in (2.53) and also insert another factor $\Lambda_+(\vec{k})$ to rewrite (2.53) in the form

$$\tilde{h}_+\tilde{\psi}(\vec{k}) = E\tilde{\psi}(\vec{k}) \tag{2.56a}$$

where

$$\tilde{h}_+ = \vec{\alpha}\cdot\vec{k} + \beta m + \Lambda_+(\vec{k}) \sum_{\vec{k}'} \langle k|V_{ext}|\vec{k}'\rangle \Lambda_+(\vec{k}') \tag{2.56b}$$

is a manifestly hermitian operator. In coordinate-space (2.56) takes the form

$$h_+\psi(\vec{r}) = E\psi(\vec{r}) \tag{2.57}$$

where

$$h_+ = \vec{\alpha}\cdot\vec{p}+\beta m + \Lambda_+ V_{ext} \Lambda_+ . \tag{2.58}$$

The trade-off in going from $\vec{k}$-space to $\vec{r}$-space is that whereas V_{ext} becomes a multiplicative function rather than an integral operator, the reverse is true for Λ_+. In $\vec{r}$-space

$$\Lambda_+ = \frac{1}{2} + \frac{1}{2E_{op}} (\vec{\alpha}\cdot\vec{p}+\beta m) \tag{2.59}$$

where E_{op}^{-1} is an integral operator.[10] The main point to notice is how Λ_+ has appeared in the configuration-space equation as a natural outcome of the split of H' into a part H'_{sc} involving only scattering of "free" electrons (and positrons) and a part H'_{pair} involving pair-creation or annihilation, which is to be treated perturbatively.

C. Electron Scattering and the Effects of H'_{pair}

It is instructive to study the scattering problem

$$|i\rangle = |\vec{k},\sigma\rangle \rightarrow |f\rangle = |\vec{k}',\sigma'\rangle, \tag{2.60}$$

with both H'_{np} and H'_{pair} treated as perturbations on H_o. With regard to H'_{np} we could go back to the field-theory formalism, or alternatively, just look at eq. (2.57) which, with H'_{pair} neglected, accounts for H'_{np} <u>fully</u>. From (2.57) we see that, with $V_+ = \Lambda_+ V_{ext} \Lambda_+$,

$$T^{np}_{fi} = \langle f|V_+ + V_+ \frac{1}{E-H_D} V_+ + V_+ \frac{1}{E-H_D} V_+ \frac{1}{E-H_D} V_+ + \ldots|i\rangle \tag{2.61a}$$

Since $\Lambda_+|i\rangle = |i\rangle$, $\Lambda_+|f\rangle = |f\rangle$, the initial and final Λ_+ factor can be removed and we get, with $E = E(\vec{k}) = E(\vec{k}')$

$$T^{np}_{fi} = \langle f|V_{ext}|i\rangle + \langle f|V_{ext} \frac{\Lambda_+}{E-H_D+i\varepsilon} V_{ext}|i\rangle + \ldots \tag{2.61b}$$

It should be clear that these terms just correspond to the sum of the time-ordered graphs shown below:

with the conventions introduced previously. As an example, on using the expression (2.54a) for Λ_+ in the second term of (2.61b) we get

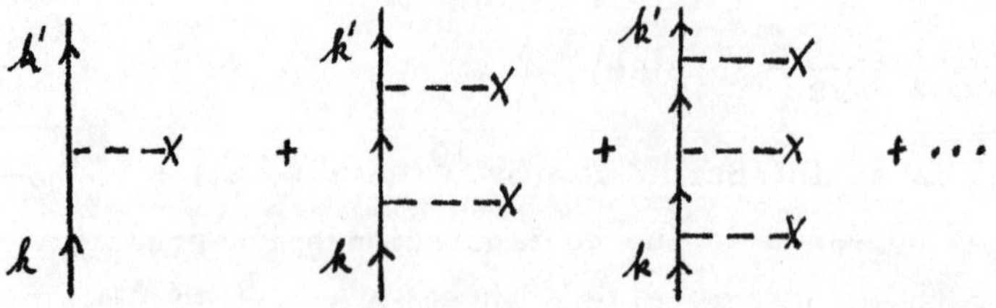

Fig. 1 Graphs for T^{np}_{fi}.

$$\sum_{\vec{k}'',\sigma''} \frac{\langle f|V_{ext}|\vec{k}'',\sigma''\rangle\langle\vec{k}'',\sigma''|V_{ext}|i\rangle}{E(\vec{k}) - E(\vec{k}'') + i\varepsilon} \; .$$

The numerator is just the product of the factors associated with the vertices in graph (b) and the denominator corresponds to a familiar rule: Supply a factor $(E - E_{int}+i\varepsilon)^{-1}$ for an intermediate state of energy E_{int}.

Now let us see how H'_{pair} acts to restore the full answer for T_{fi}, given by (2.6). Since $\langle f|H'_{pair}|i\rangle = 0$, the first chance for H'_{pair} to make a contribution is in second order:

$$T_{fi}^{(2)pair} = \langle f|H'_{pair} \frac{1}{E-H_o+i\varepsilon} H'_{pair}|i\rangle \; . \qquad (2.61c)$$

Only the creation part of H'_{pair} is relevant and on inserting a complete set of intermediate states, we see that only <u>three-body</u> intermediate states can contribute. Moreover, these must be of the form $|k,k';q\rangle$ where k and k' refer to the initial and final electrons and q to a positron. Thus we get

$$T_{fi}^{(2)pair} = \sum_q \frac{\langle f|H'_{pair}|k,k',q\rangle \; \langle k,k';q|H'_{pair}|i\rangle}{E-E_{int}}$$

corresponding to the graph

Using the explicit form of H'_{pair} one finds that, for $\vec{k}' \neq \vec{k}$,

$$T_{fi}^{(2)pair} = -\sum_q \frac{v^\dagger_{-\rho}(-\vec{q})\;\tilde{V}(-\vec{q}-\vec{k})u_\sigma(\vec{k})\;u^\dagger_{\sigma'}(\vec{k}')\tilde{V}(\vec{k}'+\vec{q})v_{-\rho}(-\vec{q})}{E-E_{int}}$$

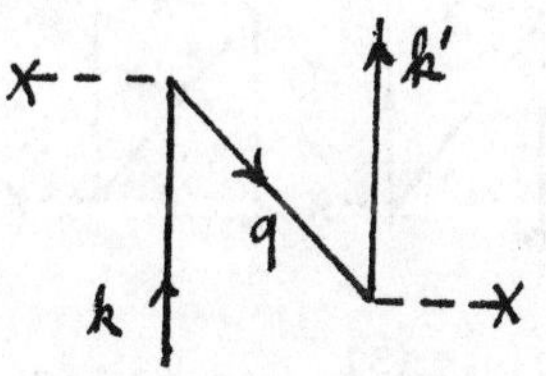

Fig. 2. Graph for the second-order contribution to T_{fi}^{pair}.

The minus sign comes from the fact that $\langle k'|b(q)a(k)|k',k;q\rangle$ and $\langle k',k;q|a^\dagger(k')b^\dagger(q)|k\rangle$ have opposite sign. The energy of the intermediate state is just $E(\vec{k})+E(\vec{k}')+E(\vec{q})$ so that the denominator is equal to $-E(\vec{k}') - E(\vec{q}) = -[E(\vec{k})+E(\vec{q})]$. With $q = (\vec{q},\rho)$, the sum on intermediate spins ρ gives

$$\sum_\rho v_{-q} v^\dagger_{-q} = \Lambda_-(-\vec{q}) \ .$$

Since the minus signs from the denominator and the pair-annihilation factor cancel, we get

$$T^{(2)pair}_{fi} = \sum_{\vec{q}} \frac{u^\dagger_{\sigma'}(\vec{k}')\tilde{V}(\vec{k}'+\vec{q})\,\Lambda_-(-\vec{q})\,\tilde{V}(-\vec{q}-\vec{k})u_\sigma(\vec{k})}{E(\vec{k}) + E(\vec{q})}$$

Because of the factor Λ_-, $E(\vec{q})$ can be replaced by $-H_D(-\vec{q})$, so that on reverting to operator language, we may write

$$T^{(2)pair}_{fi} = \langle f|V_{ext} \frac{\Lambda_-}{E-H_D} V_{ext}|i\rangle \ .$$

If we add this to $T^{(2)np}_{fi}$, and use the relation

$$\Lambda_+ + \Lambda_- = 1$$

we see that

$$T^{(2)}_{fi} = T^{(2)no\text{-}pair}_{fi} + T^{(2)pair}_{fi} \qquad (2.61d)$$

as expected.

<u>Exercise</u> The diagrams describing the third-order amplitude are given, for $\vec{k}' \neq \vec{k}$, by

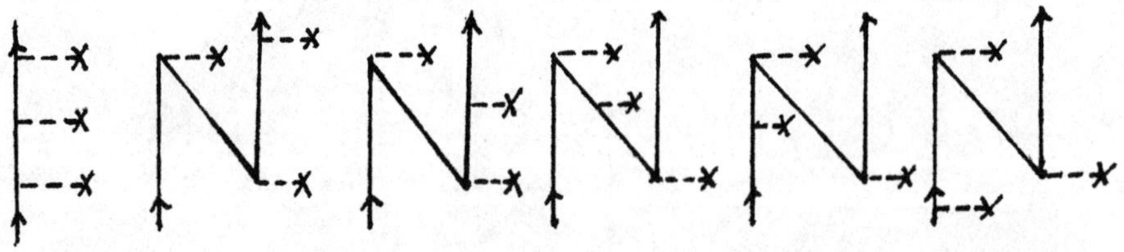

Fig. 3. Graphs for $T^{(3)}_{fi}$, the third-order part of T_{fi}.

The first diagram is just $T_{fi}^{(3)np}$ and the sum of the other five, each involving two actions of H'_{pair} and one of H'_{np}, define $T_{fi}^{(3)pair}$. Write down the contributions from each graph and show that

$$T_{fi}^{(3)} = T_{fi}^{(3)np} + T_{fi}^{(3)pair}. \tag{2.61e}$$

In summary, the amplitude for T_{fi} for an electron to scatter in an external field is given by (2.3), determined in principle by the solution $\psi_{\vec{k},\sigma}^{out}$ of the usual Dirac equation, defined by (2.5). However, we now see that the Born series (2.6) for T_{fi}, which involves contributions from negative energy intermediate states, can be thought of more physically as a sum of two kinds of terms:

$$T_{fi} = T_{fi}^{np} + T_{fi}^{pair} \tag{2.62}$$

where T_{fi}^{np} is a part which comes exclusively from terms which involve only (free) positive-energy intermediate states for the electron. T_{fi}^{pair}, on the contrary corresponds to processes which involve the virtual production of one or more electron-positron pairs and hence n-body intermediate states with n = 3,5,... . The sum of the terms which define T_{fi}^{np} is precisely the scattering amplitude which is generated by solving the equation (2.5), with V_{ext} replaced by

$$V^{np} = \Lambda_+ V_{ext} \Lambda_+ . \tag{2.63}$$

It is interesting to ask if we can find a positive-energy equation which effectively sums the contribution from many-body intermediate states also. This can be accomplished in several ways, but most easily by a sleight-of-hand as follows:

Consider

$$(H_D + V_{ext})\psi = E\psi. \tag{2.64}$$

Define positive- and negative- energy projections $\psi_\pm$ of ψ via

$$\psi_\pm = \Lambda_\pm \psi$$

Multiplying (2.64) by Λ_+ and Λ_- in turn we get

$$E_{op}\psi_+ + \Lambda_+ V_{ext}\psi_+ + \Lambda_+ V_{ext}\psi_- = E\psi_+, \quad (2.65a)$$

$$-E_{op}\psi_- + \Lambda_- V_{ext}\psi_+ + \Lambda_- V_{ext}\psi_- = E\psi_- . \quad (2.65b)$$

Solve (2.65b) for ψ_-:

$$\psi_- = (E + E_{op} - \Lambda_- V_{ext} \Lambda_-)^{-1} \Lambda_- V_{ext} \psi_+ , \quad (2.66)$$

and substitute into (2.65a) to get

$$h_+^{eff} \psi_+ = E \psi_+ , \quad (2.67)$$

where

$$h_+^{eff} = E_{op} + \Lambda_+ V_{ext} \Lambda_+ + V_E^{pair} = h_+ + V_E^{pair} \quad (2.68)$$

with

$$V_E^{pair} = \Lambda_+ V_{ext} (E+E_{op} - \Lambda_- V_{ext} \Lambda_-)^{-1} \Lambda_- V_{ext} \Lambda_+ . \quad (2.69)$$

Thus the effect of virtual pair-creation and destruction is completely included by adding the "potential" V_E^{pair} to h_+. Note that V_E^{pair} is both non-local, as is V^{np}, and explicitly energy-dependent. In practice it is of course easier to solve the Dirac equation (2.64) directly rather than (2.67), but we are in interested here in questions of principle and physical interpretation.

Magnitude of Pair Effects

Using (2.69), it easy to estimate the importance of V_E^{pair} relative to V^{np}. If V_{ext} <<m the denominator in (2.69) is > 2m so that

$$V_E^{pair} \lesssim \frac{\Lambda_+ V_{ext} \Lambda_- V_{ext} \Lambda_+}{2m} \quad (2.70)$$

Since $\Lambda_- \lesssim 1$, in norm, we see that

$$V_E^{pair} < \left(\frac{V_{ext}}{m}\right) V^{np} . \quad (2.71)$$

Actually, unless the incident electron is highly relativistic, i.e. unless $|\vec{k}| \gtrsim m$, (2.71) is an overestimate because

$$\langle u'|V_E^{pair}|u\rangle \sim \langle u'|\frac{V_{ext}\Lambda_- V_{ext}}{2m}|u\rangle$$

$$= (2m)^{-1}\langle u'|V_{ext}\left[\Lambda_-, V_{ext}\right]|u\rangle \sim \langle u'|V_{ext}\left[\vec{\alpha}\cdot\vec{p}, V_{ext}\right]|u\rangle/4m^2 \; . \tag{2.72}$$

On replacing $\vec{\alpha}$ by $\vec{p}/m$ we get

$$\langle u'|V_E^{pair}|u\rangle \sim \langle u'|V_{ext}\frac{\vec{p}^{\,2}}{4m^3}V_{ext}|u\rangle \tag{2.73}$$

so that the pair-effects are of relative order $(v^2/c^2)(V_{ext}/m)$ in a scattering situation.

With regard to bound states, the difference between the energy levels of (2.64) and (2.67) can be estimated by taking the expectation value of V_E^{pair} with a bound state solution $\psi_+^{(o)}$ of (2.57). Using (2.69) we expect, for a Coulomb binding potential, $V_{ext} = -Z\alpha/r$, that

$$\Delta E = \langle\psi_+^{(o)}|V_E^{pair}|\psi_+^{(o)}\rangle \sim \langle\psi_+^{(o)}|V_{ext}\frac{\vec{p}^{\,2}}{4m^3}V_{ext}|\psi_+^{(o)}\rangle$$

$$\sim (Z\alpha)^4(Z^2 Ry) \; . \tag{2.74}$$

Actually, this estimate is valid only for $\ell \neq 0$ states, because the operator $V_{ext}\vec{\alpha}\cdot\vec{p}\,V_{ext}$ is too singular at the origin ($\sim r^{-3}$) to have a finite expectation value for S-states. A more careful evaluation[1] shows that for such states

$$\Delta E \sim (\alpha Z)^3\;(Z^2 Ry). \tag{2.75}$$

Incidentally, we can conclude from this analysis, without further calculation, that the energy-levels of (2.57) will coincide with those of (2.64) up to terms of order $(\alpha Z)^2\; Z^2 Ry$. For $\ell \neq 0$ states the energy levels of (2.52) differ from those of (2.64) only in terms of order $(\alpha Z)^4 Z^2 Ry$. However, for $\ell = 0$, there is a spurious term of order $(\alpha Z)^3 Z^2 Ry$, in the expansion of the eigenvalues of h_+ in powers of αZ. This is canceled by the term just discussed, arising from V_E^{pair}.

III. DERIVATION OF NO-PAIR HAMILTONIANS FROM Q.E.D.

Let us now see how the hamiltonian H_+ of eq. (1.22) emerges from QED. In the radiation gauge, the total hamiltonian of QED is given, apart from renormalization counter terms, by

$$H_{QED} = H_{mat} + H_{rad} + H_T \qquad (3.1)$$

where, in the S-picture,

$$H_{mat} = H_{ext} + H_C \qquad (3.2)$$

with

$$H_C = \frac{1}{2} \iint \frac{\rho(\vec{x})\rho(\vec{x}')}{|\vec{x}-\vec{x}'|} d\vec{x}\, d\vec{x}' \qquad (3.3)$$

$$H_T = -\int \vec{j}(\vec{x})\cdot\vec{A}_T(\vec{x})\, d\vec{x} \qquad (3.4)$$

and

$$j^\mu(\vec{x}) = -e{:}\bar{\psi}_D(\vec{x})\gamma^\mu\psi_D(\vec{x}){:} \qquad (3.5)$$

We neglect H_T at first and concentrate on the matter Hamiltonian H_{mat}.

A. Main Theme

On substituting the expansion (2.19) into (3.3) one gets sixteen terms, which may be organized in the form

$$H_C = H_C^{np} + H_C^{pair} . \qquad (3.6)$$

Here H_C^{np} is the sum of four terms which correspond to scattering processes involving <u>no</u> pair-creation or destruction, and H_C^{pair} is the sum of the remaining terms. More explicitly, we have

$$H_C^{np} = \frac{1}{2} \Sigma \; [M_C(n_1'n_2';n_1n_2)A^\dagger(n_1')A(n_1)A^\dagger(n_2')A(n_2)] + \dots , \qquad (3.7a)$$

where the dots denote similar terms involving one or more $BB^\dagger$ factors and

$$M_C(n_1'n_2';n_1n_2) = \iint d\vec{x}_1 d\vec{x}_2 u^\dagger_{n_1'}(\vec{x}_1)u^\dagger_{n_2'}(\vec{x}_2)\frac{e^2}{|\vec{x}_1-\vec{x}_2|}u_{n_1}(\vec{x}_1)u_{n_2}(\vec{x}_2). \qquad (3.7b)$$

The coefficient M_C may be symbolized, in an obvious extension of the graphical notation introduced in Sec. II, by

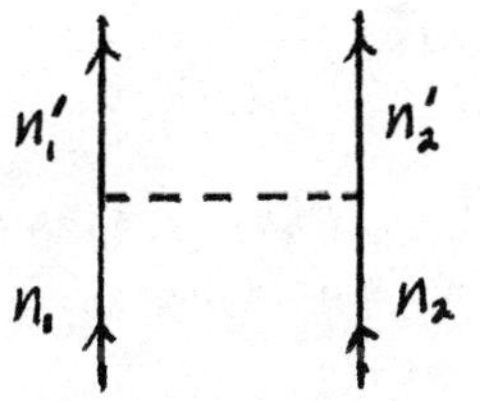

Fig. 4. Graph representing a term in H_C^{np}.

Strictly speaking, this picture should be drawn only after the A^+AA^+A product in (3.7) is normal ordered. With $A(n_1)A^+(n_2') = - A^+(n_2') A^+(n_1) + \delta(n_1,n_2')$, the δ-term gives rise to a Coulomb self-energy which must be kept for the purposes of eventual mass renormalization, when virtual photons are included. The operator factor in (3.79) now becomes

$$-A^\dagger(n_1') A^\dagger(n_2') A(n_1) A(n_2) .$$

The minus sign looks suspicious until it is remembered that the $A^\dagger$'s anticommute, so that this operator coincides with

$$[A(n_1')A(n_2')]^\dagger A(n_1)A(n_2) .$$

The coefficient M_C now looks natural again.

The coefficients of the operators in H_C^{pair} may be similarly symbolized, e.g.

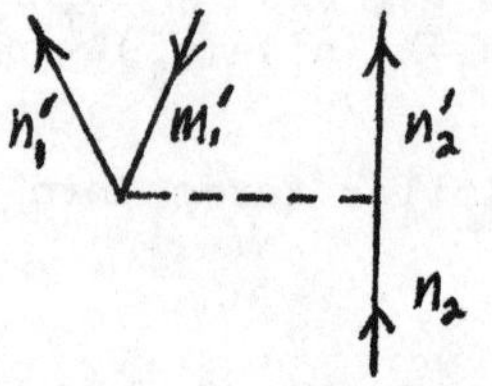

Fig. 5. Graph representing a typical term in H_C^{pair}.

represents the coefficient of the operator $A^\dagger(n_1')B^\dagger(m_1')A^\dagger(n_2')A(n_2)$, viz.

$$\iint d\vec{x}_1 d\vec{x}_2\, u^\dagger_{n_1'}(\vec{x}_1)\, u^\dagger_{n_2'}(\vec{x}_2)\, \frac{e^2}{|\vec{x}_1-\vec{x}_2|}\, v_{m_1'}(\vec{x}_1)\, u_{n_2}(\vec{x}_2) \quad (3.7c)$$

Since H_C^{np} commutes with the number operators N_e and N_p defined in Sec. II. we adjoin it to H_{ext} and define our "zero-order" matter Hamiltonian by

$$H_{mat}^{(o)} = H_{ext} + H_C^{np} . \quad (3.8)$$

Because $H_{mat}^{(o)}$ commutes with N_e and N_p, we may look for an eigenstate

$$H_{mat}^{(o)}\Psi = E\Psi \quad (3.9)$$

with $N_e\Psi = N\Psi$, $N_p\Psi = 0$. The general form of such a Ψ is

$$\Psi = \sum_{n_1\ldots n_N} f(n_1,\ldots,n_N)A^\dagger(n_1)\ldots A^\dagger(n_N)|Vac\rangle . \quad (3.10)$$

<u>Exercise</u>: Substitute (3.10) into (3.9) and show that the Fock-space wavefunction f must satisfy

$$(\varepsilon_+(n_1) + \varepsilon_+(n_2) + \ldots)\, f(n_1,n_2,\ldots) +$$

$$+ \sum_{i<j}\sum_{n_i',n_j'} \langle n_i n_j|\frac{e^2}{|\vec{x}-\vec{x}'|}|n_i'n_j'\rangle\, f(n_1,\ldots,n_i',\ldots n_N)$$

$$= E\, f(n_1, n_2, \ldots n_N). \tag{3.11}$$

To find the configuration-space equivalent of (3.11) we follow the procedure used several times in Sec. II. We define

$$\psi(\vec{x}_1, \ldots, \vec{x}_N) = \sum_{n_1 \ldots n_N} f(n_1, n_2, \ldots n_N) u_{n_1}(\vec{x}_1) \ldots u_{n_N}(\vec{x}_N), \tag{3.12}$$

multiply (3.11) by $u_{n_1}(\vec{x}_1) \ldots u_{n_N}(\vec{x}_N)$ and sum on the n_i.

<u>Exercise:</u> Show that the result of this procedure is

$$\left(\sum_{i=1}^{N} H_{D;ext}(i) + \sum_{i<j} L_+(i)\, L_+(j)\, \frac{e^2}{r_{ij}} \right)\psi = E\psi. \tag{3.13}$$

The projection operators

$$L_+(i) = \sum_n u_n(i)\, u_n^\dagger(i) \tag{3.14}$$

arise because the sum in (3.11) is only over the labels n_i associated with positive-energy one-electron states. Since, from its definition, ψ satisfies

$$L_+(i)\psi = \psi \tag{3.15}$$

we may rewrite (3.14) in the manifestly hermitian form

$$\left[\sum_{i=1}^{N} H_{D;ext}(i) + \sum_{i<j} L_+(i)L_+(j)\, \frac{e^2}{r_{ij}}\, L_+(i)L_+(j) \right]\psi = E\psi \tag{3.16}$$

Finally, since $L_+(k)$ commutes with e^2/r_{ij} if $k \neq i$ and $k \neq j$, we may rewrite (3.16) in the more compact, but equivalent form

$$H_+\psi = E\psi \tag{3.17}$$

where, as advertised in Sec. I, eq. (1.22),

$$H_+ = \sum H_{D;ext}(i) + L_+\, V_{ee}\, L_+ \tag{3.18}$$

and L_+ is defined by (1.23). This completes our study of the connection of H_+ with QED.

B. Variations

It should be noted that for the same money we can derive a whole slew of similar equations, by varying what we regard as the "external potential," provided we are willing to pay a price later on. To be explicit, we may introduce an as yet unspecified "potential" U, which need not be local, and write

$$H_{D;ext} = H_U + H'_U \tag{3.19a}$$

where

$$H_U = \int d\vec{x}\ \psi_D^\dagger(\vec{x})(\vec{\alpha}\cdot\vec{p}+\beta m+U)\psi_D(\vec{x}) \tag{3.19b}$$

and

$$H'_U = \int d\vec{x}\ \psi_D^\dagger(V_{ext}-U\ )\psi_D. \tag{3.19c}$$

We may expand ψ_D in terms of positive-energy eigenfunctions $\{\phi_n^{(+)}(\vec{x})\}$ and negative-energy eigenfunctions $\{\phi_m^{(-)}(\vec{x})\}$ of

$$H_{D;U} = \vec{\alpha}\cdot\vec{p}+\beta m+U \tag{3.20}$$

and define new creation operators $A_U^\dagger(n)$, $B_U^\dagger(m)$, via

$$\psi_D(\vec{x}) = \sum_n A_U(n)\ \phi_n^{(+)}(\vec{x}) + \Sigma\ B_U^\dagger(m)\ \phi_m^{(-)}(\vec{x}). \tag{3.21}$$

A vacuum state $\hat{\Psi}_o$ is now defined by the relations

$$A_U(n)\hat{\Psi}_o = 0, \qquad B_U(m)\ \hat{\Psi}_o = 0 \tag{3.22}$$

and number operators by

$$\hat{N}_e = \Sigma\ A_U^\dagger(n)A_U(n), \qquad \hat{N}_p = \Sigma\ B_U^\dagger\ (n)\ B_U(n). \tag{3.23}$$

We may decompose both H'_U and H_C into no-pair and pair parts, and write

$$H_{mat} = \hat{H}_{mat}^{(o)} + \hat{H}_{mat}^{pair} \tag{3.24a}$$

with

$$\hat{H}^{(o)}_{mat} = H_U + H'^{np}_U + \hat{H}^{np}_C \tag{3.24b}$$

and

$$\hat{H}^{pair}_{mat} = H'^{pair}_U + \hat{H}^{pair}_C \quad . \tag{3.24c}$$

The caret indicates that the operator in question differs from the corresponding operator defined in Sec. II, because $A_U(n)$, etc. differs from A(n) and so the meaning of "pair" and "no-pair" has changed. The eigenvalue equation

$$\hat{H}^{(o)}_{mat}\Psi = E\Psi \tag{3.25a}$$

with the constraint

$$\hat{N}_e\Psi = N\Psi, \; \hat{N}_p\Psi = 0 \tag{3.25b}$$

is now equivalent, following the steps already outlined for the case U=0, to the configuration-space equation

$$\hat{H}_+\psi = E\psi \tag{3.26a}$$

where

$$\hat{H}_+ = \sum_{i=1}^{N} (\alpha_i \cdot \vec{p}_i + \beta_i m + U(i)) + \hat{L}_+(V^{tot}_{ext} - U^{tot} + V_{ee})\hat{L}_+ \tag{3.26b}$$

with

$$\hat{L}_+ = \hat{L}_+(1) \ldots \hat{L}_+(n) \tag{3.26c}$$

and

$$\hat{L}_+(i) = \sum_n |\phi^{(+)}_n(i)\rangle\langle\phi^{(+)}_n(i)| \quad . \tag{3.26d}$$

Note that, on the one hand, for $U \to V_{ext}$, $\hat{L}_+(i) \to L_+(i)$ and $\hat{H}_+ \to H_+$, so that we recover (3.7). On the other hand, for $U \to 0$, $\hat{L}_+(i) \to \Lambda_+(i)$ and $\hat{H}_+ \to h_+$, where

$$h_+ = \sum_{i=1}^{N} H_D(i) + \Lambda_+ (V_{ext}^{tot} + V_{ee}) \Lambda_+, \tag{3.27a}$$

with

$$\Lambda_+ = \Lambda_+(1)\ldots\Lambda_+(n) \tag{3.27b}$$

the product of free positive-energy projection operators.

I will come back to the question of the optimum choice of U after a discussion of the HF-type equations associated with $\hat{H}_+$. Note that, because $\hat{L}_+\psi = \psi$, with $\langle\psi|\psi\rangle=1$ we have the relation

$$E = \langle\psi|\hat{H}_+|\psi\rangle = \langle\psi|H_{DC}|\psi\rangle \quad , \tag{3.28}$$

where H_{DC} is defined by (1.18). Thus, if one is <u>given</u> a solution ψ of (3.26) the total energy may be calculated as the expectation of value of the familiar H_{DC}. But ψ is <u>not</u> determined by H_{DC}. In fact for no choice of U is $\hat{H}_+ = H_{DC}$. Moreover, whether U = 0 or V_{ext} or something in between, $\hat{H}_+$ is not simply obtained from H_{DC} by sandwiching H_{DC} between projection operators. Thus, with

$$\hat{H}_{DC;+} \equiv \hat{L}_+ H_{DC} \hat{L}_+, \tag{3.30}$$

$\hat{H}_+ \neq \hat{H}_{DC;+}$ because $\hat{L}_+^2 = \hat{L}_+ \neq 1$. However, one can write eq. (3.26a) in the equivalent form

$$\hat{H}_{DC;+}\psi = E\psi \ . \tag{3.31}$$

This seems to be as close as one can get to the form (1.19) if one starts from field theory.

Sec. IV. RELATIVISTIC HARTREE-FOCK TYPE EQUATIONS

The simplest Hartree-Fock type of approximation $\bar{\psi}$ to a solution ψ of

$$\hat{H}_+\psi = E\psi \tag{4.1}$$

may be defined by the requirement

$$\delta \langle\tilde{\psi}|\hat{H}_{+}|\tilde{\psi}\rangle = 0 \tag{4.2}$$

with

$$\tilde{\Psi} = \frac{1}{\sqrt{N!}} \text{Det}\,(\tilde{\psi}_1(1)\ldots\tilde{\psi}_N(N)) \tag{4.3a}$$

and

$$\langle\tilde{\psi}_i|\tilde{\psi}_j\rangle = \delta_{ij}\,. \tag{4.3b}$$

The constraint

$$\hat{L}_{+}(i)\tilde{\psi} = \tilde{\psi} \tag{4.4a}$$

is satisfied if

$$\hat{L}_{+}(1)\ \tilde{\psi}_j(1) = \tilde{\psi}_j(1) \qquad (j = 1, \ldots, N). \tag{4.4b}$$

Since H_{+} differs from H_{DC} only by the presence of the projection operators, it is plausible that the HF equations differ from the usual ones only in the presence of these operators. Indeed, from

$$\delta\ [\langle\tilde{\Psi}|\hat{H}_{+}|\tilde{\Psi}\rangle - \Sigma\lambda_{ij}\ \langle\tilde{\psi}_i|\tilde{\psi}_j\rangle] = 0, \tag{4.5}$$

one finds

$$\begin{aligned}
&[\vec{\alpha}_1\cdot\vec{p}_1+\beta_1 m+U(1)]\tilde{\psi}_i(1)\\
&+ \hat{L}_{+}(1)[V_{ext}(1)-\tilde{U}(1) + \sum_{j\neq i} \langle\tilde{\psi}_j(2)|\frac{e^2}{r_{12}}|\tilde{\psi}_j(2)\rangle]\hat{L}_{+}(1)\tilde{\psi}_i(1)\\
&+ \hat{L}_{+}(1)\sum_{j\neq i} \langle\tilde{\psi}_j(2)|\frac{e^2}{r_{12}}|\tilde{\psi}_i(2)\rangle\hat{L}_{+}(1)\ \tilde{\psi}_j(1) = \sum_j \lambda_{ij}\ \tilde{\psi}_j(1)
\end{aligned} \tag{4.6}$$

Although in deriving (4.6) from (4.5), the constraint (4.4) was not explicitly imposed, the result is unchanged when this constraint is included via further Lagrangian multipliers. It should be noted, moreover, that the constraints (4.4b) are compatible with (4.6) because $\hat{L}_+(1)$ commutes with all the operators occurring on the left-hand side of (4.6).

As described in detail in Ref. 15 for the cases $U = 0$ and $U = V_{ext}$, the nonrelativistic limit of the system (4.6) coincides with the familiar nonrelativistic HF equations. If we drop the projection operators $\hat{L}_+(i)$ in (4.6), the usual DHF equations are recovered. For the case $U = V_{ext}$, the single-particle energies $\varepsilon_i = \lambda_{ii}$ determined from (4.6) will coincide with the ones determined by the DHF equations up to terms of order $(\alpha Z)^4 m$.

Sec. V. PAIR EFFECTS

Let us imagine we have found the eigenvalues, E_n of H_+ defined by

$$H_+\psi_n = E_n\psi_n , \tag{5.1}$$

corresponding to the choice $U=0$. It behooves us to try to estimate the magnitude of the terms omitted from consideration so far. The effects of repeated transverse-photon exchange can be included approximately by replacing e^2/r_{ij} by $e^2/r_{ij} + B_{ij}$, where B_{ij} is the Breit operator. This was discussed in some detail in Ref. 15, so let me focus here on the effect of H_C^{pair}. If Ψ_n denotes the Fock-space vector associated with ψ_n, the level shift arising from H_C^{pair} is, in leading order,

$$\Delta E_{n;c}^{pair} = \langle\Psi_n|H_C^{pair} \frac{1}{E_n - H_{mat}^{(o)}} H_C^{pair}|\Psi_n\rangle . \tag{5.2a}$$

Only intermediate states $\Psi_{int}^{N+1;1}$ containing N+1 electrons and one positron can contribute, so

$$\Delta E_{n;c}^{pair} = \sum_{int} \frac{\langle\Psi_n|H_C^{pair}|\Psi_{int}^{N+1;1}\rangle\langle\Psi_{int}^{N+1;1}|H_C^{pair}|\Psi_n\rangle}{E_n - E_{int}} . \tag{5.2b}$$

If we neglect the effects of electron-electron and electron-positron interaction in the intermediate states, we may visualize $\Delta E^{\text{pair}}_{n;c}$ as arising from a sum of diagrams of the type

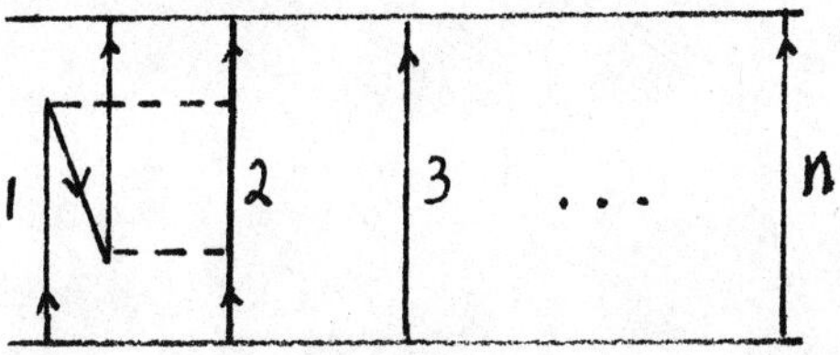

Fig. 6. Coulomb one-pair contribution to $\Delta E^{\text{pair}}_{n;c}$.

a similar one with 1 and 2 interchanged, and a diagram of the type

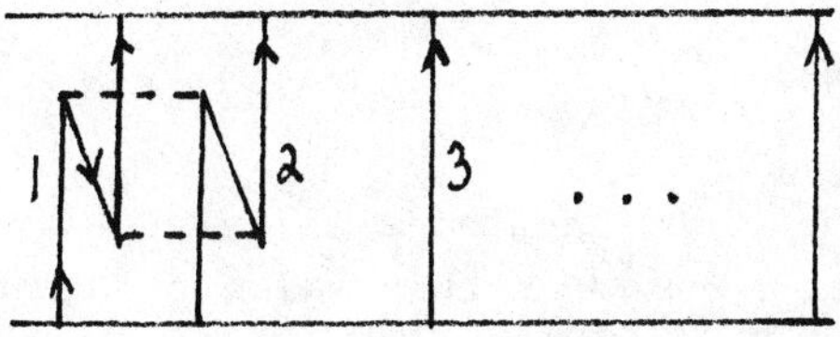

Fig. 7. Coulomb two-pair contribution to $\Delta E^{\text{pair}}_{n;c}$.

For N = 2, the leading term in $\Delta E^{\text{pair}}_{n;c}$ was calculated long ago[1], with the result that, with ϕ the n.r. limit of ψ,

$$\Delta E^{\text{pair}}_{n;c} \approx \frac{-4}{3}\frac{\alpha^2}{m^2}\langle\phi|\delta(\vec{r}_{12})|\phi\rangle. \tag{5.3}$$

Since there are N(N-1)/2 electron pairings, it follows that for $N \sim Z >> 1$,

$$\Delta E^{\text{pair}}_{n;c} \sim Z^2\frac{\alpha^2}{m^2}\langle\psi|\delta(\vec{r}_{12})|\psi\rangle\ . \tag{5.4}$$

On dimensional grounds

$$\langle\psi|\delta(\vec{r}_{12})|\psi\rangle = 1/a^3 \tag{5.5}$$

where a is a length. Now, for large Z

$$\langle\phi|\frac{1}{r_1}|\phi\rangle \sim \frac{Z^{1/3}}{a_o} \tag{5.6a}$$

and[17]

$$\langle\phi|\frac{1}{r_{12}}|\phi\rangle \sim \frac{1}{3}\langle\phi|\frac{1}{r_1}|\phi\rangle \ . \tag{5.6b}$$

This suggests that

$$a \gtrsim a_o/Z^{1/3} \tag{5.7}$$

and hence

$$\Delta E^{pair}_{n;c} \gtrsim -Z^2 \frac{\alpha^2}{m^2} \frac{Z}{a_o^3} \sim -(\alpha Z)^3 \ Ry. \tag{5.8}$$

Thus $\Delta E^{pair}_{n;c}$ is always a small fraction of the total binding energy $W_n = E_n - Nm \sim -Z^{7/3}$ Ry:

$$\Delta E^{pair}_{n;c}/W_n \lesssim \alpha^3 Z^{2/3} . \tag{5.9}$$

Even for z = 80, the right-hand side is only ~ $(4\times10^{-7})(18) \sim 10^{-5}$. However, in absolute terms, for this value of Z the bound given by the r.h.s. of (5.8) is ~ 3eV, which is not negligible in view of the accuracy with which, e.g., X-ray energies are quoted.

If we look just at the two K-shell electrons, the corresponding interaction energy δE_K is

$$\delta E_K \sim -\alpha^2/m^2 a_K^3 \tag{5.10}$$

where a_K is a length characteristic of the K-shell. Taking $a_K \sim a_o/Z$, we get

$$\delta E_K \sim -\alpha^2 Z^3/m^2 a_o^3 \ . \tag{5.11}$$

The fact that the r.h.s. of (5.11) and of (5.8) coincide is just a measure of the crudeness of our estimates. If we use a non-relativistic two-electron hydrogenic wavefunction (not a good approximation for Z ~ 80) we get $a_K \sim (16\pi)^{1/3} a_o/Z \sim 4a_o/Z$ which leads to a much smaller ΔE_K, of order .1 eV.

It should be emphasized that there are other terms of the same order of magnitude as (5.2), such as those coming from one-photon exchange combined with the action of H^{pair}_C and from two-photon exchange. Although these have been computed for He,[1,2] a similar calculation for a heavier atom has not yet been carried out.

Of course, it should be remembered that there are also radiative and vacuum-polarization effects analogous to those entering the Lamb shift in hydrogen. To the extent that the electron-electron interaction can be neglected in the relevant intermediate states, the calculation of these effects in heavy atoms is similar to that in hydrogen, with the difference that the wave funtion for which the corresponding one-body operators must be evaluated is only known approximately.

VI. APPLICATION: $2\,{}^3S_1 \rightarrow {}^1S_o + \gamma$ in He & He-like Ions

The only application which has been made so far to an actual physical problem of the H_+-type hamiltonian, involving the external-field projection operators is, to my knowledge, in the calculation of the α^3Ry corrections to the energy levels of He.[1,2] I want to describe briefly an application of more recent vintage of the formalism discussed in Sec. II. This is to radiative transitions of the type

$$2\,{}^3S_1 \rightarrow {}^1S_o + \gamma \tag{6.1}$$

in He and He-like ions. This is an M1 transition, with $\Delta J = 1$ and no change in parity. However, it is highly forbidden because in a nonrelativistic description the matrix element

$$\langle\psi_F|\vec{\sigma}_1\cdot\vec{k}\times\vec{\varepsilon}^* + \vec{\sigma}_2\cdot\vec{k}\times\vec{\varepsilon}^*|\psi_I\rangle \tag{6.2}$$

vanishes, both because the spatial wave functions associated with ψ_I and ψ_F are orthogonal, and because the transition operator is proportional to $\vec{S}\cdot\hat{k}\times\vec{\varepsilon}^*$, with $\vec{S} = (\hat{\sigma}_1+\hat{\sigma}_2)/2$, and $\vec{S}$ is zero acting on the spin-wavefunction of the ${}^1S_o$ state. The analogous transition $2\,{}^2S_{1/2} \rightarrow 1\,{}^2S_{1/2} + \gamma$ in H or H-like ions is also a forbidden M1 transition, but suppressed only by the orthogonality of the nonrelativistic spatial wavefunction.

The transition (6.1) is thus of a highly relativistic character and a relativistic framework is required to calculate the amplitude. For the purpose of this problem it is sufficient to use the decomposition of H_{QED} in which $U = 0$, i.e. $\psi_D(\vec{x})$ is expanded in terms of free creation and annihilation operators. For the case $N = 2$, one is then led to contemplate as wavefunctions for a He-like system, the eigenfunctions of the operator

$$h_+ = H_D(1) + H_D(2) + \Lambda_+(1)\Lambda_+(2)(V_{ext}(1)+V_{ext}(2)+\frac{e^2}{r_{12}})\Lambda_+(1)\Lambda_+(2), \qquad (6.3a)$$

satisfying

$$h_+\phi = E\phi, \qquad \Lambda_+(i)\psi = \phi \; . \qquad (6.3b)$$

The leading terms in the matrix element for a radiative transition

$$\phi_F \to \phi_I + \gamma \; ,$$

apart from a factor $e/\sqrt{2\omega}$, are then given as a sum

$$M = M_{dir} + M^C_{pair} + M^T_{pair} \qquad (6.4)$$

where the "direct term" is just

$$M_{dir} = \langle\phi_F|\vec{\alpha}_1\cdot\vec{\varepsilon}^*e^{-i\vec{k}\cdot\vec{r}_1} + \vec{\alpha}_2\cdot\vec{\varepsilon}^*e^{-i\vec{k}\cdot\vec{r}_2}|\phi_I\rangle \qquad (6.5)$$

and is familiar because it looks like what one would write down in Dirac-one electron theory. The quantity M^C_{pair} arises when virtual-pair production by the Coulomb field of either the nucleus or the electrons is taken into account and corresponds to the sum of the graphs shown below,

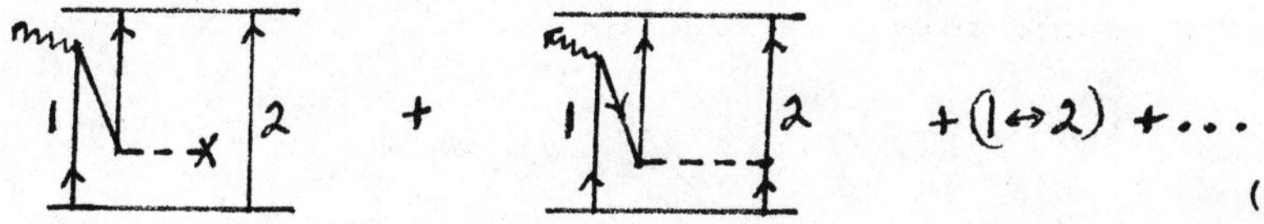

(6.6)

Fig. 8. Contributions to M^C_{pair}.

where the dots denote terms in which the time-order of the photon-emission and Coulomb interaction are reversed. M^T_{pair} denotes contributions from diagrams similar to (6.6) in which the Coulomb interaction is replaced by the exchange of a transverse-photon. If recoil and binding corrections are neglected in the energy-denominators associated with these diagrams, which is equivalent to replacing them by 3m - m = 2m, one gets for the sum $M_{pair} = M^C_{pair} + M^T_{pair}$, the result

$$M_{pair} = 2\langle\phi_F|\vec{\alpha}_1\cdot\vec{\varepsilon}^*e^{-i\vec{k}_1\cdot\vec{r}_1}\frac{\Lambda_-(1)}{2m}(V_{ext}(1)+\frac{e^2}{r_{12}}+b_{12})+r.t.|\phi_I\rangle \quad (6.7)$$

where r.t. denotes the reverse term and b_{12} is the Breit operator. The constraint $\Lambda_+(i)\phi = \phi$ lets us write

$$\phi = (1 + \frac{\vec{\alpha}_1\cdot\vec{p}_1}{m+E(\vec{p}_1)})(1 + \frac{\vec{\alpha}_2\cdot\vec{p}_2}{m+E(\vec{p}_2)})\phi^{(++)} \quad (6.8)$$

where

$$\phi^{(++)} = \frac{1+\beta_1}{2}\,\frac{1+\beta_2}{2}\,\phi \quad (6.9)$$

are the large components of ϕ and M may then be expressed in terms of the matrix elements of reduced operators taken between $\phi_F^{(++)}$ and $\phi_I^{(++)}$, which are essentially Pauli-type wavefunctions. For M1 transitions with orthogonal spatial wavefunctions $\varphi_I(\vec{r})$ & $\varphi_F(\vec{r})$ for the initial and final state one then finds

$$M \approx 2i\ \langle\chi_F|\vec{\sigma}_1\cdot\frac{\vec{k}}{m}\times\vec{\varepsilon}^*|\chi_I\rangle\ \langle\varphi_F|T(1)|\varphi_I\rangle \quad (6.10)$$

where the χ's denote spin-wave functions and

$$T(1) = \vec{p}_1^{\,2}/3m^2 + k^2 r_1^2/12 + r_1 V'_{ext}(r_1)/6m \quad . \qquad (6.11)$$

The integral may be calculated by using Hylleraas type wavefunctions for He and, for large Z, variational wavefunctions obtained by Drake in a 1/Z expansion. The theoretical values for the lifetimes of the 2^3S_1 state may then be calculated for a wide variety of ions in the He-isoelectronic sequence, and compared with experimental values, obtained largely from beam-foil experiments. The very good agreement found between theory and the experimental values, which range over fourteen orders of magnitude, from $\tau \sim 10^4$ sec in He to $\tau \sim .2 \times 10^{-9}$ sec in 34-times ionized krypton, constitutes an impressive triumph for Q.E.D. For a review, with extensive references to the original literature, see Ref. 9.

VII. SUMMARY AND CONCLUDING DISCUSSION

The main aim of the present lectures has been to describe a framework for the simultaneous treatment of relativistic effects and correlation effects in atoms or molecules. It is based on first principles, i.e. on Q.E.D. The approach outlined involves a bona-fide configuration space Hamiltonian such as H_+ (eq. 1.22), which, unlike H_{DC}, has normalizable eigenfunctions. These may be associated with bound states of an N-electron Hamiltonian in the usual way. As discussed in Ref. 15, the effect of iterated single-photon exchange between electrons can be approximately included by replacing e^2/r_{ij} by $e^2/r_{ij} + B_{ij}$, where B_{ij} is the Breit operator. The Hamiltonian

$$H'_+ = \sum_{i=1}^{N} H_{D;ext} + L_+\Big(\sum_{i=j}^{N}\Big(\frac{e^2}{r_{ij}} + B_{ij}\Big)\Big)L_+ \qquad (7.1)$$

has the virtues that an expansion of its eigenvalues in powers of

α will contain <u>all</u> the usual fine structure effects. That is, if

$$H'_{+}\psi = E\psi \ , \tag{7.2}$$

and

$$H_{nr}\phi = W\phi \tag{7.3}$$

with ϕ the extreme n.r. limit of ψ, then

$$E - Nm = W + \langle\phi|H_{fs}|\phi\rangle + \text{higher-order terms} \ . \tag{7.4}$$

If we imagine the solutions to (7.2) or (3.17) known, then it is a straightforward matter to write down expressions for further corrections arising from virtual pair-creation, multiple photon exchange, and radiative effects, because, as illustrated in Sec. V, one can return to the left-over parts $H_T + H_C^{pair}$ of H_{QED} for their consideration. The accurate evaluation of such corrections is of course another matter. Leaving this aside, the main task from the present point of view appears to be the numerical solution of, e.g., eq. (7.2) or (3.17). The most practical hope for this seems to be pursuit of the analogue of the MCDF (multi-configuration Dirac-Fock) method for obtaining approximate solutions. The H-F equations arising from (1.22) or (7.1), given in Ref. 15, differ from the usual DHF equations only in the presence of the $L_{+}(1)$ projection operator.

The presence of the projection operators in H_{+} and H'_{+} are what assures that these Hamiltonians have normalizable eigenstates. In the corresponding DHF equations, the projection operators assure that these equations are free of "variational falling," a lack-of-convergence phenomenon found in approaches to the usual DHF equations which use analytic basis sets. However, it should be emphasized that this is a purely technical problem, quite distinct from the absence of bound states for H_{DC}. It does not arise when the DHF equations are solved numerically, with the constraint that the single-particle energies ε_i are positive.[18]

Alternatives to (1.22) or (7.1) are easily written down, as shown in Sec. IV. That is, by writing $V_{ext} = U + (V_{ext}-U)$ in H_{QED} and using eigenfunctions of H_D+U for the expansion of $\psi_D(\vec{x})$ one can get a continuous class of no-pair Hamiltonians of the form

$$\hat{H}_+ = \sum_{i=1}^{N} [H_D(i)+U(i)] + \hat{L}_+[V_{ext}^{tot}-U^{tot} + \sum_{i<j} \frac{e^2}{r_{ij}}]\hat{L}_+ \qquad (7.5)$$

or a similar operator $\hat{H}'_+$ which includes the B_{ij}. These operators are equivalent to but not identical with the operators $\hat{L}_+H_{DC}\hat{L}_+$, and $\hat{L}_+H'_{DC}\hat{L}_+$, respectively.

In connection with the choice of U, it has recently been argued by Mittleman,[19] that the "best choice" for U is the H-F potential found by solving the usual DHF equations. With this choice the modified equations (4.6) will coincide with the usual ones; this is readily verified from (4.6) if the constraint (4.4b) is taken into account and use is made of the fact that $\hat{L}_+(1)$ commutes with $H_D(1) + U(1)$ to remove the factor $\hat{L}_+(1)$. However, it must be remembered that if V_{ext} is replaced by such a U, there will appear new terms in the pair-part of the matter Hamiltonian, involving pair creation or destruction by the left-over external potential $\delta U = V_{ext} - U$. In the absence of concrete calculations, it is hard to say how much will be gained or lost by such a replacement.

In any case, the central message of these lectures has been a methodological one: If one wants to go beyond a central field approximation in the treatment of relativistic effects in atoms, one needs to have a configuration space hamiltonian which includes at least some of the physical electron-electron interaction from the outset, as one does in the nonrelativistic case. The Hamiltonians we have considered here do include such interactions without falling prey to continuum dissolution, and have a clear-cut origin in QED, as do their eigenfunctions. While approximation schemes for finding eigenvalues and eigenfunctions of such Hamiltonians will, generally speaking, have to deal with the projection operators, the complication presented by these do not appear to be insuperable. I hope to return to this problem elsewhere.

Acknowledgements

I would like to thank J. P. Desclaux, W. E. Schwarz and S. Vosko for stimulating discussions, which influenced the written version of these lectures. I am grateful to Professor G. Malli for his warm hospitality. This work was supported in part by the National Science Foundation.

APPENDIX A: Nonrelativistic Q.E.D. and the Concept of a Relativistic Effect in Atomic Physics

1. Definition of nonrelativistic Q.E.D.

In the Schrödinger picture, the quantized transverse radiation field $\vec{A}_T(\vec{x})$, may be written in the form

$$A_T(\vec{x}) = \frac{1}{(2\pi)^{3/2}} \int \frac{d\vec{k}}{\sqrt{2\omega}} \sum_{\lambda=1}^{2} [\vec{\varepsilon}_\lambda(\vec{k}) e^{i\vec{k}\cdot\vec{x}} a_\lambda(\vec{k}) + \text{h.c.}], \qquad (A.1)$$

where the annihilation and creation operators $a_\lambda(\vec{k})$ and $a_\lambda^\dagger(\vec{k})$ satisfy

$$[a_\lambda(\vec{k}), a_{\lambda'}(\vec{k}')] = \delta_{\lambda\lambda'} \delta(\vec{k}-\vec{k}') \ . \qquad (A.2)$$

The energy operator describing free radiation is, with $\omega \equiv |\vec{k}|$,

$$H_{rad} = \int d\vec{k} \sum_\lambda \omega \, a_\lambda^\dagger(\vec{k}) a_\lambda(\vec{k}) \ . \qquad (A.3)$$

With $\vec{E}_T(\vec{x})$ and $\vec{B}_T(\vec{x})$ <u>defined</u> by

$$\vec{E}_T(\vec{x}) = -i\left[H_{rad}, A_T(\vec{x})\right], \quad \vec{H}_T(\vec{x}) = \vec{\nabla}\times A_T(\vec{x}) \ . \tag{A.4}$$

H_{rad} may be rewritten in the classical-looking form

$$H_{rad} = \frac{1}{2}\int :\vec{E}_T^2(\vec{x}) + \vec{H}_T^2(\vec{x}): d\vec{x} \tag{A.5}$$

where the colons denote normal ordering. (Note that $\vec{A}_T(\vec{x})$ is independent of the time in the S-picture so that it is somewhat confusing to denote $\vec{E}$ by $-\dot{\vec{A}}_T$, as is often done.)

A matter Hamiltonian which neglects relativistic kinematics and pair creation, but which includes electron-spin and some v^2/c^2 effects is given, say for a single atom for ease of writing, by

$$H'_{mat} = H_{nr}^{atom} + H'_{fs} \tag{A.6}$$

where H_{nr}^{atom} is the completely nonrelativistic Hamiltonian. The operator H'_{fs} refers to the part of the fine structure operator H_{fs} which does not come from the exchange of a transverse photon between the electrons. Most of the terms in H'_{fs} can be written down on a semi-classical basis; they include the spin-self-orbit interaction, the analogous terms with an electron replacing the nucleus, and the p^4/m^3 kinetic correction. Still, great care is needed in the correct interpretation of the singular operators entering H'_{fs} (see Ref. 12 and papers referred to therein).

A useful "nonrelativistic QED" for atomic or molecular physics may be obtained by defining

$$H_{qed} = H_{rad} + H'_{mat} + H'_T \tag{A.7}$$

where H'_T is the usual interaction with the transverse field obtained by the replacement $\vec{p} \to \vec{p} + e\,\vec{A}_T$, in H_{nr}^{atom}, augmented by a magnetic moment term:

$$H'_T = \sum_{i=1}^{N} [(\vec{p}_i+e\vec{A}_T(\vec{x}_i))^2-\vec{p}_i^{\,2}]/2m - \Sigma\vec{\mu}_i\cdot\vec{H}_T(i) \ . \tag{A.8}$$

Here $\vec{\mu}_i$ is the magnetic moment operator for electron "i" and $\vec{H}_T \equiv \vec{\nabla} \times \vec{A}_T$.

2. Virtues of H_{qed}

Let us see to what accuracy H_{qed} is capable of describing the properties of atoms or molecules. Let ϕ_n denote a bound eigenstate of H_{nr}^{atom},

$$H_{nr}^{atom} \phi_n = W_n \phi_n \quad . \tag{A.9}$$

The fine-structure level shift from H'_{fs} is is given by

$$\Delta W'_n = \langle \phi_n | H'_{fs} | \phi_n \rangle \equiv \langle n | H'_{fs} | n \rangle \tag{A.10}$$

An additional shift ΔW_n^T of comparable magnitude comes from using H'_T in second order,

$$\Delta W_n^T = \langle n; vac | H'_T (W_n - H_{rad} - H_{nr}^{atom})^{-1} H'_T | n; vac \rangle \quad . \tag{A.11}$$

Here the symbol "vac" denotes the photon vacuum state. Insertion of a complete set of intermediate states , antisymmetric under exchange of electron variables, leads to

$$\Delta W_n^T = \sum_m \int \frac{d\vec{k}}{(2\pi)^3 2\omega} \sum_{\lambda=1}^{2} \frac{\langle n | \sum_i \Gamma_{\lambda i}^{\dagger}(\vec{k}) e^{-i\vec{k}\cdot\vec{r}_i} | m \rangle \langle m | \sum_j \Gamma_{\lambda j}(\vec{k}) e^{i\vec{k}\cdot\vec{r}_j} | n \rangle}{W_n - W_m - \omega} \tag{A.12}$$

where

$$\Gamma_{\lambda j}(\vec{k}) = \vec{\varepsilon}_\lambda \cdot \left(\frac{e}{m}\vec{p}_j - i\vec{\mu}_j \times \vec{k}\right) \equiv \vec{\varepsilon}_\lambda \cdot \vec{\Gamma}_j(\vec{k}) \tag{A.13}$$

is a nonrelativistic analogue of the vertex operator $e\,\gamma_j^\mu\,\varepsilon_\mu$ familiar from standard QED. Now comes a tricky point: It seems that the i=j terms in the above sum correspond to electron self-energy effects and should be considered separately. But this is an illusion. Because of the invariance (up to a sign) of the states $|n\rangle$ and $|m\rangle$ under permutations of electron coordinates the $i=j$ and $i\neq j$ terms in (A.12) are all equal! So how can we disentangle the electron self-energy from interaction between electrons? The answer was given long ago by Feynman: Without changing the left-hand side we may extend the sum on "m" to all eigenstates of H_{nr}^{atom}, not just those which satisfy the antisymmetry principle. With this understanding, the $i=j$ terms, while independent of i, and the $i\neq j$ terms, while independent of the pair (i,j) are no longer equal; the $i\neq j$ terms <u>then</u> correspond to genuine interactions between electrons, whereas the $i=j$ terms correspond to self-energy effects similar to those giving rise to the major part of the Lamb shift in hydrogen.

It is an instructive exercise to calculate approximately the sum of the $i\neq j$ terms. For N = 2,

$$\Delta W_n = 2\int \frac{d\vec{k}}{(2\pi)^3\, 2\omega}$$

$$\times \sum_{\lambda=1}^{2} \sum_{\text{all } m} \frac{\langle n|\Gamma_{\lambda 1}(\vec{k})e^{i\vec{k}\cdot\vec{r}_1}|m\rangle\langle m|\Gamma^{\dagger}_{\lambda 2}(\vec{k})e^{-i\vec{k}\cdot\vec{r}_2}|n\rangle}{E_n-E_m-\omega} \quad . \qquad \text{(A.14)}$$

The dominant contributions come from $kr \tilde{<} 1$ or, since $r \sim a_o = (\alpha m)^{-1}$, from $k \tilde{<} \alpha m$. But $E_n-E_m \sim \alpha^2 m$ for the nearest lying states. Neglecting E_n-E_m relative to ω, i.e. neglecting "recoil," and using closure, we get

$$\Delta W_n \approx \Delta W'' \equiv \langle n|b_{12}|n\rangle \ . \qquad \text{(A.15)}$$

where

$$b_{12} = -\,2\int \frac{d\vec{k}}{(2\pi)^3\, 2\omega^2}\, e^{i\vec{k}\cdot\vec{r}_{12}} \sum_{\lambda} \Gamma_{1\lambda}(\vec{k})\, \Gamma^{\dagger}_{2\lambda}(\vec{k}) \ . \qquad \text{(A.16)}$$

A straightforward calculation then gives b_{12} as the sum of the familiar orbit-orbit, spin-other-orbit and spin-spin interaction terms. For particles with no anomalous moment, this coincides with the result obtained originally by Breit from the reduction of the Breit operator, except for the spin-spin contact term, which arises as a surface term from an integration by parts used in the course of the reduction. This contact interaction is just the electron -electron analogue of the electron-proton contact interaction (Fermi term). It is rather amazing that it was overlooked for almost twenty years, despite the fact that the Fermi term had been on the market all the time. The upshot of all this is that with

$$H''_{fs} \equiv \sum_{i<j} b_{ij} \tag{A.17a}$$

and

$$H_{fs} \equiv H'_{fs} + H''_{fs} , \tag{A.17b}$$

the quantity

$$\Delta W^{(2)} = \langle\phi_n | H_{fs} | \phi_n\rangle \tag{A.18}$$

is the complete expression for the order α^2Ry level shift.

The amplitude for one-photon decay of an excited state is given in this framework by

$$M = \langle\phi_F ; 1\gamma | H_T | \phi_I\rangle = (2\omega)^{-1/2} \langle\phi_F | \sum_i \vec{\Gamma}_i^{op} \cdot \vec{\varepsilon}^* e^{-i\vec{k}\cdot\vec{r}_i} | \phi_I\rangle . \tag{A.19}$$

Then

$$M_{el}^{(o)} = (2\omega)^{-1/2} \langle\phi_F | \sum_i \frac{e\vec{p}_i^{op} \cdot \vec{\varepsilon}^*}{m} | \phi_I\rangle \tag{A.20}$$

is the leading term in parity-changing transitions and

$$M_{mag}^{(o)} = (2\omega)^{-1/2} \langle\phi_F | \sum_i \vec{\mu}_i \cdot \vec{k}\times\vec{\varepsilon}^* | \phi_I\rangle \tag{A.21}$$

is <u>nominally</u> the leading term in a transition in which the parity does not change. I emphasize the word nominally because there are cases of interest in which $M_{mag}^{(o)}$ vanishes, although the transition still takes place.

3. Limitations of nonrelativistic Q.E.D.

Let us now ask how much further the hybrid theory described by (A.6) may be taken. With regard to energy levels, we may imagine that (A.9) is solved exactly and the ϕ_m are known. One might think that one could go beyond perturbation theory in the treatment of H_{fs} by including it in the zero-order Hamiltonian and studying the eigenfunctions of

$$H = H_{nr}^{atom} + H_{fs} . \qquad (A.22)$$

This fails because H_{fs} is too singular to lead to a well-defined eigenvalue problem. This is reflected in the circumstance that the attempt to calculate some corrections, nominally of order α^4Ry, via

$$\Delta W^{(4)} = \langle\phi_n|H_{fs} \; (W_n - H_{nr}^{atom})^{-1} \; H_{fs}|\phi_n\rangle \qquad (A.23)$$

founders on the fact that some of the integrals in (A.23) are divergent. Thus (A.22) is only a pseudo-Hamiltonian. In fact, the next corrections to $\Delta W^{(2)}$ are of order α^3Ry, not α^4Ry. This can be seen by returning to the perturbation theory based on H_T itself. As already pointed out, ΔW_n^T contains some of the radiative corrections corresponding to the Lamb shift which involves terms order $\alpha^3 \log \alpha$ Ry. and α^3Ry. The major (lower-energy) part of this shift can still be gotten from the hybrid theory.Some of the α^3Ry corrections without a logarithm come from the $\alpha/2\pi$ correction to the electron g-factor and can be grafted onto the fine-structure operator by replacing $\mu_o\vec{\sigma}_i$ by $\mu_o(1+\frac{\alpha}{2\pi})\vec{\sigma}_i$. However, there are other α^3Ry corrections such as those coming from two-photon exchange, obtained by using H_T in fourth-order perturbation theory, which are not fully obtainable from the hybrid theory.

What about decay amplitudes? It would seem that at least the leading term for M is obtainable from (A.19)for any one-photon transition. However, there is even an exception to this, for transitions of the magnetic type. If (A.21) vanishes, the next term, obtained from the quadratic term in the expansion of $\exp(i\vec{k}\cdot\vec{r})$ in powers of $\vec{k}\cdot\vec{r}$, gives a non-vanishing contribution, of order α^4. However, the coefficient of this term is not given correctly by the hybrid theory: There are terms of the same order of magnitude which arise because the electon must be treated relativistically, i.e. described by the Dirac equation. (For a review, see Ref. 9.)

In summary, the usual framework for atomic or molecular physics, which is basically equivalent to that described above, cannot, strictly speaking, go beyond terms of order α^2Ry in the calculation of energy levels and fails even in leading order for certain transitions of magnetic type. One might therefore define a "relativistic effect" or better a "truly relativistic effect" in atomic physics as an effect which is not described correctly in leading order by the hybrid theory outlined. This definition has a slight drawback: It lumps together effects arising from diverse physical sources: radiative effects, multi-photon exchange effects, and effects arising from the creation and destruction of virtual electron-positron pairs, together with effects associated with the relativistic description of an electron, through the Dirac equation. Its main merit is that it is brief.

REFERENCES

1. J. Sucher, "α^3Ry Corrections to the energy levels of helium," Columbia University Ph.D. Thesis, 1958; available from University Microfilms, Ann Arbor, Michigan.
2. J. Sucher, Phys. Rev. 109, 1010 (1958).

3. E. E. Salpeter, Phys. Rev. 87, 328 (1952).

4. In Ref. 1, eq. (1.1) was called the "no-pair Coulomb ladder equation," in view of its connection with the so-called "ladder approximation" to the Bethe-Salpeter equation, in which only the repeated exchange of a single boson is taken into account.

5. M. Douglas and N. M. Kroll, Ann. Phys. N.Y. 82, 89 (1974).

6. G. E. Brown and D. G. Ravenhall, Proc. Roy. Soc. A208, 552 (1951).

7. G. Feinberg and J. Sucher, Phys. Rev. Lett. 26, 681 (1971).

8. M. H. Mittleman, Phys. Rev. A 4, 893 (1971).

9. J. Sucher, Rep. Prog. Phys. 41, 1781 (1978).

10. J. Sucher, Proceedings of the Workshop on the Foundations of the Relativistic Theory of Atomic Structure, edited by H. G. Berry, K. T. Cheng, W. R. Johnson and Y.-K. Kim (Argonne National Laboratory Report, 1981).

11. G. Breit, Phys. Rev. 34, 553 (1929); 36, 383 (1930); 39, 616 (1932).

12. In Ref. 11, the spin-spin contact interaction was overlooked. For a review, see H. A. Bethe and E. Salpeter, Quantum Theory of One- and Two-Electron Atoms, (Academic Press, New York, 1971).

13. For reviews of DHF and related calculations for heavy atoms, see the lectures of J.-P. Desclaux and of I. P. Grant, in these Proceedings.

14. E. E. Salpeter, Phys. Rev. 87, 328 (1952); R. Karplus and A. Klein, Phys. Rev. 87, 848 (1952); T. Fulton and P. C. Martin, Phys. Rev. 93, 903 (1954).

15. J. Sucher, Phys. Rev. A22, 348 (1980).

16. W. H. Furry, Phys. Rev. 81, 115 (1951); for a review see S. Schweber, An Introduction to Relativistic Quantum Field Theory, (Row, Peterson and Company, Evanston 1961), p. 566.

17. J. Sucher, J. Phys. B: Atom. Molec. Phys. 11, 1515 (1978).

18. For a clear discussion of this problem and a remedy for it, see W. H. E. Schwarz and E. Wechsel-Trakowski, University of Siegen report (1981); F. Mark and W. H. E. Schwarz, ibid (1981).

19. M. H. Mittleman, Phys. Rev. A 24, 1167 (1981).

INCIDENCE OF RELATIVISTIC EFFECTS IN ATOMS

I.P. Grant

Theoretical Chemistry Department
1 South Parks Road, Oxford, OX1 3TG, England
and
Instituto de Fisica
Universidad Nacional Autonoma de Mexico
Mexico 20, D.F., Mexico

INTRODUCTION

From the explosive growth of publications dealing with the study of relativistic effects in atoms, molecules and solids in the last decade, it is clear that the subject has come of age. Some ten years ago, it was possible to write a review of relativistic atomic structure theory[1] that could present a reasonably comprehensive survey of what had been done to date. This is no longer possible, and it will be necessary for us to be selective in this series of lectures.

Before embarking on the more technical aspects of relativistic atomic structure theory, it is useful to ask some simple questions about the incidence of relativistic effects in atoms. Theory will help us to see qualitatively how relativity modifies electron charge distributions, and calculation should reveal trends which will, we hope, be followed by observational data. We shall begin where the textbooks usually leave off.

THE ONE ELECTRON PROBLEM

We start with the classical relativistic energy-momentum four-vector of a free particle moving with velocity $\vec{v}$,

$$p^{\mu} = (p^{o}, \vec{p}), \quad (\mu = 0,1,2,3)$$

where, if $\gamma = (1 - v^2/c^2)^{-\frac{1}{2}}$

$$cp^{o} = E = \gamma mc^2 = mc^2 + \tfrac{1}{2} mv^2 + \frac{3}{8} m \frac{v^4}{c^4} + \ldots$$

$$\vec{p} = \gamma m\vec{v}.$$

Departures from non-relativistic values clearly depend upon the size of $(v/c)^2$, and it seems reasonable to take $(v/c)^2 > 0.1$ as a criterion for saying that relativistic corrections are appreciable. It is straightforward to verify (perhaps using the virial theorem) that the non-relativistic energy E_{nr}, mean kinetic energy T_{nr} and mean potential energy V_{nr} of the n-th bound state of a hydrogenic atom of nuclear charge Z are related by*

$$\langle E_{nr} \rangle = - \langle T_{nr} \rangle = \tfrac{1}{2} \langle V_{nr} \rangle = - \tfrac{1}{2} \left(\frac{\alpha Z}{n}\right)^2,$$

where α is the fine-structure constant. Thus

$$\langle v^2/c^2 \rangle = 2\langle T_{nr} \rangle/mc^2 = (\alpha Z/n)^2,$$

and our criterion leads us to expect relativistic effects to be important if, on average,

$$Z/n \geqq 40.$$

Of course, we may expect more localized effects which will not be predicted by an average argument. Thus an electron in a nearly bound classical orbit ($E_{nr} \uparrow 0$) can still acquire a relativistic velocity if it enters the region near the nucleus in which V_{nr} is strongly negative. We find the critical distance is

$$r \leqq 10^{-3} Z \text{ a.u. (approximately)}$$

and we shall find that low angular momentum orbitals will always have appreciable density within this radius.

Dirac central field problem

We shall follow the notation of reference 1, in which Dirac's equation for an electron moving in a well- defined central force field is written

*We shall use Hartree atomic units (a.u.) throughout, so that $m = 1$, $\hbar = 1$, $e^2/4\pi\varepsilon_o = 1$, $c = \alpha^{-1} \cong 137$. The a.u. of energy is 27.2 eV, or 2Ry.; the a.u. of length is the hydrogen Bohr radius, $a_o = 0.529.10^{-10}$ m.

$$(H\psi)(x) \equiv [c\vec{\alpha}.\vec{p} + (\beta - 1)c^2 + V(r)]\,\psi(x) = i\,\frac{\partial\psi}{\partial t} \tag{1}$$

The symbols $\vec{\alpha}$ and β stand for the usual 4x4 Dirac matrices,[1] and the identity matrix should be inserted where necessary in the obvious way. The symbol $x \equiv x^{\mu} = (ct, \vec{x})$, $\mu = 0,1,2,3$, and the metric (which we shall not use in this lecture) is given by $g_{00} = +1;\ g_{11} = g_{22} = g_{33} = -1;\ g_{\mu\nu} = 0$ if $\mu \neq \nu$. For the Coulomb case,

$$V(r) = -Z/r, \tag{2a}$$

but it is not necessary to restrict ourselves in this way, and we may for example wish to use a nucleus with a distributed charge rather than a mathematical point charge, for which

$$V(r) = -Z(r)/r \tag{2b}$$

where $V(r)$ is now bounded on $[0,\infty]$, and where

$$Z(r) \to Z \quad \text{as} \quad r \to \infty. \tag{2c}$$

The electron rest energy has been subtracted in (1).

Most standard texts (and reference 1) show that eigensolutions of (1) may be classified as simultaneous eigenfunctions of H, $\vec{J}^2$, J_3 and βK (defined below) which can be written

$$\psi(x) = e^{-i\varepsilon t}\,\frac{1}{r}\begin{bmatrix} P_{\varepsilon,\kappa}(r)\ \chi_{\kappa,m} \\ iQ_{\varepsilon,\kappa}(r)\ \chi_{-\kappa,m} \end{bmatrix}. \tag{3}$$

ε is the eigenvalue of H, and $\chi_{\pm\kappa,m}$ are two-component spinors belonging to the irreducible representations of SU(2) formed by the Clebsch-Gordan expansion of the product

$$D^{(\ell)} \times D^{(\frac{1}{2})} = D^{(\ell+\frac{1}{2})} \oplus D^{(\ell-\frac{1}{2})}, \quad \ell = 0,1,2,\ldots$$

so that if $a(= \pm 1)$ is used to label the two representations in the direct sum, the eigenvalues of $\vec{J}^2$ have (half-integral) quantum number $j = \ell + \frac{1}{2}a$, and we write h

$$\kappa = -(j + \tfrac{1}{2})\,a \tag{4}$$

The basis functions $\chi_{\kappa,m}$ are defined by

$$\chi_{\kappa,m}\ \sum_{\sigma} \langle \ell, m-\sigma, \tfrac{1}{2}, \sigma | \ell, \tfrac{1}{2}, \ell+\tfrac{1}{2}a, m\rangle\ Y_{\ell}^{m-\sigma}(\theta,\phi)\phi^{\sigma} \tag{5}$$

where $\phi^{\sigma}, \sigma = \pm\frac{1}{2}$ form a basis for $D^{(\frac{1}{2})}$; in addition to being eigenfunctions of $\vec{J}^2$ with eigenvalue $j(j+1)$, they are eigenfunctions of J_3 with eigenvalue m. Explicitly

$$\chi_{\kappa,m} = \begin{cases} (2j+2)^{-\frac{1}{2}} \begin{bmatrix} -\sqrt{j+1-m} & Y^{m-\frac{1}{2}}_{j+\frac{1}{2}} \\ \sqrt{j+1+m} & Y^{m+\frac{1}{2}}_{j+\frac{1}{2}} \end{bmatrix} & a = -1, \ell=j+\frac{1}{2} \quad (6a) \\ (2j)^{-\frac{1}{2}} \begin{bmatrix} \sqrt{j+m} & Y^{m-\frac{1}{2}}_{j-\frac{1}{2}} \\ \sqrt{j-m} & Y^{m+\frac{1}{2}}_{j-\frac{1}{2}} \end{bmatrix} & a = +1, \ell=j-\frac{1}{2} \quad (6b) \end{cases}$$

We define the operator K, mentioned above, by

$$K = \vec{L}^2 + \vec{S}^2 - \vec{J}^2 - 1 = -1 - \vec{\sigma}.\vec{L}, \quad \vec{S} = \tfrac{1}{2}\vec{\sigma}$$

so that

$$K\chi_{\kappa,m} = \kappa\chi_{\kappa,m} .$$

From (6a) and (6b), we see that under the parity operators, $\vec{x}\rightarrow-\vec{x}$ we have, using the properties of spherical harmonics,

$$\chi_{\kappa,m} \rightarrow (-1)^{\ell}\chi_{\kappa,m}$$

By invoking the identity[1]

$$(\vec{\sigma}.\vec{p}) \left[\frac{F(r)}{r}\chi_{\kappa,m}\right] = \frac{i}{r}\left[\frac{dF}{dr} + \frac{\kappa F}{r}\right]\chi_{-\kappa,m} \qquad (7)$$

we obtain the radial wave equations*

*The signs here, and elsewhere, are a matter of convention. They depend upon the order in which orbital and spin functions are coupled in (5): $\vec{j}=\vec{\ell}+\vec{s}$; and on the position of the factor i in (5), inserted so that P(r) and Q(r) can be chosen to be real. Some authors couple $\vec{\ell}$ and $\vec{s}$ in the reverse order, and i can be placed in front of P(r) rather than Q(r). This is a fruitful source of confusion. The precise convention chosen does not matter as long as <u>all</u> formulas are internally consistent.

$$\begin{bmatrix} \varepsilon - Z(r)/r & c(\frac{d}{dr} - \frac{\kappa}{r}) \\ c(-\frac{d}{dr} - \frac{\kappa}{r}) & 2c^2 + \varepsilon - Z(r)/r \end{bmatrix} \begin{bmatrix} P(r) \\ Q(r) \end{bmatrix} = 0 \tag{8}$$

As in (3) their solutions may be classified by (ε,κ), or, for bound solutions, in terms of a principal quantum number n and κ. The following labels are equivalent: (n,k), (n,j,a), and (n,ℓ,j); we also use (n,ℓ) for the case $j = \ell + \frac{1}{2}$ and $(n,\bar{\ell})$ for the case $j = \ell - \frac{1}{2}$ where the value of ℓ is to be associated with the P-component which, as we shall see shortly, approaches the non-relativistic radial amplitude in the formal limit $c \to \infty$. Table 1 makes these relations clear.

Angular Densities

The electron density is obtained from $\psi(x)$, equation (3), and its adjoint

$$\rho_{\varepsilon\kappa m}(x) = \psi^+(x)\, \psi(x)$$

$$= \frac{1}{r^2} [A_{\kappa m}(\theta) |P_{\varepsilon\kappa}(r)|^2 + A_{-\kappa,m}(\theta) |Q_{\varepsilon\kappa}(r)|^2]$$

where

$$A_{\kappa m}(\theta) = \chi^{\dagger}_{\kappa,m}\, \chi_{\kappa,m}$$

is easily seen to be independent of the azimuth ϕ. Hartree[2] showed that it is also independent of the sign of κ, so that the density factorizes - a point text books seem to take for granted - into a radial part and an angular part:

$$r^2 \rho_{\varepsilon\kappa m}(x) = D_{\varepsilon\kappa}(r) . A'_{\kappa m}(\theta) . \tag{9}$$

Table 1

κ:	-1	+1	-2	+2	-3	....
j:	1/2	1/2	3/2	3/2	5/2	....
ℓ:	0	1	1	2	2	....
a:	1	-1	1	-1	1	....
Label:	s	$\bar{p}$	p	$\bar{d}$	d	....

Table 2, which displays the Dirac and Schrödinger angular densities, shows some resemblances between $A_{\kappa m}(\theta)$ and $A'_{\ell m}(\theta)$ for $\ell = |\kappa| - 1$, though the Dirac distributions have less distinct lobes in general,

Table 2. Dirac and Schrödinger angular densities

Dirac ($A_{\kappa m}(\theta)\times 4\pi$)			Schrödinger ($A'_{\ell m}(\theta)\times 4\pi$)		
$\lvert\kappa\rvert$	$\lvert m\rvert$		ℓ	$\lvert m\rvert$	
1	1/2	1	0	0	1
2	3/2	$3/2\ \sin^2\theta$	1	1	$1/2\ \sin^2\theta$
	1/2	$1/2\ (1 + 3\cos^2\theta)$		0	$\cos^2\theta$
3	5/2	$15/8\ \sin^4\theta$	2	2	$15/8\ \sin^4\theta$
	3/2	$3/8\ \sin^2\theta(1 + 15\cos^2\theta)$		1	$15/2\ \sin^2\theta\cos^2\theta$
	1/2	$3/4(3\cos^2\theta-1)^2+3\sin^2\theta\cos^2\theta$		0	$1/4\ (3\cos^2\theta-1)^2$

which could have an influence on chemical binding in the heaviest atoms. The sum rules

$$\sum_{m=-j}^{j} A_{\kappa m}(\theta) = \frac{2j+1}{4\pi}$$

$$\sum_{m=-\ell}^{\ell} A'_{\ell m}(\theta) = \frac{2\ell+1}{4\pi}$$

show that filled Dirac subshells, like Schrödinger ones, have a spherically symmetric charge distribution.

Energy levels and radial density distributions

Standard texts give solutions of the radial equations for a Coulomb potential. We quote here only results for bound states; see Johnson and Cheng[3] for continuum states and an extension to quantum defect theory. The energy levels are given by Sommerfeld's formula

$$\begin{aligned}\varepsilon_{n\kappa} &= c^2\{[1 - (\frac{\alpha Z}{N})^2]^{\frac{1}{2}} - 1\} \\ &= -\frac{Z^2}{N^2}\{[1 - (\frac{\alpha Z}{N})^2]^{\frac{1}{2}} + 1\}^{-1} \qquad (10)\end{aligned}$$

where the apparent principal quantum number, N, is given by

$$N^2 = (n_r+\gamma)^2+(\alpha Z)^2 , \quad \gamma = + [\kappa^2 -(\alpha Z)^2]^{\frac{1}{2}}, \ \alpha Z < |\kappa|$$

and the principal quantum number, n, is given by

$$n = n_r + |\kappa| , \quad n_r = 0,1,2,....$$

This definition ensures that[4]

$$N = [n^2 - 2n_r (|\kappa| - \gamma)]^{\frac{1}{2}} \leqq n.$$

Evidently

$$\varepsilon_{n\kappa} \leqq -\frac{Z^2}{2n^2}$$

so that one-electron theory gives relativistic energy levels below the non-relativistic ones, characterized by the magnitude of the angular quantum number, $|\kappa| = j + \frac{1}{2}$, and so by j. Expanding (10) in powers of αZ gives the familiar fine structure formula

$$\varepsilon_{n\kappa} = -\frac{Z^2}{2n^2}\{1 + \frac{(\alpha Z)^2}{n}(\frac{1}{|\kappa|} - \frac{3}{4n}) + 0\ (\alpha Z)^4\}.$$

A similar argument[4] demonstrates the relativistically induced contraction of hydrogenic radial distributions. Consider, for example the mean radius

$$<r_{n\kappa}> = \frac{(\gamma+n_r)(3N^2-\kappa^2) - \kappa N}{2ZN}$$

$$= \frac{1}{2Z}\{\frac{\gamma+n_r}{N}(3N^2 - \kappa^2) - \kappa\}$$

$$< \frac{1}{2Z}\{(3N^2-\kappa^2) - \kappa\}.$$

Since $\kappa(\kappa+1) = \ell(\ell+1)$, and $N < n$, we get

$$<r_{n\kappa}> < \frac{1}{2Z}\{3n^2 - \ell(\ell+1)\} = <r_{n\ell}>,$$

the non-relativistic value. Other expectation values give similar results. Qualitatively, the most important relativistic effects on the radial distribution[4] are:

a) Contributions are most marked for penetrating low angular momentum orbitals, even if the binding energy is small.

b) The "large" component P(r) tends smoothly to the corresponding Schrödinger amplitude in the formal non-relativistic limit $c \to \infty$. To lowest order, (8) shows that

$$cQ(r) \sim \tfrac{1}{2}\left(\frac{dP}{dr} + \frac{\kappa P}{r}\right)\left(1 + 0\left(\frac{1}{c^2}\right)\right)$$

Near the origin, the ratio $Q/P \sim [\varepsilon/(2c^2+\varepsilon)]^{\frac{1}{2}} \approx Z/2n$, so that Q is certainly not small near the origin for high Z atoms.

THE MANY ELECTRON PROBLEM

Now that we have some idea of the behaviour of electron density distributions in the one electron case, it is not difficult to predict what will happen in the many electron atom. As was first observed by Mayers[5], there will be two sorts of effect to study. The inner shells will be hydrogen-like, and we expect them to contract; this modifies the field experienced by valence electrons, giving rise to the indirect effect of relativity. There is also a direct effect due to the need to describe the dynamics of such electrons relativistically when they approach the nucleus.

The inner shell behaviour is dominated by the direct effect, and needs little explanation. As shown by Rose[6], the behaviour of valence electrons is a good deal more subtle. He carried out SCF calculations on three simple neutral atom ground configurations having just one valence electron: $_{71}$Lu 5d; $_{79}$Au 6s; and $_{81}$Tℓ 6p. Each fine structure component was computed separately using a Dirac-Fock program; then the velocity of light was increased by a factor of 1000 to give an effectively non-relativistic solution and the calculation was repeated. Table 3 shows the computed splittings; in the "non-relativistic"case, the values reflect the self-consistency criterion of 1 part in 10^9 of total energy.

Table 3. Calculated and experimental fine structure (cm^{-1}) in the ground terms of $_{71}$Lu and $_{81}$Tℓ

	Expt.	Relativistic	"Non-relativistic"
Lu	1994†	1409	20
Tℓ	9972‡	7684	12

†Reference 7 ‡ Reference 8.

(The calculations were done in double precision on an IBM 360/195, giving about 16 decimals). Only a single CSF was considered, and Breit, QED and core polarization corrections, which have little influence on the effect we wish to study, were ignored. The core electron distributions were frozen, and the calculations repeated so that relativistic dynamics could be used with the non-relativistic core and non-relativistic dynamics with the relativistic core. Ignoring technical complications[6], the results appear in Table 4, which shows the effective one electron energy e in the screened nuclear potential, and the mean radius r. Notice that, for example, $e(6\bar{p})-e(6p)$ is not the splitting reported in Table 3 for the pure relativistic (R) case, since core relaxation in omitted. For the 6s in Au, the calculations with relativistic dynamics give closely similar orbitals, as do the pair calculated with non-relativistic dynamics. The core potential has only marginal effects. The $6\bar{d}$ and 6d orbitals in Lu show the opposite behaviour dominated by the changes in core potential. The contracted relativistic core, from which d orbitals are excluded by an angular momentum barrier, leads to increased screening of the nuclear charge and so to orbital expansion. A small "spin-orbit"splitting due to direct relativistic dynamics is superimposed. The $6\bar{p}$ and 6p orbitals in Tℓ occupy an intermediate position. The $6\bar{p}$ orbital, which has appreciable density near the nucleus, shows a large direct effect and signs of a large indirect effect. The less penetrating 6p orbital has a larger indirect effect but also a non-negligible direct effect.

Table 4. Effective one-electron energy e(cm^{-1}) and mean radius r(a.u.) in various SCF calculations - see text

	Core potential:	R	NR	R	NR
	Valence dynamics:	R	NR	NR	R
$_{79}Au$	e(6s)	-64071	-48455	-49858	-64308
	r(6s)	3.059	3.701	3.655	3.017
$_{81}T\ell$	$e(6\bar{p})$	-46826	-42262	-36959	-54748
	e(6p)	-38595	-42236	-36007	-45395
	$r(6\bar{p})$	3.502	3.925	4.170	3.289
	r(6p)	4.021	3.926	4.249	3.731
$_{71}Lu$	$e(6\bar{d})$	-42372	-53488	-38260	-59073
	e(6d)	-40444	-53447	-38826	-55666
	$r(6\bar{d})$	2.685	2.485	2.876	2.348
	r(6d)	2.784	2.485	2.862	2.428

Thus all valence electron motion is best described by relativistic dynamics. Direct relativistic effect are never completely negligible, and are very important for s and $\bar{p}$ orbitals. The indirect effects of the relativistic contraction of core orbitals dominates the behaviour of electrons with higher angular momenta, and acts in the opposite direction to direct effects. These subtleties are often ignored in perturbation treatments of relativistic effects, and can be responsible for some failures of perturbation approaches which are otherwise attributed to unspecified "correlation". (See lectures by Pyper in this volume[9])

COMPARISON OF DIRAC-FOCK AND HARTREE-FOCK SOLUTIONS

Extensive Hartree-Fock calculations of the LS-coupled ground states of atoms have been published by Fischer[10] and by Mann[11] and corresponding relativistic (DF) solutions using the LS-average of configuration method have been published by Desclaux[12] and Mann and Waber[13]. These tables make extensive comparisons possible, from which I have selected the elements of Groups II and VI of the Periodic Table as representatives.

Each vertical line in Figure 1 is marked with DF (relativistic) eigenvalues on the left, and HF (non-relativistic) equivalents on the right. The main features are as predicted in the last section. Inner shells behave as predicted for hydrogenic systems, and show the same relativistic decrease in eigenvalue, the same ordering in terms of quantum numbers. Radial mean values (not shown) display similar hydrogenic behaviour. Valence shells show the competition between direct and indirect effects. Thus s electrons always have more negative eigenvalues in DF calculations, and the trend for their magnitudes to diminish towards the high-Z end of a Group is ultimately reversed by relativistic effects. This is also true for $\bar{p}$ electrons, whilst p, $\bar{d}$, d, $\bar{f}$, f, ... electrons are usually less tightly bound. Shell ordering is often disturbed by relativity at high Z, for example, 4f and 5s in Hg, and such adjustments can affect the interaction between configurations and modify spectra, both in the optical region and at higher energies.

RELATIVITY AND THE PERIODIC TABLE

So far, we have revealed certain trends which can be attributed to relativistic effects by comparing HF and DF eigenvalues. We may ask: do these accord with observation?

We can first compare MCDF single manifold predictions of first ionization potentials with observed values[14]. (A single manifold, say $(np)^q$, is the set of all jj-coupled CSF of the relativistic configurations $(n\bar{p})^r(np)^{q-r}$, $0 \leq r \leq \min(2,q)$. The atomic levels are thus properly described in intermediate coupling but no further correlation is admitted).

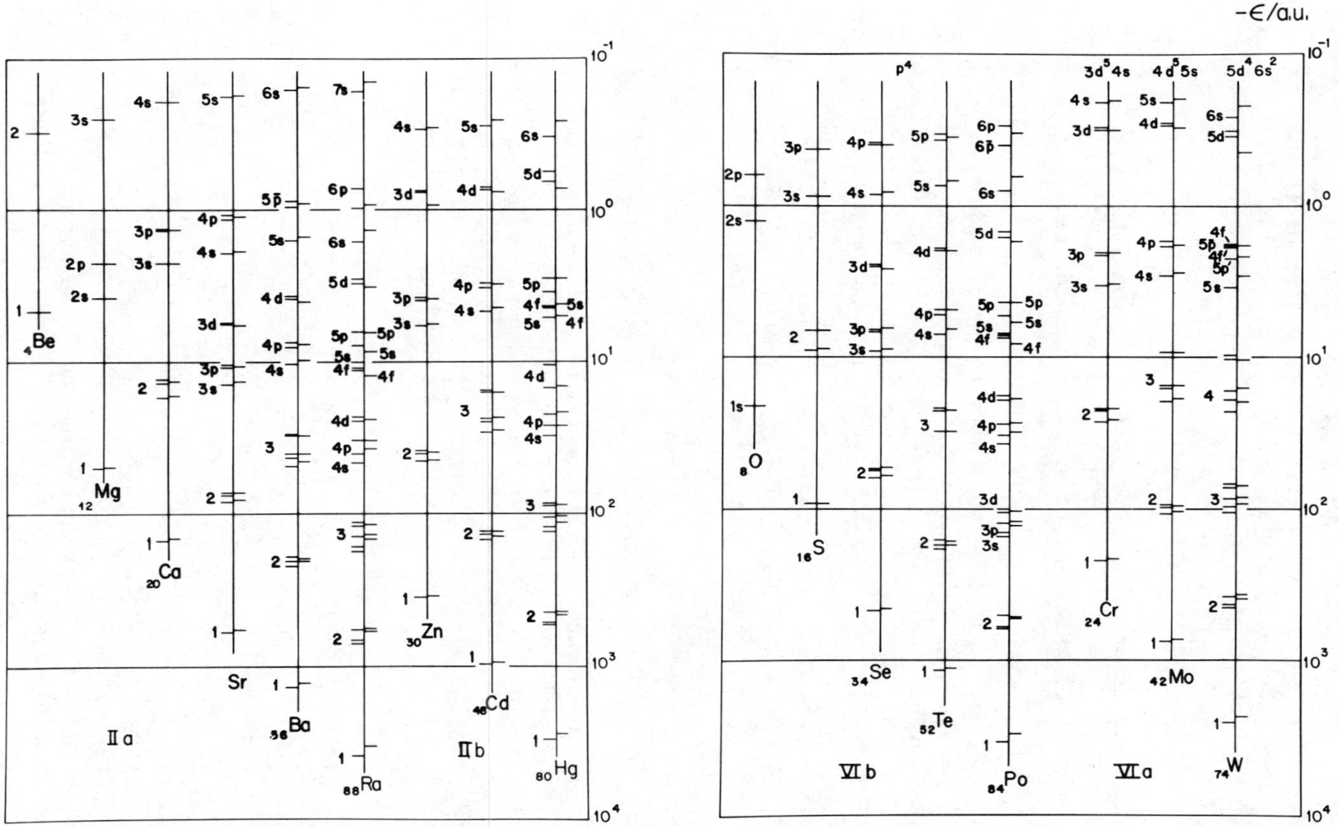

Fig. 1. Comparison of orbital eigenvalues from DF calculations (left side of each vertical line) with those from HF calculations (right side). a) Elements of Group II, b) Elements of Group IV.

Figures 2 a, b, c show theoretical values due to Pyper and Grant[15] obtained as differences of ground state energies of the ion and the neutral atoms. Computed and measured values are linked by straight-lines to bring out trends. Qualitatively, theoretical and observed polygonal traces have much the same shape, the error attributable to neglect of correlation being roughly constant. Whilst there is a general decrease in ionization potential down a group of the Periodic Table, there are some exceptions (the rise in Group IIa from Ba to Ra, in Group IIIb from In to Tℓ, and in Group IVb from Sn to Pb) which are consistent with relativistic predictions.

Although one would like more data to justify the assertion that such simple MCDF calculations are capable of reporducing ionization potential trends down groups, we have found little evidence to the contrary. It is therefore interesting to adopt a different viewpoint and ask if observed ionization potentials show any trace of relativistic effects. The simple model of ionization used by Pyper and Grant[15] motivates our presentation of the data. We simply proposed that the energy of the ground state of the n-th positive ion of an element be written

$$E^{(n)} = E_c + q_a^{(n)}\, e(a) + \tfrac{1}{2}\, q_a^{(n)}(q_a^{(n)} - 1)F(a) \tag{11}$$

Here $q_a^{(n)}$ is the number of electrons, $q_a^{(n)} = N_a - n$, in the valence shell a from which ionization is taking place; e(a) is a mean effective one electron energy in the nuclear potential screened by core electrons; F(a) is a mean pair repulsion energy of two valence electrons; and E_c is the core energy. All parameters are assumed to be independent of n as the shell a is ionized. Thus the n-th ionization potential is

$$I_n = E^{(n)} - E^{(n-1)} = -\, e(a) - q_a^{(n)} F(a)$$

and, in particular,

$$I_o = -\, e(a) - N_a F(a).$$

If the n = 0 configuration is that of the neutral atom, I_o is the electron affinity, which is usually relatively small, and may be replaced by zero as a first approximation. The above equations can be rearranged to give

$$\begin{aligned} I_n &= \frac{n}{N_a}\,(|e(a)| - I_o) + I_o \\ F(a) &= (I_o - |e(a)|)/N_a \end{aligned} \tag{12}$$

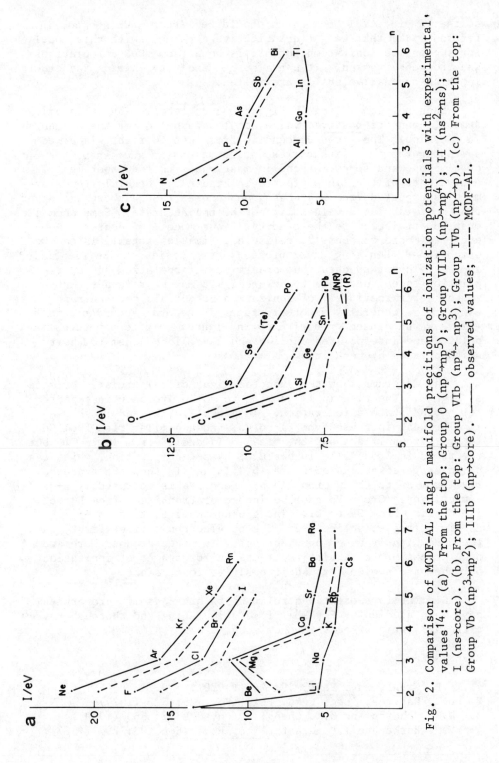

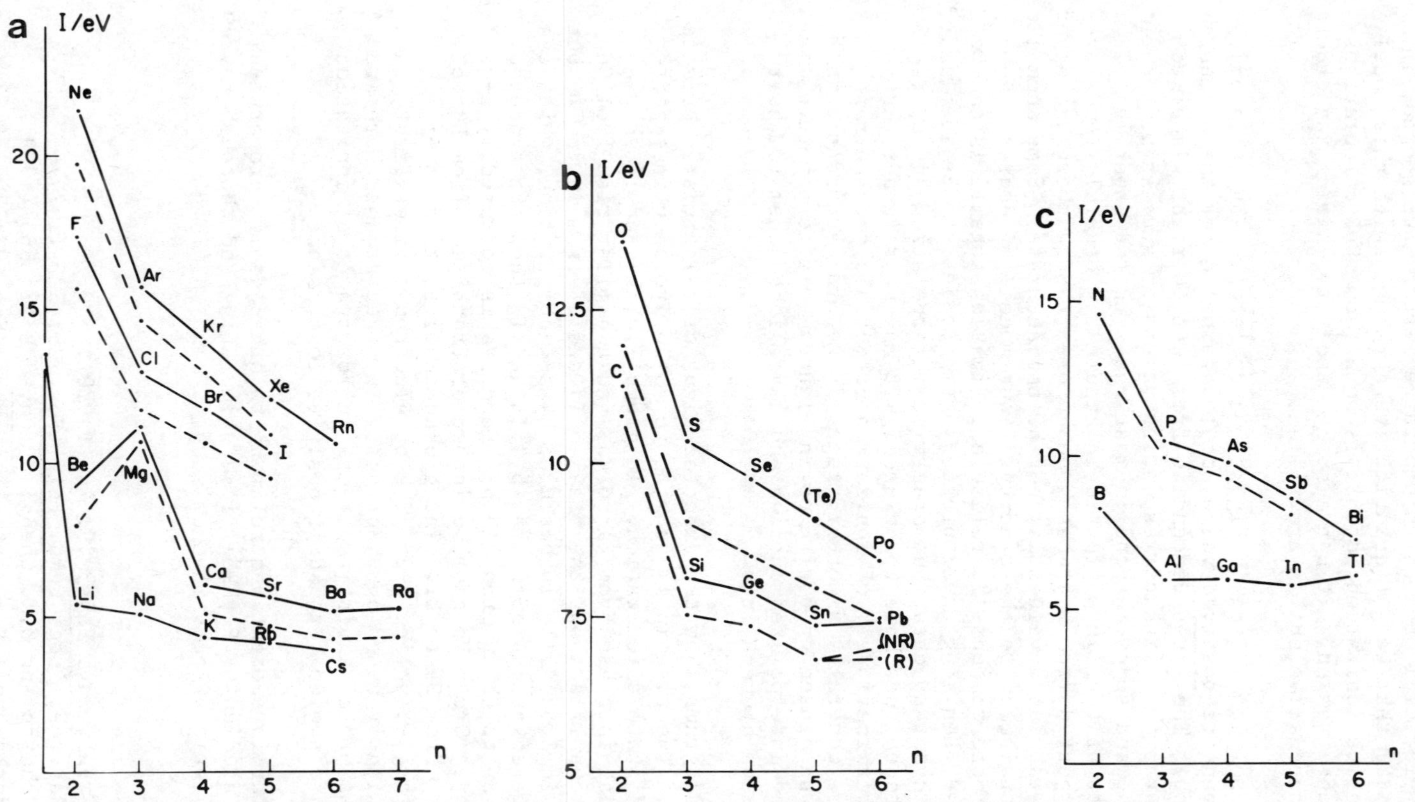

Fig. 2. Comparison of MCDF-AL single manifold precitions of ionization potentials with experimental values[14]: (a) From the top: Group 0 ($np^6 \rightarrow np^5$); Group VIIb ($np^5 \rightarrow np^4$); II ($ns^2 \rightarrow ns$); I (ns→core). (b) From the top: Group VIb ($np^4 \rightarrow np^3$); Group IVb ($np^2 \rightarrow p$). (c) From the top: Group Vb ($np^3 \rightarrow np^2$); IIIb (np→core). ——— observed values; ---- MCDF-AL.

so that a plot of I_n against n should be linear. We showed, in reference 15, that we get quantitative agreement with this model for successive ionization potentials both from MCDF calculations and from observation. It can also be used[15] to examine the order in which successive shells are filled.

Each polygonal line in Figures 3a-d, labelled by its atomic symbol, connects observed values of I_n against n for the element in question. The lines descending from left to right, labelled by configurations, link corresponding values in successive elements of a group, and so correspond in most cases to the experimental equivalents of the lines plotted in Figure 2. General trends are much as expected from what has gone before, though there are signs that some tabulated values may not be too accurate. Some anomalies must be attributed to the crudity of our model, of course. However, as Moore comments[14], reliable ionization potentials can be established when long unperturbed Rydberg series can be identified in the relevant spectra, but otherwise interpolation and extrapolation techniques have to be used. It may be relevant that recent determinations[16] of ionization potentials for the rare earths give generally smoother traces in general accordance with (12). We have used open circles in Figures 3a-d to denote extrapolated and interpolated entries and have linked them to their neighbours by broken lines.

The diagrams show the expected signs of relativistic effects at high Z. The rise in I_1 from Ag to Au in Group Ib is expected for ground state s ionization. In group IIa, I_1 and I_2 show a similar weak increase from Ba to Ra, and a similar rise from Cd to Hg. Group IIIa shows the rise in I_2 and I_3 from In to Tℓ, but the trend in Group IIIb is obscured by s-d competition, which can be interpreted as in reference 15. I_3 and I_4 show the expected rise from Sn to Pb in Group IV b; Group IVa is confused by s-d competition. Group Vb shows evidence of transition from LS to jj-coupling from Sb to Bi. The ground configuration in Bi is nearly pure jj-coupling $\bar{p}^2 p$. The 6p electron ionizes first, then the more tightly bound $6\bar{p}$ electrons. In other groups there are some traces of relativistic effects in low stages of ionization, but they are two weak to admit positive conclusions.

Similar investigations relating to other atomic properties have been studied by Pyykkö[17] in a stimulating and thorough review of relativistic effects in chemistry.

REFERENCES

1. I.P. Grant, Adv. Phys. 19: 747 (1970).
2. D.R. Hartree, Proc. Camb. Phil. Soc. 25: 225 (1929).
3. W.R. Johnson and K.T. Cheng, J. Phys. B 12: 863 (1979),
4. V.M. Burke and I.P. Grant, Proc. Phys. Soc. 90: 297 (1967).

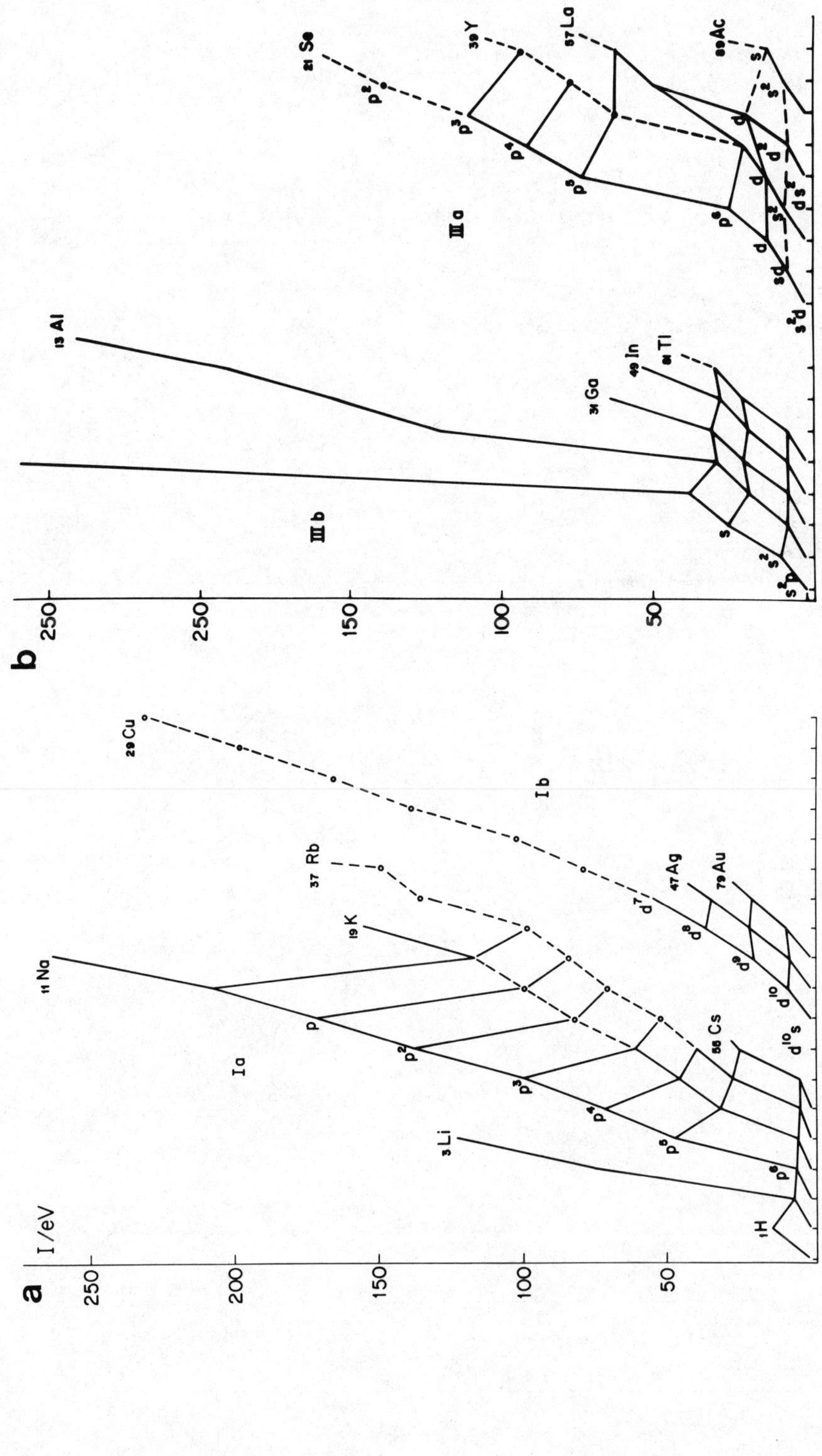

Fig. 3. Trends of successive ionization potentials down some main chemical groups in the Periodic Table (a) Groups Ia, Ib, (b) Groups IIIa, IIIb.

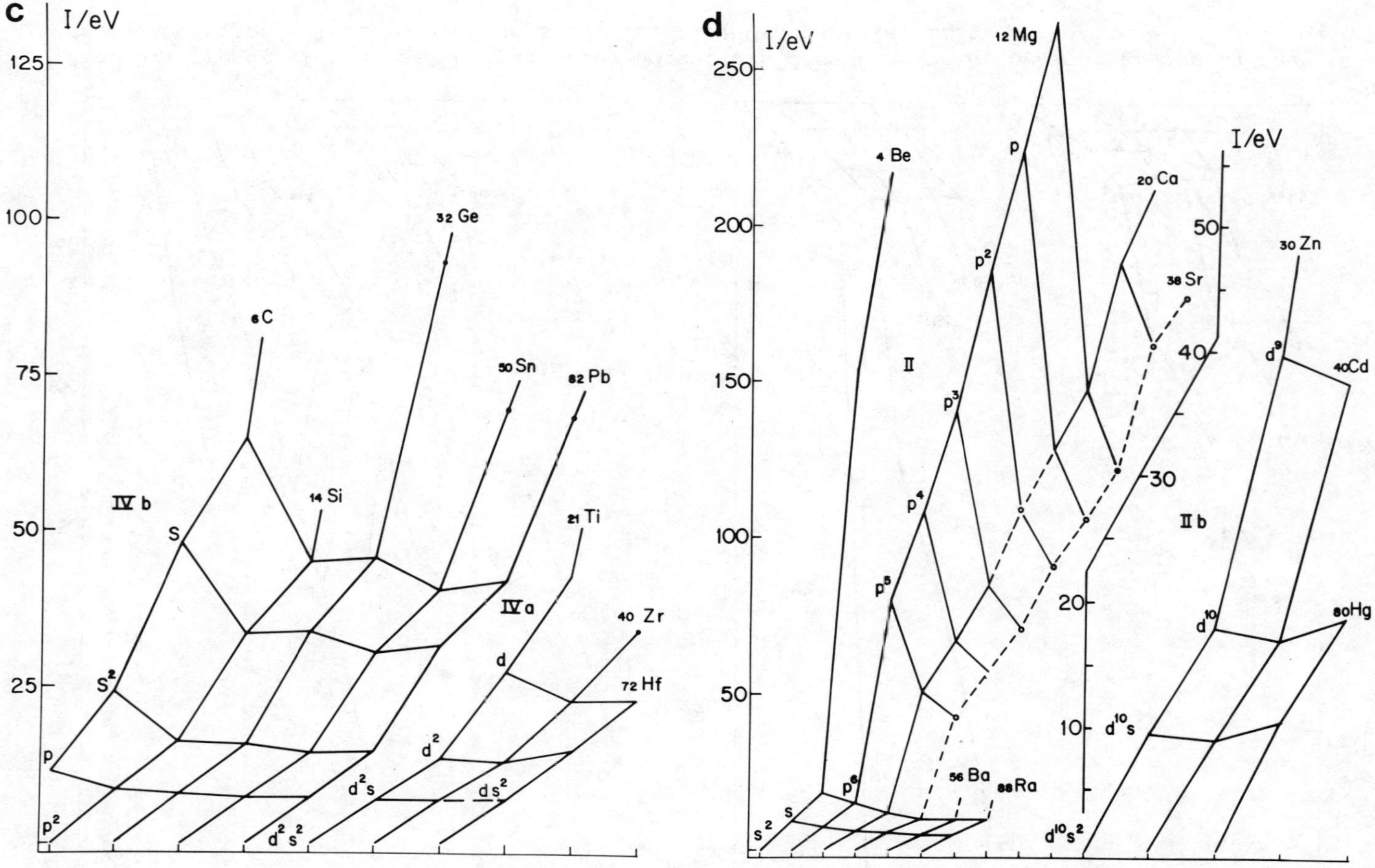

Fig. 3. (continued) (c) Groups IVa, IVb, (d) Groups IIa, IIb.

5. D.F. Mayers, Proc. R. Soc. London. A241: 93 (1957).
6. S.J. Rose, I.P. Grant and N.C. Pyper, J. Phys. B 11: 1171 (1978).
7. P.F.A. Klinkenberg, Physica 21: 53 (1955).
8. C.E. Moore, "Atomic Energy Levels" (NBS Circular No. 467, Vol. 3,), U.S. Government Printing Office, Washington, D.C., (1958).
9. N.C. Pyper, these proceedings.
10. C.F. Fischer, Atomic Data 4: 302 (1972); At. Data. Nucl. Data. Tab. 12: 87 (1973).
11. J.B. Mann, At. Data Nucl. Data. Tab. 12: 1 (1973).
12. J.P. Desclaux, At. Data. Nucl. Data. Tab. 12: 311 (1973).
13. J.B. Mann and J.T. Waber, Atomic Data, 5: 201 (1973).
14. C.E. Moore, "Ionization potentials and ionization limits derived from the analysis of optical spectra"(NSRDS-NBS 34) U.S. Government Printing Office, Washington D.C. (1971).
15. N.C. Pyper and I.P. Grant, Proc. R. Soc. Lond. A359: 525 (1978).
16. W.C. Martin, R. Zalubas, L. Hagan, "Atomic energy levels - the rare earth elements". (NSRDS-NBS 60), U.S. Government Printing Office, Washington D.C. (1978).
17. P. Pyykkö, Adv. Quant. Chem. 11 : 353 (1978).

FORMULATION OF THE RELATIVISTIC N-ELECTRON PROBLEM

I.P. Grant

Theoretical Chemistry Department
1 South Parks Road, Oxford, OX1 3TG, England
and
Instituto de Fisica
Universidad Nacional Autonoma de Mexico
Mexico 20, D.F., Mexico

OBJECTIVES

The first of my lectures examined relatively general features of the relativistic quantum mechanics of atoms in order to show that there is good evidence for taking relativistic effects seriously in atomic and molecular physics. It is therefore disturbing that the methods employed, both in the simple calculations described in the first lecture (and in high precision calculations reported elsewhere in this volume) have often been criticized as lacking proper foundations. In the past, the argument has often gone by default, those concerned being too deeply involved in calculations to devote much time to the tricky arguments needed, and it is now time to remedy the deficiency. Indeed, while the viewpoint is very different from that adopted elsewhere in this volume by Sucher, the final conclusions are not all that different.

PROPERTIES OF THE DIRAC OPERATOR

There are comparatively few calculations in atomic and molecular structure that do not built upon a set of suitably defined one-body wavefunctions. The relativistic theory requires that these should represent states of the Dirac operator, so that an understanding of its properties is an essential preliminary. Some properties have been known for many years, and feature in all the standard text book treatments[1,2]; but others, such as the relativistic orbital contraction[3], and the rigorous study of the self-adjointness of the Dirac operator[4], are more recent, and less well-known to physicists. A particularly lucid and concise account of the self-adjointness properties of the Dirac Coulomb operator has recently been given by Richtmyer[4], and I shall begin by trying to outline its main features.

The appropriate Hilbert space to describe the 4-component solutions of Dirac's equation we denote by $h = (L^2(\mathbb{R}^3))^4$; thus if $\psi(\vec{x})$, $\vec{x}\epsilon\mathbb{R}^3$ is in h, each of its four components, $\psi_\sigma(\vec{x})$ is in $L^2(\mathbb{R}^3)$ as a distribution. The space h has the usual norm $\|\psi\|^2 = \sum_{\sigma=1}^{4}\int_{\mathbb{R}^3}|\psi_\sigma|^2$, and inner product, $(\psi,\phi) = \sum_{\sigma=1}^{4}\int_{\mathbb{R}^3}(\overline{\psi}_\sigma,\phi_\sigma)$, both of which must be bounded. Since the Dirac operator is unbounded, it can only be defined on a dense subset in h. For example, the free-particle Dirac operator can be shown[4] to be self-adjoint on the domain

$$D(T) = \{\psi\epsilon h, \vec{\alpha}\cdot\nabla\psi\epsilon h\}$$

$$T\psi = [-ic\vec{\alpha}\cdot\vec{\nabla}+(\beta-1)c^2]\psi \tag{1}$$

The simple proof uses the fact that T becomes a matrix-valued multiplicative operator when one takes the Fourier transform.

It has been conjectured[4] that when $V(\vec{r}) = -Z/r$, the Dirac operator denifed formally by

$$H = T + V(\vec{r}) \tag{2}$$

is self-adjoint for $Z \leqq 137$ on the space $(W^1(\mathbb{R}^3))^4$, where

$$W^1(\mathbb{R}^3) = \{\psi_\sigma(\vec{x}), \vec{x}\epsilon\mathbb{R}^3 : \psi_\sigma,\ \partial\psi_\sigma/\partial x_i, \text{ are in } L^2(\mathbb{R}^3)\}$$
$$(i = 1,2,3)$$

with the norm

$$\|\psi_\sigma\|_1^2 = \|\psi_\sigma\|^2 + \|\vec{\nabla}\psi_\sigma\|^2 .$$

However, it is simpler and, for our purposes, more illuminating to discuss in more detail the reduced Dirac operator obtained by separating angular radial coordinates. We write the vector transpose $u^\dagger(r) = [P(r),Q(r)]$, and define $h = (L^2(\mathbb{R}))^2$. Then the reduced operator T_κ is self-adjoint on

$$D(T_\kappa) = \{u\epsilon h, \frac{du}{dr}\epsilon h\}\ , \kappa = \pm1, \pm2, \ldots..$$

where

$$T_\kappa u = \begin{bmatrix} 0 & c(-\frac{d}{dr}+\frac{\kappa}{r}) \\ c(\frac{d}{dr}+\frac{\kappa}{r}) & -2c^2 \end{bmatrix} u. \tag{3}$$

One next shows that on $D(T_\kappa)$,

$$H_\kappa^{(Z)} = T_\kappa - Z/r$$

is self-adjoint for $(Z/c)^2 < 3/4$, $Z \leqq 118$. If $u(r)$ is to be an eigenfunction of $H_\kappa^{(z)}$ with eigenvalue ε, then u must be in $D(T_\kappa)$, and in particular, its norm

$$||u||^2 = \int_0^\infty (|P|^2 + |Q|^2)\, dr \tag{4}$$

must be finite. It is now necessary to examine the behaviour at the end points, both of which are singular points of the differential operator. As $r \to \infty$, both $P(r)$ and $Q(r)$ have the same behaviour, either $e^{\lambda r}$ or $e^{-\lambda r}$, where

$$\lambda = + \sqrt{[-2\varepsilon(1 + \varepsilon/2c^2]} \tag{5}$$

is real only when $-2c^2 < \varepsilon < 0$; only one of these solutions at most will be quadratically integrable on, say, $(1,\infty)$ (for any complex ε) and one has to deal with a limit point case. No boundary condition is needed in addition to the condition $u \varepsilon D(T_\kappa)$. Similarly, as $r \to 0$, $P(r)$ and $Q(r)$ both have the same behaviour, varying like r^γ or $r^{-\gamma}$, $\gamma = + \sqrt{[\kappa^2 - Z^2c^2]}$. Only one of these solutions is quadratically integrable on $(0,1)$ for $1 \geqq \gamma > \frac{1}{2}$, and we again have a limit point case. The lowest value of Z for which $\gamma = \frac{1}{2}$ occurs when $|\kappa| = 1$, $Z^2/c^2 = 3/4$ or $Z = 118$. When $\frac{1}{2} \geqq \gamma > 0$, both solutions, behaving like r^γ and $r^{-\gamma}$, are square integrable on $(0,1)$ and we have a limit circle case. A further boundary condition is then needed to define the solution, and it is conventional to require the potential energy to have a finite expectation

$$\int_0^\infty (|P|^2 + |Q|^2) \frac{dr}{r} < \infty . \tag{6}$$

This condition is satisfied automatically for $\gamma > \frac{1}{2}$.

The spectrum of the Coulomb Dirac operator (on its domain of self-adjointness) is shown in Figure 1. There are two continua, $-\infty < \varepsilon < -2c^2$, $0 < \varepsilon < \infty$, together with a point spectrum in the gap $-2c^2 < \varepsilon < 0$, the usual Sommerfeld eigenvalues, all of which are non-degenerate. Only the eigenfunctions of the point spectrum are square integrable, and we see that when this is the case, the boundary conditions at $r = 0$ and $r = \infty$ are uniquely defined.

These results are not sufficiently general for application to

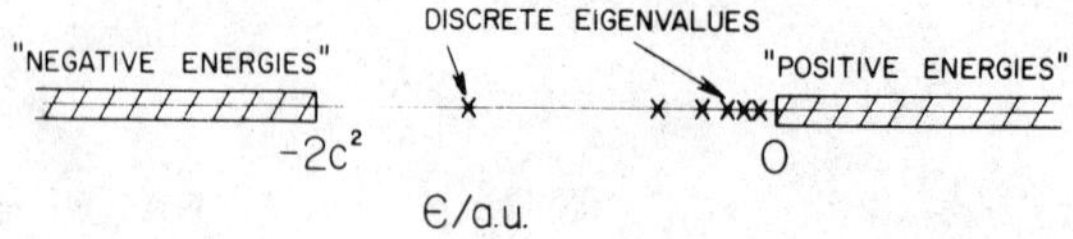

Fig. 1. Spectrum of the Dirac Coulomb Hamiltonian

atomic and molecular structure, though we are unlikely to go far wrong if we take them as guidelines We need to generalize in two directions. The first is to remove the Coulomb singularity at the origin, for the finite radius of the nucleus has an important influence on inner shell binding energies for high -Z elements. This makes $V(r) \to V_1$, $r \to 0$, where V_1 depends on the model charge distribution chosen; for example, $V_1 = -3Z/2r_o$ for a uniform charge distribution of radius r_o. For $\kappa = -1, -2, \ldots$ $\ell = 0, 1, 2, \ldots$ the regular solution is

$$\begin{bmatrix} p_o\ r^{\ell+1} \\ q_1\ r^{\ell+2} \end{bmatrix} (1 + O(r^2))$$

and the irregular solution is

$$\begin{bmatrix} p_1\ r^{-\ell} \\ q_o\ r^{-\ell-1} \end{bmatrix} (1 + O(r^2))$$

so that only one solution is square integrable in a neighbourhood of the origin for any value of ℓ and we always have a limit point case. For $\kappa = 1, 2, \ldots,$ $\ell = 1, 2, \ldots$ the regular solution is

$$\begin{bmatrix} p_1\ r^{\ell+1} \\ q_o\ r^{\ell} \end{bmatrix} (1 + O(r^2))$$

and the irregular solution is

$$\begin{bmatrix} p_o\ r^{-\ell} \\ q_1\ r^{-\ell+1} \end{bmatrix} (1 + O(r^2))$$

so that we have the limit circle case when $\ell = 0$. Solutions are then obtainable for Z > 137, and as shown in Figure 2, the 1s, $2\bar{p}$ and 2s eigenvalues reach the upper edge of the lower continuum for Z ∿ 170, 185 and 245 respectively[5]. Such large charges may effectively be formed in heavy ion collisions as quasi-superheavy atoms, a topic of interest for understanding QED in high fields[5].

The second type of generalization is to relax the restriction to a Coulomb central potential to permit, say, effective atomic central potentials defined by self-consistent fields or molecular potentials. The conditions on the potentials are rather complicated, and we refer the interested reader to the literature[6]. We shall rely on the conjecture that the results outlined above are valid in practice for the atomic case in what follows.

THE PROBLEM OF NEGATIVE ENERGY STATES

From the earliest days, the existence of "negative energy" states - a phrase we shall use to indicate states in the lower continuum, $\varepsilon < -2c^2$ - has posed a problem. It is well-known that among the motives for Dirac's introduction of his wave equation, the avoidance of negative energy states ranked high, along with the need to construct a conserved charge density that would have definite sign. It is therefore ironic that not only do negative energy states appear to be inevitable, but that their interpretation requires the equation to be seen in a many-particle context - as a field equation. For that is essentially the implication of Dirac's notion[7] that in the ground, or vacuum state, all the negative energy levels are filled, in accordance with Pauli's exclusion principle, up to a Fermi level usually taken as $\varepsilon = -2c^2$ on our energy

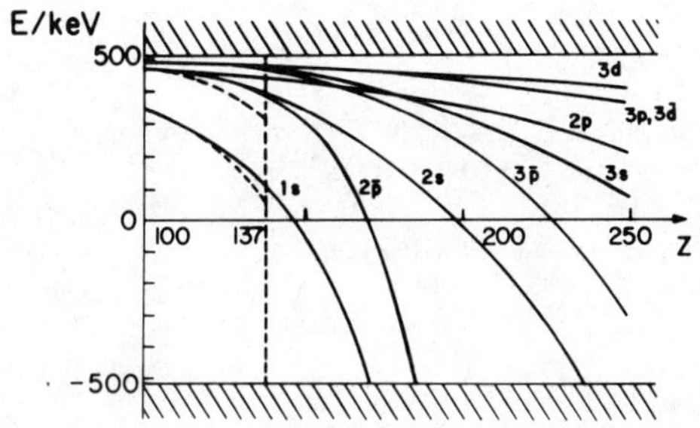

Fig. 2. Schematic variation of hydrogenic eigenvalues as a function of atomic number Z. Dashed lines show point nucleus values, solid lines values for distributed nuclear change.

scale. The reinterpretation of "holes" in this "negative energy sea" as positrons give a different way of looking at this: the vacuum state is the state with no electrons and no positrons.

It is hardly surprising that this problem does not disappear when we have two or more electrons which are allowed to interact. Soon after Dirac's equation was introduced, Breit[8] tried to model the two-electron system by using the classical analogy with Darwin's[9] two-electron hamiltonian function. In more convenient notation, his equation was

$$[H^\circ + V]\psi(x_1,x_2) = i\frac{\partial}{\partial t}\psi(x_1,x_2), \quad x_i = (ct,\vec{x}_i) \tag{7}$$

where

$$H^\circ = h_D(1) + h_D(2), \qquad h_D = c\vec{\alpha}\cdot\vec{p} + c^2(\beta-1) - \frac{Z}{r}$$

a sum of non-interacting Dirac Coulomb operators, and

$$V = \frac{1}{R} + B(0,\vec{R}), \quad \vec{R} = \vec{x}_1 - \vec{x}_2$$

is the sum of interelectronic Coulomb repulsion and another term, the Breit potential, we shall study later. At the time, the only method available to solve (1) was to eliminate the small components of the two-electron function and pass to the non-relativistic limit. This gave the well-known Breit-Pauli hamiltonian[1,2]. First order perturbation theory gave good results for the 2^3P fine structure, provided a term of order higher than $1/c^2$ was discarded. Second order perturbation theory gave a correction as large as the first. Something was badly wrong.

It apparently took nearly 20 years before it was realized, by Brown and Ravenhall[10], that unless something is done in the formalism to cope with the presence of negative energy states, equation (7) has no stationary states. The demonstration is elementary. Suppose first that V = 0; then (7) has a zero order solution

$$\psi_0(x_1,x_2) = \phi_1(x_1)\phi_2(x_2)$$

where ϕ_1 and ϕ_2 are square integrable one electron functions with energies ε_1 and ε_2 in the point spectrum of h_D. Now let V be <u>any</u> two electron interaction. We expand ψ as a linear combination of product states,

$$\psi = C_o\,\psi_o + C_1\psi_1 + \ldots, \qquad \sum |C_1|^2 = 1;$$

can find a term in the sum, say

$$\psi_n(x_1,x_2) = \chi_1(x_1)\chi_2(x_2), \quad \langle\psi_n|V|\psi_1\rangle \neq 0$$

where χ_1 is a Coulomb solution with $\varepsilon = \varepsilon_1 + \Delta$, and χ_2 is a similar solution with $\varepsilon = \varepsilon_2 - \Delta$. For large enough Δ, one or both of these must belong to the continuous spectrum, and ψ can no longer be a bound stationary state. This argument was quoted by Bethe and Salpeter[1] in the context of a discussion of the Breit potential, $B(0,\vec{R})$, though it is apparent that the detailed nature of the interaction is not important. The trouble is that the formalism does not prevent variational collapse into negative energy states, and the remarks by Brown and Ravenhall[10] and Bethe and Salpeter[1] can be interpreted in this sense, although they have usually been represented only as a commandment not to use the Breit potential save in first order perturbation theory. There is a subsidiary question that the Breit potential is not appropriate for large excitation energies, but this is easily dealt with.

It is clear that we need an appropriate formalism to handle the problem, and indeed this is already available in the standard texts on quantum electrodynamics,[2,11]. One method, first used in this context by Brown and Ravenhall[10] and followed by Sucher[11] and Mittleman[12], uses projection operators to eliminate the unwanted contributions. Another approach, which fits naturally into atomic and molecular structure calculations, is to assume an underlying one-electron potential, $U(\vec{x})$, satisfying reasonable conditions to ensure a spectrum of the type of Figure 1, with a complete set of states in h

$$\psi_p(x) = \phi_p(\vec{x})e^{-i\varepsilon_p t} \quad (0 > \varepsilon_p > -2c^2) \tag{8}$$

and such continuum states as may be necessary. We identify states as "electron" ($\varepsilon > -2c^2$) or "positron" ($\varepsilon < -2c^2$), and use the notions of second quantization and normal ordering to achieve the same effects. The underlying potential, $U(x)$, cna be a prescribed model potential or, in the manner of many-body perturbation theory[14] it can be chosen <u>a posteriori</u> to annihilate certain diagrams as in the Hartree-Fock approximation. We must now explain this in more detail.

SECOND QUANTIZATION AND NORMAL ORDERING

Fock space can be described as the direct sum of product spaces such as h^n, $n = 1,2,\ldots$. We define operator-valued solutions of Dirac's equation

$$h_D\psi(x) = i\frac{\partial}{\partial t}\psi(x), \quad h_D = c\vec{\alpha}\cdot\vec{p}+c^2(\beta-1) + U(\vec{x}), \tag{9}$$

of the form

$$\psi(x) = \sum_p^{(+)} a_p\psi_p(x) + \sum_q^{(-)} b_q^{+}\psi_q(x), \tag{10}$$

where the first sum runs over all solutions of (9) with $\varepsilon > -2c^2$, and the second over the remainder. A discrete basis is implied here for simplicity; a more exact notation, involving integration over spectral projections, would complicate the presentation without much benefit to the reader. The operators a_p, a_p^+, b_q, b_q^+ satisfy a set of anticommutation relations

$$\{a_p, a_{p'}^+\} = \delta_{pp'}, \qquad \{b_q, b_{q'}^+\} = \delta_{qq'},$$

$$\{\text{all others}\} = 0. \tag{11}$$

It is standard manipulation to show[2,11] that: a_p^+ creates an electron in state ψ_p; a_p destroys an electron in state ψ_p; b_q^+ creates a positron in state ψ_q; b_q destroys a positron in state ψ_q. The vacuum, having no electrons or positrons, is that state $|0\rangle$ for which

$$a_p|0\rangle = 0, \quad \forall p$$

$$b_q|0\rangle = 0, \quad \forall q \tag{12}$$

Then the operator $n_p^{(+)} = a_p^+ a_p$ has eigenvalues 0,1, and can be interpreted as the number of electrons present in the (non-degenerate) state ψ_p. A similar interpretation holds for $n_q^{(-)} = b_q^+ b_q$

The notion of normal ordering of operators eliminates the infinite zero-point energy of the vacuum and of some other observables. A product of any number of creation and annihilation operators is said to be in normal order if is arranged so that all annihilation operators stand to the right of all creation operators. If it is necessary to permute operators to put a product in normal order, this is to be done as if <u>all</u> anticommutators vanish. Thus if N(...) is used to denote a normally ordered product,

$$N(a_p^+ a_p) = a_p^+ a_p, \qquad N(a_p a_p^+) = -\, a_p^+ a_p$$

and so on. Assume now that all basis functions are orthonormal; indeed this is essential for consistency with the relations (11). Then the total charge-current-density 4 vector is

$$j^\mu(x) = -\, ecN(\tilde{\psi}(x)\gamma^\mu\psi(x)), \quad \tilde{\psi}(x) = \psi^+(x)\gamma^o \tag{13}$$

in terms of the Dirac γ matrices: $\gamma^o = \beta$, $\vec{\gamma} = \beta\vec{\alpha}$; it is easy to show that this is a conserved quantity

$$\partial_\mu j^\mu(x) = 0 \tag{14}$$

(this was an essential feature of Dirac's original construction), or, in more familiar language

$$\frac{\partial\rho}{\partial t} + \mathrm{div}\,\vec{j} = 0, \qquad \rho(x) = \frac{1}{c}\, j^{\circ}(x)$$

The total charge of the system is the expectation of

$$Q = \int d^3x\rho(x) = -\, e\left(\sum_p^{(+)} n_p^{(+)} - \sum_q^{(-)} n_q^{(-)}\right)$$

so that the vacuum expectation value, $<0|Q|0>$, vanishes as expected. Similarly, the energy of the system of non-interacting electrons is the expectation of

$$\begin{aligned} H^D &= \int d^3x\; N(\psi^+(x)\; h_D\psi(x)) \\ &= \sum_p^{(+)} n_p^{(+)}\varepsilon_p - \sum_q^{(-)} n_q^{(-)}\;\varepsilon_q \\ &= \sum_p^{(+)} n_p^{(+)}\varepsilon_p + \sum_q^{(-)} n_q^{(-)}\;|\varepsilon_q| \end{aligned} \tag{16}$$

This is clearly a positive definite operator, with vanishing vacuum expectation $<0|H^D|0> = 0$, as expected.

Similar benefits accrue in this formalism when two-particle interactions are invoked. We discuss how such terms arise in quantum electrodynamics in the next section. We shall find that, to lowest non-vanishing order in the coupling constant, $\alpha = 1/c$, between the electron and photon fields, we shall get an effective atomic hamiltonian.

$$H_{eff} = H^D + H' + H^{(2)} \tag{17}$$

where

$$H' = \int d^3x\; N(\psi^+(x)\;[-\frac{z}{r} - U(\vec{x})]\;\psi(x))$$

is a counterterm to eliminate the potential $U(\vec{x})$ and replace it by the correct electron-nucleus interaction, and $H^{(2)}$ is a two-electron term having the general form

$$\iint d^3x d^4x'\; j^\mu(x)\;\; v_{\mu\nu}(x,x')\; j^\nu(x') \tag{18}$$

where $j^\mu(x)$ is the operator defined in (13), and $v_{\mu\nu}(x,x')$ will turn out to be the configuration space representation of the photon propagator in some gauge. We shall also find divergent and higher order terms describing electron self-energy (Lambshift[15]) and vacuum polarization as well as 3-body[13] and higher order terms. However, as discussed by Desclaux[16], it appears that the operator (17), supplemented by relatively simple corrections for Lamb shift and vacuum polarization, is sufficient in practice to predict most atomic properties at the current level of experimental precision. A definitive test for many-electron systems has still to be made.

How, then, may we use H_{eff}? We first need to construct suitable n-electron states; in general, these can be written[17]

$$|\Psi_n\rangle = \sum c_{\alpha_1 \ldots \alpha_n} a^+_{\alpha_n} \ldots a^+_{\alpha_1} |0\rangle \qquad (19)$$

where the coefficients may be chosen so that the state $|\Psi_n\rangle$ is a basis function for a suitable symmetry group[18]. Because of the anticommutation relations (11) each term in the sum of (19) can be put into one-to-one correspondence with a Slater determinant, so that in general (19) represents a symmetry-adapted, antisymmetric, superposition of configurational wave-functions. If we now consider (17) as a trial wave-function, we can find the expectation of H_{eff}, which will be a sum of terms of the form

$$\langle 0|(\text{Product of } a_\beta) \;\; N(\ldots) \; (\text{Product of } a^+_\alpha)|0\rangle.$$

In effect, we now rearrange this product, using (11) to get pairs, $a^+_p a_p = n_p^{(+)}$, which we can think of as projection operators onto the one-dimensional subspace spanned by the single state ψ_p:

$$n_p^{(+)} (1 - n_p^{(+)}) = 0$$

Only when no creation or annihilation operators are left over do we get a non-null result. As shown by Judd[17], this gives the standard expectation of the energy in terms of Slater integrals, which we shall examine in more detail elsewhere in this volume[18]. It is at this point, though, that we make contact with the approach advocated by Sucher[12]. By restricting (19) to contain electron states only, we have effectively introduced "positive energy projectors" as he advocates. Since parts of the normal products involving positron states cannot enter into the final result, we may as well drop them, this gives his "no pair" hamiltonian. Of course, if we are willing to replace (19) with a wave function that merely conserves charge rather that electron number, we shall introduce terms involving "pair" creation and destruction[12]. The two approaches at this level appear completely equivalent.

To complete this section, we note that we now have an explanation of why Dirac-Fock programs using numerical integration methods are not troubled by variational collapse or "continuum dissolution" in Sucher's phrase[12]. For we have seen that all states in the Hilbert space $\mathcal{h}$ must have specific behaviour as $r \to 0$ and $r \to \infty$ when h_D is self-adjoint, and these boundary conditions are necessary to properly define the numerical orbitals[19]

$\psi_p(x) = \phi_p(\vec{x})\, e^{-i\varepsilon_p t}$. Thus the boundary conditions are equivalent to imposing a positive energy restriction on the terms of (19).

THE ELECTRON-ELECTRON INTERACTION

In deriving an effective interaction between electrons from quantum electrodynamics, most texts[1,2,11] start with the interaction hamiltonian

$$H_{int} = \int d^3x\, j_\mu(x)\, A^\mu(x) \tag{20}$$

for minimal coupling of the electron and photon fields and write down the lowest order contribution to the S-matrix in perturbation theory that conserves the number of particles. This comes from the diagram of Figure 3, and may be written

$$S^{(2)} = \int d^4x_1 \int d^4x_2\, j^\mu(x_1) v_{\mu\nu}(x_1-x_2) j^\nu(x_2) \tag{21}$$

We can think of this formula as the interaction of an electron at x_1 with the field of another electron

$$S^{(2)} = \int d^4x_1 j^\mu(x_1)\, A_\mu(x_1);$$

of course $A_\mu(x)$ must satisfy Maxwell's equations

$$\Box A_\mu(x) = \frac{1}{\varepsilon_o}\, j_\mu(x)$$

where the source on the right hand side is the current of the second electron. This can be solved by standard Green's function methods, and we can then identify $v_{\mu\nu}(x_1-x_2)$ through

$$v_{\mu\nu}(x) = \frac{i}{2\varepsilon_o}\, D_{F\mu\nu}(x) \tag{22}$$

where $D_{F\mu\nu}(x)$ is Feynman's causal propagator for the photon, the solution of

$$\Box\, D_{F\mu\nu}(\gamma) = -\, 2ig_{\mu\nu}\delta^4(x) \tag{23}$$

$g_{\mu\nu}$ being the usual metric in Minkowski's space. Equation (23) is most simply solved by Fourier transformation; standard prescriptions[2,11] give

$$D_{F\mu\nu}(x) = \frac{1}{(2\pi)^4} \int e^{-ik\cdot x}\, \hat{D}_{F\mu\nu}(k)\, d^4k \tag{24}$$

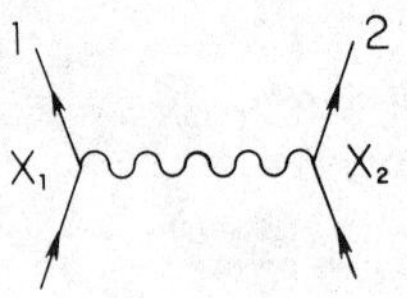

Fig. 3

where

$$\hat{D}_{F\mu\nu}(k) = \frac{2ig_{\mu\nu}}{k^2+i\varepsilon}$$

The small positive imaginary quantity $i\varepsilon$ is here added to displace the poles of $\hat{D}_{F\mu\nu}$ in the complex k^o-plane off the real axis. More generally, one notes that this solution is only unique up to a gauge transformation, and we express this by writing

$$\hat{D}_{F\mu\nu}(k) = \frac{2ig_{\mu\nu}}{k^2+i\varepsilon} + \chi_\mu k_\nu + \chi_\nu k_\mu \tag{25}$$

where χ_μ is an arbitrary 4-vector, and the additional terms maintain the symmetry in the indices μ,ν.

In the original gauge, $\chi_\mu = 0$, the result is

$$V_{\mu\nu}(x) = g_{\mu\nu} \cdot \frac{1}{4\pi\varepsilon_o r} \cdot \frac{1}{2\pi} \int_{-\infty}^{\infty} \exp\left[-i(k_o x^o - |k_o| r)\right] \cdot dk_o \tag{26}$$

We extract the time dependence by using the spectral decomposition of $j^\mu(x)$, using (10) and (13); a typical term will have the form

$$\begin{aligned} j^\mu_{pq}(x) &= -\, ec\psi_p(x)\, \gamma^\mu \psi_q(x) \\ &\equiv j^\mu_{pq}(\vec{x})\, e^{-i\omega_{qp} x^o} \end{aligned} \tag{27}$$

where

$$\omega_{qp} = -\,\omega_{pq} = \frac{\varepsilon_q - \varepsilon_p}{c}$$

Thus, the vector potential due to the second electron has contributions

$$A^{\mu}_{pq}(x_1) = \int d^3x_2 \, \frac{j^{\mu}_{pq}(\vec{x}_2)}{4\pi\varepsilon_o R} \cdot \exp\left[-i\omega_{qp}(x_1^o - R \,\mathrm{sgn}\, \omega_{qp})\right], \quad (28)$$

$$R = |\vec{x}_1 - \vec{x}_2|.$$

When $\omega_{qp} > 0$, this describes a retarded potential, for the exponent has the time-dependence due to a harmonic component associated with the transition frequency ω_{qp} at a time $t_2 = t_1 - R/c$. This corresponds to a photon emission. When $\omega_{qp} < 0$, we get the reverse process and an advanced potential.

We can now rewrite a typical amplitude in $S^{(2)}$ in terms of a matrix element of electron amplitudes $\phi_p(\vec{x})$ and an effective interaction:

$$S^{(2)}_{pp';qq'} = \int d^3x_1 \int d^3x_2 \, \phi^+_p(\vec{x}_1)\phi^+_{p'}(\vec{x}_2) . V(\omega,R) . \phi_q(\vec{x}_1)\phi_{q'}(\vec{x}_2) \quad (29)$$

where

$$V(\omega,R) = \frac{e^2}{4\pi\varepsilon_o} \, \frac{1 - \vec{\alpha}(1) \cdot \vec{\alpha}(2)}{R} \, e^{i|\omega|R/c}$$

The integrations that were performed over the time components x_1^o, x_2^o force conservation of energy, so that

$$\omega = ck_o = \varepsilon_p - \varepsilon_q = \varepsilon_{q'} - \varepsilon_{p'}$$

or

$$\varepsilon_p + \varepsilon_{p'} = \varepsilon_q + \varepsilon_{q'}.$$

Since $e^2/4\pi\varepsilon_o$ is equal to unity in atomic units, we see that $V(\omega,R)$, the Møller interaction, gives relativistic corrections to Coulomb's law.

In fact, virtually all atomic and molecular structure calculations are done in Coulomb gauge, defined by the analogue of the static field condition div $\underset{\sim}{A} = 0$,

$$\sum_{i=1}^{3} k^i \hat{D}_{F\mu i}(k) = 0$$

This splits $S^{(2)}$ into two parts. The first is the (instantaneous) Coulomb potential

$$S^{(2)}_{c;pp',qq'} = \int d^3x_1 \int d^3x_2 \, \phi^+_p(\vec{x}_1)\phi^+_{p'}(\vec{x}_2) \, \frac{e^2}{4\pi\varepsilon_o R} \, \phi_q(\vec{x}_1)\phi_{q'}(\vec{x}_2) . \quad (30)$$

The second part, the contribution of transverse photon components, is more complicated. When the states $\phi_p(\vec{x})$ are sufficiently localized, and ω is small enough so that only regions with $|\omega R|/c \ll 1$ contribute, we get

$$S^{(2)}_{tr;pp',qq'}(\omega\to 0) = \int d^3x_1 \int d^3x_2 \left(\frac{1}{8\pi\varepsilon_o Rc^2}\right)[\vec{j}(\vec{x}_1)\cdot\vec{j}(\vec{x}_2) + (\vec{j}(\vec{x}_1)\cdot\vec{R})(\vec{j}(\vec{x}_2)\cdot\vec{R})/R^2]$$

which is formally equal to the classical expression for the interaction of two currents, giving a correction of relative order $1/c^2$. The corresponding effective interaction is the Breit potential

$$B(0,\vec{R}) = -\frac{e^2}{8\pi\varepsilon_o R}[\vec{\alpha}(1)\cdot\vec{\alpha}(2) + (\vec{\alpha}(1)\cdot\vec{R})(\vec{\alpha}(2)\cdot\vec{R})/R^2] \tag{31}$$

introduced in equation (7). This is sometimes split into two parts, the dominant Gaunt (or magnetic) interaction

$$H^G = -\frac{e^2}{4\pi\varepsilon_o R}\vec{\alpha}(1)\cdot\vec{\alpha}(2),$$

which is the $\omega = 0$ correction to Coulomb's law in the Møller interaction (29) and the residue (the so-called retardation correction)

$$H^{Ret} = \frac{e^2}{8\pi\varepsilon_o R}[\vec{\alpha}(1)\cdot\vec{\alpha}(2)-(\vec{\alpha}(1)\cdot\vec{R})(\vec{\alpha}(2)\cdot\vec{R})/R^2].$$

The full effective interaction due to transverse photons ($\omega \neq 0$) is

$$B_{tr}(\omega,\vec{R}) = -\frac{e^2}{4\pi\varepsilon_o}\{\vec{\alpha}(1)\cdot\vec{\alpha}(2)e^{i|\omega|R/c}-(\vec{\alpha}(1)\cdot\vec{\nabla})(\vec{\alpha}(2)\cdot\vec{\nabla})f(\frac{|\omega|}{c}, R)\}$$

where (32)

$$f(\nu,R) = (1 + i\nu R-e^{i\nu R})/\nu^2 R.$$

We expect quantum electrodynamics to give results that are independent of the gauge chosen for evaluating the potentials. This has never been seriously tested.

The derivation given here emphasizes the fact that both the Coulomb potential and its lowest order relativistic corrections arise from the same diagram in QED, and should be treated together for consistency, contrary to what is usually claimed with the authority of Bethe and Salpeter[1]. One sees also that the Breit potential, $\omega = 0$, is quite inappropriate for matrix elements with large energy differences such as those linking positive and negative energy states. The rapidly oscillating exponentials will then reduce the magnitudes of such matrix elements. There is still one deficiency; the derivation is restricted to two electrons, and it is no longer obvious what ω to use in many-electron cases. However, this can be dealt with by adapting a procedure suggested by Brown and Ravehall[10] and by Mittleman[13]. The idea is to start with a hamiltonian describing the interacting electron and photon fields in quantum electrodynamics

$$H^{QED} = H_d + H_{rad} + H_{int}$$

where H_{int} is given by (20), and to seek a new representation defined by a sequence of unitary transformations, in which H_{int} appears only in higher order in the coupling constant, α. This procedure yields an effective interaction to order $(1/c^2)$ of the form

$$H^{(2)} = \tfrac{1}{2} \sum_{pqrs} N\,(a_p^+ a_q^+ a_r a_s)\{ \langle p,q|\tfrac{1}{R}|sr\rangle + \langle p,q|[\tfrac{1}{2}B(\tfrac{\omega_{ps}}{c}, \vec{R}) + \tfrac{1}{2}(\tfrac{\omega_{qr}}{c},\vec{R})]|sr\rangle\}$$

where we have reinstated the annihilation and creation operators, and p,q,r,s run over a complete set of basis functions in the one electron potential $U(\vec{x})$. The expression $B(\nu,\vec{R})$ is just the real part of (32)

$$B(\nu,\vec{R}) = \text{R.P. } B_{tr}(\nu,\vec{R}).$$

It is this expression we had in mind for equation (17). Mittleman[13] gives further terms corresponding to three-body potentials and radiative corrections, as well as attempting to define a self-consistent potential. The radiative corrections have, of course, been considered in atomic calculations[15,16], though as yet no results are available on the effect of three-body potentials. Mittleman's formalism is a good deal more complicated than that presented in this lecture.

Finally we should remember that our theory is in some respects incomplete. External potentials (Stark effect, Zeeman effect) have not been incorporated, nor have we dealt with the problem of centre of mass motion in the atom. Recent work[20] offers some progress in that direction.

REFERENCES

1. H.A. Bethe and E.S. Salpeter, "Quantum mechanics of One- and Two-Electron Systems" [Handbuch der Physik, Vol XXXV, pp. 88-436], Berlin, Springer-Verlag (1957).
2. V.B. Berestetskii, E.M. Lifshitz and L.B. Pitaevskii, "Relativistic Quantum Theory" Parts 1 and 2. Pergamon Press, Oxford (1971 and 1974).
3. V.M. Burke and I.P. Grant, Proc. Phys. Soc. 90: 297 (1967)

4. R.D. Richtmyer, "Principles of Advanced Mathematical Physics", Vol. 1, Sec.10.17,11.6 Springer-Verlag, New York (1978).
5. J. Reinhardt and W. Greiner, Rep. Prog. Phys. 40: 219 (1977); J. Rafelski, L.P. Fulcher, A. Klein, Phys. Reports 38: 227 (1978).
6. W. Hunziker, Commun Math. Phys. 40: 215 (1975); H. Kalf, U.-W. Schmincke, J. Walter and R. Wüst, Springer Lecture Notes in Mathematics, 448: 182 (1975). see also ref. 4.
7. P.A.M. Dirac, Proc. Roy. Soc. Lond. A126: 360 (1930).
8. G.E. Breit, Phys. Rev. 34: 553 (1929); ibid, 36:383 (1930); ibid, 39:616 (1932).
9. C.G. Darwin, Phil. Mag. 39: 537 (1920)
10. G.E. Brown and D.G. Ravenhall, Proc. Roy. Soc. Lond. A208:552 (1951).
11. J.D. Bjørken and S.D. Drell, "Relativistic Quantum Mechanics" McGraw Hill, New York (1964); S.S. Schweber, "An Introduction to Relativistic Quantum Field Theory", Harper and Row, New York (1964); J.M. Jauch and F. Rohrlich, "Theory of photons and electrons" (2nd ed.) Springer-Verlag, New York (1976); and many other texts.
12. J. Sucher, Phys. Rev. A22: 348 (1980); and elsewhere in these proceedings.
13. M.H. Mittleman, Phys. Rev. A4: 893 (1971); ibid, A5, 2395 (1972).
14. H.P. Kelly, Phys. Scripta 21: 448 (1980); and other references cited therein.
15. P.J. Mohr, these proceedings.
16. J.P. Desclaux, these proceedings.
17. B.R. Judd, "Second Quantization in Atomic Spectroscopy", Baltimore, Johns Hopkins Press (1967).
18. I.P. Grant, these proceedings.
19. I.P. Grant, Adv. Phys. 19: 747 (1970).
20. R.A. Krajcik and L.L. Foldy, Phys. Rev. D10: 1777 (1974); K. Sebastian and D. Yun, Phys. Rev. D19: 2509 (1979).

TECHNIQUES FOR OPEN SHELL CALCULATIONS FOR ATOMS

I.P. Grant

Theoretical Chemistry Department
1 South Parks Road, Oxford, OX1 3TG, England
and
Instituto de Fisica
Universidad Nacional Autonoma de Mexico
Mexico 20, D.F., Mexico

ANGULAR MOMENTUM IN ATOMIC STRUCTURE

As in nonrelativistic atomic structure theory[1-6], angular momentum plays a big part in the construction of n-electron states and of simplifying the matrices of operators in the relativistic theory [7]. Racah algebra, second quantization, and diagram techniques, discussed in the references just mentioned, all play a part in deriving general purpose expressions which are suitable for implementation in a computer program. Some aspect of this work, which have been important in developing one system of relativistic atomic structure programs[8] incorporating radiative corrections[9] will be described in this lecture. Of course, this approach, which depends heavily on the use of Dirac central field orbitals, has little to offer those wishing to study molecular structure, where symmetry has less of a dominant role. It is also fair to point out that Desclaux[10] uses algorithms based on somewhat different principles; it is encouraging that nevertheless, results from both systems of programs are in close agreement.

SECOND QUANTIZATION, OPERATORS AND STATES

Tensor operators and their uses

A set of operators $\{T_q^{(k)}, q = -k, -k+1,\dots,k\}$ is said to constitute the components of an irreducible tensor operator $\vec{T}^{(k)}$ if its members satisfy commutation relations

$$[J_3, T_q^{(k)}] = q\, T_q^{(k)}$$

$$[J_\pm, T_q^{(k)}] = [k(k+1) - q(q\pm1)]^{\frac{1}{2}}\, T_{q\pm1}^{(k)} \tag{1}$$

where $J_\pm = J_1 \pm iJ_2$, J_3 are the components of the angular momentum operator $\vec{J}$ (the infinitesimal generators of the rotation group SO(3)). I shall assume some familiarity with the properties of irreducible tensor operators such as may be found in well-known texts[1-5].

Dirac central field orbitals, for which we shall use orbital labels $\alpha,\beta,\gamma,\delta,\ldots$, are characterized by quantum numbers, say $(n_\alpha,\kappa_\alpha,j_\alpha,m_\alpha)$; here j_α is redundant, since $\kappa_\alpha = -(j_\alpha + \frac{1}{2})a_\alpha$, $\ell_\alpha = j_\alpha - \frac{1}{2}a_\alpha$, $a_\alpha = \pm 1$, but we shall need to use it in formulas. The operators of interest will be either single particle, having matrices $\langle\alpha|f|\beta\rangle$, say, or two particle $\langle\alpha\beta|g|\gamma\delta\rangle$. The classification of orbitals makes it convenient to expand f in terms of irreducible tensor operator components $\vec{f}^{(k)}$. Similarly, two-particle operators, g, are always scalar (tensor operators of rank 0) and can always be expanded in a sum of scalar products of irreducible tensor operators

$$g(\vec{r}_1,\vec{r}_2) = \sum_h g_k(r_1,r_2)\, (\vec{T}^{(k)}(1).\vec{T}^{(k)}(2)) \tag{2}$$

where $g_k(r_1,r_2)$ carries all the dependence on radial coordinates and the scalar product of irreducible tensors is defined by

$$\vec{T}^{(k)}(1).\vec{T}^{(k)}(2) = \sum_q (-1)^q\, T_q^{(k)}(1) T_{-q}^{(k)}(2) \tag{3}$$

Standard techniques[7] relying on Racah algebra and the Wigner-Eckart theorem, enable us to write, using Wigner's covariant notation[11] for 3j-symbols*

---

*[The relation of Wigner's covariant notation to the ordinary 3j-symbol is given by

$$\begin{pmatrix} m & \cdot & \cdot \\ j & \cdot & \cdot \end{pmatrix} = (-1)^{j-m} \begin{pmatrix} j & \cdot & \cdot \\ m & \cdot & \cdot \end{pmatrix}$$

The projection number appears on the top line for a covariant argument, and on the bottom line for a contravariant argument. As legal sums over projection quantum numbers always involve a matched covariant/contravariant pair, the use of this notation minimizes the number of phase factors appearing, and simplifies book-keeping].

$$\langle n\ \kappa\ m|f_q^{(k)}|n'\kappa'm'\rangle = \begin{pmatrix} m & k & j' \\ j & q & m' \end{pmatrix} X^k(n\kappa,n'\kappa') \tag{4}$$

and

$$\langle\alpha\beta|g|\gamma\delta\rangle = \sum_{k,q} \begin{pmatrix} m_\alpha & k & j_\gamma \\ j_\alpha & q & m_\gamma \end{pmatrix}\begin{pmatrix} m_\beta & q & j_\delta \\ j_\beta & k & m_\delta \end{pmatrix} X^k(\alpha,\beta,\gamma,\delta) \tag{5}$$

For the purposes of the present discussion, the details of the X^k factors, which we call effective interaction/transition strengths are unimportant. We give two important examples. First, let

$$f_q^{(k)} = V_k(r)\ C_q^{(k)},$$

where

$$C_q^{(k)} = \left(\frac{4\ \pi}{2k+1}\right)^{\frac{1}{2}} Y_{kq}(\theta,\phi);$$

then[7]

$$X^{(k)}(n\kappa,n'\kappa') = \Pi^e(\kappa,\kappa',k)\langle j\|\vec{C}^{(k)}\|j'\rangle[n\kappa|V_k|n'\kappa'] \tag{6}$$

The terms are, respectively: a parity factor; the jj-coupling reduced matrix element

$$\langle j\|\vec{C}^{(k)}\|j'\rangle = (-1)^{j+j'}[j,j']^{\frac{1}{2}} \begin{pmatrix} j & k & -\frac{1}{2} \\ \frac{1}{2} & 0 & j' \end{pmatrix};$$

and a radial integral

$$[n\kappa|V_k|n'\kappa'] = \int_0^\infty \rho_{n\kappa,n'\kappa'}(r).V_k(r)dr,$$

where

$$\rho_{n\kappa,n'\kappa'}(r) = P_{n\kappa}(r)P_{n'\kappa'}(r) + Q_{n\kappa}(r)Q_{n'\kappa'}(r)$$

is r^2 times a radial overlap density of the participating orbitals, Similarly, for the Coulomb repulsion of two electrons, $g = 1/R$, $R = |\vec{x}_1 - \vec{x}_2|$, we get[7]

$$X^k(\alpha,\beta,\gamma,\delta) = \Pi^e(\kappa_\alpha,\kappa_\gamma,k)\Pi^e(\kappa_\beta,\kappa_\delta,k)$$
$$\times(-1)^k\langle j_\alpha\|\vec{C}^{(k)}\|j_\gamma\rangle\langle j_\beta\|\vec{C}^{(k)}\|j_\delta\rangle R^k(\alpha,\beta,\gamma,\delta) \tag{7}$$

where

$$R^k(\alpha,\beta,\gamma,\delta) = \int_0^\infty \int_0^\infty \rho_{\alpha,\gamma}(r)\, U_k(r,s)\, \rho_{\beta,\delta}(s) dr\, ds,$$

with

$$U_k(r,s) = r^k/s^{k+1} \quad \text{if } r < s,$$
$$= s^k/r^{k+1} \quad \text{if } s < r$$

is a relativistic Slater integral. Comparable formulas for the Breit and transverse interactions are derived in reference 12.

Specification of states

It has been pointed out for example by Judd[2], that there is a one-to-one correspondence between Slater determinants $\{\alpha_1, \ldots, \alpha_n\}$ and Fock space states $a_{\alpha_n}^+ \ldots a_{\alpha_1}^+|0\rangle$, where $\{a_{\alpha_i}^+\}$ denote some set of anticommuting creation operators. This follows because the state labels $\{\alpha_i\}$ are associated with the columns of the Slater determinant, whilst the rows are labelled by the corresponding coordinates $\vec{x}_1, \ldots, \vec{x}_N$. The interchange of two adjacent columns of the Slater determinant introduces a factor (-1), exactly what we find on exchanging neighbouring creation operators. For this reason, the Pauli exclusion principle is built in automatically.

It proves fruitful to exploit the angular momentum characteristics of Dirac central field orbitals. To do this, we observe that the states $\{|n\kappa m\rangle, m = -j, \ldots j\}$ are a basis for the irreducible representation $D^{(j)}$ of SO(3), and so the operator $\vec{J}$ can be represented on the span of these states by

$$\vec{J} := \sum_m a_{n\kappa m}^+ \langle\kappa m|\vec{j}|\kappa' m'\rangle\, a_{n\kappa m} \tag{8}$$

where

$$\langle\kappa m|j_3|\kappa' m'\rangle = m\delta_{mm'}$$
$$\langle\kappa m|j_\pm|\kappa' m'\rangle = [j(j\pm1) - m'(m'\pm1)]^{\frac{1}{2}}\, \delta_{m,m'\pm1}.$$

There are, of course, no matrix elements of $\vec{J}$ or $\vec{j}$ linking different representations $D^{(j)}$. Dropping the label κ, which in this context is redundant, we now find

$$[J_3, a_m^+] = m\, a_m^+$$
$$[J_\pm, a_m^+] = [j(j\pm1) - m(m\pm1)]^{\frac{1}{2}}\, a_{m\pm1}^+ \tag{9}$$

so that, by (1) we can interpret the set $\{a^+; m = -j, \ldots, j\}$ as the components of a rank j tensor operator $\vec{a}^{+}$. Similarly, the set $\{\tilde{a}_m = (-1)^{j-m} a_{-m} : m = -j, \ldots, j\}$ can be regarded as components of the adjoint tensor operator $\vec{\tilde{a}}$. Creation and destruction operators therefore carry information on symmetry, and by using rules for coupling irreducible tensor operators, taking care over anticommutation properties, we can construct states with well-defined angular symmetry.

It is best to start with antisymmetric states for equivalent electrons, the degenerate states with N electrons ($N \leq 2j + 1$) in a given n,κ subshell. Commuting tensor operators can be combined by repeated use of the formula

$$[\vec{T}^{(k_1)}\,\vec{U}^{(k_2)}]^{(k)}_q = \sum_{q_1,q_2} T_{q_1}^{(k_1)}\, U_{q_2}^{(k_2)} \langle k_1 q_1, k_2 q_2 | kq\rangle \qquad (10)$$

and this serves for combining the $a^{+}_{n\kappa m}$, provided we take care with ordering. States of the jj-coupling configuration j^N (configurational state functions, of CSF) can therefore be formally written

$$|j^N, \Gamma JM\rangle = (N!)^{-\frac{1}{2}}\,[(\vec{a}^{+})^N]^{(\Gamma,J)}_M |0\rangle \qquad (11)$$

where J is the rank of the final tensor, and Γ defines a sequence of couplings, the leading factor being necessary for normalization. Table 1 lists the states obtained in this way; it turns out that the label Γ, which may serve to distinguish degenerate CSF with the same value of J, is not required for subshells with $j \leq 5/2$ (s, $\bar{p}$, p, $\bar{d}$, d or $\bar{f}$ electrons). This is much simpler than in the corresponding LS-coupling case[1]. The first degeneracies occur for $(7/2)^4$ configurations, where there are 2 states with J=4 and 2 states with J=2. For $(9/2)^N$, there are similar degeneracies for N=3, 4, 5, 6, 7, 8. Several possible group theoretical schemes are available for classifying the degenerate states[13,14] which it is unnecessary to describe here. The simplest, which is quite adequate for most applications of atomic structure programs is the seniority scheme, which can be formulated in terms of the concept of quasispin[13-15]. The three operators

$$Q_+ = \tfrac{1}{2}\,[j]^{\frac{1}{2}}\,[\vec{a}^{+}\,\vec{a}^{+}]^{(o)}_o$$

$$Q_- = -\tfrac{1}{2}\,[j]^{\frac{1}{2}}\,[\vec{\tilde{a}}\,\vec{\tilde{a}}]^{(o)}_o$$

$$Q_3 = -\tfrac{1}{4}\,[j]^{\frac{1}{2}}\,\{[\vec{a}^{+}\,\tilde{a}]^{(o)}_o + [\vec{\tilde{a}}\,\vec{a}^{+}]^{(o)}_o\}$$

satisfy the commutation relations

$$[Q_+, Q_-] = 2Q_3, \quad [Q_3, Q_\pm] = \pm\, Q_\pm \qquad (12)$$

Table 1. Allowed states of j^N with their multiplicity

j	ℓ	N*	J
1/2	s,$\overline{p}$	1	1/2
3/2	p,$\overline{d}$	1	3/2
		2	2,0
5/2	d,$\overline{f}$	1	5/2
		2	4,2,0
		3	9/2,5/2,3/2
7/2	f,$\overline{g}$	1	7/2
		2	6, 4, 2, 0
		3	15/2, 11/2, 9/2, 7/2, 5/2, 3/2.
		4	8, 6, 5, $(4)^2$, $(2)^2$, 0
9/2	g,$\overline{h}$	1	9/2
		2	8, 6, 4, 2, 0
		3	21/2,17/2.15/2,13/2,11/2,$(9/2)^2$,7/2,5/2,3/2
		4	12, 10, 9, $(8)^2$, 7, $(6)^3$, 5, $(4)^3$, 3, $(2)^2$, $(0)^2$
		5	25/2,21/2,19/2,$(17/2)^2$,$(15/2)^2$,$(13/2)^2$,$(11/2)^2$, $(9/2)^3$, $(7/2)^2$, $(5/2)^2$, 3/2,1/2.

*[States with occupation 2j+1-N are the same as those with occupation N]

characterizing generators of SO(3), just as does ordinary spin. Their usefulness stems from the fact that on the space spanned by states of j^N, Q_3 has eigenvalues

$$M_Q = -\frac{1}{4}(2j + 1 - 2N) = \frac{1}{2}(N - |\kappa|) \tag{13}$$

whilst Q_+ and Q_- are stepping operators which connect states of j^N with j^{N+2} and j^{N-2} respectively. The construction of $\vec{Q}$ ensures that its components are scalar with respect to ordinary rotations, so that they are diagonal with respect to J. Thus there will be some minimum value of N, denoted by ν, the seniority number, for which any given J,M value first appears. If then, we use ν as a classifying label in place of Γ, the definition of ν is such that

$$Q_-|j^\nu, \nu, J, M\rangle = 0,$$

since there is no state of $j^{\nu-2}$ with the same J,M values. In the usual way, we now find that $\vec{Q}^2$ has eigenvalues Q(Q + 1), where

$$Q = \tfrac{1}{2}|\kappa| - \tfrac{1}{2}\nu \tag{14}$$

So the seniority number, ν, is related to Q in much the same way as N is to Q_3. Table 2 shows how this solves the classification problem for $j \leqq 9/2$, except for the J = 4,6 states of $(9/12)^{4,6}$, which are all doubly degenerate. In this case, it is not worth the trouble to seek further classifying variables, and one can just pick two orthonormal pairs in each case.

Table 2. States of j^N in the seniority scheme

j	N	ν	J	j	N	ν	J
1/2	0,2	0	0	9/2	0,10	0	0
	1	1	1/2		1,9	1	9/2
3/2	0,4	0	0		2,8	0	0
	1,3	1	3/2			2	2,4,6,8
	2	0	0		3,7	1	9/2
		2	2			3	3/2, 5/2, 7/2, 9/2,
5/2	0,6	0	0				11/2, 13/2, 15/2,
	1,5	1	5/2				17/2. 21/2
	2/4	0	0		4,6	0	0
		2	2,4			2	2,4,6,8
	3	1	5/2			4	$0,2,3,4^2,5,6^2,7,8$
		3	3/2, 9/2				9,10,12
7/2	0,8	0	0		5	1	9/2
	1,7	1	7/2			3	3/2. 5/2. 7/2. 9/2,
	2,6	0	0				11/2. 13/2, 15/2,
		2	2,4,6				17/2, 21/2
	3,5	1	7/2			5	1/2. 5/2. 7/2, 9/2.
		3	3/2, 5/2, 9/2,				11/2, 13/2, 15/2.
			11/2. 15/2				17/2. 19/2, 25/2
	4	0	0				
		2	2,4,6				
		4	2,4,5,8				

States of j^N can therefore be generated by recursion:

$$a_m^+|j^{N-1},\bar{\nu},\bar{J},\bar{M}\rangle = \sum_{\nu,J,M} |j^N,\nu,J,M\rangle\langle j^N,\nu,J,M|a_m^+|j^{N-1},\bar{\nu},\bar{J},\bar{M}\rangle \qquad (15)$$

The matrix elements on the right-hand side can be simplified using the Wigner-Eckart theorem

$$\langle j^N\nu JM|a_m^+|j^{N-1}\overline{\nu JM}\rangle = \begin{pmatrix} M & j & \bar{J} \\ J & m & \bar{M} \end{pmatrix} \langle j^N\nu J\|\vec{a}^+\|j^{N-1}\overline{\nu J}\rangle;$$

whenever they are non-zero one says that $(j^{N-1}\overline{\nu J})$ and $(j^N\nu J)$ have a parent-daughter relation. It is convenient to express the reduced matrix elements in terms of coefficients of fractional parentage (cfp):

$$\langle j^N\nu J\|\vec{a}^+\|j^{N-1}\overline{\nu J}\rangle = (-1)^N N^{\frac{1}{2}}[J]^{\frac{1}{2}}(j^N\nu J\{|j^{N-1}\overline{\nu J})$$

and (16)

$$\langle j^{N-1}\overline{\nu J}\|\vec{a}\|j^N\nu J\rangle = (-1)^N(-1)^{\bar{J}-j-J}N^{\frac{1}{2}}[J]^{\frac{1}{2}}(j^{N-1}\overline{\nu J}|\}j^N\nu J)$$

where

$$(j^N \nu J\{|j^{N-1}\bar{\nu}\bar{J}) = (j^{N-1}\bar{\nu}\bar{J}|\}j^N \nu J)$$

There are a number of sum rules for cfp which are useful for checking purposes: for example, the orthonormality of states in the seniority scheme gives

$$\sum_{\bar{\nu}\bar{J}} (j^N \nu J\{|j^{N-1}\bar{\nu}\bar{J})(j^{N-1}\bar{\nu}\bar{J}|\}j^N \nu' J') = \delta_{\nu\nu'}\delta_{JJ'}$$

Construction of cfp is described, for example, by Judd[2].

The main use of cfp appears when we consider the inverse relation of (15) in configuration space

$$\langle \vec{x}_1,\ldots,\vec{x}_{N-1},\vec{x}_N|j^N \nu JM\rangle = \sum_{\bar{\nu}\bar{J}\bar{M}m} \langle \vec{x}_1,\ldots,\vec{x}_{N-1}|j^{N-1}\bar{\nu}\bar{J}\bar{M}\rangle\langle \vec{x}_N|jM\rangle$$

$$\times \langle \bar{J}\bar{M},jm|JM\rangle \; (j^{N-1}\bar{\nu}\bar{J}|\}j^N \nu JM) \tag{17}$$

We see that the antisymmetric state of j^N is expressed a sum of products of antisymmetric states of j^{N-1} for the first N-1 particles with states of the N-th particle. The cfp project out the antisymmetric combination from the set of vector-coupled products.

The final step, to construct general open shell CSF, is now easy; we need to modify the notation so that $\vec{a}_1$, $\vec{a}_2$,..., are spherical tensor creation operators for subshells 1,2,... respectively and label the CSF 1^{N_1}, 2^{N_2},... by $T_1, T_2, \ldots$, where, $T_i = (J_i, \nu_i, \ldots)$ including any additonal classifying labels needed. Thus

$$|1^{N_1}, T_1 M_1\rangle = [\vec{a}_1{}^+]^{N_1,T_1}_{M_1} |0\rangle \tag{18}$$

where the definition implicitly includes the normalizing factor $(N_1!)^{-\frac{1}{2}}$. We now define a <u>standard CSF</u>,

$$|T,M\rangle = ([\vec{a}_k]^{N_k,T_k} \ldots [\vec{a}_2{}^+]^{N_2,T_2}[\vec{a}_1{}^+]^{N_1,T_1})^{N,T}_M |0\rangle \tag{19}$$

in which the subshell operators are to be coupled in a standard order from right to left according to the coupling scheme

$$((\ldots((J_1J_2)X_1J_3)X_2\ldots)X_{k-2}J_k)J \tag{20}$$

where $\vec{X} = (X_1, X_2, \ldots, X_{k-2})$ is a set of resultant intermediate angular momenta. The adoption of a standard order simplifies the subsequent organization. If desired, other orders can be related through standard recoupling techniques. Thus the label T for a standard CSF will consist of the list

$$T \equiv (T_1N_1,\ T_2N_2,\ldots,\ T_kN_k;\ X_1,\ldots,X_{k-2};J) \tag{21}$$

which uniquely defines the method of construction.

CONSTRUCTION OF MATRIX ELEMENTS. COMPUTER PROGRAMS

The formalism of the last section, combined with the operator representation

$$H_{eff} = H^{(1)} + H^{(2)}$$

$$H^{(1)} = \sum_{pq} N(a_p^+ a_q)\langle p|c\vec{\alpha}.\vec{p}+(\beta-1)c^2 - \frac{Z}{r}|q\rangle \tag{22}$$

$$H^{(2)} = \tfrac{1}{2} \sum_{pqrs} N(a_p^+ a_q^+ a_s a_s)\langle pq|\frac{1}{R} + \tfrac{1}{2}B(\frac{\omega_{ps}}{c},\vec{R}) + \tfrac{1}{2}B(\frac{\omega_{qr}}{c},\vec{R})|sr\rangle$$

deduced in another lecture[16] makes it quite straightforward to construct the matrix of H_{eff} in the span of any chosen set of CSF. For each standard CSF (19), we can use (17) to split off one or more "active" electrons which will link with the terms of the operator in non-zero matrix elements, $\langle TM|H_{eff}|T'M'\rangle$. The derivation is too lengthy to give here; one can either use the algebraic properties of the creation and anmihilation operators alone, or assisted by the use of diagram techniques as pioneered by Yutsis and his collaborators[5] and developed by El Baz and Castel[4]. The flavour of this kind of analysis will be found in an article by Huang[17], whose approach is very similar to ours.

We shall quote only a typical result for a two-particle interaction, $H^{(2)}$. The general formula may be written[18]

$$\langle T,M|H^{(2)}|T',M'\rangle = \delta_{JJ'}\delta_{MM'} \sum_{p,q,r,s} (-1)^{\Delta}[N_p(N_p-\delta_{pq})N_s'(N_r'-\delta_{sr}]^{\frac{1}{2}}$$

$$\times \sum_k \sum_{\{T\}} \left(\prod_{i\neq p,q,r,s} \delta_{T_i T_i'}\right)(T_p\{|T_p')(T_q\{|T_q')(T_s|\}T_s')(T_r|\}T_r')$$

$$\times\{R_d(1+\delta_{pq}\delta_{sr})^{-1}\ [j_p,j_r]^{-\frac{1}{2}}\ X^k(p,q,s,r)$$

$$- R_e\ (1-\delta_{pq})(1-\delta_{sr})[j_p,j_s]^{-\frac{1}{2}}\ X^k(p,q,r,s)\}, \tag{23}$$

where p,q run over orbitals occupied in CSF,T,r,s over orbitals occupied in T'. The terms on the first line arise from the antisymmetry of the wave-functions, and if p,q,r,s are ordered so that p<q, r<s,

$$\Delta = \sum_{p+1}^{q} N_\lambda - \sum_{r+1}^{s} N'_\mu$$

is a phase-factor originating in the anticommutations carried out in the course of the evaluation. Those on the second express the equivalence, in T and T', of the "inactive" shell CSF ($i \neq p,q,r,s$), and the parent-daughter relations, through equation (17) in the "active" subshells. The quantity in brackets is a product of a recoupling coefficient and an effective interaction strength, given by equation (7) in the case of the Coulomb interaction, one for the "direct" term, R_d, one for exchange. Symbolically

$$R_d = (\ldots(\bar{J}_q j_q)J_q \ldots '\bar{J}_p (j_s k)j_p)J_p \ldots ; \vec{X}|$$

$$\ldots(\bar{J}_r (kj_q)j_r)J'_r \ldots (\bar{J}_s j_s)J_s{}' \ldots ; \vec{X}')$$

where $\bar{J}_q = J_q{}'$, $\bar{J}_p = J_p{}'$, $\bar{J}_r = J_r$, $\bar{J}_s = J_s$. The roles of r,s, are interchanged in R_e. The form (23) is, in fact, quite general; to deal with other interactions, say $B(\nu,\vec{R})$, it is only necessary to replace X^k by the appropriate expression, with its selection rules[12,8,9]. The general one-particle matrix element has similar structure.

Finally, we arrive at an energy expression which has the form

$$\langle TM|H|T'M'\rangle = \sum_{pq} A_{T,T'}(p,q)I(p,q) + \sum_{pqrs\,k} B_{T,T'}(p,q,r,s)R^k(p,q,r,s)$$

$$+ \text{ transverse energy terms}$$

Both the one-electron angular coefficients, $A_{T,T'}(p,q)$, and the two-electron coefficients $B_{T,T'}(p,q,r,s)$ have the structure implied by (23), so that essentially the same methods can be employed to construct all of them. The programs require: a list of configurations, {T}; an analysis to determine which radial integrals will appear in (24); and numerical evaluation of the coefficients. The first part can be laborious, and we have now designed the method of input to minimize the amount of data that has to be supplied manually[8]. Indeed it is now possible to give a list of configurational states in LS-coupling, and the program generates those jj-coupled CSF which will be needed to give a proper description in intermediate coupling[6]. A recent addition, due to K.G. Dyall (not yet published) allows the final ASF (atomic state functions) to be expressed in terms of LS-coupled CSF. The analysis of CSF must determine what coupling schemes are allowed, and provide data for the calculation of recoupling coefficients. This is done using the well-known package due to Burke[19]. Fractional parentage coefficients are all tabulated quantities of the form $\pm a\sqrt{b}/c\sqrt{d}$, a,b,c,d integers, so that a simple table layout permitting output of cfp to machine precision is easy to devise[20].

These energy expressions, (24), can be used in many different ways. The programs[8,9] are presented as a multiconfiguration Dirac-Fock package, but they can easily be segmented so that modules can be used for other purposes. Nor is it necessary to do self-consistent calculations; the package can, if desired, be used to study CI. Some aspects of SCF theory are discussed elsewhere in this volume[21].

REFERENCES

1. B.R. Judd, "Operator techniques in atomic spectroscopy", McGraw Hill, New York (1963).
2. B.R. Judd, "Second quantization in atomic spectroscopy", Johns Hopkins Press, Baltimore (1967).
3. D.M. Brink and G.R. Satchler "Angular momentum" (2nd edition) Clarendon Press, Oxford (1968).
4. E. El Baz and B. Castel, "Graphical methods of spin algebras in atomic, nuclear and particle physics". Marcel Dekker, New York (1972).
5. A.P. Yutsis, I.B. Levinson and V.V. Vanagas "Mathematical Apparatus of the Theory of Angular Momentum" (Translated by A. Sen and R.N. Sen), Israel Program for Scientific Translations, Jerusalem (1962).
6. E.U. Condon and G.H. Shortley, "Theory of atomic structure", Cambridge University Press, Cambridge (1935).
7. I.P. Grant, Adv. Phys. 19: 747 (1970)
8. I.P. Grant, B.J. McKenzie, P.H. Norrington, D.F.Mayers and N.C. Pyper, Computer Phys. Commun. 21:207 (1980); and earlier programs referenced therein.
9. B.J. McKenzie, I.P. Grant and P.H. Norrington, Computer. Phys. Commun. 21: 233 (1980).
10. J.P. Desclaux, Computer Phys. Commun. 9:31 (1975).
11. E.P. Wigner, "Group Theory", Academic Press, New York (1959).
12. I.P. Grant and N.C. Pyper, J. Phys. B 9: 761 (1976); I.P. Grant, and B.J. McKenzie, ibid, 13: 2671 (1980).
13. B.R. Judd and J.P. Elliott, "Topics in atomic and nuclear theory" New Zealand, University of Canterbury (1970).
14. B.G. Wybourne, "Classical Groups for Physicists", Wiley, New York (1974).
15. B.H. Flowers and S. Szpikowski, Proc. Phys. Soc. (Lond) 84: 193 (1964); R.D. Lawson and M.H. Macfarlane, Nucl. Phys. 66: 80 (1965).
16. I.P. Grant, these proceedings.
17. K.N. Huang, Rev. Mod. Phys. 51: 215 (1979).
18. I.P. Grant, Computer Phys. Commun. 5: 263 (1973).
19. P.G. Burke, Computer Phys. Commun. 1: 241 (1970); ibid 2: 173 (1971); ibid 5: 161 (1973); ibid 8: 151 (1974).
20. I.P. Grant, Computer Phys. Commun. 4: 377 (1972); ibid 14:311 (1978).
21. See articles by J.P. Desclaux, and by the author, these proceedings.

SELF-CONSISTENCY AND NUMERICAL PROBLEMS

I.P. Grant

Theoretical Chemistry Department
1 South Parks Road, Oxford, OX1 3TG, England
and
Instituto de Fisica
Universidad Nacional Autonoma de Mexico
Mexico, 20. D.F., Mexico

THE VARIATIONAL METHOD IN RELATIVISTIC CALCULATIONS

In previous lectures, I have discussed how one can extract from QED an effective relativistic hamiltonian and shown how this can be expressed in terms of Slater integrals involving only radial functions defined in terms of a suitable one-body potential. From this point on, the process is very close to that employed in non-relativistic calculations of similar type.

We suppose that each atomic state (ASF), say $\Psi_\alpha(J^\pi,M)$ where α distinguishes states having the same total angular momentum parameters J,M and parity π, can be expressed as a linear superposition of CSF, say $\Phi_r(J^\pi,M)$ having the same symmetry (the CSF denoted $|\Gamma M\rangle$ on page). Write

$$\Psi_\alpha(J^\pi,M) = \sum_r c_{r\alpha}\Phi_r(J^\pi,M) \tag{1}$$

and let $\hat{H} = \{H_{ij}\}$ be the real symmetric matrix with elements

$$H_{ij} = \langle\Phi_i(J^\pi,M)|H|\Phi_j(J^\pi,M)\rangle \tag{2}$$

Then the expectation of H in the ASF (1) can be written compactly

$$W(\vec{c}_\alpha) = c_\alpha^+ \ \hat{H}\ \vec{c}_\alpha \tag{3}$$

where $\vec{c}_\alpha^{\,+} = \{\bar{c}_{i\alpha}\,;\ i = 1, 2, \ldots\}$ If the matrix elements H_{ij} are kept fixed, the requirement that $W(\vec{c}_\alpha)$ be stationary with respect to variation of the mixing coefficients c_α leads to the familiar secular equation

$$(\hat{H} - E_\alpha^{(N)}\,\hat{I})\,\vec{c}_\alpha = 0, \quad \alpha = 1,2,\ldots, N, \tag{4}$$

if the basis of CSF has dimension N. Order the eigenvalues so that $E_1^{(N)} \leqslant E_2^{(N)} \leqslant \ldots \leqslant E_N^{(N)}$; since $\hat{H}$ is symmetric, the eigenvectors will be orthonormal, $\vec{c}_\alpha^{\,+}\,\vec{c}_\beta = \delta_{\alpha\beta}$, whenever $E_\alpha \neq E_\beta$, and we can always choose an orthonormal set when two or more eigenvalues are degenerate. Suppose now we add another CSF to the basis; then we have the well-known Hylleraas-Undheim theorem[1], which states that

$$E_k^{(N+1)} \leqslant E_k^{(N)} \leqslant E_{k+1}^{(N+1)}, \quad k = 1, 2, \ldots, N. \tag{5}$$

Provided $\hat{H}$ is bounded below, one of the aims of our construction of an effective bound state Hamiltonian, the sequence $\{E_k^{(N)} : N = k,k+1\ldots\}$ must converge monotonically from above to a limit E_k, which we hope will be an eigenvalue of the hamiltonian H if the CSF set is complete in the space on which H acts.

This is the technique used in configuration interaction (CI) calculations; the orbital basis is fixed, the matrix elements H_{ij} are thereby determined, and the only question is the choice of CSF, and whether the set is sufficiently large to approximate E_k to the desired precision. Whilst many papers discuss the choice of CSF for atomic and molecular problems[2] it often seems as if there is little but consistency of successive approximations allied to physical insight to help decide if convergence has been obtained, with rare exceptions[3].

Greater flexibility, and, one hopes, an optimization of the number of CSF needed for convergence at a given precision, can be obtained by permitting more flexibility in the underlying orbital basis, either by incorporating adjustable parameters into some model potential[4], or by the SCF method. To explain the latter, we need more explicit expressions for the matrix elements H_{ij} given in the previous lecture[5]. Ignoring transverse energy terms and QED corrections, we write

$$H_{ii} = \sum_a q_a(i)\, I(a,a) + \sum_{\substack{a,b\\ b\geqslant a}} \sum_k f_i^{\,k}(a,b)\, F^k(a,b) + \sum_{\substack{a,b\\ b>a}} \sum_k g_i^{\,k}(a,b)\, G^k(a,b) \tag{6}$$

$$H_{ij} = \sum_{a,b} T_{ij}(a,b)\, I(a,b) + \sum_{a,b,c,d} \sum_{k} V_{ij}{}^{k}(a,b,c,d) R^{k}(a,b,c,d),$$

$$\text{for } i \neq j.$$

The details of the angular coefficients are not required in this context ; clearly it is sufficient, for programming purposes, to have a list of the coefficients $f_i{}^k(a,b)$, $g_i{}^k(a,b)$, $T_{ij}(a,b)$, and $V_{ij}{}^k(a,b,c,d)$ from which the rest of the expressions required can be constructed. However, the structure of the radial integrals is important. Let

$$u_a(r) = \begin{bmatrix} P_a(r) \\ Q_a(r) \end{bmatrix} \tag{7}$$

denote the Dirac radial amplitudes for orbital a in the chosen central potential. Then the one-electron integrals are defined by

$$I(a,b) = \delta_{\kappa,\kappa_a} \delta_{\kappa,\kappa_b} \int_0^\infty u_a{}^+(r) . h_\kappa{}^o . u_b(r) dr \tag{8}$$

where

$$h_\kappa{}^o = \begin{bmatrix} -Z(r)/r & c(-\frac{d}{dr} + \frac{\kappa}{r}) \\ c(+\frac{d}{dr} + \frac{\kappa}{r}) & -2c^2 - Z(r)/r \end{bmatrix} \tag{9}$$

is the radial Dirac operator, and the nuclear-electron potential energy is written $-Z(r)/r$ to allow for the possibility of a finite nuclear charge distribution. The diagonal integrals, I(a,a) are each weighted by $q_a(i)$, the occupation of orbital a in CSFi. The Slater integrals are defined by

$$R^k(a,b,c,d) = \int_0^\infty \rho_{ac}(r) \frac{Y^k_{bd}(r)}{r} dr \tag{10}$$

where

$$Y^k_{bd}(r) = r \int_0^\infty u_k(r,s)\, \rho_{bd}(s)\, ds$$

with

$$\rho_{ac}(r) = P_a(r)\, P_c(r) + Q_a(r)\, Q_c(r)$$

and

$$U_k(r,s) = r^k/s^{k+1} \quad \text{if } r < s$$
$$= s^k/r^{k+1} \quad \text{if } s < r$$

In particular

$$F^k(a,b) = R^k(a,b,a,b) = R^k(b,a,b,a)$$
$$G^k(a,b) = R^k(a,b,b,a) = R^k(b,a,a,b)$$

are the relativistic versions of the familiar direct and exchange integrals.

We can now write down our energy functional, which should take account of the orthonormality constraints

$$N(a,b) = \delta_{\kappa_a,\kappa_b} \int_0^\infty \rho_{ab}(r)\, dr = \delta_{a,b}\, \delta_{\kappa_a,\kappa_b} \tag{11}$$

We incorporate these into the energy expression to give the functional

$$W(\vec{c}_\alpha, u_1, \ldots, u_{NW}, \vec{\varepsilon}) = \vec{c}_\alpha^{\,\dagger} \hat{H} \vec{c}_\alpha + \sum_a q_{a\alpha} \varepsilon_a N(a,a)$$
$$+ \sum_{\substack{a,b \\ a \neq b}} \delta_{\kappa_a,\kappa_b} \varepsilon_{ab} N(a,b) \tag{12}$$

where $\vec{\varepsilon}$ represents the set of Lagrange multipliers ε_a, $\varepsilon_{ab} = \varepsilon_{ba}$ $(a \neq b)$ and,

$$q_{a\alpha} = \sum_i c_{i\alpha}^2\, q_a(i) \tag{13}$$

is the generalized occupation number of orbital a in ASF α. Variation of W with respect to the mixing coefficients and orbitals now gives the secular equations (4) together with nonlinear integro-differential equations for the orbitals we must consider presently, which depend on which of the solutions, α, of the secular equation we choose. This is therefore called an optimal level (OL) - MCDF calculation. Experience shows that OL calculations are often difficult to converge, and are very time consuming. A simplification of this scheme may therefore be worth considering in many circumstances, the average level (AL)-MCDF scheme. In it, we merely average (12) over all ASF α; since all CSF appearing have the same J (H is a scalar operator), all ASF have the same weight, and we get

$$\overline{W}^{(J)}(u_1,\ldots,u_{NW},\vec{\varepsilon}) = \text{trace } \hat{H}^{(J)} + \sum_a \overline{q}_a \varepsilon_a N(a,a)$$

$$+ \sum_{\substack{a,b \\ b\neq a}} \delta_{\kappa_a \kappa_b} \varepsilon_{ab} N(a,b) \tag{14}$$

where we have made the (J) value explicit on the right-hand side. This restricts the derived orbitals to be the same for all ASF, unlike (12), but in many cases gives as good energies at a fraction of the computing effort.[6] A further variant,[6] the extended average level (EAL) - MCDF scheme, defined by

$$\overline{W} = \sum_J \omega_J \; \overline{W}^{(J)}, \quad \sum_J \omega_J = 1 \tag{15}$$

for some suitable J values[7] is extremely useful for transition probability calculations, since the same orbital basis is used for different ASF. The best choice is usually $\omega_J \propto 2J + 1$, though others are possible.

The EAL and AL schemes are clearly cheap and useful, though there can be cases in which the choice of the same orbitals for all ASF can be a distinct disadvantage, for example, when a transition is known to involve substantial orbital relaxation. On the other hand, EAL calculations seem to give very reliable fine structures[7]. We have already seen a connection of the AL and EAL schemes with transition state methods[6], and we have recently noted a connection with similar ideas in molecular MCSCF theory[8].

FINITE DIFFERENCE FORM OF MCDF EQUATIONS

The deduction of SCF equations from any of the functionals (12), (14) or (15) is a simple extension of the procedure for a single configuration[9] and does not need a detailed description. The equations that appear have the general form

$$(h_{\kappa_a} - \varepsilon_a)u_a(r) = -\frac{1}{r}X_a(r) + \frac{1}{\overline{q}_{a\alpha}} \sum_{b\neq a} \delta_{\kappa_a,\kappa_b} \varepsilon_{ab} u_b(r) \tag{16}$$

where we have exploited the symmetry afforded by (9) to write the equations more compactly than in the earlier reference.[9] Here h_κ is the same as $h_\kappa{}^o$ in equation (9), with the replacement

$$Z(r) \to Y_a(r) = Z(r) - \sum_b \sum_k y^k(a,b) Y^k_{bb}(r) \tag{17}$$

to account for orbital screening. Similarly

$$X_a(r) = \sum_k \{ \sum_{b \neq a} x^k(a,b) Y_{ab}{}^k(r) u_b(r) + \sum_{b,c,d} x^k(a,b,c,d) Y_{bd}{}^k(r) u_c(r) \} \quad (18)$$

The coefficients $y^k(a,b)$, $x^k(a,b)$ and $x^k(a,b,c,d)$ are easily constructed from the $f_i{}^k(a,b)$, $g_i{}^k(a,b)$ and $v_{ij}{}^k(a,b,c,d)$ coefficients; for example,[5]

$$y^k(a,b) = \frac{1 + \delta_{ab}}{\bar{q}_{a\alpha}} \sum_i c_{i\alpha}{}^2 f_i{}^k(a,b)$$

We have seen that the bound state solutions of (16), in the no exchange case, will be in a domain[10] in which $u_a(r)$, du_a/dr will be square integrable, and we conjecture the same will be true when exchange is included. We also recall [10] that the boundary conditions $u_a(r)$ at $r = 0$ and $r = \infty$, previously defined, are sufficient to ensure this.

The numerical difference schemes used in reference 5 are very close to those described in reference 9, and will not be repeated here. There has been little analysis of the accuracy and convergence of such numerical schemes; some unpublished work has been done, however, on the single Dirac equation, in which the method of deferred correction[9] is analysed. A typical result is exploited in reference 11. This has not be extended to the SCF case, though it is likely that some clues will be found to do this by writing (16) as a compact 2×NW dimensional self-adjoint system. Define

$$\mathbb{U}^+ = (\vec{u}_1{}^+, \ldots, \vec{u}_{NW}{}^+);$$

$$\mathcal{J} = \text{diag}\,(J, \ldots, J), \quad J = \begin{bmatrix} 0 & 1 \\ -1 & 0 \end{bmatrix};$$

$$\mathbb{Z}(\mathbb{U}) = [\mathbb{Z}_{ab}],$$

where

$$\mathbb{Z}_{ab} = \sum_k [x^k(a,b) Y_{ab}{}^k(r) + \sum_{cd} x^k(a,c,b,d) Y_{cd}{}^k(r)]\, I_2,$$

$$I_2 = \begin{bmatrix} 1 & 0 \\ 0 & 1 \end{bmatrix},$$

when $a \neq b$, and

$$\mathbb{Z}_{aa} = \begin{bmatrix} -Y_a(r) & c\kappa \\ c\kappa & -2rc^2 - Y_a(r) \end{bmatrix}$$

We also write

$$\boldsymbol{\varepsilon} = [\boldsymbol{\varepsilon}_{ab}],\ \boldsymbol{\varepsilon}_{ab} = \varepsilon_{ab} I_2,\ a \neq b,\ \boldsymbol{\varepsilon}_{aa} = \bar{q}_{a\alpha}\varepsilon_a I_2$$

$$\mathcal{Q} = \mathrm{diag}\,(\bar{q}_{a\alpha} I_2, \ldots, \bar{q}_{NW,\alpha} I_2)$$

Then the system (16) becomes

$$c\,\mathcal{J}\frac{d\mathbb{U}}{dr} + \frac{1}{r}\,\mathbb{Z}(\mathbb{U}).\mathbb{U} = \mathcal{Q}^{-1}\,\boldsymbol{\varepsilon}\,\mathbb{U} \tag{19}$$

The non-linearity of the system appears in $\mathbb{Z}(\mathbb{U})$, a matrix functional of $\mathbb{U}$. In the OL case, $\mathcal{Q}$ depends on the mixing coefficients, and so will change at each iteration; in AL and EAL cases it is fixed throughout the SCF process. The boundary values $\mathbb{U}(0)$ and $\mathbb{U}(\infty)$ are, of course, the usual ones for L^2 solutions.

DISCRETE BASIS FORM OF MCDF EQUATIONS

An alternative approach, expansion of each atomic or molecular spinor in terms of a set of finite basis functions is more appropriate for molecular calculations[12,13], though the most successful calculations of this type so far have been for atoms[14]. We shall present a discussion here in terms of one possibility, the formal expansion

$$u_a(r) = \begin{bmatrix} P_a(r) \\ Q_a(r) \end{bmatrix} = \sum_{j=1}^{NB} \phi_j(r)\, u_{ja},\quad u_{ja} = \begin{bmatrix} p_{ja} \\ q_{ja} \end{bmatrix} \tag{20}$$

where $\{\phi_j(r)\}$ is a suitably chosen scalar basis set. In molecular problems, where the central dominance of the nuclear Coulomb potential cannot be used, four components may be needed. Likewise, one might use a spinor (2- or 4-) component basis and scalar coefficients. Apart from the practical arguments,[12,13], the structures obtained are very similar, and the present analysis may be imitated to study the other possibilities. For simplicity, we shall also assume that the set $\{\phi_j(r)\}$ is orthonormal.

When we substitute (20) into one of the functionals (12), (14) or (15), the result is a nonlinear algebraic expression to be made stationary with respect to the expansion coefficients, which we can group first in vectors

$$P_a^+ = (p_{ia}, \ldots, p_{NB,a}) \qquad a = 1,2,\ldots, NW$$

$$Q_a^+ = (q_{ia}, \ldots, q_{NB,a}) \qquad a = 1,2,\ldots, NW$$

and then in supervectors of dimension M = NB * NW

$$\boldsymbol{P}^+ = (P_1^+, \ldots, P_{NW}^+)$$

$$\boldsymbol{Q}^+ = (Q_1^+, \ldots, Q_{NW}^+)$$

The functionals can be written compactly in the form

$$W(\vec{c},\boldsymbol{P},\boldsymbol{Q}) = (\boldsymbol{P}^+,\boldsymbol{Q}^+) \begin{bmatrix} -E+NZ(\boldsymbol{P},\boldsymbol{Q}) & cN(-D+KW) \\ cN(\ D+KW) & -2c^2N-E+NZ(\boldsymbol{P},\boldsymbol{Q}) \end{bmatrix} \begin{bmatrix} \boldsymbol{P} \\ \boldsymbol{Q} \end{bmatrix} \qquad (21)$$

where each of the submatrices is M×M, divided into square blocks of dimension NB×NB. We use a circumflex accent over an operator to indicate that it is the matrix of order NB×NB of that operator with respect to the basis $\{\phi_j(r): j = 1,\ldots,NB\}$. Thus

$$D = \text{diag}\ \{\hat{D}, \ldots, \hat{D}\}$$

where

$$\hat{D} = [D_{ij}], \quad D_{ij} = -D_{ji} = \int_0^\infty \phi_i(r) \frac{d}{dr} \phi_j(r)\, dr \qquad (22)$$

Similarly,

$$W = \text{diag}\ \{\hat{W}, \ldots, \hat{W}\},$$

where

$$\hat{W} = [W_{ij}], \quad W_{ij} = W_{ji} = \int_0^\infty \phi_i(r) \frac{1}{r} \phi_j(r)\, dr; \qquad (23)$$

$$Z = [\hat{Z}_{ab}], \qquad (24)$$

where

$\hat{Z}_{aa}$ is the matrix of $-Y_a(r)/r$

$\hat{Z}_{ab}$ is the matrix of $\frac{1}{r} \sum_k [x^k(a,b)Y_{ab}{}^k(r) + \sum_{cd} x^k(a,c,b,d)Y_{cd}{}^k(r)]$;

$$N = \text{diag}\ \{\bar{n}_i \hat{I}, \ldots, \bar{n}_{NW,\alpha}\hat{I}\}, \quad \hat{I} = \text{identity}, \qquad (25)$$

where the notation $\bar{n}_{a\alpha}$ is here used for $\bar{q}_{a\alpha}$ in earlier equations to avoid confusion; and

$$\hat{E}_{a\alpha} = \bar{n}_{a\alpha}\varepsilon_a \hat{I}, \quad \hat{E}_{ab} = \varepsilon_{ab} \hat{I}. \tag{26}$$

Of course $\hat{Z}_{ab}$ can be expressed in terms of density matrices and elementary integrals, $<\phi_i \phi_j|O|\phi_k\phi_l>$, in the usual way [14].

STUDY OF THE ONE ELECTRON PROBLEM

The nature of the difficulties encountered with (21) becomes clearer if we consider the case of a single function in a prescribed potential, for which the equation to be solved is just

$$\begin{bmatrix} -\hat{E}+\hat{Z} & c(-\hat{D}+\kappa\hat{W}) \\ c(\hat{D}+\kappa\hat{W}) & -2c^2\hat{I}-\hat{E}+\hat{Z} \end{bmatrix} \begin{bmatrix} P \\ Q \end{bmatrix} = 0 \tag{27}$$

where, now, $\hat{E} = \varepsilon\hat{I}$, and $\hat{Z}$ is just the matrix of $-Z(r)/r$ with respect to the chosen basis set. We shall show that provided $-\hat{Z}$ is positive definite, and has a sufficiently small spectral radius $\rho_{\hat{Z}}$ (eigenvalue of maximum modulus), the eigenvalues of (27) divide into two classes: the set $E^+ = (\varepsilon_1^+, \ldots \varepsilon_{NB}^+)$ linked to eigenvalues of some nonrelativistic operator in this limit $c \to \infty$, and the set $E^- = (\varepsilon_1^-, \ldots, \varepsilon_{NB}^-)$ which are all less than $-2c^2$, and so provide L^2 approximations to states in the lower continuum.

We first perform the matrix equivalent of "elimination of small components". We may write the result

$$\{\hat{T}_\ell + \hat{Z} - \varepsilon\hat{I} - \hat{C}_N^{(Z)}(\varepsilon)\} P = 0 \tag{28}$$

Let

$$\hat{\Pi}_\kappa = -\hat{D} + \kappa\hat{W}, \quad \hat{\Pi}_\kappa^{+} = -\hat{\Pi}_{-\kappa} = \hat{D} + \kappa\hat{W}\ ; \tag{29}$$

then we define

$$\hat{T}_\ell = \tfrac{1}{2}\hat{\Pi}_\kappa\hat{\Pi}_\kappa^{+} = \tfrac{1}{2}\,(\kappa^2\hat{W}^2 - \kappa[\hat{D},\hat{W}] - \hat{D}^2),$$

and

$$\hat{T}_{\bar{\ell}} = \tfrac{1}{2}\hat{\Pi}_\kappa^{+}\hat{\Pi}_\kappa = \tfrac{1}{2}(\kappa^2\hat{W}^2 + \kappa[\hat{D},\hat{W}] - \hat{D}^2) \tag{30}$$

The matrices $\hat{T}_\ell$ and $\hat{T}_{\bar{\ell}}$ can be regarded as matrix representations of the Schrödinger radial kinetic energy operators

$$T_\ell = -\tfrac{1}{2}\frac{d^2}{dr^2} + \frac{\ell(\ell+1)}{2r^2}, \quad T_{\bar{\ell}} = -\tfrac{1}{2}\frac{d^2}{dr^2} + \frac{\bar{\ell}(\bar{\ell}+1)}{2r^2}$$

where ℓ and $\bar{\ell}$ are, respectively, the positive roots of $\ell(\ell+1)=\kappa(\kappa\pm1)$ For $[\hat{D},\hat{W}]$ should be a matrix representative of $[\frac{d}{dr}, \frac{1}{r}] = -\frac{1}{r^2}$, so that we immediately get a formal justification of our interpretation We shall return later to the problem of defining conditions for this to be correct.

The matrix $\hat{C}_\kappa^{(Z)}(\varepsilon)$ contains all the relativistic corrections; writing $-Z(r)/r = -ZU(r)$, we have

$$\hat{C}_\kappa^{(Z)}(\varepsilon) = \tfrac{1}{2}\hat{\Pi}_\kappa \left(\hat{I} + \frac{Z\hat{U}+\varepsilon\hat{I}}{2c^2}\right)^{-1} \left(\frac{Z\hat{U}+\varepsilon\hat{I}}{2c^2}\right) \hat{\Pi}_\kappa^{+} ; \tag{31}$$

clearly $\hat{C}_\kappa^{(Z)}(\varepsilon)$ is analytic in $1/c^2$, $c \to \infty$, so that $\lim \hat{C}_\kappa^{(z)}(\varepsilon)=0$, and P and ε can be approximated by regular perturbation theory using a power series in $1/c^2$. The eigenvalues of

$$\hat{H}_\ell^{(Z)} = \hat{T}_\ell + Z\hat{U}$$

say $\{\varepsilon_{n\ell}^{(o)} : n = 1,2\ldots, NB\}$ should therefore be approximations to the discrete nonrelativistic eigenvalues in the central potential $-ZU(r)$, and one requirement is therefore that the basis chosen should provide a good approximation at this level. Clearly, the relativistic correction matrix will change the eigenvalues, and we should like to ensure that the eigenvalues $\{\varepsilon_{n\ell} : n = 1,2\ldots, NB\}$ remain in the appropriate interval, $-2c^2 < \varepsilon < 0$, for bound states. A sufficient condition, though possibly not a necessary one, is that the spectral radius, $\rho_{\hat{Z}}$, should be less than $2c^2$, independent of the size, NB, of the basis set. To see this, rewrite (28) as a sort of Rayleigh quotient

$$\varepsilon = \{P^+\hat{H}_\ell^{(z)}P - P^+C_\kappa^{(z)}(\varepsilon)P\}/P^+P$$

Let $t = \hat{\Pi}_\kappa P$ and assume (a condition on the basis set) that $t \neq 0$; then $P^+\hat{T}_\ell P = \tfrac{1}{2}t^+t > 0$. Also

$$\frac{P^+C_\kappa^{(z)}(\varepsilon)P}{P^+P} = \left[\tfrac{1}{2}t^+\left(\hat{I} + \frac{Z\hat{U}+\varepsilon\hat{I}}{2c^2}\right)^{-1} \left(\frac{Z\hat{U}+\varepsilon\hat{I}}{2c^2}\right) t\right]/(P^+P)$$

$$< \left[\tfrac{1}{2}\, t^+\left(\hat{I} + \frac{Z\hat{U}}{2c^2}\right)^{-1} \left(\frac{Z\hat{U}}{2c^2}\right) t\right] /(P^+P)$$

on $-2c^2 < \varepsilon < 0$, and this can be bounded by

$$\max_{\|P\|=1} \left(\tfrac{1}{2}\, t^+t\right) \quad \rho_{\hat{Z}}/(2c^2 + \rho_{\hat{Z}})$$

where $\rho_{\hat{Z}}$ is the largest eigenvalue of the positive definite matrix ZU(r). Under these conditions, all eigenvalues of E^+ are bounded below by

$$\varepsilon_{min} = - \rho_{\hat{Z}} > -2c^2 \tag{32}$$

If we now "eliminate large components" and, at the same time, set $\varepsilon = -2c^2 + \varepsilon'$ to take a new energy zero at the upper edge of the lower continuum, we find in place of (28)

$$\{\hat{T}_{\overline{\ell}} - \hat{Z} + \varepsilon'\hat{I} - \hat{C}^{(-Z)}_{-\kappa}(-\varepsilon')\}\ Q = 0 \tag{33}$$

In the nonrelativistic limit, this gives

$$-\varepsilon' = Q^+\hat{H}^{(-Z)}_{\overline{\ell}} Q/Q^+Q > 0$$

since $-\hat{Z}$ represents the repulsive ZU(r), consistent with the interpretation of the lower continuum in terms of positron states. Furthermore, the relativistic correction will not increase ε', provided $-\hat{Z}$ is positive definite, so that all eigenvalues ε_i' remain negative. Restoring the energy zero, $-2c^2$, we see that every member of the set $E^- = (\varepsilon_1^-, \ldots, \varepsilon_{NB}^-)$ is below $-2c^2$, and, of course, disappears to $-\infty$ when $c \to \infty$.

There are a good many hypotheses here which, if the matrices have the relevant properties, will suffice to ensure that the eigenvalues of the set E^+ will converge from above to fixed limits as the basis set is enlarged. For, as in the Hylleraas-Undheim theorem[1], the symmetric matrix (27) will have the separation property, and E^+ is bounded below for all NB. It is not at all obvious that the eigensolutions approach those of the original Dirac equation, and it is necessary to look a bit more carefully at properties of the basis set.

It is clear, first of all, that we shall get nowhere unless each $\psi_j(r)$ is in $L^2(\mathbb{R})$, and moreover, that $d\psi_j/dr$ is also in $L^2(\mathbb{R})$, and that $\{\psi_j(r)\}$ is a dense set. However, this is by no means sufficient. Not all choices of basis will ensure $\rho_{\hat{Z}} < 2c^2$; it may even be unbounded as $NB \to \infty$. Also, the conditions for $[\hat{D},\hat{W}]$ to approximate $-W^2$ accurately, and $\hat{D}^2$ to approximate d^2/dr^2 are by no means obvious [15]. If the formal relations are poorly represented, it is by no means clear that results making physical sense will be obtained, as has been observed in many of the calculations reported[12,13]. Clearly, a good deal of practical numerical work to test these questions on selected basis sets would be worthwhile. For the moment, we can only speculate. However, there exists at least one basis set with finite $\rho_{\hat{Z}}$: the nonrelativistic Coulomb functions. These have expectation values of 1/r proportional to $1/n^2$, for the basis function $P_{n\ell}$, so that trace $\hat{W} < Z \sum_1^\infty 1/n^2 = \pi^2 Z/6$, and for a Coulomb potential, Z/r, it follows that $\rho_{\hat{z}} < \pi^2 Z^2/6$. Drake abd Goldman[16] studied the Coulomb case using a basis set $\psi_j(r) = r^{\gamma+j-1}e^{-\lambda r}$, $\gamma = \sqrt{[\kappa^2 - \alpha^2 Z^2]}$, $j = 1,2\ldots$ and found behaviour consistent with a lower bound of E^+ independent of NB. However, in this case, $\rho_{\hat{Z}}$ appears to be unbounded as $NB \to \infty$, so that whilst the condition that $\rho_{\hat{Z}}$ be finite is sufficient, it cannot be necessary. They also showed that the complete set of 2NB eigenfunctions gave a useful representation of the Coulomb-Dirac Green's function which could be used to calculate accurately a number of expectation values.

EXTENSION TO THE SCF CASE

It is clear that the formal arguments of the last section can be repeated for the more general system (21) under very similar hypotheses. The most dubious here is that $-Z(\mathbb{P},\mathbb{Q})$ be positive definite, though a plausible physical argument suggests it is not unreasonable for neutral systems or positive ions. The analogue of (32) will be

$$\mathbb{P}^+ \, E \, \mathbb{P} > \mathbb{P}^+ \, NZ(\mathbb{P},\mathbb{Q})\mathbb{P}$$

Now choose $\mathbb{P}^+ = (0,0,\ldots,0, P_a^+,0,\ldots)$; the inequality reduces to

$$\varepsilon_a > P_a^+ \hat{Z}_{aa} P_a / P_a^+ P_a \qquad a = 1,2,\ldots, NW$$

$$> -\bar{\rho}, \text{ say,}$$

where $\bar{\rho}$ is the equivalent of $\rho_{\hat{Z}}$, (32), for a bare nucleus. Of course, all the questions connected with the choice of basis sets described in the one-electron case remain open, and give some idea of the reasons why so many recent molecular calculations with finite basis sets have given results of dubious value [12,13]. In particular, the absence of a lower bound on E^+ or one that is too low, will imply no guarantee that computed eigenvalues give upper bounds to the correct eigenvalues. For example, if the representation $\hat{H}_\ell^{(Z)}$ is poor, so that T_ℓ is underestimated, the non-relativistic limit will lie too low, and there is no hope that any relativistic eigenvalue connected to it will always be an upper bound. Criteria for the choice of suitable basis sets for relativistic calculations are urgently needed.

REFERENCES

1. J.K.L. MacDonald, Phys. Rev. 43: 830 (1933); the theorem, in a different guise is familiar to numerical analysts, e.g. A. Ralston "A first course in numerical analysis", Theorem 10.15, page 495. McGraw Hill, New York (1965).
2. See, for example, articles in Proceedings of Nobel Symposium 46; Phys. Scripta, 21, nos. 3 - 4 (1980).
3. D.P. Carroll, H.J. Silverston and R.M. Metzger, J. Chem. Phys. 71: 4142 (1979).

4. W. Eissner, M. Jones, H. Nussbaumer, Computer Phys. Commun. 8: 270 (1974); M. Klapisch, ibid, 2: 239 (1971); M. Klapisch, J.L. Schwob, B.S. Fraenkel and J. Oreg, J. Opt. Soc. Amer 67: 148 (1977); and subsequent papers.
5. I.P. Grant, these proceedings ; we use here the notation of I.P. Grant, B.J. McKenzie, P.H. Norrington, D.F. Mayers and N.C. Pyper, Computer Phys. Commun. 21: 207 (1980).
6. I.P. Grant, D.F. Mayers and N.C. Pyper, J. Phys. B. 9: 2777 (1976).
7. J. Hata and I.P. Grant, J. Phys. B. 14: 2111 (1981).
8. K.K. Docken and J. Hinze, J. Chem. Phys. 57 : 4928 (1972).
9. I.P. Grant, Adv. Phys. 19: 747 (1970).
10. I.P. Grant, these proceedings.
11. I.P. Grant, Phys. Scripta , 21: 443 (1980).
12. G.L. Malli and J. Oreg, J. Chem. Phys. 63: 830 (1975); J. Oreg and G.L. Malli, ibid, 65: 1746, 1755 (1976); O. Matsuoka, N. Suzuki, T. Aoyama and G.L. Malli, ibid, 73: 1320, 1329 (1980); J.C. Barthelat, M. Pelissier and Ph. Durand, Phys. Rev. A21: 1773 (1980): G.L. Malli, these proceedings, and many others.
13. F. Mark, H. Lischka and F. Rosicky, Chem. Phys. Letts.71: 507 (1980); F. Mark and F. Rosicky, ibid, 74: 562 (1980).
14. Y.K. Kim, Phys. Rev. 154: 17 (1967); ibid 159: 190 (1967); T. Kagawa, ibid, A12: 2245 (1975); ibid, A22: 2340 (1980).
15. M. Reed and B. Simon, "Functional Analysis" Vol. 1., Sec VIII.5 Academic Press, New York (1972).
16. G.W.F. Drake and S.P. Goldman, Phys. Rev. A23: 2093 (1981)

NUMERICAL DIRAC-FOCK CALCULATIONS FOR ATOMS

J.P. Desclaux

Centre d'Etudes Nucléaires de Grenoble
Département de Recherche Fondamentale
LIH 85 X
38041 GRENOBLE, France

INTRODUCTION

If judged from the number of papers dealing with the subject, the interest in relativistic corrections to the theoretical determination of electronic properties of atoms, molecules and solids has been growing up very rapidly during the last years. There is much to be said for explaining it : increase in the experimental accuracy especially when dealing with heavy atoms, access to more exotic systems like very stripped ions, theoretical developments which make pratical calculations manageable, availability of fast computers, etc... But despite (or because !) the fact that the importance of relativistic effects is now well established and that sophisticated models have been built up to handle them, it is necessary to question the validity and accuracy of these calculations. This question is not a simple one because of the complexity of the problem. To perform these calculations we have, as everyone knows, to use an approximate Hamiltonian and there is still questions concerning the most appropriate choice of it. Furthermore, as we try to improve our methods, we have to worry more and more about various contributions. Examples are : QED corrections, finite size of the nucleus or non additivity of relativistic and many-body effects. My purpose here is neither to try to cover all these subjects (I will be quite unable to reach this goal) nor to present an exhaustive review of the importance of relativistic effects on various properties. More simply I will describe one of the most elaborate method available to carry out atomic calculations for bound systems and illustrate the results by few examples.

The first section is thus devoted to a brief discussion of the theoretical fundations of the method, more fundamental presentations will appear during this school. Next I describe in some detail the hardware necessary to reach numerical values. In the last part selected examples are given to illustrate the order of magnitude of the various corrections, the importance of a relativistic treatment or the limitations of present day state of the art in atomic calculations.

I. THEORETICAL ASPECTS

I.1 Hamiltonian

In all relativistic Dirac-Fock calculations performed up to now, the Hamiltonian for the N electron system was taken to be the Breit-Dirac Hamiltonian :

$$H = \sum_{i=1}^{N} h_D(i) + \sum_{i<j=1}^{N} g(i,j) \tag{I.1}$$

where, in atomic units ($e=\hbar=m=1$), h_D is the one electron Dirac operator :

$$h_D(i) = c\,\vec{\alpha}_i \cdot \vec{p}_i + \beta_i c^2 + V_N(r_i) \tag{I.2}$$

c is the velocity of light and $\vec{\alpha}$, β are the usual fourth order Dirac matrices. V_N is a realistic electron nucleus potential energy. The electron-electron interaction energy $g(i,j)$ is approximated by the instantaneous Coulomb repulsion corrected up to a certain order in 1/c for the dynamical part of the interaction (the Breit operator). Various approximations have been used as we shall see later but we first start by discussing the validity of the above Hamiltonian I.1. Beside the fact that in all practical calculations the electron-electron interaction taken in account was not fully Lorentz invariant, Eq. I.1 has the further disavantage of having no eigenfunctions which can describe bound states. This problem, first pointed out by Brown and Ravenhall[1], is obviously related to the existence of the negative energy (including the rest energy) states but arises only because the electrons are interacting whatever is the interaction. In the one-electron case Dirac assumed that all these negative energy states are filled but this constraint is not explicitly included in the Dirac equation itself. To illustrate how this problem show up let us first neglect the electron-electron interaction $g(i,j)$ and consider the eigenstates of :

$$H_o = \sum_{i=1}^{N} h_D(i) \tag{I.3}$$

To any solution of total energy E and including only bound wave functions we can associate an infinite number of states with the same total energy but with one electron being in a positive energy continuum state and a second electron in a negative energy one. We have thus to deal with the problem of a bound state coupled to an infinity of continuum states and, when the interaction $g(i,j)$ is switched on, the bound state will dilute into the sea of continuum states giving rise to "autionization". Obviously we know that this does not occur in reality. Furthermore, Dirac-Fock calculations have produced quite reasonable results and we may, from the preceding comments, worry why ? To describe bound states we should in principle not used the Breit-Dirac Hamiltonian given by Eq. I.1 but a corrected one such that the positive energy states will be decoupled from the negative ones. The correct derivation of the proper operator must be done using quantum electrodynamics (QED) and we refer to the paper of Sucher[2] for this purpose. We outline here only the main points and discuss why Dirac-Fock calculations are nevertheless justified. A corrected Hamiltonian will be of the form :

$$H^{+} = \sum_{i} h_D(i) + \Lambda_+ \; V_{ee} \; \Lambda_+ \qquad (I.4)$$

where

$$\Lambda_+ = \Lambda_+(1)\ \Lambda_+(2)\ \ldots.\ \Lambda_+(N) \qquad (I.5)$$

is a positive energy projection operator which ensures that the electron-electron interaction V_{ee} will connect positive energy states only among themselves. Then we need to solve :

$$H^{+}\psi = E\ \psi \qquad (I.6)$$

with

$$\Lambda_+\psi = \psi \qquad (I.7)$$

The projection operator Λ_+ can be defined with respect to a given one particle Hamiltonian h_o by :

$$\Lambda_+(i) = \sum_{n} |u_n^{+}\ (i)\rangle\langle\ u_n^{+}\ (i)| \qquad (I.8)$$

where the u_n^+ are the positive energy eigenfunctions (both bound and continuum) of h_o. We immediately notice at this point that H^+ depends, via Λ_+, on the choice made for h_o and consequently we may ask about the best possible choice for it. Mittleman[3] has considered this question from a variational point of view by imposing the condition :

$$\partial E/\partial\Lambda_+ = 0 \qquad (I.9)$$

which obviously must be understood as a functional derivative. Restricting the total wave function ψ to be a single Slater determinant, Mittleman claimed that h_o should be the Hartee-Fock operator. As far as I know this proof has not be extended yet to more complex wave functions but it nevertheless provides a theoretical justification for carrying out Dirac-Fock calculations for simple systems. In fact in these calculations one never really solves equations for the many body problem but only a set of one-electron differential equations in a self-consistent manner. Thus from a practical point of view there is no difficulty to restrict oneself to only positive energy solutions. Furthermore if we assume that the Hamiltonian H^+ of Eq. I.4 is built up with Λ_+ defined with respect to the one-electron Dirac-Fock Hamiltonian, then we have, in a perturbative treatment, no first order correction to the total energy since Λ_+ becomes unity when acting on a positive energy Dirac-Fock orbital. This choice may not be the best one for other quantities than energy but no numerical investigations have been done yet. For other possible choices of h_o we again refer to Sucher paper. To conclude let us just point out that the Dirac-Fock projector, which is a non local operator, is rather complicated and that using it with anything else than Dirac-Fock orbitals would be difficult. Furthermore Mittleman has also shown that the Hamiltonian should include three, four etc... body terms. These terms describe an effective interaction due to the fact that we have restricted the variational space to the subspace of positive energy states while in a perturbative treatment the negative ones must appear. Again no investigation of their numerical importance has been performed up to now.

I.2 Breit Interaction

We come now to the discussion of which electron-electron interaction has been taken into account in practical relativistic calculations of N-electron systems. Again the correct description should be given by QED. This interaction must describe the emission or absorption of photons by electrons. It turns out that the lowest order terms correspond to a two body interaction where the photon emitted by an electron is absorbed by an other electron. When the photon is absorbed by the same electron this gives rise to the Lamb shift corrections which will be considered in the next section. To give an explicit form to the electron-electron interaction we have, as usual, to make the choice of a given gauge and as we will never obtain the exact solution, the numerical results will be gauge dependent (like for the "length" and "velocity" forms of dipole transition probabilities). The interaction energy being given by :

$$\iint \varphi_A^*(1)\,\varphi_B^*(2)\; g(1,2)\; \varphi_C(1)\; \varphi_D(2)\; dv_1 dv_2 \tag{I.10}$$

In the Coulomb gauge the operator g(1,2) is :

$$g^C(1,2) = \frac{1}{R} - \frac{\vec{\alpha}_1 . \vec{\alpha}_2}{R}\, e^{i\omega R} + (\vec{\alpha}_1 . \vec{\nabla}_1)(\vec{\alpha}_2 . \vec{\nabla}_2)\, \frac{e^{i\omega R} - 1}{\omega^2\, R} \tag{I.11}$$

while in the Feynman gauge we have :

$$g^F(1,2) = \frac{1 - \vec{\alpha}_1 . \vec{\alpha}_2}{R}\, e^{i\omega R} \tag{I.12}$$

where in both of the above equation $R = |\vec{r}_1 - \vec{r}_2|$ is the interelectronic distance and ω the frequency of the virtual exchange photon given by :

$$\omega = \omega_{AC} = |\varepsilon_A - \varepsilon_C|/c = \omega_{BD} = |\varepsilon_B - \varepsilon_D|/c \tag{I.13}$$

the ε's being the electron energies. The two expressions are identical only at the low frequency limit, i.e. if we take $e^{i\omega R} = 1$, and give :

$$(1 - \vec{\alpha}_1.\vec{\alpha}_2)/R \tag{I.14}$$

known as the Gaunt interaction. The next non vanishing term corresponds to the retardation in the interaction. Note that its contribution to a direct type integral (A=C, B=D) will be zero since ω=0 and this is just what is expected for the interaction between two static charges. The lowest order terms are different in the two gauges and are given by :

$$\frac{1}{2R}\ \vec{\alpha}_1\ \vec{\alpha}_2 - [\frac{(\vec{\alpha}_1.\vec{R})(\vec{\alpha}_2.\vec{R})}{R^2}] \qquad \text{Coulomb} \tag{I.15}$$

$$-\ \omega^2 R/2 \qquad \text{Feynman} \tag{I.16}$$

All atomic calculations have been done in the Coulomb gauge and in the most elaborate ones the following form was used for the electron-electron interaction :

$$\frac{1}{R} - \frac{\vec{\alpha}_1.\vec{\alpha}_2}{R}\cos\omega R + (\vec{\alpha}_1.\vec{\nabla}_1)(\vec{\alpha}_2.\vec{\nabla}_2)\ \frac{\cos\omega R - 1}{\omega^2 R} \tag{I.17}$$

As given by Eq. I.17 the operator is strictly applicable only to diagonal matrix elements (i.e. for Coulomb and exchange integrals). For the general case where $\omega_{AC} \neq \omega_{BD}$, Mittleman[3] suggested that one should use one half of the sum of an operator in which $\omega = \omega_{AC}$ and a second on in which $\omega = \omega_{BD}$. Before concluding this section we would like to make two comments. First, expressions like I.11 and I.12 or even the approximations I.16 or I.17 are not really operators since they depend on the wave functions due to the presence of the energies in the definition of the ω's. Second in nearly all calculations only the instantaneous Coulomb repulsion was introduced in the self-consistent process for the determination of the wave functions. The reason advocated comes from the derivation of the Breit interaction from second order perturbation theory[4]. From the latest derivations[2,3] of the relativistic Hamiltonian there seem to be no reasons, except practical ones, of not including magnetic and retardation corrections (at least at the low frequency limit) in the self-consistent process providing that the positive energy operator is explicitly or, as in the Dirac-Fock method, implicitly included.

I.3 Lamb-shift Corrections

As already stated they correspond to the process involving a virtual photon emitted and absorbed by the same electron and are not included in the Dirac-Fock Hamiltonian discussed in the previous section. Up to now only few calculations of the self energy, which together with the vacuum polarization gives the Lamb shift, have been performed for many electron atoms. If we except two electron systems[5], the only other results are the screened self energy results for the K shell of few atoms[6,8]. Consequently, in most cases, the self energy was just evaluated from the hydrogenic results of Mohr[9] by using an effective nuclear charge. In most of the published results the screening coefficients were obtained by comparing the mean radius of Dirac-Fock orbitals to the one of their hydrogenic counterparts. As the Lamb-shift corrections increased rapidly with the atomic number Z (Z^4) and are sensitive to the behaviour of the wavefunctions near the nucleus, this procedure may not be too reliable. If the calculations of the Lamb-shift for two electron systems seems straightfoward it will be quite time consuming and extension to N-electron systems may be practically untractable in the framework of S formulation of QED due to the need of including diagrams which involve also the electron-electron interaction[5].

The vacuum polarization contribution is easier to obtain since it has been shown by Wichmann and Kroll[10] that the Uehling potential[11] corresponds to the first order term. In the case of a spherical nuclear charge distribution $\rho_N(r)$, this potential is given by :

$$V(r) = -\frac{2Xe}{3cr}\int_0^\infty u\rho_N(u)\,[K_o(\frac{2}{\lambda_e}|r-u|) - K_o(\frac{2}{\lambda_e}|r+u|)]du \quad (I.18)$$

with λ_e the electron Compton wavelength and :

$$K_o(x) = \int_1^\infty (\frac{1}{t^3}+\frac{1}{2t^5})(t^2-1)^{1/2}\,e^{-xt}\,dt \quad (I.19)$$

If higher order terms are needed it is also possible to calculate them as the expectation value of the next order potential given by Källén and Sabry[12].

II. THE MULTICONFIGURATION DIRAC-FOCK METHOD

II.1 Wave functions

From now on we shall use, according to the previous section, the following Hamiltonian :

$$H = \sum_i h_D(i) + \sum_{i<j}\,[\frac{1}{r_{ij}} + B(i,j)] \quad (II.1)$$

with $B(i,j)$ given by Eq. I.16 or I.17. Thus compared to the non-relativistic case, we introduce mass variation, Darwin and spin-orbit corrections by using Dirac one-electron Hamiltonian h_D to describe the kinematics and the electron-nucleus interaction. In the non-relativistic limit the Breit interaction gives the Breit-Pauli operator containing terms known as spin-other-orbit, spin-spin and orbit-orbit interaction. Also, as we have seen before, retardation in the electron-electron interaction is taken care of up to a certain order. The total wave function which must be an eigenstate of the total angular momentum J and of the parity P (the only two good quantum numbers in the present case) is writen for a given atomic state as :

$$\psi(PJM) = \sum_{\nu=1}^{NCF} W_\nu \, \phi_\nu \, (\gamma_\nu PJM) \qquad (II.2)$$

i.e. as a linear combination of configuration state functions (CSF) ϕ. These CSF are simultaneous eigenstates of P, J^2 and J_Z with J(J+1) and M respectively as eigenvalues of the two last operators. The label γ_ν stands for all other values (angular momentum recoupling scheme, seniority numbers etc...) necessary to define unambigously the νth CSF. Each of these CSF is a fully antisymmetric product of fourth order central field Dirac spinors :

$$\varphi_{n\kappa m}(r,\theta,\varphi) = \frac{1}{r} \begin{vmatrix} P_{n\kappa}(r) & \chi_{\kappa m}(\theta,\varphi) \\ iQ_{n\kappa}(r) & \chi_{-\kappa m}(\theta,\varphi) \end{vmatrix} \qquad (II.3)$$

where n is the principal quantum number. The one-electron orbitals $\varphi_{n\kappa m}$ are eigenvectors of the total angular momentum $\vec{j} = \vec{\ell} + \vec{s}$ and of its projection j_Z and are of given parity. Here we use the relativistic quantum number "kappa" defined by :

$$\kappa = \ell \quad \text{if } j = \ell - 1/2 \quad \text{and} \quad \kappa = -(\ell+1) \quad \text{if } j = \ell + 1/2 \qquad (II.4)$$

which includes both j and the parity. The two component Pauli spinors χ are simultaneous eigenfunctions of ℓ^2, s^2, j^2 and j_Z with eigenvalues $\ell(\ell+1)$, $s(s+1)$, $j(j+1)$ and m respectively. As already pointed out we have to use jj coupling to construct the CSF's. Should we consider a single jj coupled state, such a description would be unphysical for most of atomic systems since we know that Coulomb repulsion is never negligible compared to spin-orbit interaction. In this respect the multiconfiguration method gives the opportunity to handle intermediate coupling and, if we restrict the basis functions to the jj ones arising from a single LS state, we are still almost at the Hartree-Fock level but in intermediate coupling. Note nevertheless that, even for light elements, this is not completly analogous to a non-relativistic treatment including spin-orbit interaction since we allow the radial functions for $j = \ell - 1/2$ and $j = \ell + 1/2$ to vary independently. Extension of the basis set will introduce correlation corrections as in the non-relativistic case.

II.2 Matrix elements

As only the radial function P and Q of the one-electron orbitals and the mixing coefficients w between the CSF depend on the state under consideration and are optimized during the self-consistent process, the mean value of any operator is given, in the CSF basis, as a quadratic expression of the w's with radial integrals as coefficients. We can thus write the total energy in terms of the following radial integrals : the rest mass energy associated to the βc^2 operator is given by :

$$\langle n\kappa m|\beta c^2|n'\kappa' m'\rangle = \delta_{\kappa,\kappa'}\delta_{m,m'}\, c^2 \int_0^{\infty} (P_{n\kappa}P_{n'\kappa} - Q_{n\kappa}Q_{n'\kappa})\,dr \qquad \text{(II.5)}$$

The kinetic energy operator $c\vec{\alpha}.\vec{p}$ results in :

$$c\langle n\kappa m|\vec{\alpha}.\vec{p}|n'\kappa' m'\rangle = \delta_{\kappa,\kappa'},\ \delta_{m,m'},\, c\int_0^{\infty}\{Q_{n\kappa}[\frac{dPn'\kappa}{dr} + \frac{\kappa}{r}Pn'\kappa]$$
$$-P_{n\kappa}[\frac{dQn'\kappa}{dr} - \frac{\kappa}{r}Qn'\kappa]\}\,dr \qquad \text{(II.6)}$$

If the nuclear potential is assumed to be spherically symmetric, the electron-nucleus interaction is :

$$\langle n\kappa m|V_N(r)|n'\kappa' m'\rangle = \delta_{\kappa,\kappa'}\ \delta_{m,m'} \int_{\alpha}^{\infty} (P_{n\kappa}P_{n'\kappa} + Q_{n\kappa}Q_{n'\kappa})V_N(r)\,dr \qquad \text{(II.7)}$$

For the instantaneous Coulomb repulsion the usual multipole expansion :

$$\frac{1}{r_{12}} = \sum_k U^k(1,2) \sum_q (-1)^q\, C^k_q(1)\, C^k_{-q}(2) \qquad \text{(II.8)}$$

with

$$U^k(1,2) = r_<^k / r_>^{k+1}$$
$$C^k_q = (4\pi/2k+1)^{1/2}\, Y^k_q\,(\theta,\varphi) \qquad \text{(II.9)}$$

leads after straightfoward angular momentum algebra to the result :

$$\langle AB|\frac{1}{r_{12}}|CD\rangle = \delta_{m_A+m_B,m_C+m_D} \sum_k d^k(j_Cm_C,j_Am_A)\, d^k(j_Bm_B,j_Dm_D) \qquad \text{(II.10)}$$
$$\times \int_0^{\infty}\int_0^{\infty} [P_A(1)P_C(1)+Q_A(1)Q_C(1)]U^k(1,2)[P_B(2)P_D(2)+Q_B(2)Q_D(2)]\,dr_1\,dr_2$$

where the d^k coefficients are given in terms of 3j symbols by:

$$d^k(jm,j'm') = (-1)^{m'+1/2}[(2j+1)(2j'+1)]^{1/2}\begin{pmatrix} j & k & j' \\ 1/2 & 0 & -1/2\end{pmatrix}\begin{pmatrix} j & k & j' \\ -m & q & m'\end{pmatrix} \qquad \text{(II.11)}$$

and the following triangular conditions must hold :

$$\mathrm{Max}\{|j_A-j_C|,\ |j_B-j_D|\} \leq k \leq \mathrm{Min}\{j_A+j_C,\ j_B+j_D\} \qquad \text{(II.12)}$$

$\ell_A + \ell_C + k$ and $\ell_B + \ell_D + k$ must both be even

The Breit interaction is more tedious to handle and we give the main points of the derivation. We first consider the magnetic part and notice that we can expand it in terms of tensor operators as :

$$\frac{\vec{\alpha}_1 . \vec{\alpha}_2}{r_{12}} = \sum_{k,L,M} (-1)^{L+k+M}\, U^k(1,2) X^L_M(1) X^L_{-M}(2) \qquad \text{(II.13)}$$

where the X^L_M tensors are given by the direct product :

$$X^L_M = [C^k_q \times \alpha^1_p]^L_M \qquad \text{(II.14)}$$

Evaluation of X^L_M is obtained from standard formula of angular momentum theory i.e. :

$$\langle \ell\tfrac{1}{2}jm|X^L_M|\ell'\tfrac{1}{2}j'm'\rangle = (-1)^{j-m}\begin{pmatrix} j & L & j' \\ -m & M & m' \end{pmatrix}\langle \ell\tfrac{1}{2}j||X^L||\ell'\tfrac{1}{2}j'\rangle \qquad \text{(II.15)}$$

and the reduced matrix element is given by :

$$\langle \ell\tfrac{1}{2}j||\ X^L\ ||\ell'\tfrac{1}{2}j'\rangle = [j,L,j']^{1/2}\begin{Bmatrix} \ell & 1/2 & j \\ \ell' & 1/2 & j' \\ k & 1 & L \end{Bmatrix}\langle \tfrac{1}{2}||\sigma||\tfrac{1}{2}\rangle\langle \ell||C^k||\ell'\rangle \qquad \text{(II.16)}$$

where $\{\}$ is a 6j symbol and :

$$[j,L,j'] = (2j+1)(2L+1)(2j'+1) \qquad \text{(II.17)}$$

Recombining II.15 and II.16 and using the well known results :

$$\langle \tfrac{1}{2}||\sigma||\tfrac{1}{2}\rangle = \sqrt{6}$$

$$\langle \ell||C^k||\ell'\rangle = (-1)^{\ell}[\ell,\ell']^{1/2}\begin{pmatrix} \ell & k & \ell' \\ 0 & 0 & 0 \end{pmatrix} \qquad \text{(II.18)}$$

we reach the final result :

$$\langle \ell\tfrac{1}{2}jm|X^L_M|\ell'\tfrac{1}{2}j'm'\rangle = (-1)^{j+\ell-m}\sqrt{6}[j,L,j',\ell,\ \ell']^{1/2}\begin{pmatrix} j & L & j' \\ -m & M & m' \end{pmatrix}$$
$$\times \begin{pmatrix} \ell & k & \ell' \\ 0 & 0 & 0 \end{pmatrix}\begin{Bmatrix} \ell & 1/2 & j \\ \ell' & 1/2 & j' \\ k & 1 & L \end{Bmatrix} \qquad \text{(II.19)}$$

Using formulae given in Brink and Satchler[13], the last product of the 3j and 6j can be reduced to a single 3j multiplied by a simple factor. We thus obtain for the general matrix element of the magnetic interaction :

$$\langle AB|\frac{-\vec{\alpha}_1\cdot\vec{\alpha}_2}{r_{12}}|CD\rangle = (-1)^q d^L(j_Am_A,j_Cm_C)d^L(j_Bm_B,j_Dm_D)X\sum_k(-1)^{L+k}$$

$$\{\eta_k^L(\ell_Aj_A,\overline{\ell}_Cj_C)[\eta_k^L(\ell_Bj_B,\overline{\ell}_Dj_D)R_M^k(AC,BD)-\eta_k^L(\overline{\ell}_Bj_B,\ell_Dj_D)R_M^k(AC,DB)]$$

$$-\eta_k^L(\overline{\ell}_Aj_A,\ell_Cj_C)[\eta_k^L(\ell_Bj_B,\overline{\ell}_Dj_D)R_M^k(CA,BD)-\eta_k^L(\overline{\ell}_Bj_B,\ell_Dj_D)R_M^k(CA,DB)]\}$$

$$q = j_C + j_D + m_A - m_C \qquad \text{(II.20)}$$

where $\overline{\ell}$ is the orbital angular momentum of the small component and because the α_i matrices are antidiagonal, the large and small components are mixed yielding a radial integral of the form :

$$R_M^k(AC,BD) = \iint P_A(1)Q_C(1)U^k(1,2)P_B(2)Q_D(2)dr_1dr_2 \qquad \text{(II.21)}$$

the d^L coefficients are those given by Eq.II.11 and the η ones are defined by :

$$\eta_k^L\ (\ell j,\ell' j') =$$

$$(-1)^{j+\ +\ell'+1/2}\{(\ell-j)(2j+1)+(\ell'-j')(2j'+1) + \xi_k^L\ \}/\Delta_k^L \quad \text{if } L=k\pm 1$$

$$\xi_k^{k-1} = L + 1\ ;\ \xi_k^{k+1} = -L \qquad \text{(II.22)}$$

$$\Delta_k^{k-1} = [(L+1)(2L+3)]^{1/2}\ ;\ \Delta_k^{k+1} = [L(2L-1)]^{1/2}$$

$$(-1)^{\ell+1}\ \frac{(2j'+1) + (-1)^{j+j'+k}\ (2j+1)}{2\ [L(L+1)]^{1/2}} \quad \text{if } L = k$$

The following triangular relations must be satisfied :

$$\text{Max}\ \{|j_A-j_C|,|J_B-j_D|\} \lesssim L \lesssim \text{Min}\ \{j_A+j_C,\ j_B+j_D\}$$

$$L = k - 1,\ k,\ k + 1 \qquad \text{(II.23)}$$

$$\ell_A + \ell_C + k \quad \text{and} \quad \ell_B + \ell_D + k \text{ must both be odd}$$

The reduction of the retardation term to radial integrals is similar to that of the magnetic interaction but the angular momentum algebra is more tedious. Results can be found in reference 14 and will not be given here. If the derivation is carried out in the same spirit as done above for the magnetic term, it is possible for the low frequency limit (Eq. I.15) to express all the angular coefficients in terms of the d^L and η coefficients. But, because of the gradient operator, some of the integrals will not extend over all the domain of integration and will be of the type :

$$\int_0^\infty [P_B(2)Q_D(2)\int_0^{r_2} P_A(1)Q_C(1)U^k(1,2)dr_1]dr_2 \qquad \text{(II.24)}$$

As can be seen from the above expressions, the complexity of the angular coefficients makes highly desirable to calculate them automatically. Two computer programs have been developped for this purpose. One by Pyper et al.[15] uses Racah techniques to construct the CSF. Each subshell configuration $(n_i\kappa_i)^{q_i}$ is characterized by its total angular momentum quantum numbers J_i, M_i and its seniority number V_i. Such a characterization is complete up to f electrons. The various CSF are then generated by coupling in all possible ways the angular momentum of the various subshells to a given total value J. The second program written by the author is working in terms of Slater determinants and is the relativistic counterpart of the program described by Eissner and Nussbaumer[16]. The N electron wave function of each CSF is given as the linear combination of determinants having all the same M value and which is an eigenvector of J^2. These eigenstates are obtained by the diagonalization of the J^2 matrix constructed on the basis of all possible determinants associated to the given M value. It is then straightforward to calculate the angular coefficients needed for the energy expression. It is also possible to obtain the expectation value of other operators like for oscillator strengths or Landé g_J values.

II.3 Average energy

As in the non-relativistic case, it is useful to introduce the concept of the average energy defined as the centre of gravity of all the levels belonging to a given configuration and defined by :

$$E_{AV}(jj) = \frac{\Sigma(2J+1)\ E(J)}{\Sigma(2J+1)} \tag{II.25}$$

The usefulness of the average energy is due first to the fact that the energy of a given level is just the sum of the E_{AV} plus a small number of extra two electron integrals. As the expression of E_{AV} is very easy to incorporate in any program this makes the imput data for a given level rather convenient to specify. Second, except in special cases like the sp configuration, the radial functions do not depend strongly on the specific level considered and as a first approximation they can be determined for the average energy only. To obtain E_{AV} we have to derive the electron interaction with a closed shell. For the Coulomb repulsion we have to perform the following summations :

$$\sum_m (-1)^m \begin{pmatrix} j & k & j \\ -m & 0 & m \end{pmatrix} = [j]^{-1/2}\ \delta_{k,0}$$
$$\sum_{m,m_k} \begin{pmatrix} j & k & j' \\ -m & m_k & m' \end{pmatrix}^2 = [j']^{-1} \tag{II.26}$$

So that we obtain :

$$\sum_m d^k(jm,jm)\, d^k(j'm',j'm') = [j]\, \delta_{k,0} \qquad \text{(II.27)}$$

for direct integrals and for exchange ones :

$$\sum_m [d^k(jm,j'm')]^2 = [j]\, \Gamma_{jkj'} \qquad \text{(II.28)}$$

with

$$\Gamma_{jkj'} = \begin{pmatrix} j & k & j' \\ \frac{1}{2} & 0 & -\frac{1}{2} \end{pmatrix}^2 \qquad \text{(II.29)}$$

For the magnetic interaction the triangular relations II.23 combined with the result II.27 shows that there is no direct contribution as expected since a closed shell has no net magnetic moment. For the exchange integrals we obtain :

$$\langle AB|\frac{-\vec{\alpha}_1\,\vec{\alpha}_2}{r_{12}}|BA\rangle = -[j_A,j_B] \sum_{k,\gamma,L} \Gamma_{j_A L j_B}\, \xi^{\gamma}_{kL}(\kappa_A,\kappa_B)\, G^k_M(A,B;\gamma) \qquad \text{(II.30)}$$

$$\gamma = -1,0,1$$

where using the definition II.21 of the R integrals :

$$\begin{aligned} G^k_M(A,B;1) &= R^k_M(AB,AB) \\ G^k_M(A,B;0) &= R^k_M(AB,BA) \qquad \text{(II.31)} \\ G^k_M(A,B;-1) &= R^k_M(BA,BA) \end{aligned}$$

and

$$\begin{aligned} \xi^{\pm1}_{k,k-1}(\kappa_A,\kappa_B) &= \frac{(\kappa_A-\kappa_B+k)^2}{k(2k+1)} & \xi^{0}_{k,k-1} &= \frac{2[(\kappa_A-\kappa_B)^2-k^2]}{k(2k+1)} \\ \xi^{\pm1}_{k,k}(\kappa_A,\kappa_B) &= \frac{(\kappa_A+\kappa_B)^2}{k(k+1)} & \xi^{0}_{k,k} &= 2\xi^{\pm1}_{k,k} \qquad \text{(II.32)} \\ \xi^{\pm1}_{k,k+1}(\kappa_A,\kappa_B) &= \frac{(\kappa_A-\kappa_B-k-1)^2}{(k+1)(2k+1)} & \xi^{0}_{k,k+1} &= \frac{2\,[(\kappa_A-\kappa_B)^2-(k+1)^2]}{(k+1)(2k+1)} \end{aligned}$$

With the help of the above relations we can now give the explicit expression of the average energy which, restricted to Coulomb interaction for simplicity, reads:

$$E_{AV} = \sum_A q_A I(A,A) + \frac{1}{2}\sum_A q_A\, \{(q_A-1)F^0(A,A)$$

$$-(q_A-1)/2j_A \sum_{k>0} (2j_A+1)\, \Gamma_{j_A k j_A}\, F^k(A,A)\} \qquad \text{(II.33)}$$

$$+\frac{1}{2} \sum_{\substack{A,B \\ A\neq B}} q_A q_B \{F^0(A,B) - \sum_k \Gamma_{j_A k j_B}\, G^k(A,B)\}$$

the q's are the occupation numbers of the orbitals and the one-electron integrals I are the sum of II.5, II.6 and II.7 which after subracting the rest mass energy becomes :

$$I(A,B) = c \int_0^\infty \{Q_A \frac{dP_B}{dr} - P_A \frac{dQ_B}{dr} + \frac{\kappa_A}{r} (P_A Q_B + P_B Q_A) + \qquad \text{(II.34)}$$

$$\frac{V_N}{c}(P_A P_B + Q_A Q_B) - 2cQ_A Q_B\}\, dr$$

As already pointed out, considering a single jj configuration will be quite unphysical in most atomic cases and it is thus necessary to extend the above definition of the average energy. This is easily achieved by introducing a new averaging over all the jj configurations associated to a single LS one[17,18].

II.4 Radial equations

In the multiconfiguration Dirac-Fock (MCDF) method we solve the two following problems sequentially :

1) given a set of mixing coefficients w between the CSF's we minimize the total energy with respect to the radial orbitals P and Q.

2) with the radial functions thus determined the energy matrix is diagonalized and a new set of w coefficients determined.

This two-step process is repeated until convergence is hopefully reached. From the preceeding section it is obvious that the total energy can be writen as :

$$E_T = \sum_\nu w_\nu^2\, E_{AV}^\nu + \sum_{\nu,\mu} w_\nu\, w_\mu \{\sum e_n(A,B)\, I_n(A,B) + \qquad \text{(II.35)}$$

$$\sum_m c_m^k\, (AC,BD)\, R_m^k\, (AC,BD)\}$$

where again for simplicity we have omitted the Breit interaction. The Coulomb integrals R^k have been defined in Eq. II.10 and the one-electron integrals I by II.34. $e_n(A,B)$ and $c_m^k(AC,BD)$ are numerical coefficients obtained from integration over angular variables. Requiring that the total energy is stationary we obtain the Dirac-Fock equations which for a given orbital A read :

$$\frac{d}{dr}\begin{pmatrix}P_A\\ Q_A\end{pmatrix}=\begin{Bmatrix}-\kappa_A/r & 2c+\{\varepsilon_A-V_A(r)\}/c\\ -\{\varepsilon_A-V_A(r)\}/c & \kappa_A/r\end{Bmatrix}\begin{pmatrix}P_A\\ Q_A\end{pmatrix}+\begin{pmatrix}X_{Q_A}(r)\\ X_{P_A}(r)\end{pmatrix} \quad \text{(II.36)}$$

where V_A is the direct Coulomb potential and the X's are generalized exchange terms containing contributions from the extra integrals of Eq.II.35.

III. NUMERICAL METHODS

We now consider the solution of the pair of differential equations II.36 in which V(r), $X_P(r)$ and $X_Q(r)$ are known functions and ε is an eigenvalue to be determined. For the same reasons as in the non-relativistic case it is convenient to make the change of variable

$$t = \log(r) \quad \text{(III.1)}$$

and to tabulate all the radial functions on this grid of points equally spaced with respect to the new variable t. Making this transformation, the equations become :

$$\begin{aligned} P' + \kappa P &= r\{2c+(\varepsilon-V)/c\}Q + rX^Q \\ Q' - \kappa Q &= -r\{(\varepsilon-V)/c\}P - rX^P \end{aligned} \quad \text{(III.2)}$$

where the prime indicates differentiation with respect to t. In the non-relativistic case we have to deal with a second order differential equation of a particularly simple type for which the Numerov method[19] is particularly well adapted and very efficient. Here we have a pair of first order coupled differential equations and a wide variety of methods can be used.

III.1 Predictor-corrector methods

In most of the first relativistic atomic calculations a predictor-corrector type of method was used. As an example we consider the five point Adams method which has been widely selected because of its stability properties[20]. The predictor is given by :

$$p_{n+1}=Y_n+(1901Y'_n-2774Y'_{n-1}+2616Y'_{n-2}-1274Y'_{n-3}+251Y'_{n-4})/720 \quad \text{(III.3)}$$

and the corrector by :

$$c_{n+1}=Y_n+(251p'_{n+1}+646Y'_n-264Y'_{n-1}+106Y'_{n-2}-19Y'_{n-3})/720 \quad \text{(III.4)}$$

The final value is taken to be :

$$Y_{n+1} = (475c_{n+1}+27p_{n+1})/502 \quad \text{(III.5)}$$

This linear combination is selected in order to cancel the term of order h^6 (h being the constant interval with respect to the variable t). In the expressions above Y stands for either P or Q radial functions. As in the Schrödinger case, the integration strategy involves both an outward and an inward integration for stability purposes. The outward integration is started by means of power series near the origin in order to obtain the five initial values required by the Adams method. For the inward integration a practical infinity has to be determined and the empirical relation :

$$\{V(r)+\varepsilon\}r^2>700 \tag{III.6}$$

gives a reasonable estimate in most of the cases. Then the asymptotic relations:

$$P(r) = a_P\exp(-\mu r) \qquad Q(r) = a_Q\exp(-\mu r) \tag{III.7}$$

$$\text{where} \quad \mu = \{-\varepsilon(2+\varepsilon/c^2)\}^{\frac{1}{2}} \quad \text{and} \quad a_Q = -\{-\varepsilon/(\varepsilon+2c^2)\}^{\frac{1}{2}}a_P \tag{III.8}$$

can be used to determine the first five points of the inward integration. Nevertheless note that these asymptotic forms are exact only for homogeneous systems (i.e. when the X^P and X^Q potentials terms are missing), while in the Hartree-Fock case the long range behaviour is always determined by the outermost orbitals. It would thus be necessary to consider a linear combination of exponential functions with the energies of the various orbitals as exponents. But in practice the approximation above does not seem to introduce significant numerical errors if the method is stable which is indeed the case for the Adams one. For arbitrary values of the slope at origin and infinity the outward and inward integrations will not match at the matching point. In order to get a solution which is continuous everywhere we proceed in the following way: we notice that the solution of an inhomogeneous system can be written as the linear combination of a particular solution of the inhomogeneous system and of the general solution of the homogeneous associated system. Thus denoting by P^i and P^o the inward and outward solutions for the large component of the inhomogeneous system and by P^i_H and P^o_H their corresponding ones of the homogeneous system, it comes, with similar notations for the small component Q :

$$\begin{aligned} \{P^o + \alpha P^o_H\}_{r=R^-} &= \{P^i + \beta P^i_H\}_{r=R^+} \\ \{Q^o + \alpha Q^o_H\}_{r=R^-} &= \{Q^i + \beta Q^i_H\}_{r=R^+} \end{aligned} \tag{III.9}$$

where R is the matching point. The coefficients α and β can be determined from these equations and a solution continuous everywhere is obtained. Obviously for an arbitrary value of the energy ε this solution will not be normalized. This is just due to the fact that the asymptotic expressions (III.7) are valid only for the exact eigenvalue while for an arbitrary value one obtains a linear combination of increasing and decreasing exponentials. The default in the norm can then be used to modify the starting value of ε until the proper eigenvalue is found. In practice it is not always necessary to enforce the normalization constraint at each step of the calculation as long as one converges to a normalized solution (this has been used in some computer codes in order to save time).

III.2 Finite difference methods

It is obvious from the preceding section that the predictor-corrector methods require the solution of at least part of the homogeneous system to obtain continuity in one of the component (for example the homogoneous system can be integrated only from infinity to the matching point in order to insure continuity in the large component while the small one remains discontinuous). This is in contradiction with the Numerov method which, associated with the tail correction[19], gives always a continuous radial function in the non-relativistic case (the derivative remains discontinuous until the proper eigenvalue is found). We consider now the other type of methods used in relativistic numerical programs and which allow easily to enforce the continuity of one of the two components.

We may define an approximation to the solution by writing :

$$Y_{n+1} = Y_n + h(Y'_{n+1} + Y'_n) + \Delta_n \qquad \text{(III.10)}$$

where Δ_n is a difference correction given in terms of central differences by :

$$\Delta_n = \frac{-1}{12}\delta^3 Y_{n+\frac{1}{2}} + \frac{1}{120}\delta^5 Y_{n+\frac{1}{2}} \qquad \text{(III.11)}$$

with :

$$\begin{aligned} \delta^3 Y_{n+\frac{1}{2}} &= Y_{n+2} - 3Y_{n+1} + 3Y_n - Y_{n-1} \\ \delta^5 Y_{n+\frac{1}{2}} &= Y_{n+3} - 5Y_{n+2} + 10Y_{n+1} - 10Y_n + 5Y_{n-1} - Y_{n-2} \end{aligned} \qquad \text{(III.12)}$$

At this stage we may notice that precision in the resolution is required only when we are about to reach self-consistency and not each time we are solving the Eq.(III.2). Consequently the difference corrections Δ_n may be calculated very efficiently (with respect to computer time) by using the wave functions from the previous iteration as it is done for the potential terms. Let:

$$\begin{aligned} a &= 1 + h\kappa/2 & u_n &= \Delta_n^P + h(r_n X_n^Q + r_{n+1} X_{n+1}^Q)/2 \\ b &= -1 + h\kappa/2 & v_n &= \Delta_n^Q + h(r_n X_n^P + r_{n+1} X_{n+1}^P)/2 \\ \theta_n &= hr_n(\varepsilon - V_n)/(2c) & \phi_n &= hcr_n + \theta_n \end{aligned} \qquad \text{(III.13)}$$

where all the quantities (except for ε) are calculated as just said with the help of the estimates available from the preceding iteration. Then the system of algebraic equations :

$$
\begin{aligned}
aP_{n+1} - \theta_{n+1}Q_{n+1} + bP_n - \theta_n Q_n &= u_n \\
\phi_{n+1}P_{n+1} - bQ_{n+1} + \phi_n P_n - bQ_n &= v_n
\end{aligned}
\tag{III.14}
$$

determines P_{n+1} and Q_{n+1} if P_n and Q_n are known. For the outward integration this system is solved step by step from near the origin to the matching point after getting the solution at the first point by series expansion. For the inward integration an elimination process is used by expressing the solution in the matrix form:

$$
\begin{vmatrix}
-a & \phi_{n+1} & -b & 0 & \cdots & & & & 0 \\
-\theta_n & a & -\theta_{n+1} & 0 & \cdots & & & & 0 \\
0 & \phi_{n+1} & -a & \phi_{n+2} & -b & 0 & \cdots & & 0 \\
0 & b & -\theta_{n+1} & a & -\theta_{n+2} & 0 & \cdots & & 0 \\
\cdots & \cdots & \cdots & \cdots & \cdots & \cdots & \cdots & \cdots & \cdots \\
0 & \cdots & 0 & b & -\theta_{N-2} & a & -\theta_{N-1} & & 0 \\
0 & \cdots & & & 0 & \phi_{N-1} & -a & & \phi_N \\
0 & \cdots & & & 0 & b & -\theta_{N-1} & & a
\end{vmatrix}
\times
\begin{Vmatrix}
Q_n \\ P_{n+1} \\ Q_{n+1} \\ P_{n+2} \\ \cdots \\ P_{N-1} \\ Q_{N-1} \\ P_N
\end{Vmatrix}
=
\begin{Vmatrix}
v_n - \phi_n P_n \\ u_n - bP_n \\ v_{n+1} \\ u_{n+1} \\ \cdots \\ u_{N-2} \\ v_{N-1} + bQ_N \\ u_{N-1} + \theta_N Q_N
\end{Vmatrix}
\tag{III.15}
$$

or :
$$A(PQ) = (uv) \tag{III.16}$$

where each row of the matrix A has at most four non zero elements. It is easy to prove that the matrix A can be expressed as the product:

$$A = L.T \tag{III.17}$$

in which L is a lower triangular matrix with only three non zero elements on each row and T an upper triangular matrix with the same property. Introducing an intermediate vector (pq) we first solve L(pq)=(uv) for n,n+1,....,N and then T(PQ)=(pq) for N,N-1,....,n. n is the index of the matching point and N is determined by requiring that P_N should be smaller than a specified very small value when assuming Q_N=0. Thus the number of tabulation points of each orbital is determined automatically without the help of an empirical relation like (III.6). This elimination process is quite similar to the tail procedure[19] used in the non-relativistic case and as written above, produces a large component which is everywhere continuous. The discontinuity in the small component can then be used to adjust the eigenvalue ε. If we compare with the predictor-corrector methods we see that, if it is not necessary to enforce the normalization condition at each step, we avoid the need to integrate the homogeneous system.

III.3 Orthonormality constraints.

When it is necessary to enforce the orthonormality constraints each time the radial equations are solved, we proceed in the following way: the first order variation of the large component due to a change $\Delta\varepsilon_{AB}$ of one of the Lagrange multipliers ($\varepsilon_{AA}=\varepsilon_A$) is given by :

$$P(\varepsilon^0_{AB} + \Delta\varepsilon_{AB}) = P(\varepsilon^0_{AB}) + \Delta\varepsilon_{AB}\left(\frac{\partial P}{\partial\varepsilon_{AB}}\right)_{\varepsilon_{AB}=\varepsilon^0_{AB}} \quad \text{(III.18)}$$

and an equivalent relation will hold for the small component. Let :

$$p_{AB} = \partial P/\partial\varepsilon_{AB} \quad \text{and} \quad q_{AB} = \partial Q/\partial\varepsilon_{AB} \quad \text{(III.19)}$$

then p and q satisfy the system :

$$\begin{aligned} p'_{AB} + \kappa p_{AB} &= r\{2c+(\varepsilon_{AA}-V)\}q_{AB} + rQ_B/c \\ q'_{AB} - \kappa q_{AB} &= -r\{(\varepsilon_{AA}-V)/c\}p_{AB} - rP_B/c \end{aligned} \quad \text{(III.20)}$$

i.e. exactly the same system as (III.2) with the substitution of P_B/c and Q_B/c for X^P and X^Q (note that we may have B=A if $\varepsilon_{AB}=\varepsilon_{AA}$). Consequently the methods described in the previous sections can be used to solve for p_{AB} and q_{AB} which in turn determine the variation $\Delta\varepsilon_{AB}$ from the first order relation :

$$2\Delta\varepsilon_{AB} \int(P_B p_{AB} + Q_B q_{AB})dr = \delta_{AB} - \int(P_B P_A + Q_B Q_A)dr \quad \text{(III.21)}$$

or in a more general way from a system of linear equations if more than one Lagrange multiplier have to be changed simultaneously. In most of the cases we found that only the eigenvalue (i.e. ε_{AA}) needs to be adjusted when solving for the radial equations while the off-diagonal Lagrange multipliers can just be calculated with the help of the solutions from the previous iteration.

As in the non-relativistic case, it is always possible to set all off-diagonal Lagrange multipliers among the closed shells equal to zero since this amounts performing a unitary transformation in the subspace of the filled orbitals. If the occupation numbers n_A and n_B of the two orbitals A and B are different, we can obtain the ε_{AB} parameter from the equation:

$$\begin{aligned} \varepsilon_{AB}(n_B - n_A)/n_B = {} & \int(V_A - V_B)(P_A P_B + P_A P_B)dr \\ & - c\int(X^Q_A Q_B - X^Q_B Q_A + X^P_A P_B - X^P_B P_A)dr \end{aligned} \quad \text{(III.22)}$$

obtained with the help of :

$$\varepsilon_{AB}/n_A = \varepsilon_{BA}/n_B \quad \text{(III.23)}$$

Equation (III.22) shows that many terms will cancel out (for example the closed shell contribution to both V_A and V_B) and provides an accurate way of calculating these parameters by explicitly retaining only the non zero contributions. If $n_A = n_B$ then equations (III.20) and (III.21) can be used to determine the off-diagonal Lagrange multipliers.

IV. ILLUSTRATIVE EXAMPLES

Before considering examples of relativistic effects in many-electron systems, it may be worthwhile to give orders of magnitude for the various corrections.

Table I. Hydrogenic Energy Levels (in eV)

		a	b	c	d	e	f
Z=20	1s	-5442.32	-29.29	1.75	-0.13	-	0.01
	2s	-1360.58	-9.16	0.24	-0.02	-	-
	2p*	-1360.58	-9.16	-0.01	-	-	-
	2p	-1360.58	-1.82	0.01	-	-	-
Z=40	1s	-21769.29	-484.57	18.40	-2.09	-0.04	0.50
	2s	-5442.32	-151.76	2.64	-0.28	-0.01	0.07
	2p*	-5442.32	-151.76	-0.03	-0.01	-	-
	2p	-5442.32	-29.29	0.19	-	-	-
Z=60	1s	-48980.89	-2603.67	73.45	-11.70	-0.23	6.16
	2s	-12245.22	-817.89	11.42	-1.68	-0.03	0.92
	2p*	-12245.22	-817.89	0.30	-0.10	-	0.04
	2p	-12245.22	-150.34	1.21	-0.01	-	-
Z=80	1s	-87077.15	-9039.48	207.24	-45.52	-0.67	54.30
	2s	-21769.29	-2853.08	35.59	-7.33	-0.12	9.27
	2p*	-21769.29	-2853.08	3.25	-0.89	-0.01	0.75
	2p	-21769.29	-484.57	4.60	-0.06	-	-
Z=100	1s	-136058.04	-25557.02	515.55	-157.80	-5.43	451.45
	2s	-34014.51	-8126.92	100.43	-29.83	-1.06	93.01
	2p*	-34014.51	-8126.92	19.06	-6.58	-0.21	13.93
	2p	-34014.51	-1214.35	13.21	-0.21	-0.01	-

a: Non-relativistic value.
b: Dirac correction.
c: Self energy correction[5].
d: Vacuum polarization as given by the Uehling potential[11].
e: Higher order vacuum polarization corrections[12].
f: Nuclear size correction.

Empty entries indicate that the correction is less than 0.01 eV. p* stands for $p_{1/2}$ and p for $p_{3/2}$.

Table I gives the lowest energy levels of hydrogen-like atoms for a few atomic numbers and for the sake of completeness we have listed, in the last column, the finite nuclear size correction. This correction was calculated by assuming an uniform proton charge distribution inside a sphere of radius $1.2A^{1/3}$ Fm (A being the atomic mass). Looking at the numbers two comments emerge: first if the Dirac correction is by far the dominant one, none of the other corrections can really be neglected for nuclear charges greater than 80. For lower atomic numbers and unless very high precision is required it seems enough to consider only the contributions of terms up to order $(Z\alpha)^4$, i.e. the solutions of the Dirac equation corrected up to that order for QED. Note that the very recently calculated self energy correction[21] of $2p_{3/2}$ electrons dominates that of $2p_{1/2}$ ones for medium Z values. Our second comment is related to the finite nuclear correction. For very high Z values and j=1/2 levels, this correction is as important as the sum of the QED contributions and consequently it is necessary in actual calculations to rely on experimental determination of the nuclear parameters. To illustrate this point let us just mention that for Fermium (Z=100) the nuclear radius is about 7.6 Fm and that a change of 0.1 Fm results in a variation of 8 eV of the 1s electron binding energy[22]. This change has to be compared with the typical few eV value of the correlation energy for K electrons. The almost zero contribution of the finite size of the nucleus to energy levels with angular momenta greater than 1/2 can be related to the fact that only j=1/2 wavefunctions display a s character either in their large or small component and are thus the only ones to have a finite amplitude at origin.

We now turn to the more complex problem of trying to assess the quality of present days atomic calculations. The first thing to realize is that a meaningful comparison between theory and experiment is not always easy to achieve. The underlying reason is that we have to deal with two conflicting constraints: on one side it would be desirable to calculate the non-relativistic value of the quantity of interest with a precision at least as good as the experimental uncertainty in order to be able to extract the relevant relativistic corrections. On the other side, we would like to consider large nuclear charges so that the relativistic contributions are important. As up to now rather few high quality experiments exist for systems with a very small number of electrons and very high Z values, we will always have to face the problem of the poorly known many-body corrections. Furthermore it should be remembered that correlation and relativistic corrections are not additive.

IV.1 Inner-shell properties

One of the first candidates to try to solve the dilemma above where the binding energies of K electrons of heavy atoms for which it can be expected that relativistic corrections will dominate the many-body effects. To illustrate the kind of agreement achieved

Table II. Mercury (all energies in eV)

	1s energy	K_{α_1} energy
DF finite nucleus	83 559.1	71 231.8
Breit interaction ω=0	-303.4	-267.8
Transverse correction	6.0[a]	3.4[b]
QED corrections	-152.8	-154.4
Total	83 108.9	70 813.0
Experiment[c]	83 102.3	70 819.0

(a): Ref. 23 (b): Ref. 24 (c): Ref. 25

between theory and experiment we give in Table II the example of mercury K electron binding energy for which we list the various contributions. The kind of agreement displayed here is typical of this part of the periodic table if certainly not the best one but may be in some sense misleading since not only many-body effects are not included but also because comparison is done between atomic calculations and experiments generally performed on solid samples, some of them of uncharacterized chemical composition. Besides chemical shifts, the influence of the solid will results in a screening of the localized K hole by conduction electrons. This screening may be more or less efficient depending on the metallic character of the sample but will shift the K energy by at least a few eV[26] compared to the value one should get for monoatomic gases.

As we have just discussed, inner-shell binding energies are certainly not the best possible choice to check high order relativistic corrections and one thing to be minimized is the influence of the sample chemical composition. K_α X-ray energies are undoubtedly a better candidate since for the inner electrons involved in these transitions, the modification in the valence orbitals between the atom and the solid will be reflected as only a change in the outer screening and consequently will be of minor importance in the calculation of these quantities from pure atomic results. Taking again mercury as a prototype we also list in Table II the various contributions to the K_{α_1} X-ray energy. They are seen to be quite comparable to the one found for the K binding energy and to contribute for less than one percent of the total value. Consequently none of these quantities can be used for a very sensitive test and we will come back latter to this point, for the time being let us concentrate on how a systematic comparison can be done. To avoid any calibration problem between different experiments, X-ray energies for low Z and high Z elements have been measured simultaneously[27]. The low Z X-ray energy was measured at the first order Bragg reflection while the high Z one was obtained at the second or third order. The results for five pairs of atoms are summarized in Table III and clearly no

Table III

Comparison between X-ray energies of low Z and high Z elements (all energies in eV)

El.	Trans.	Order	Energy *	Δ exp.	Δ the.
Pt	K_{β_1}	3	75 750	-22.06	-23.77
Sn	K_{α_1}	1	25 271		
Pt	K_{α_1}	2	66 832	-26.45	-26.00
La	K_{α_1}	1	33 442		
Bi	K_{α_2}	3	74 815	-3.68	-4.08
Ag	K_{β_1}	1	24 943		
Th	K_{α_1}	2	93 350	-26.29	-26.37
Ho	K_{α_2}	1	46 700		
U	K_{β_1}	2	111 300	-23.49	-24.25
Er	K_{β_1}	1	55 674		

(*) Ref. 28. Δ is defined as the X-ray energy of the high Z element divided by the order minus the X-ray energy of the low Z element.

systematic deviation can be observed. A serie of high precision measurements of X-ray energies for heavy atoms is presently being done and will be more than useful to establish the degree of agreement between experiment and theory[29].

As already said, the high order relativistic corrections to either the K binding or X-ray energies amount to only a small fraction of the total amount and as we pointed out some years ago[30] they can be tested in a more crucial way by looking at the shift between the normal X-ray and its hypersatellite, i.e. the X-ray emitted in

Table IV

Hypersatellite $K^h_{\alpha_1}$ shifts for Thulium and Mercury (in eV)

	Tm	Hg	
	Present		Ref 24
DF finite nucleus	804.3	991.5	992.9
Breit interaction ω=0	113.9	185.3	187.0
Transverse correction	–	–	0.1
QED corrections	-1.2	-1.3	-1.3
Total	917.0	1175.5	1178.7
Experiment	902(9)$^+$	1145(12)*	

(+) Ref. 31 (*) Ref. 32

the presence of a cacancy in one of the inner-shells. Taking the case of a K spectator vacancy, the shift of the K_α line is given, in terms of total energies, by:

$$\Delta(K_\alpha - K_\alpha^h) = \{E(1s^1 2p^6) - E(1s^2 2p^5)\} - \{E(1s^0 2p^6) - E(1s^1 2p^5)\} \quad \text{(IV.1)}$$

where we have listed only the electrons participating in the transitions. As seen from counting the number of electrons (or holes), this shift will be zero in a frozen core approximation. Consequently the non zero observed value can arise from only three contributions: the relaxation of the orbitals between the various states, the change in the correlation energy and variations in the contributions of the high order relativistic corrections. For these latter ones it is expected that the contribution of the Breit interaction will be large because only one of the configurations (the $1s^1 2p^5$) contains two open shells which will give a direct spin-spin interaction not even partially cancelled out by contributions from the other three configurations. On the other hand as the numbers of 1s and 2p electrons add separately to zero, the one-electron contributions to the Lamb-shift (the only ones considered up to now) will result in a small value given only by the relaxation of the orbitals. Indeed the values given in Table IV show that the Breit interaction contributes to a large amount of the shift (12% for thulium and 16% for mercury) while QED corrections are almost negligible and less than the contribution of 2eV due to the mixing between the $^3P_1$ and $^1P_1$ levels of the $1s^1 2p^5$ configuration and included in the DF result. The agreement with experimental results is rather poor and much worse than for X-ray energies or binding energies. The reasons for this disagreement are still not clearly understood. The almost perfect agreement between the two theoretical values for mercury obtained with two completely independent computer codes seems to exclude any large numerical error. Obviously correlation corrections are missing in both calculations but it is unlikely that both the magnitude of the discrepancy for mercury and the increase in the disagreement when going from thulium to mercury can be attributed only to the many-body effects. It would be of great interest to investigate on the theoretical side the influence of including the Breit operator in the determination of the radial functions while at the same time new experiments should be done.

IV.2 Outer-shell properties

In the previous comparisons inner shells were obviously selected because of their extreme relativistic behaviour and if on one side we do not observe serious disagreement with experiment (except may be in the case of hypersatellite shifts), on the other side we always run into the problem of the missing many-body effects. To include these contributions we have to rely on much simpler systems and good candidates are highly stripped ions since there exist very accurate beam-foil spectroscopy measurements of the transition wave-

Table V. Some energy levels of Fe XXI (in cm^{-1})

Configuration	State	MCDF	Breit	QED
$2s^2 2p^2$	$^3P_0$	0	0	0
	$^1S_0$	372 932	-4 564	-92
$2s2p^3$	$^3D_1$	784 303	-1 192	-4 896
	$^1P_1$	1 283 066	-4 841	-5 250
$2p^4$	$^3P_2$	1 666 615	-768	-8 325
	$^1S_0$	2 076 194	-2 365	-8 116

lengths. For these ionized systems the correlation effects are dominated by intra-shell contributions since the nuclear Coulomb field remains very strong even for the outer electrons. This configuration interaction among orbitals with the same principal quantum number can be handled rather efficiently by the multiconfiguration Dirac-Fock method (MCDF). Examples of the results obtained in the study of the first eighteen energy levels levels of carbon-like iron[33] are given in Table V. They confirm the prime importance, already pointed out for lithium-like ions[34], of the Breit term to describe accuratly the fine structure splittings within a given configuration while the Lamb-shift correction is very important for transitions between configurations with different occupations of the 2s orbitals. In such cases the QED contribution may even be greater than that of the Breit operator. Comparison with experiment[35] shows that the agreement is in the 1% range and the remaining discrepancy may quite well be attributed to the missing inter-shell correlation contributions. An estimate of these contributions is highly desirable in order to definitively assess the limits of the Breit and one-electron Lamb-shift approximations for medium Z elements.

The stripped ion results are also of great help in studying systematic trends along a given isolelectronic sequence. We consider here the case of the allowed transition from the ground state $1s^2 2s^2$ ($^1S_0$) to the excited $1s^2 2s2p$ ($^1P_1$) one along the beryllium sequence as an example. To include intra-shell correlation, the ground state wave function is written as :

$$(1s^2)\{\alpha(2s^2) + \beta(2p_{1/2}^2) + \gamma(2p_{3/2}^2)\} \quad \text{(IV.2)}$$

while the excited state wave function is simply :

$$(1s^2)\{(1-\delta^2)^{\frac{1}{2}} 2s2p_{3/2} - \delta 2s2p_{1/2}\} \quad \text{(IV.3)}$$

and the $^1P_1$ state is associated with the upper J=1 level (obviously

Table VI. Mixing coefficients for the beryllium sequence

Z	α	β	γ	$(\gamma/\beta)^2$	δ
10	0.968	0.147	0.205	1.94	0.572
42	0.986	0.141	0.089	0.40	0.200
74	0.992	0.126	0.025	0.04	0.038

we need to take intermediate coupling into account). The values of the α,β,γ and δ coefficients[36] are given in Table VI for some atomic numbers as well as the value of the ratio $(\gamma/\beta)^2$.
In the relativistic case only the 2s and $2p_{1/2}$ levels are degenerate in the hydrogenic limit and Table VI displays the expected result that the weight β of the $(2p_{1/2})^2$ configuration remains essentially constant throughout the sequence. On the other hand the increase in the spin-orbit splitting causes the contribution of the $(2p_{3/2})^2$ configuration to decrease when the atomic number Z increases. The influence of the spin-orbit interaction is also reflected in the variation of the ratio $(\gamma/\beta)^2$, for Z=10 this ratio is almost equal to its non-relativistic limit of 2 while for Z=74 the very small value of 0.04 indicates that one has reached the jj coupling scheme limit. The same type of evolution is observed for the upper excited state as indicated by the variation of the mixing coefficient δ. We discuss now the behaviour of the transition probabilities: when plotted as a function of Z, the relativistic results for the dipole oscillator strengths (f values) display a quite different behaviour than the one obtained in the non-relativistic case. While the latter one gives a monotonically decreasing function of Z, the relativistic f values of the beryllium sequence present a minimum around Z=35 and then begin to increase with Z. Furthermore, starting from Z$\sim$20, the relativistic results are significantly larger than the non-relativistic ones and are in better agreement with the recommended values of Smith and Wiese[37]. To analyze the relativistic

Table VII. f values for the beryllium sequence

Z	E(eV)		SZ^2		f	
	R	NR	R	NR	R	NR
10	28.3	28.1	57.44	57.32	0.398	0.395
42	251.9	136.0	40.16	41.90	0.140	0.079
74	1758.0	243.0	33.80	40.33	0.266	0.044

effects it is convenient to consider separately the two factors contributing to the f value, i.e. the excitation energy E and the line strength S. In terms of these, f is defined in the length form as:

$$f = ES(3gR)^{-1} \qquad \text{(IV.4)}$$

where R=13.6 eV and g the degeneracy of the initial state. Results for E, s and f are given in Table VII. It can be seen that E increases much more rapidly with Z in the relativistic case while S remains close to the non-relativistic throughout the sequence (note that in the non-relativistic hydrogenic approximation SZ^2 will be constant). Similar behaviour was found for the magnesium sequence by Aymar and Luc-Koenig[38] and explained in the framework of the theory of Layzer and Bahcall[39]. In this theory the energy is expressed as the double serie expansion:

$$E_T = Z^2 \sum_{n=0} \sum_{m=0} E_{nm} (Z^2\alpha^2)^n Z^{-m} \qquad \text{(IV.5)}$$

where the n series is associated to the relativistic corrections and the m one corresponds to the non-relativistic expansion of the total energy of the system. For a $\Delta n=0, \Delta j=0$ transition like the one considered here, the transition energy E is given by :

$$E = \Delta E_{01} Z + \Delta E_{10} Z^4\alpha^2 + \Delta E_{11} Z^3\alpha^2 + \dots \qquad \text{(IV.6)}$$

thus because of roughly Z^{-2} dependence of the line strength S:

$$f \sim \Delta E_{01}/Z + \Delta E_{10} Z^2\alpha^2 + \Delta E_{11} Z\alpha^2 + \dots \qquad \text{(IV.7)}$$

which explains that for low Z values the f value begins to decrease because the relativistic corrections are small while for high Z values when these corrections become important we see a rapid increase with the atomic number. As observed from Table VII, S decreases more rapidly in the relativistic case (R) than in the non-relativistic one (NR). This behaviour is a consequence of the relativistic contraction of the radial orbitals. For transitions to the lowest J=1 state (corresponding to the $^3P_1$ state in the LS limit) it was found that the f values are smaller by a factor of 10^9 (low Z-ions) to 20 (high Z-ions) compared to the allowed transition to the upper state. Thus the "forbidden" transition remains weak in the jj coupling limit for the beryllium sequence. Before leaving the subject of transition probabilities let us point out another consequence of the transition from LS to jj coupling: as the level ordering between two states may change when going from the low Z to the high Z region, these two levels have to cross at some value of the atomic number. Even if true crossings are forbidden, the two levels will interact strongly around this Z value and the oscillator strength may be either largely enhanced or almost suppressed depending upon the coherent or destructive character of the interference. Examples of such behaviours have been found in many isoelectronic sequences (see for example the case of the magnesium one[40]).

Not to leave the reader with the feeling that relativistic corrections are important only for more or less exotic systems, we consider few examples involving neutral atoms. Even if the outer orbitals may be considered as "non-relativistic" in the sense that their small components are negligible compared to the large ones over most of the range of the radial variable, these orbitals are substantially changed due to the self-consistent relativistic effects and their charge distributions are shifted from those obtained in non-relativistic calculations. Up to now the relativistic results have always been confirmed by experiments. For example the binding energy of the $4f_{5/2}$ electron becomes lower than that of the 5s electron for bismuth (Z=83) and heavier atoms according to relativistic calculations (and in agreement with experiment) while this crossover is predicted to occur at lower atomi number (Z=78,Pt) by non-relativistic calculations. The changes in the charge distribution are greatly dependent on the total angular momentum j of the electron and this dependence will be reflected in the oscillator strength. Indeed it was shown[41] that the dipole oscillator strength of mercury for the reonance transition $6s^2$ to 6s6p is decreased by one third due to the changes in the 6s and 6p orbitals.

The last example we want to discuss is related to the importance of relativistic corrections in predicting the ground state configuration of very heavy atoms. This example will provide an easy transition to the discussion of the influence of relativistic effects on chemial properties which will be the subject of the last lecture. Every Periodic Table gives $5f^{14}7s^26d$ as the ground state configuration of lawrencium (Z=103) and this is what is expected from the lighter elements of the same column. It was pointed out rather early in the beginning of relativistic calculations that the $5f^{14}7s^27p$ configuration gives in fact a slightly lower Dirac-Fock energy. Because the splitting between these two configurations is small, single configuration calculations may give the wrong ordering if the neglected correlation contributions are substantially different for the two configurations. We have recently investigated this problem[42] in the framework of the multiconfiguration Dirac-Fock method and came to the conclusion that the $^2P_{1/2}$ level of the 7p configuration was indeed lower than the $^2D_{3/2}$ level of the 6d one by about 2000 cm^{-1}. To reach this conclusion it was necessary to include most of the intra-shell correlation of the three valence electrons $7s^2$ and 6d or 7p and part of the inter-shell correlation with the n=6 other orbitals. Furthermore it was also necessary to assume that the missing correlation contributions (mainly the core polarization) remains the same as that estimated for the lighter elements. This assumption was supported by the fact that the difference between theory and experiment was almost constant for the other atoms (Y, La, Lu and Ac) studied at the same level of approximation.

ACKNOWLEDGEMENTS

Most of the results reported here have been obtained in collaboration with Drs K.T. Cheng and Y.K. Kim and with Professor B. Fricke. I wish also to thank Dr. P.J. Mohr for making available before publication his results for the self-energy of 2p electrons.

REFERENCES

1) G.E. Brown and D.G. Ravenhall Proc. R. Soc.(London) A208, 552 (1951)

2) J. Sucher Phys. Rev. A 22, 348 (1980) and these proceedings

3) M.H. Mittleman Phys. Rev. A 4, 893 (1971); A 5, 2395 (1972)

4) H.A. Bethe and E.E. Salpeter, *Quantum Theory of One- and Two-electron Atoms*, Springer-Verlag, Berlin (1957)

5) P.J. Mohr in *Proceedings of the Workshop on Foundations of the Relativistic Theory of Atomic Structure* ANL Report ANL-80-126 (1981) and references therein.

6) G.E. Brown and D.F. Mayers Proc. R. Soc. (London) A251, 105 (1951)

7) A.M. Desiderio and W.R. Johnson Phys. Rev. A 3, 1267 (1971)

8) K.T. Cheng and W.R. Johnson Phys. Rev. A 14, 1943 (1976)

9) P.J. Mohr Ann. Phys. (New-York) 88, 52 (1974); Phys. Rev. Lett. 34, 1050 (1975)

10) E.H. Wichmann and N.M. Kroll Phys. Rev. 101, 843 (1956)

11) E.A. Uehling Phys. Rev. 48, 55 (1935)

12) G. Källen and A. Sabry K. Dan. Vidensk. Selsk. Mat. Fys. Medd. 29, 17 (1955) see also J. Blomqvist Nucl. Phys. B 48, 95 (1972)

13) D.M. Brink and G.R. Satchler, *Angular Momentum*, Clarendon Press, Oxford (1968)

14) I.P. Grant and B.J. McKenzie J. Phys. B 13, 2671 (1980)

15) N.C. Pyper, I.P. Grant and N. Beatham Comp. Phys. Comm. 15, 387 (1978)

16) W. Eissner and H. Nussbaumer J. Phys. B 2, 1028 (1969)

17) D.F. Mayers J. Phys. 31, C4-213 (1970)

18) J.P. Desclaux, C.M. Moser and G. Verhaegen J. Phys. b 4, 296 (1971)

19) C.F. Fisher, *The Hartree-Fock Method for Atoms*, Wiley, New-York, (1977)

20) F.B. Hildebrand, *Introduction to Numerical Analysis*, Mc Graw Hill, New-York (1956)

21) P.J. Mohr to be published

22) B. Fricke, J.P. Desclaux and J.T. Waber Phys. Rev. Lett. 28, 714 (1971)

23) J.B. Mann and W.R. Johnson Phys. Rev. A 4, 41 (1971)

24) N. Beatham, I.P. Grant, B.J. McKenzie and S.J. Rose Phys. Scripta 21, 423 (1980)

25) J.A. Beardeen and A.F. Burr Rev. Mod. Phys. 39, 78 (1967)

26) L. Ley, S.P. Kowalczyk, F.R. Mcfeely, R.A. Pollak and D.A. Shirley Phys. Rev. B 8, 2392 (1973)

27) G.L. Borchert, P.G. Hansen, B. Jonson, H.L. Ravn and J.P. Desclaux in *Atomic Masses and Fundamental Constants-6* p189, J. Nolen and W. Benenson, Eds., Plenum Press, New-York (1979)

28) C.M. Lederer and V.S. Shirley *Tables of Isotopes*, J. Wiley New-York (1978)

29) R.D. Deslattes private communication.

30) J.P. Desclaux, Ch. Briançon, J.P. Thibaud and R.J. Walen Phys. Rev. Lett. 32, 447 (1974)

31) C.N.E. Van Eijk and J. Wijnhorst Phys. Rev. A 15, 1794 (1977)

32) K. Schrekenbach, H.G. Börner and J.P. Desclaux Phys. Lett. 63A, 330 (1977)

33) J.P. Desclaux, K.T. Cheng and Y.K. Kim J. Phys. B 12,3819 (1979)

34) K.T. Cheng, J.P. Desclaux and Y.K. Kim J. Phys. B 11,L359 (1978)

35) K. Nori, M. Otsuka and T. Kato Institute of Plasma Physics, Nagoya University Report IPPJ-AM-3 (1977)

36) Y.K. Kim and J.P. Desclaux Phys. Rev. Lett. 36, 139 (1976)

37) M.W. Smith and W.L. Wiese Astrophys. J. Suppl. Ser. 23,103 (1971)

38) M. Aymar and E. Luc-Koenig Phys. Rev. A 15, 821 (1977)

39) D. Layzer and J. Bahcall Ann. Phys.(New-York) 17, 177 (1962)

40) K.T. Cheng and W.R. Johnson Phys. Rev. A 16, 263 (1977)

41) J.B. Mann private communication, 1971.

42) J.P. Desclaux and B. Fricke J. Phys. 41, 943 (1980)

LAMB SHIFT IN HIGH-Z ATOMS

Peter J. Mohr

Yale University
Gibbs Laboratory, Physics Department
New Haven, CT 06520

I INTRODUCTION

Most of the qualitative features of the structure of atoms, molecules, and solids are well described by the many-electron Schrödinger equation with an instantaneous Coulomb interaction between electrons and nuclei. However, relativistic effects can play an important role in the quantitative description of such systems, as discussed in the accompanying papers in these proceedings. For a many-electron system, the relativistic description is formulated in terms of a sum of one-electron Dirac Hamiltonians with an effective instantaneous external potential and a suitable form of the relativistic electron-electron interaction, such as the Breit interaction. Questions concerning the detailed formulation, methods of calculation, and associated problems are addressed in the accompanying papers.

In some cases, it is necessary to further generalize to include purely quantum electrodynamic (QED) effects such as self-energy and vacuum polarization corrections in order to understand the experimental data. These effects are primarily responsible for the well-known Lamb shift in ordinary hydrogen, and they can play an important role in more complicated systems.

In high-Z one-electron atoms, quantum electrodynamics provides a fully relativistic framework for calculations of radiative corrections, and has yielded precise theoretical results.[1] In high-Z few-electron atoms, a straightforward generalization of the fully relativistic formulation can be made that, to the accuracy of the calculations, gives results consistent with experiment. For high-Z many-electron atoms, radiative corrections calculated in a one-electron model correctly explain the dominant level shifts of inner-shell electrons.

As will be seen in the following, the one-electron Coulomb field radiative corrections give the dominant effect at high-Z for both few-electron atoms and inner shells of many-electron atoms. Hence, in this paper, the emphasis will be on the one-electron corrections and their role in more complicated high-Z systems.

There are various aspects of strongly-bound high-Z systems that distinguish them qualitatively from weakly-bound systems such as ordinary hydrogen or helium. i. In high-Z systems, electrons with low principal quantum number n have relativistic velocities, as indicated by the order of magnitude estimate $v \sim (Z\alpha/n)c$. ii. The electron binding energy is high, of order $BE \sim (Z\alpha/n)^2 m_e c^2$, which is comparable to the electron rest energy $m_e c^2$ at high Z. iii. The relative magnitude of the radiative corections = (radiative corrections)/(binding energy) $\sim\alpha(Z\alpha)^2$ is greater at high Z than at low Z. iv. The radiative corrections are strongly Z dependent and must be calculated nonperturbatively (in $Z\alpha$) to achieve good accuracy at high Z. v. The strong Z-dependence invalidates weak-field based ideas, such as the anomalous magnetic moment correction to the spin-orbit interaction. At high Z, the correction is not a constant multiplicative factor as it is at low Z; in fact, the correction for the hydrogenic $2P_{1/2}$-$2P_{3/2}$ splitting changes sign at $Z \approx 90$.

II QED EFFECTS

In high-Z atoms, the dominant radiative corrections are the lowest-order (in α) self energy and vacuum polarization. The self-energy correction dominates over the vacuum-polarization correction for electrons in a strong Coulomb field, although historically, the vacuum-polarization effect was understood much earlier. In 1935, Serber[2] and Uehling[3] discussed vacuum polarization, to lowest order in the external Coulomb field, and in 1956 Wichmann and Kroll examined vacuum polarization to all orders in the external field.[4] In hole theory, vacuum polarization has the following interpretation. Near a positive nucleus, there is an accumulation of the electrons that fill the negative energy states of the vacuum, and the effective charge of the nucleus for a bound electron sufficiently far from the nucleus is the net charge +Ze of the nucleus and the accumulated vacuum electrons. Inside the characteristic distance associated with this screening, $\lambda\!\!\!^{-}_e = 4 \times 10^{-11}$ cm, the net effective charge is larger and approaches the (infinite) bare nuclear charge $+Ze_0$ at the nucleus. Hence, s electrons are more tightly bound when vacuum polarization is taken into account; higher-l states are affected less, because the electron is less likely to be near the nucleus. From the point of view of field theory, the vacuum polarization correction is the result of the process in which the photon that mediates the Coulomb interaction between the bound electron and the nucleus produces a virtual electron-positron pair in the vacuum. The corresponding Feynman diagram appears in Fig. 1. In that diagram, the single line represents a free electron or positron, the double line represents the bound electron, the wavy line symbolizes the photon, and the × denotes interaction with the nuclear Coulomb field. This diagram gives a correction to the Coulomb potential V(r) known as the Uehling potential $\delta V(r)$:

$$V(r) \to V(r) + \delta V(r) \tag{1}$$

In the Coulomb field of a nucleus of charge Ze, $\delta V(r)$ is given by

$$\delta V(r) = -\frac{\alpha}{\pi}(Z\alpha)\, f(r/\lambda\!\!\!^-_e)\, m_e c^2 \tag{2}$$

where

$$f(r/\lambda\!\!\!^-_e) = \frac{\lambda\!\!\!^-_e}{3r}\int_1^\infty dt\,(t^2-1)^{1/2}(2t^{-2}+t^{-4})\,e^{-2tr/\lambda\!\!\!^-_e} \tag{3}$$

The range of the Uehling potential is exhibited by the asymptotic form for $r \gg \lambda\!\!\!^-_e$

$$f(r/\lambda\!\!\!^-_e) \sim \frac{\pi^{1/2}}{4}\left(\frac{\lambda\!\!\!^-_e}{r}\right)^{5/2} e^{-2r/\lambda\!\!\!^-_e} \tag{4}$$

The range of δV is short on the atomic scale, so the energy shift is proportional to the bound-electron wave function at the origin, to

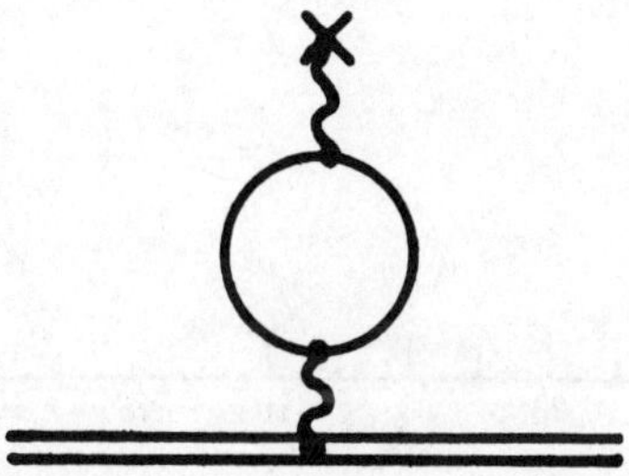

Fig. 1. Feynman diagram for the lowest-order vacuum-polarization correction for a bound electron.

lowest order in $Z\alpha$:

$$\Delta E_U = \langle \delta V \rangle = \frac{\alpha}{\pi} \frac{(Z\alpha)^4}{n^3} G_n(Z\alpha) m_e c^2$$
$$\approx -\frac{\alpha}{\pi} (Z\alpha) |\psi(0)|^2 \int d^3r \, f(r/\lambda\!\!\!^{-}_e) m_e c^2 \tag{5}$$

In (5), the function $G_n(Z\alpha)$ gives the exact expectation value of δV calculated with Dirac wave functions. For small Z, we have

$$G_n(Z\alpha) \sim \begin{cases} -\frac{4}{15} & \text{s states} \\ 0 & \text{p,d,... states} \end{cases} \tag{6}$$

Figure 2 shows $G_n(Z\alpha)$ over a wide range of Z. The $2P_{1/2}$-state values for $G(Z\alpha)$ grow rapidly as Z increases, because the small components of the wave function, which are smaller than the large components by a

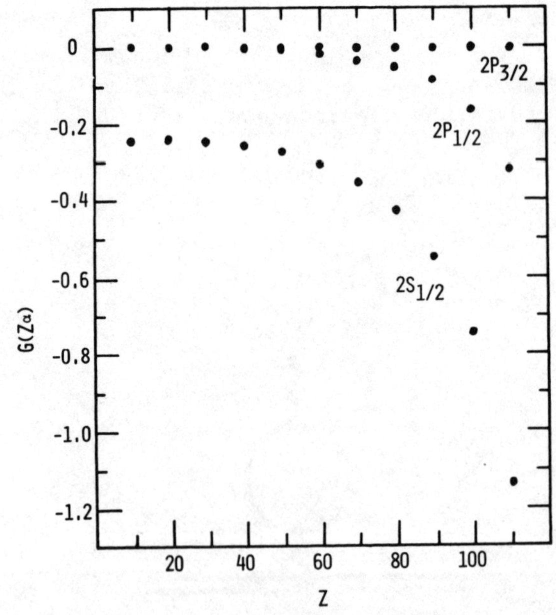

Fig. 2. Graph of the function $G(Z\alpha)$ that appears in Eq. (5).

factor ($Z\alpha$), are non-vanishing at the nucleus. All components of the $2P_{3/2}$ wave function vanish at the nucleus, so the corresponding values of $G(Z\alpha)$ do not grow as rapidly at high Z.

The self energy (radiative correction) is the dominant QED correction in ordinary atoms and ions. This is in contrast to muonic atoms where the vacuum polarization dominates.[1] The first self-energy calculation was done by Bethe in 1947 based on the nonrelativistic formulation of radiation theory, including mass renormalization.[5] The expression he evaluated is

$$\Delta E_{SE}^{B} = \frac{\alpha}{4\pi^2 c^2} \sum_{\lambda=1}^{2} \int d^3k\, k^{-1} \langle n| \hat{\varepsilon}_\lambda \cdot \vec{v} \sum_i \frac{|i\rangle\langle i|}{E_n - E_i - k} \hat{\varepsilon}_\lambda \cdot \vec{v} |n\rangle + \frac{1}{2}\delta m_e \langle v^2\rangle \tag{7}$$

The process described by this formula is the virtual emission and reabsorption of a photon by the bound electron, which occurs in second order perturbation theory. In Eq. (7), $|n\rangle$ is the bound-state vector, $e\hat{\varepsilon} \cdot \vec{v}/\sqrt{k}$ is the photon emission or absorption operator in the dipole approximation, and the intermediate states are summed over electron states i and photon states of energy k less than K. The intermediate state energy is $E_i + k$. Taking into account the order of magnitude of atomic velocities and binding energies, one can show that Eq. (7) gives a level shift of leading order $(\alpha/\pi)(Z\alpha)^4$. Hence, we write

$$\Delta E_{SE}^{(2)} = \frac{\alpha}{\pi} \frac{(Z\alpha)^4}{n^3} F_n(Z\alpha) m_e c^2 \tag{8}$$

where F_n is a relatively slowly varying function of Z. The small-Z behavior of F_n is given by

$$F_n(Z\alpha) \sim \begin{cases} \frac{4}{3} \ln(Z\alpha)^{-2} + C_n & \text{s states} \\ C_n & \text{p,d,... states} \end{cases} \tag{9}$$

where C_n is a state-dependent constant. To determine the complete Z-dependence of F_n, a relativistic formulation of the interaction of radiation with atoms is necessary, as provided by the Furry bound-interaction picture of quantum electrodynamics.[6]

III THE FURRY PICTURE

In QED in the Furry picture, the electron-positron field operator

$$\psi(x) = \sum_{n+} b_n \phi_n(\vec{x}) e^{-iE_n t} + \sum_{n-} d_n^* \phi_n(\vec{x}) e^{-iE_n t} \tag{10}$$

is expanded in terms of electron destruction operators b_n and positron creation operators $d_n{}^*$ multiplying positive (n+) and negative (n−) energy eigenfunctions of the Dirac equation in an external Coulomb field (in this section, units in which $\hbar = m_e = c = 1$ are employed)

$$[-i\vec{\alpha}\cdot\vec{\nabla} + V(\vec{x}) + \beta - E_n]\phi_n(\vec{x}) = 0 \tag{11}$$

The creation and destruction operators obey the usual Fermi anticommutation rules. Interaction of the electromagnetic current of the electron-positron field with the radiation field is provided by the interaction Hamiltonian density:

$$H_I(x) = -\tfrac{1}{2}e\,[\bar{\psi}(x)\gamma^\mu,\ \psi(x)]A_\mu(x) - \tfrac{1}{2}\delta m[\bar{\psi}(x),\ \psi(x)] \tag{12}$$

where −e is the electron charge and A_μ is the vector potential of the radiation field. The second term in (12) is the mass renormalization counter term. Energy level shifts are calculated in a perturbation expansion in H_I by applying the Gell-Mann and Low theorem[7] in the symmetric form discussed by Sucher[8]

$$\Delta E_n = \lim_{\substack{\varepsilon\to 0\\ \lambda\to 1}} \frac{i\varepsilon}{2}\,\frac{\frac{\partial}{\partial\lambda}\langle n|S_{\varepsilon,\lambda}|n\rangle}{\langle n|S_{\varepsilon,\lambda}|n\rangle} \tag{13}$$

where

$$S_{\varepsilon,\lambda} = \sum_{j=0}^{\infty} S_{\varepsilon,\lambda}^{(j)} \tag{14}$$

and where

$$S_{\varepsilon,\lambda}^{(j)} = \frac{(-i\lambda)^j}{j!}\int d^4x_j \ldots \int d^4x_1\, e^{-\varepsilon|t_j|-\ldots-\varepsilon|t_1|} \times T[H_I(x_j)\ldots H_I(x_1)] \tag{15}$$

In (13), the vector $|n\rangle$ for a state with k electrons is a linear combination of products of k creation operators acting on the vacuum

$$|n\rangle = \sum_{i_1\ldots i_k} c_{i_1\ldots i_k}\, b_{i_1}^{*}\ldots b_{i_k}^{*}|0\rangle \tag{16}$$

The contributions to the level shift ΔE_n of order j are represented by Feynman diagrams with j vertices as in ordinary free-electron QED, except that here, the electron lines represent bound electrons with the appropriate bound-electron propagation function

$$S_F(x_2,x_1) = \begin{cases} \sum\limits_{n+} \phi_n(\vec{x}_2)\bar{\phi}_n(\vec{x}_1)e^{-iE_n(t_2-t_1)} & t_2 > t_1 \\ -\sum\limits_{n-} \phi_n(\vec{x}_2)\bar{\phi}_n(\vec{x}_1)e^{-iE_n(t_2-t_1)} & t_2 < t_1 \end{cases} \tag{17a}$$

$$= \frac{1}{2\pi i}\int_{C_F} dz \sum_n \frac{\phi_n(\vec{x}_2)\bar{\phi}_n(\vec{x}_1)}{E_n - z} e^{-iz(t_2-t_1)} \tag{17b}$$

where C_F denotes the Feynman contour. The lowest-order diagrams for the QED corrections to the energy levels of one electron in a Coulomb field are shown in Fig. 3. The vacuum-polarization diagram, corresponding to the generalization of the Uehling potential correction, appears in Fig. 3a, and the self-energy diagram, corresponding to the relativistic generalization of Bethe's calculation, appears in Fig. 3b.

Wichmann and Kroll have examined the complete vacuum polarization in Fig. 3a, and have shown that the effect of the correction to the Uehling potential included in the exact result is very small in ordinary atoms over a wide range of Z.[4] In muonic atoms, on the other hand, this correction is of measurable magnitude.[4,1]

To evaluate the self-energy level shift at high Z, a complete evaluation of the expression corresponding to Fig. 3b is necessary, because the power series in $Z\alpha$ for this correction converges too slowly numerically to give accurate results. The level shift is given by

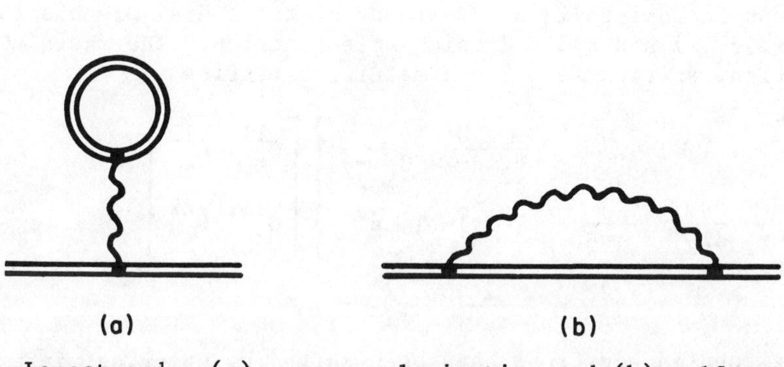

Fig. 3. Lowest-order (a) vacuum-polarization and (b) self-energy corrections to bound-electron energy levels.

$$\Delta E_{SE}^{(2)} = -4\pi i\alpha \int d(t_2 - t_1) \int d^3x_2 \int d^3x_1 \; D_F(x_2 - x_1) \times\bar{\phi}_n(\vec{x}_2)e^{iE_nt_2} \gamma_\mu S_F(x_2,x_1)\gamma^\mu \phi_n(\vec{x}_1)e^{-iE_nt_1} - \delta m \int d^3x \bar{\phi}_n(\vec{x})\phi_n(\vec{x}) \tag{18}$$

where $D_F(x_2-x_1)$ is the photon propagator

$$D_F(x_2-x_1) = \frac{2i}{(2\pi)^4} \int d^4k \frac{e^{-ik\cdot(x_2-x_1)}}{k^2 + i\varepsilon} \tag{19}$$

To completely evaluate $\Delta E_{SE}^{(2)}$ at high Z, it is useful to write the bound-electron popagation function S_F in (17b) in terms of the Coulomb Green's function

$$G(\vec{x}_2,\vec{x}_1,z) = \sum_m \frac{\phi_m(\vec{x}_2)\phi_m^\dagger(\vec{x}_1)}{E_m - z} \tag{20}$$

which satisfies the inhomogeneous Dirac equation

$$[-i\vec{\alpha}\cdot\vec{\nabla}_2 + V(\vec{x}_2) + \beta - z]G(\vec{x}_2,\vec{x}_1,z) = I\delta(\vec{x}_2 - \vec{x}_1) \tag{21}$$

The Coulomb Green's function can be written as an expansion in eigenfunctions of the Dirac angular momentum operator $K = \beta(\vec{\sigma}\cdot\vec{L} + I)$

$$G(\vec{x}_2,\vec{x}_1,z) = \sum_\kappa G_\kappa(\vec{x}_2,\vec{x}_1,z) \tag{22}$$

Each term G_κ can be written as a 2×2 matrix, in which each matrix element factorizes into an element of the radial Green's function $G_\kappa^{ij}(r_2,r_1,z)$ and a 2×2 spin-angle function. The radial Green's function, written as a 2×2 matrix, satisfies

$$\begin{bmatrix} V(r_2) + 1 - z & -\frac{1}{r_2}\frac{d}{dr_2}r_2 + \frac{\kappa}{r_2} \\ \frac{1}{r_2}\frac{d}{dr_2}r_2 + \frac{\kappa}{r_2} & V(r_2) - 1 - z \end{bmatrix} \begin{bmatrix} G_\kappa^{11} & G_\kappa^{12} \\ G_\kappa^{21} & G_\kappa^{22} \end{bmatrix} = I\frac{1}{r_2r_1}\delta(r_2 - r_1) \tag{23}$$

These coupled equations have been solved by numerical integration, or by taking advantage of the fact that for a Coulomb potential the solutions are known functions. For a Coulomb potential

$$G_\kappa(r_2,r_1,z) = \theta(r_2 - r_1)F_>(r_2)F_<^T(r_1) + \theta(r_1 - r_2)F_<(r_2)F_>^T(r_1) \qquad (24)$$

where $F_<(F_>)$ is the 2-component solution of the homogeneous version of (21) that is regular at zero(infinity)

$$F_<(r) = \begin{bmatrix} \dfrac{\sqrt{1+z}}{2ar^{3/2}}\,[(\lambda-\nu)M_{\nu-\frac{1}{2},\lambda}(2ar)-(\kappa-\frac{\gamma}{a})M_{\nu+\frac{1}{2},\lambda}(2ar)] \\ \dfrac{\sqrt{1-z}}{2ar^{3/2}}\,[(\lambda-\nu)M_{\nu-\frac{1}{2},\lambda}(2ar)+(\kappa-\frac{\gamma}{a})M_{\nu+\frac{1}{2},\lambda}(2ar)] \end{bmatrix} \qquad (25a)$$

and

$$F_>(r) = \frac{\Gamma(\lambda-\nu)}{\Gamma(1+2\lambda)}\begin{bmatrix} \dfrac{\sqrt{1+z}}{2ar^{3/2}}\,[(\kappa+\frac{\gamma}{a})W_{\nu-\frac{1}{2},\lambda}(2ar)+W_{\nu+\frac{1}{2},\lambda}(2ar)] \\ \dfrac{\sqrt{1-z}}{2ar^{3/2}}\,[(\kappa+\frac{\gamma}{a})W_{\nu-\frac{1}{2},\lambda}(2ar)-W_{\nu+\frac{1}{2},\lambda}(2ar)] \end{bmatrix} \qquad (25b)$$

In Eq. (25), $a = (1-z^2)^{1/2}$; $\gamma = Z\alpha$; $\nu = \gamma z/a$; $\lambda = (\kappa^2-\gamma^2)^{1/2}$; the branches of the square roots are chosen to have positive real parts. The Whittaker functions M and W can be accurately evaluated by the known power series and asymptotic expansions. This method has been applied to evaluate ΔE_{SE} in Ref. 9.

In any method of evaluation of the self energy, it is necessary to deal with the mass renormalization term proportional to δm in Eq. (18). The magnitude of δm is fixed by the condition

$$\lim_{Z\alpha\to 0} \Delta E_{SE}^{(2)} = 0 \qquad (26)$$

The value of δm, so defined, is formally infinite, i.e., with the conventional regularization prescription

$$\frac{1}{k^2+i\varepsilon} \to \frac{1}{k^2+i\varepsilon} - \frac{1}{k^2-\Lambda^2+i\varepsilon} \qquad (27)$$

in the definition of D_F in Eq. (19), the two terms in Eq. (18) are separately of order $\ln\Lambda$. The energy shift is then the limit

$$\Delta E_{SE}^{(2)} = \lim_{\Lambda\to\infty} \Delta E_{SE}^{(2)}(\Lambda) \qquad (28)$$

In Ref. 9, an auxiliary term is subtracted from the first term in (18) to cancel the divergence, and added to the second term. This subtraction term is constructed by replacing the operator $G(z) = [\vec{\alpha}\cdot\vec{p} + V + \beta - z]^{-1}$, corresponding to the Green's function in Eq. (20), by the operator

$$G_A(z) = \frac{1}{\vec{\alpha}\cdot\vec{p}+\beta-z} - V\frac{1}{p^2+1-z^2} - V\frac{2z(\beta+z)}{[p^2+1-z^2]^2} \tag{29}$$

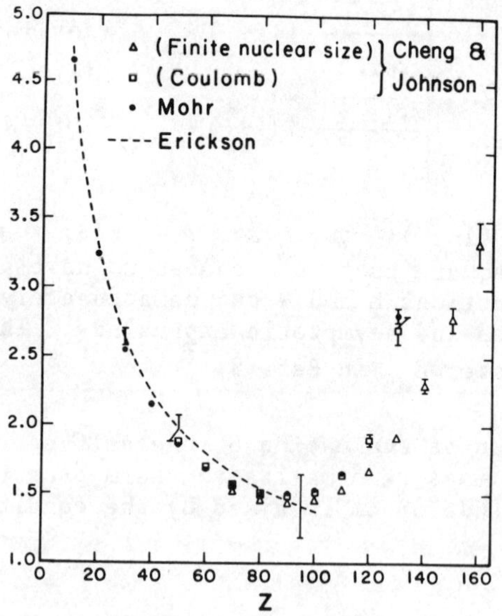

Fig. 4. Results of calculations of the $1S_{1/2}$-state self-energy level shift from Refs. 9-11.

in the first term of Eq. (18). The auxiliary term is similar in form to the first term in (18); it is also expanded in angular momentum eigenfunctions and subtracted from the unrenormalized level shift before any integration or summation is carried out, leaving a finite, numerically tractable expression for the difference. On the other

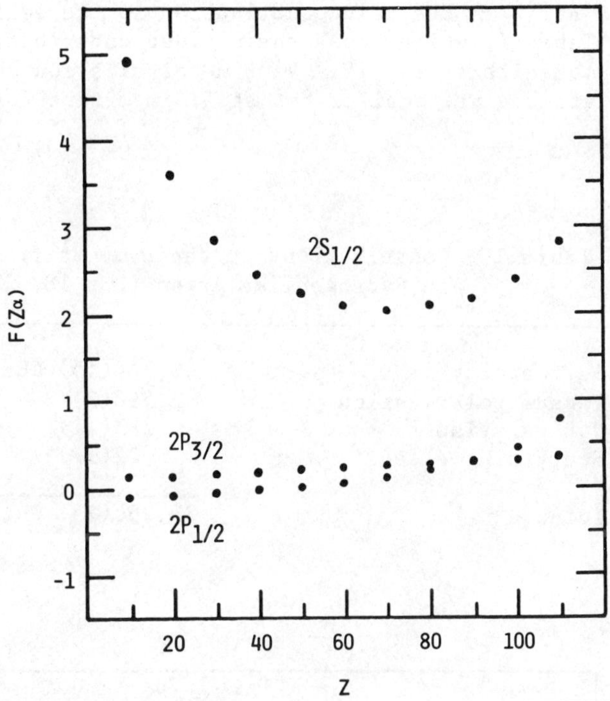

Fig. 5. Results of a calculation of the self-energy level shift for n = 2 states.[12,13]

hand, the auxiliary term is sufficiently simple that the divergent part (ln Λ) can be isolated analytically, and cancelled against the second term on the right-hand side in (18). The finite remainders give the physical level shift.

The results of various methods of calculation of ΔE_{SE} for the $1S_{1/2}$ state appear in Fig. 4.[9-11] Results of a calculation for the n = 2 states appear in Fig. 5.[12,13]

IV ONE-ELECTRON ATOMS

The theory of radiative corrections in high-Z systems is, of course, best understood in one-electron atoms. Quantitative predictions for the Lamb shift have been made for these systems over a wide range of Z.[11,14] The main contribution is the self energy as indicated in Table I, which shows the various contributions to the hydrogenic Lamb shift at Z = 18. Vacuum polarization and finite nuclear size effects are smaller but still important for an accurate prediction.

Table I. Contributions to the Lamb Shift in Hydrogenlike Argon (Z = 18)

Self energy	40,544(15) GHz
Vacuum polarization	-2,598(3)
Nuclear size	283(12)
Everything else	22(17)
Total	38,250(25) GHz

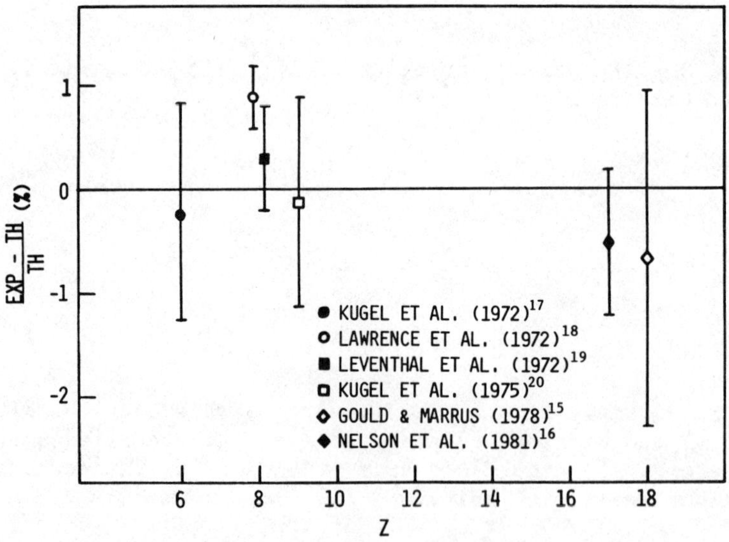

Fig. 6. Comparison of experiment and theory for the Lamb shift in high-Z hydrogenlike atoms.

Measurements of the Lamb shift in one-electron atoms have been extended to Z = 17 and 18 in recent experiments.[15,16] In both experiments, the hydrogenic charge state is created by passing a fast beam of partially ionized atoms through a thin foil. Some of the atoms in the beam are in the one-electron metastable $2S_{1/2}$ state after passing through the foil. In the experiment at Z = 18 by Gould and Marrus, the metastable atoms are passed through a magnetic field.[15] This increases the radiative decay rate of the metastable state, because the $2S_{1/2}$ and $2P_{1/2}$ states are mixed in the motional electric field. Measurement of the lifetime of the atoms as a function of field strength yields a value for the Lamb shift according to the Bethe-Lamb theory. The experiment at Z = 17 by Nelson et al. is based on a doppler-tuned laser resonance method.[16]

A comparison of theory[14] and experiment[15-20] for the one-electron Lamb shift for the range Z = 6-18 is shown in Fig. 6. There is good agreement between theory and experiment at about the 1% level of accuracy.

V TWO-ELECTRON ATOMS

The completely relativistic formulation of QED in the Furry picture can be extended in a straightforward way to high-Z systems with more than one electron. Consequently, the fundamental theory can be unambiguously tested in these systems. As noted in Section I, radiative corrections are relatively more important at high Z than at low Z. This is particularly true for the two-electron $2^3S_1-2^3P_0$ splitting, the analog of the Lamb shift, and for the two-electron $2^3S_1-2^3P_2$ splitting. These splittings are dominated at low Z by the Coulomb interaction of the electrons, of order $\alpha(Z\alpha)m_ec^2$, but the radiative correction, of order $\alpha(Z\alpha)^4m_ec^2$, increases comparatively rapidly as Z increases. This is illustrated for the $2^3S_1-2^3P_0$ splitting in Fig. 7 where the ratio of the radiative correction to the total splitting is plotted as a function of Z.

The physical idea behind the extended Furry picture theory is the fact that for a few electrons bound to a high-Z nucleus, the relativistic atomic structure is dominated by the Coulomb field of the nucleus. Electron-electron interactions are smaller by a factor 1/Z. Hence, perturbation theory can be applied, where in zero order, each electron is regarded as a Dirac hydrogenic electron. Electron-electron interactions and radiative corrections are taken into account as perturbations. This procedure produces a perturbation series in powers of 1/Z, and is the completely relativistic generalization of the familiar nonrelativistic 1/Z expansion methods.

Historically, relativistic perturbation theory has been applied in various forms and contexts. An early application by Christy and

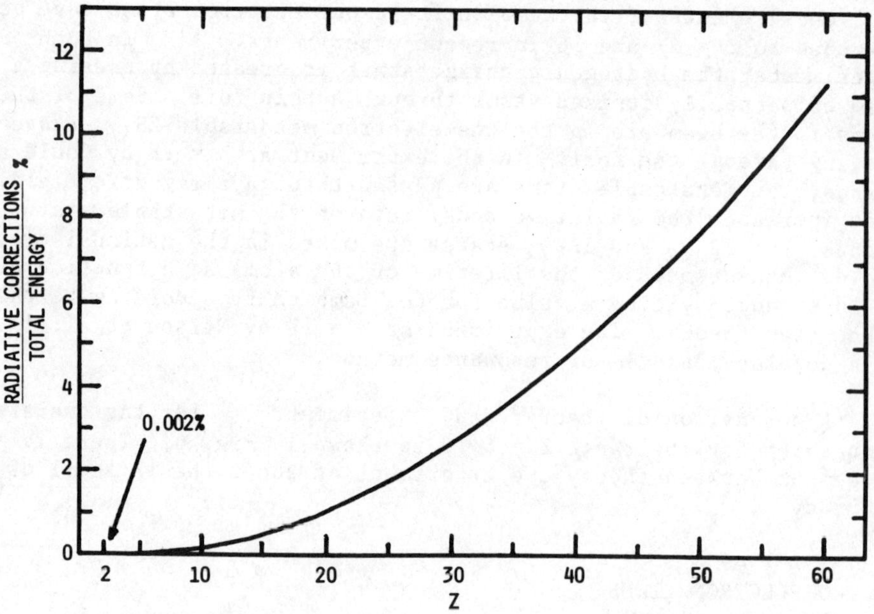

Fig. 7. The ratio (radiative corrections)/(total energy difference) for the $2\,^3S_1$–$2\,^3P_0$ splitting as a function of Z.

Keller was an evaluation of the one-photon electron-electron interaction with Dirac hydrogenic wave functions to estimate the fine structure in the L shell of heavy atoms.[21] Bethe and Salpeter discussed relativistic perturbation theory for high-Z two-electron atoms, although experiments with Z greater than 10 were not foreseen at the time.[22] Sucher noted that the Gell-Mann Low theorem would provide a basis for such calculations.[8] Relativistic 1/Z expansion methods that include the Breit interaction have been applied by Layzer and Bahcall[23] and by Dalgarno and Stewart.[24] The fully relativistic perturbation approach has been examined by Labsovsy,[25] Klimchitskaya and Labsovsy,[26] and by Ivanov, Ivanova and Safonova.[27]

The perturbation approach is complementary to the methods that accurately predict the fine structure in neutral helium.[28] In that case, the electron-electron interaction is large and is taken into account exactly in the nonrelativistic wave function of the electron. Relativistic corrections are small and are treated perturbatively. In the Furry picture formulation, by contrast, relativistic corrections are taken completely into account, and the electron-electron interactions are treated as perturbations.

In relativistic perturbation theory, the zero-order configurations of the $2\,^3S_1$, $2\,^3P_0$, and $2\,^3P_2$ states are $1s_{1/2}2s_{1/2}$, $1s_{1/2}2p_{1/2}$, and $1s_{1/2}2p_{3/2}$, respectively. The zero-order energies are just the sums of the Dirac energies of the states in the zero-order configurations

$$\Delta E^{(0)}(2^3S_1 - 2^3P_0) = 0 \tag{30}$$

$$\Delta E^{(0)}(2^3S_1 - 2^3P_2) = m_e c^2 \left\{ \left[1 - \frac{(Z\alpha)^2}{4}\right]^{\frac{1}{2}} - \left[\tfrac{1}{2} + \tfrac{1}{2}(1 - (Z\alpha)^2)^{\frac{1}{2}}\right]^{\frac{1}{2}} \right\}$$

The leading (second order in e) corrections are given by the Feynman diagrams shown in Figs. 8a-c. The one-photon-exchange diagram corresponds to both the static Coulomb interaction and the retarded transverse photon interaction that together contribute[29-31]

$$\Delta E^{(2)}_{pe}(2^3S_1 - 2^3P_0) = \alpha(Z\alpha)m_e c^2 \left[\frac{248}{6561} + 0.1428(Z\alpha)^2 + 0.1(Z\alpha)^4 + \ldots\right] \tag{31}$$

$$\Delta E^{(2)}_{pe}(2^3S_1 - 2^3P_2) = \alpha(Z\alpha)m_e c^2 \left[\frac{248}{6561} - 0.0363(Z\alpha)^2 - 0.03(Z\alpha)^4 + \ldots\right]$$

The diagrams in Figs. 8b-c represent the self-energy and vacuum-polarization corrections. These diagrams give exactly the same contribution to the energy levels, from each member of the configuration, as the hydrogenic corrections in Fig. 3, so the level differences are just the one-electron splittings discussed in Section IV.

The next order (fourth order in e) corrections correspond to the Feynman diagrams in Figs. 9-11. They are grouped according to the number of photon lines joining the two electron lines, which will be

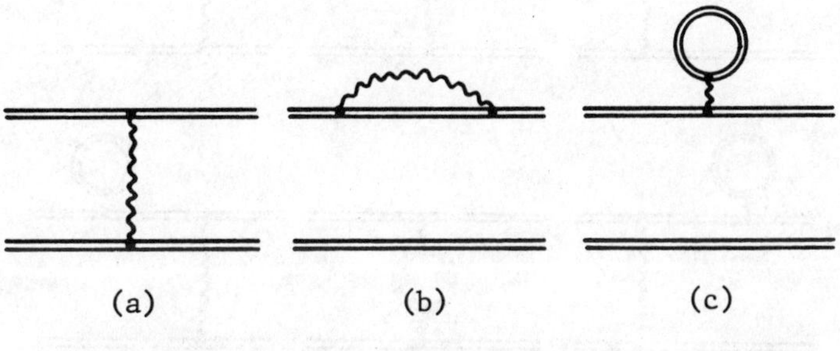

(a) (b) (c)

Fig. 8. Second-order Feynman diagrams for a two-electron atom: (a) photon exchange, (b) self-energy, and (c) vacuum polarization.

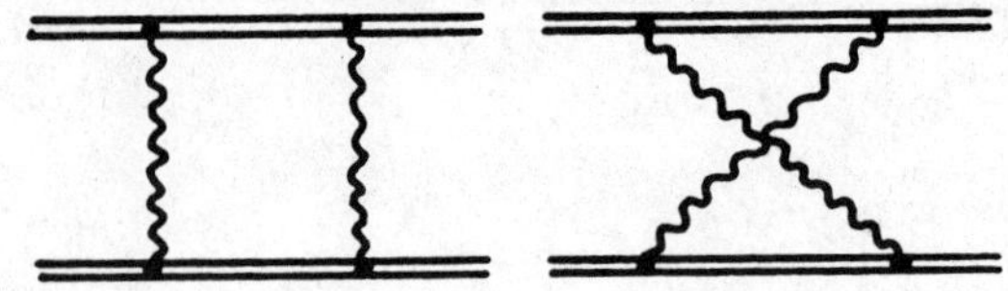

Fig. 9. Fourth-order Feynman diagrams with two exchanged photons.

seen to be related to the order of magnitude in $(Z\alpha)$ for the diagrams. The dominant corrections correspond to the two-photon exchange diagrams in Fig. 9. The corrections to the splittings are

$$\Delta E^{(4)}_{pe}(2^3S_1 - 2^3P_0) = \alpha^2 m_e c^2[-0.0256 - 0.27(Z\alpha)^2 + \ldots]$$
$$\Delta E^{(4)}_{pe}(2^3S_1 - 2^3P_2) = \alpha^2 m_e c^2[-0.0256 + 0.002(Z\alpha)^2 + \ldots] \qquad (32)$$

The leading terms in (32) are known from nonrelativistic 1/Z-expansion calculations,[32-34] and the next corrections have been extracted from a calculation of the expectation value of the Breit operators with variational wave functions.[30,35-37]

Graphs with one photon joining the two electron lines, comprising the second group, appear in Fig. 10. These graphs represent screening corrections to the radiative corrections, or equivalently radiative corrections to the electron-electron interaction. A calculation of these corrections at high Z has not been carried out, but they can be expected to be of order $(1/Z)\Delta E^{(2)}_{SE}$.

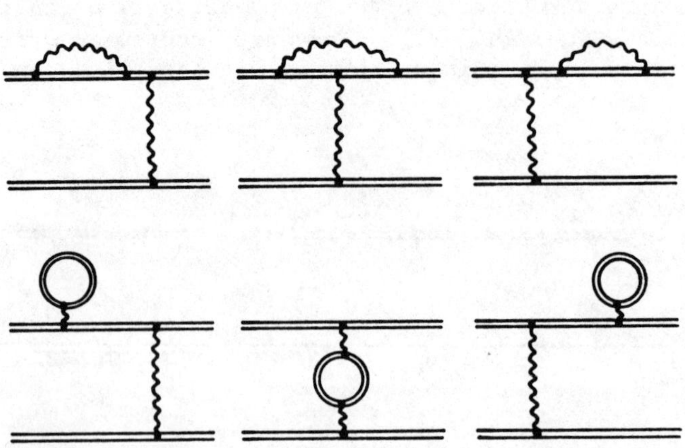

Fig. 10. Fourth-order Feynman diagrams with one exchanged photon.

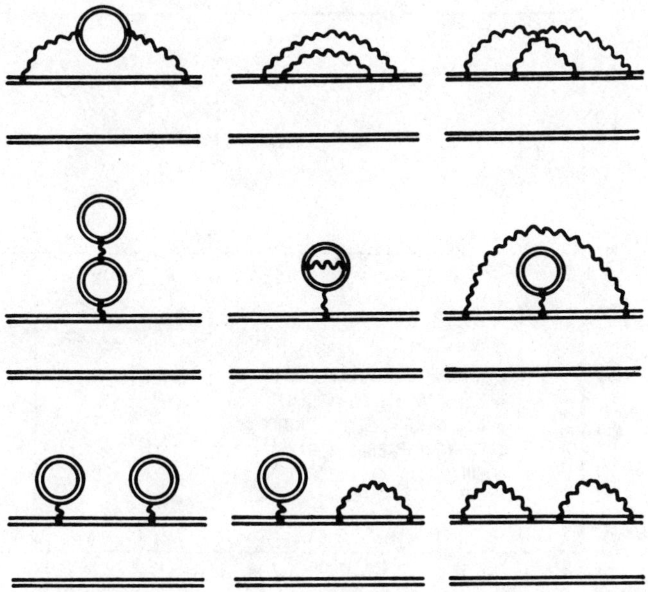

Fig. 11. Fourth-order Feynman diagrams with no exchanged photon.

The third class of fourth-order diagrams consists of diagrams with no photons joining the two electron lines, shown in Fig. 11. These diagrams are just the fourth-order hydrogenic radiative corrections, of order $(\alpha/\pi)\Delta E_{SE}^{(2)}$, that are known to lowest order in $Z\alpha$.[38] Scaling the known term to higher Z indicates that these diagrams are negligible at the present level of theoretical accuracy.

In sixth order, in analogy with fourth order, the dominant correction can be expected to arise from diagrams with three exchanged photons. To lowest order in $Z\alpha$, this correction is the third-order Coulomb interaction of the electrons, that is known from variational 1/Z-expansion calculations[32-34]

$$\begin{aligned} \Delta E_{pe}^{(6)}(2^3S_1 - 2^3P_0) &= \alpha^3(Z\alpha)^{-1}m_ec^2[-0.0117 + \ldots] \\ \Delta E_{pe}^{(6)}(2^3S_1 - 2^3P_2) &= \alpha^3(Z\alpha)^{-1}m_ec^2[-0.0117 + \ldots] \end{aligned} \tag{33}$$

The total splittings are the sums of the contributions described above

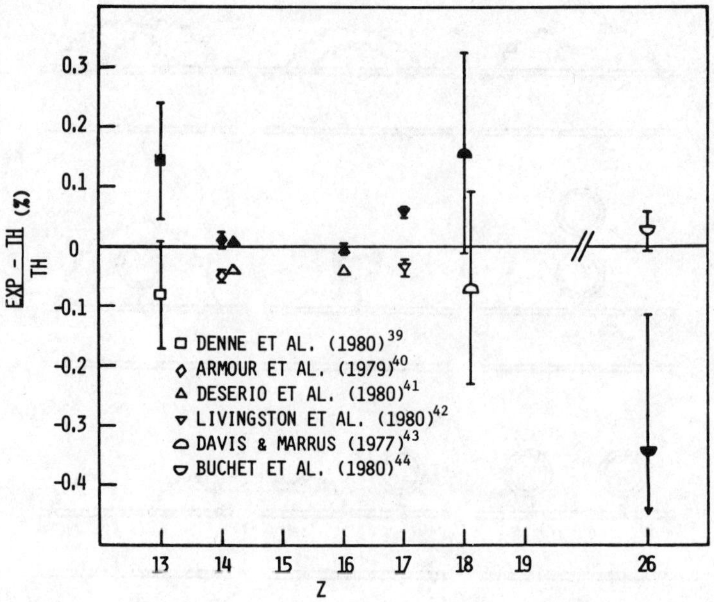

Fig. 12. Comparison of experiment and theory for the $2\,^3S_1$–$2\,^3P_0$ and $2\,^3S_1$–$2\,^3P_2$ energy level separations in high-Z two-electron atoms. The filled(open) shapes correspond to the $2\,^3S_1$–$2\,^3P_0$ ($2\,^3S_1$–$2\,^3P_2$) transitions.

$$
\begin{aligned}
\Delta E(2\,^3S_1 - 2\,^3P_0) &= -S + \Delta E_{pe}^{(2)} + \Delta E_{pe}^{(4)} + \Delta E_{pe}^{(6)} + \ldots \\
\Delta E(2\,^3S_1 - 2\,^3P_2) &= E_f - S + \Delta E_{pe}^{(2)} + \Delta E_{pe}^{(4)} + \Delta E_{pe}^{(6)} + \ldots
\end{aligned}
\tag{34}
$$

The zero- and second-order one-electron corrections are combined in the terms $-S$ and $E_f - S$, where $S = \Delta E(2P_{1/2}-2S_{1/2})$ and $E_f = \Delta E(2P_{1/2}-2P_{3/2})$ are the hydrogenic Lamb shift and fine-structure, respectively. It is clear from comparison of Eqs. (30)-(33) that the perturbation series described here gives the relativistic generalization of the nonrelativistic 1/Z expansion of energy levels. In Fig. 12, the splittings in Eq. (34) are compared to experiment.[39-44] The level of agreement is better than 1%, which is the magnitude of the one-electron radiative correction. The remaining differences between theory and experiment are of the order of the screening corrections to the radiative corrections that are not included in the theoretical values.

VI MANY-ELECTRON ATOMS

It is only in the last decade that radiative corrections to energy levels have been shown to make an improvement in the comparison of theory and experiment for energy levels of many electron atoms. Although there has been no systematic development of the treatment of radiative corrections in relativistic many-electron systems, the fact that the nuclear Coulomb field dominates over the electron-electron interactions in inner shells of high-Z atoms means that a hydrogenic or screened hydrogenic approximation can give meaningful results. Some examples of comparisons of theory and experiment are given here to show the general picture. A more detailed discussion is given by Desclaux in these proceedings.

An impressive example is the work of Desiderio and Johnson[45] and of Mann and Johnson[46] on K-electron binding energies in high-Z atoms. They were first to show that including the Hartree-Fock self energy of the $1s_{1/2}$ electron gives a substantial improvement in the theoretical binding energy as illustrated in Table II.

A comparison of theory and experiment for the K-$L_{II}(1s_{1/2}$-$2p_{1/2})$ x-ray transitions can be made over a wide range of Z. To illustrate the relative magnitude of the radiative corrections, we consider separately T_0 = (theory without self energy) and T_R = (self energy), where the total theoretical x-ray energy is $T = T_0 + T_R$. Values for T_0 have been calculated over a wide range of Z by Huang et al.[47] Fig. 13 shows a graph of the quantity $(T_0-E)/T_0$, where E is the experimental value for the K-L_{II} x-rays as a function of Z. The experimental values are based on the tabulation by Bierden and Burr[48] (circles in Fig. 13) and recent accurate values reported by Deslattes et al.[49] (squares in Fig. 13). Also shown in that figure is the ratio $-T_R/T_0$ where the self energy T_R is based on the Coulomb field

Table II. K-electron Energy Levels [Ry] from Mann and Johnson[46]

Atomic number	Self-energy and vacuum polarization Desiderio and Johnson[45]	E(th)	E(expt)
74	8.65	-5110.50	-5110.46 ± 0.02
80	11.28	-6108.52	-6108.39 ± 0.06
82	12.27	-6468.79	-6468.67 ± 0.05
86	14.43	-7233.01	-7233.08 ± 0.90

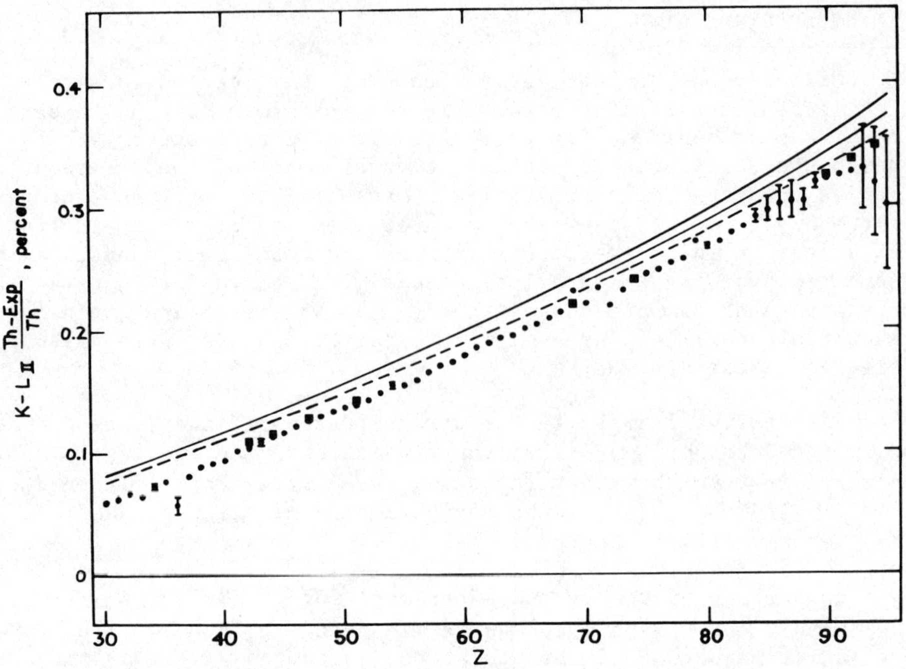

Fig. 13. Comparison of theory and experiment for the self-energy correction to the K-L_{II} x-ray energy in heavy atoms.

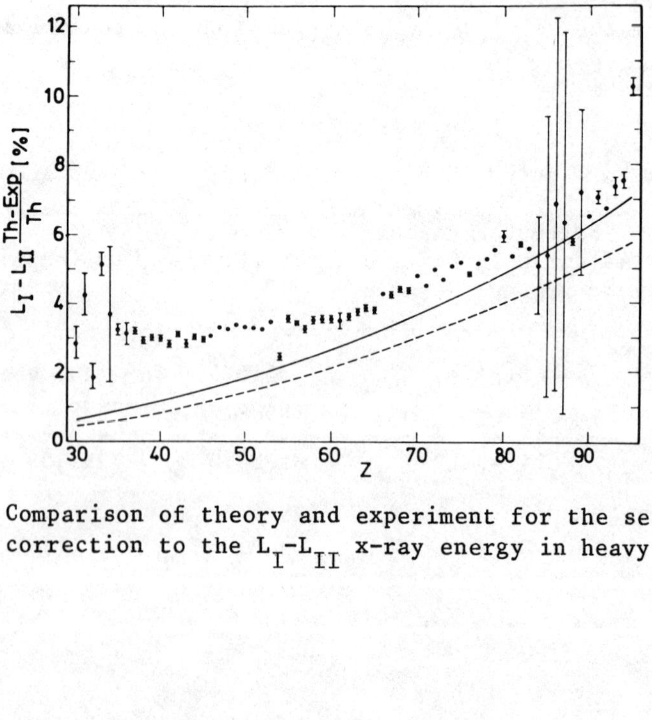

Fig. 14. Comparison of theory and experiment for the self-energy correction to the L_I-L_{II} x-ray energy in heavy atoms.

values[9,12] (solid line in Fig. 13) or on Coulomb field values modified by a phenomenological screening correction[47] (dashed line in Fig. 13). The solid line segment from Z = 70 to 95 shows values for $-T_R/T_0$ that result when the Coulomb self energy of the $1s_{1/2}$ state is replaced by the self energy of the $1s_{1/2}$ state in a Hartree-Fock potential as calculated by Cheng and Johnson.[10] From the figure, it is clear that the self energy accounts for the main part of the difference, i.e., $(T_0-E)/T_0 \approx -T_R/T_0$ or $T_0 + T_R \approx E$.

A similar comparison may be made for excited states. Figure 14 shows the analogous comparison of theory and experiment for the L_I-L_{II} $(2s_{1/2}-2p_{1/2})$ splitting, the many-electron analog of the Lamb shift. Here again, the self energy improves the agreement between theory and experiment, particularly at high Z.

VII CONCLUSION

From the comparisons of theory and experiment described here, it is clear that Lamb shift effects play an important role in the structure of high-Z atoms. These effects are quantitatively confirmed in a variety of systems.

The direction for future theoretical work seems well defined. In highly-relativistic systems with more than one electron, the theoretical predictions are limited in accuracy at the level where effects of the interaction of the electrons, strong binding effects, and radiative corrections simultaneously come into play. For two-electron atoms, the formulation of the theory described above seems to provide a suitable approach to this problem. For many-electron atoms, the fundamental problem of relativistic electron interactions itself is more complex, and the best approach to the combined problem of relativistic electron interactions and radiative corrections is not clear at present.

Research supported by the National Science Foundation, Grant No. PHY80-26549.

REFERENCES

1. For a review of strong field QED, see S.J. Brodsky and P.J. Mohr, in Structure and Collisions of Ions and Atoms, ed. by I.A. Sellin (Springer, Berlin, 1978), Topics in Current Physics, Vol. 5, p.3.
2. R. Serber, Phys. Rev. 48, 49 (1935).
3. E.A. Uehling, Phys. Rev. 48, 55 (1935).
4. E.H. Wichmann, N.M. Kroll, Phys. Rev. 96, 232 (1954); 101, 843 (1956).

5. H.A. Bethe, Phys. Rev. 72, 339 (1947).
6. S.S. Schweber, An Introduction to Relativistic Quantum Field Theory, Harper and Row, New York, 1961).
7. M. Gell-Mann and F. Low, Phys. Rev. 84, 350 (1951).
8. J. Sucher, Phys. Rev. 107, 1448 (1957).
9. P.J. Mohr, Ann. Phys. (N.Y.) 88, 26, 52 (1974).
10. K.T. Cheng and W.R. Johnson, Phys. Rev. A 14, 1943 (1976).
11. G.W. Erickson, Phys. Rev. Lett. 27, 780 (1971); J. Phys. Chem. Ref. Data 6, 831 (1977).
12. P.J. Mohr, Phys. Rev. Lett. 34, 1050 (1975).
13. P.J. Mohr, to be published.
14. P.J. Mohr, in Beam-Foil Spectroscopy, eds. I.A. Sellin and D.J. Pegg, (Plenum Press, New York, 1976), p. 89.
15. H. Gould and R. Marrus, Phys. Rev. Lett. 41, 1457 (1978).
16. E.T. Nelson, O.R. Wood II, C.K.N. Patel, M. Leventhal, D.E. Murnick, H.W. Kugel, and Y. Niv, Abstract submitted to the Second Int. Conf. on Precision Meas. and Fund. Const., Gaithersburg, MD, June 1981, p. 92.
17. H.W. Kugel, M. Leventhal, D.E. Murnick, Phys. Rev. A 6, 1306 (1972).
18. G.P. Lawrence, C.Y. Fan, S. Bashkin, Phys. Rev. Lett. 28, 1612 (1972).
19. M. Leventhal, D.E. Murnick, H.W. Kugel, Phys. Rev. Lett. 28, 1609 (1972).
20. H.W. Kugel, M. Leventhal, D.E. Murnick, C.K.N. Patel, O.R. Wood II, Phys. Rev. Lett. 35, 647 (1975).
21. R.F. Christy and J.M. Keller, Phys. Rev. 61, 147 (1942).
22. H.A. Bethe and E.E. Salpeter, Quantum Mechanics of One- and Two-Electron Atoms (Springer, Berlin, 1957).
23. D. Layzer and J. Bahcall, Ann. Phys. (N.Y.) 17, 177 (1962).
24. A. Dalgarno and A.L. Stewart, Proc. Phys. Soc. 75, 441 (1960).
25. L.N. Labzovskii, Zh. Eksp. Teor. Fiz. 59, 168 (1970), [Sov. Phys. JETP 32, 94 (1971)].
26. G.L. Klimchitskaya and L.M. Labzovskii, Zh. Eksp. Teor. Fiz. 60, 2019 (1971), [Sov. Phys. JETP 33, 1088 (1971)].
27. L.N. Ivanov, E.P. Ivanova, and U.I. Safronova, J. Quant. Spectrosc. Radiat. Transfer. 15, 553 (1975).
28. J. Daley, M. Douglas, L. Hambro, and N.M. Kroll, Phys. Rev. Lett. 29, 12 (1972).
29. H.T. Doyle, Advances in Atomic and Molecular Physics, (Academic, New York, 1969), Vol. 5, p. 337.
30. P.J. Mohr, unpublished (1975).
31. K.T. Cheng, unpublished (1978).
32. R.E. Knight and C.W. Scherr, Rev. Mod. Phys. 35, 431 (1963).
33. F.C. Sanders and C.W. Scherr, Phys. Rev. 181, 84 (1969).
34. K. Aashamar, G. Lyslo, and J. Midtdal, J. Chem. Phys. 52, 3324 (1970).
35. Y. Accad, C.L. Pekeris, and B. Schiff, Phys. Rev. A 4, 516 (1971).

36. P.J. Mohr, in Beam-Foil Spectroscopy, eds. I.A. Sellin and D.J. Pegg, (Plenum Press, New York, 1976), p. 97.
37. A.M. Ermolaev and M. Jones, J. Phys. B 7, 199 (1974).
38. T. Appelquist and S.J. Brodsky, Phys. Rev. Lett. 24, 562 (1970).
39. B. Denne, S. Huldt, J. Pihl, and R. Hallin, Physica Scripta 22, 45 (1980).
40. I.A. Armour, E.G. Myers, J.D. Silver, and E. Träbert, Phys. Lett. 75A, 45 (1979).
41. R. DeSerio, H.G. Berry, and R.L. Brooks, in Proceedings of the Workshop on Foundations of the Relativistic Theory of Atomic Structure, Argonne National Laboratory, Dec., 1980, Argonne Report ANL-80-126, p. 240.
42. A.E. Livingston, S.J. Hinterlong, J.A. Poirier, R. DeSerio, and H.G. Berry, J. Phys. B 13, L139 (1980).
43. W.A. Davis and R. Marrus, Phys. Rev. A 15, 1963 (1977).
44. J.P. Buchet, M.C. Buchet-Poulizac, P. Ceyzeriat, A. Denis, J. Désesquelles, M. Druetta, and K. Tran Cong, in Proceedings of the Workshop on Foundations of the Relativistic Theory of Atomic Structure, Argonne National Laboratory, Dec., 1980, Argonne Report ANL-80-126, p. 236.
45. A.M. Desiderio, W.R. Johnson, Phys. Rev. A 3, 1267 (1971).
46. J.B. Mann, W.R. Johnson, Phys. Rev. A 4, 41 (1971).
47. K. Huang, M. Aoyagi, M.H. Chen, B. Crasemann, H. Mark, At. Data and Nucl. Data Tables 18, 243 (1976).
48. J.A. Bearden, A.F. Burr, Rev. Mod. Phys. 39, 125 (1967); Atomic Energy Levels (U.S. Atomic Energy Commission, Oak Ridge 1965).
49. R.D. Deslattes, E.G. Kessler, Jr., L. Jacobs, and W. Schwitz, Phys. Lett. 71A, 411 (1979).

CALCULATION OF RELATIVISTIC EFFECTS IN ATOMS AND MOLECULES FROM THE SCHRÖDINGER WAVE FUNCTION

John H. Detrich and Clemens C. J. Roothaan*

NASA Langley Research Center
Mail Stop 283, Hampton, Virginia 23665
*Department of Physics and Chemistry
University of Chicago, Chicago, Illinois 60637

I. INTRODUCTION

The traditional method for dealing with relativistic effects in atoms and molecules consists of a somewhat heuristic combination of quantum electrodynamics and a many-electron generalization of the one-electron Dirac theory. On the whole, results calculated from this theory agree with experimental data. Nevertheless, the theory is by no means entirely satisfactory; in its development, certain ambiguities and divergencies are resolved by somewhat arbitrary or questionable means. In this paper we attempt to illuminate - and sidestep - some of the more questionable aspects of the traditional method, by reformulating electromagnetic interactions between particles in a different way. In the following we shall be concerned mostly with electrons, although in many instances our considerations will apply equally well to other electromagnetic particles, in particular to nuclei.

A major difficulty we encounter is already present in the classical electromagnetic theory of electrons; it stems from the fact that the very notion of mutually interacting point particles, as well as that of rigid bodies, is not compatible with the principles of relativity. Thus, it becomes necessary to distribute the charge of an electron over a finite, albeit a very small, volume. The electrons generate, and are acted upon by, electric as well as magnetic fields, and the finite velocity of propagation of these fields manifests itself through retardation and/or advancement effects; it is through these fields that the electrons interact with one another. In the limit of infinitesimally small electrons, their mutual interaction remains finite, and leads to the well-known Lienard-Wiechert potentials; these potentials correctly describe the electromagnetic interaction of electrons, at least to order $(v/c)^2$, and they depend only on the total charge of the electron, rather than the precise distribution of that charge.

Once the charge of an electron is distributed over a finite volume, one must also consider the interaction of an electron with itself through its own electromagnetic field. The energy associated with this field, which is of course the electromagnetic self-energy, obviously depends strongly on the details of the charge distribution of the electron, and becomes infinite for a point-electron. Furthermore, in order to have stable electrons, one must now introduce additional forces which are non-electromagnetic in nature; these forces only serve to maintain the internal structure of the electrons. As is well known, all attempts to formulate such stabilizing forces have been unsuccessful. Thus, while the assumption of small finite electrons renders a satisfactory account of the classical interaction between electrons, this same assumption leads to serious inconsistencies for the self-interaction and self-energy of an electron. Usually this problem is set aside by the implicit assumption that the self-energy is rigorously constant, regardless of the environment in which the electron finds itself. The self-energy then becomes an additive constant which drops out from all practical energy balance formulas.

It is to be expected that the quantization of a relativistic system of particles and field, using the classical scheme just described as the point of departure, will be plagued by similar ambiguities and inconsistencies. An additional complication arises from the requirement that a proper quantum mechanical description of a system of particles and field must also allow for the creation/annihilation of particles. Actually, the problems of particle self-energy and particle creation/annihilation are intimately connected. These questions are addressed in Quantum Electrodynamics and Quantum Field Theory. However, as long as the theory does not deliver finite particles with an appropriate internal structure, or at least permits such a structure to be imposed as an acceptable constraint, one has in fact point particles, which cause divergencies as in the classical case.

In the framework of Quantum Electrodynamics and Quantum Field Theory the Hamiltonian for a system of particles and electromagnetic field consists of three parts, describing respectively the particles, the field, and the interaction between particles and field. The interaction part of the Hamiltonian serves two functions: it governs how the field is being generated by the particles acting as sources, and how the particles are being acted upon by the forces the field represents. As in the classical picture, the field can be considered as the carrier of the mutual interaction and self-energy of the particles. But while in the classical realm one can easily identify the terms representing the particle self-energy, in a field theoretical description this is not the case, thus rendering the removal of the consequent divergencies much more difficult.

Another difficulty associated with a field-theoretical picture is that we are not longer dealing with a wave function as the sole or primary representation of an atom of molecule. Nevertheless, when relativistic effects constitute only small corrections to the non-relativistic results, it is obviously necessary to recast the field-theoretical formalism as a perturbation of ordinary many-electron quantum mechanics. This is a formidable task, since the mathematical constructs used in the two regimes are radically different. In addition, field theory is only simple for non-interacting particles, which is of course a very inappropriate first approximation for an atom or molecule.

The recasting of field-theoretical mechanisms as a perturbation of non-relativistic quantum mechanics can be viewed as an elimination of the dynamical variables representing the electromagnetic field. After this elimination one has in hand a perturbation scheme from which the wave functions and energies can be calculated systematically to any order. The lowest wave functions and energies follow from an eigenvalue equation; the corresponding operator can be identified as the particle Hamiltonian, which is of course related to the field theoretical Hamiltonian, but not identical with it or any part thereof. From a certain order onward, the wave functions and energies do not follow from eigenvalue equations. In this sense, relativistic many-electron quantum mechanics is not a Hamiltonian problem; this is the price one has to pay for having eliminated the electromagnetic field variables.

Thus, the particle-particle interactions are the consequences of the electromagnetic field, which is itself no longer explicitly present in many-electron quantum mechanics. The leading term in the particle-particle interaction is of course the non-relativistic Coulomb energy. The next few terms, of order $(v/c)^2$, are the lowest order relativistic effects; they represent magnetic interaction and retardation/advancement due to the finite velocity of propagation of electromagnetic radiation.

It is our contention that if we eliminate the electromagnetic field from the description of a system of particles and field <u>before</u> making the transition to quantum mechanics, we should obtain the same results. As we shall see below, this can be done conveniently and unambiguously for the stationary states of an electronic system. The labor involved in this process is considerably less than what is required in the traditional method based on Quantum Electrodynamics and Quantum Field Theory. Furthermore, in our framework the removal of the particle self-energy is trivially simple, and relativistic effects are calculated as expectation values of appropriate operators over ordinary wave functions.

II. THE ENERGY EXPRESSION IN TERMS OF WAVE FUNCTIONS

One of our priorities is a formulation in terms of ordinary N-electron wave functions, where N is the number of electrons in the atom or molecule. These wave functions are envisioned as superpositions of antisymmetrized products of 4-component Dirac spinors. We characterize the electronic system in terms of stationary states with wave functions denoted by

$$\tilde{\Psi}_{m\alpha}(\vec{r}_1,\vec{r}_2,\ldots,\vec{r}\ ;t) = \Psi_{m\alpha}(\vec{r}_1,\vec{r}_2,\ldots,\vec{r}_N)e^{-iE_m t/\hbar} \quad . \tag{1}$$

Here, m labels distinct energies E_m, while α refers to degenerate wave functions if necessary. We shall assume that m is a discrete index; since our interest is in bound states, there is no need for an explicit generalization to handle possible continuum states.

We assume that the time-independent wave functions $\Psi_{m\alpha}$ constitute a complete orthonormal set; this is expressed by

$$\langle\Psi_{m\alpha}|\Psi_{n\beta}\rangle = \delta_{mn}\delta_{\alpha\beta} \tag{2}$$

and

$$\Psi = \Sigma_{m\alpha}\Psi_{m\alpha}\langle\Psi_{m\alpha}|\Psi\rangle \ , \tag{3}$$

where Ψ is an arbitrary wave function within certain reasonable constraints.

The completeness property of our set of wave functions permits characterization of operators in terms of their matrix elements, which will prove very convenient. Among our operators are the position operators $\vec{r}_\mu$, the momentum operators $\vec{p}_\mu$, and the usual Dirac matrix operators $\vec{\alpha}_\mu$ and β_μ for each of the electrons, $\mu=1,2,\ldots,N$. We shall also need operators representing the charge and current densities of the electrons. We denote the charge density of the μ-th electron by $\rho_\mu(\vec{r}\,')$; in the case of a point electron this becomes

$$\rho_\mu(\vec{r}\,') = e\delta(\vec{r}\,'-\vec{r}_\mu) \ , \tag{4}$$

where e is the charge of the electron, and $\delta(\vec{r}\,'-\vec{r}_\mu)$ indicates the three-dimensional Dirac delta-function at the point $\vec{r}\,'=\vec{r}_\mu$. Here, the parameter space indicated by $\vec{r}\,'$ can be regarded as the position of an external probe, or the observer's space.

It is well known that the operator for the velocity of a Dirac particle is given by $c\vec{\alpha}_\mu$; accordingly we define the current density operator by

$$\vec{j}_\mu(\vec{r}') = \vec{\alpha}_\mu \rho_\mu(\vec{r}') \ . \quad (5)$$

Note that we have used the electromagnetic rather than the electrostatic definition of the current density; $\vec{j}_\mu(\vec{r}')$ has the same dimensions as $\rho_\mu(\vec{r}')$.

We now proceed to an electrodynamical description of our system of electrons. Our first task is to write down a valid quantum mechanical expression for the energies E_m. We partition E_m according to

$$E_m = E_{D,m} + E_{I,m} \ , \quad (6)$$

where $E_{D,m}$ is the many-electron generalization of the Dirac energy, and $E_{I,m}$ is the electron-electron interaction energy. The Dirac energy is given by

$$E_{\Delta,m} = \langle\Psi_{m\alpha}|H_D|\Psi_{m\alpha}\rangle \ , \quad (7)$$

where

$$H_D = \Sigma_\mu [\beta_\mu m_e c^2 + \vec{\alpha}_\mu \cdot \vec{p}_\mu c + U(\vec{r}_\mu)] \quad (8)$$

is simply the sum of the individual particle Dirac Hamiltonians. Here m_e denotes the electron mass and U represents the electrostatic potential due to atomic nuclei and (time-independent) external fields. Introduction of U is the customary simplification of the electron-nuclear interaction; it is also possible (and more correct) to include nuclear coordinates in the wave function, in which case $E_{I,m}$ would include electron-nuclear and nuclear-nuclear interactions.

The direct particle-particle interaction of classical electrodynamics was developed by many workers, culminating in the work of Wheeler and Feynman[1]. We follow their development in order to formulate the electromagnetic interaction energy of the N-electron system. The deviations from conventional electrodynamics may be summarized as follows.

1) There is no such concept as "the" electromagnetic field with degrees of freedom of its own. Instead, there is a collection of adjunct fields, each produced by an individual particle, and completely determined by the motion of that particle.
2) The prevailing field on a given particle is determined by the sum of the fields produced by every particle other than the given particle. The interaction of a particle with its own

field does not occur.

3) The fields produced by the particles are taken half-retarded and half-advanced. We note that this is the necessary and sufficient condition that energy and momentum are conserved - and therefore defined - within a finite, but perhaps very large volume. In classical electrodynamics, the half-retarded and half-advanced solution describes a system that neither emits nor absorbs radiation.

In the light of these remarks, we now write down the classical electromagnetic interaction energy

$$E_I = \frac{1}{2}\Sigma_\mu\Sigma_{\nu\neq\mu}\int d\vec{r}'[\rho_\mu(\vec{r}',t)\phi_\nu(\vec{r}',t) - \vec{j}_\mu(\vec{r}',t)\cdot\vec{A}_\nu(\vec{r}',t)] \ , \tag{9}$$

where $\phi_\mu(\vec{r}',t)$ and $\vec{A}_\mu(\vec{r}',t)$ are the scalar and vector potentials associated with the μ-th particle, namely

$$\left.\begin{aligned} \phi_\mu(\vec{r}',t) &= \frac{1}{2}\int d\vec{r}''R^{-1}[\rho_\mu(\vec{r}'',t-R/c) + \rho_\mu(\vec{r}'',t+R/c)] \ , \\ \vec{A}_\mu(\vec{r}',t) &= \frac{1}{2}\int d\vec{r}''R^{-1}[\vec{j}_\mu(\vec{r}'',t-R/c) + \vec{j}_\mu(\vec{r}'',t+R/c)] \ , \end{aligned}\right\} \tag{10}$$

with the abbreviation

$$R = |\vec{r}'-\vec{r}''| \ . \tag{11}$$

In quantum mechanics, the quantities in Eq. (9) are reinterpreted as operators. The time-dependent charge and current densities are handled in terms of operator matrix elements evaluated with respect to the time-dependent wave functions $\Psi_{m\alpha}$. This immediately yields quantized versions of Eq. (10) which can be reduced to relations between time-independent matrix elements by dividing by the exponential time factor; the result is

$$\left.\begin{aligned} \phi_{\mu,m\alpha,n\beta}(\vec{r}') &= \int d\vec{r}''R^{-1}\cos(k_{mn}R)\rho_{\mu,m\alpha,n\beta}(\vec{r}'') \ , \\ \vec{A}_{\mu,m\alpha,n\beta}(\vec{r}') &= \int d\vec{r}''R^{-1}\cos(k_{mn}R)\vec{j}_{\mu,m\alpha,n\beta}(\vec{r}'') \ , \end{aligned}\right\} \tag{12}$$

where

$$k_{mn} = (E_m-E_n)/\hbar c \ . \tag{13}$$

The right-hand side of Eq. (9) is now interpreted in terms of operator products, using time-dependent matrix elements. The exponential time factors cancel, so we obtain for the quantum mechanical interaction energy

$$E_{I,m} = \frac{1}{2}\sum_{\mu}\sum_{\nu\neq\mu}\sum_{n\beta}\int d\vec{r}\,'[\rho_{\mu,m\alpha,n\beta}(\vec{r}\,')\phi_{\nu,n\beta,m\alpha}(\vec{r}\,') - \vec{j}_{\mu,m\alpha,n\beta}(\vec{r}\,')\cdot A_{\nu,n\beta,m\alpha}(\vec{r}\,')] \ . \tag{14}$$

We can eliminate the potentials using Eq. (12) to obtain

$$E_{I,m} = \frac{1}{2}\sum_{\mu}\sum_{\nu\neq\mu}\sum_{n\beta}\int d\vec{r}\,'\int d\vec{r}\,''R^{-1}\cos(k_{mn}R) \times [\rho_{\mu,m\alpha,n\beta}(\vec{r}\,')\rho_{\nu,n\beta,m\alpha}(\vec{r}\,'')-\vec{j}_{\mu,m\alpha,n\beta}(\vec{r}\,')\cdot\vec{j}_{\nu,n\beta,m\alpha}(\vec{r}\,'')] \ . \tag{15}$$

In this expression the symmetry of the interaction between electrons is apparent.

We now combine the Dirac and interaction energies to obtain the total energy. Since we will soon consider the energies as functionals of trial wave functions, we display the wave functions and energies explicitly wherever they occur. We also generalize slightly to obtain

$$\begin{aligned} E_m\delta_{\alpha\beta} &= \langle\Psi_{m\alpha}|H_D|\Psi_{m\beta}\rangle \\ &+ \frac{1}{2}\sum_{\mu}\sum_{\nu\neq\mu}\sum_{n\gamma}\int d\vec{r}\,'\int d\vec{r}\,''R^{-1}\cos[E_m-E_n)R/\hbar c] \\ &\times [\langle\Psi_{m\alpha}|\rho_\mu(\vec{r}\,')|\Psi_{n\gamma}\rangle\langle\Psi_{n\gamma}|\rho_\nu(\vec{r}\,'')|\Psi_{n\beta}\rangle \\ &- \langle\Psi_{m\alpha}|\vec{j}_\mu(\vec{r}\,')|\Psi_{n\gamma}\rangle\cdot\langle\Psi_{n\gamma}|\vec{j}_\nu(\vec{r}\,'')|\Psi_{m\beta}\rangle] \ . \end{aligned} \tag{16}$$

If the factor $\cos[(E_m-E_m)R/\hbar c]$ were absent in Eq. (16), the summation over n could be carried out using the completeness property indicated in Eq. (3); E_m would then reduce to the diagonal matrix element of an operator. Hence, it is the retardation/advancement effect of the electromagnetic interaction, expressed in the factor $\cos[(E_m-E_n)/\hbar c]$, which prevents a Hamiltonian formulation here.

In the non-relativistic formulation, the wave function and energy can be determined by application of the variation principle to the energy expression considered as a functional of a trial wave function; this approach applies in any Hamiltaonian case. We wish to apply a similar approach here, but this is of course complicated by our lack of a Hamiltonian formulation: Eq. (16) references <u>all</u> energies and <u>all</u> wave functions. We consider that Eq. (16)

nevertheless constitutes a system of equations which implicitly defines each E_m as a functional. Evidently, we should demand that all E_m are stationary simultaneously to first order for any change in the set of trial wave functions $\Psi_{m\alpha}$ which satisfies the requirements of orthonormality and completeness. We use the variation principle in this form to formulate the perturbation scheme from which the wave functions and energies are to be calculated.

III. THE PAULI PERTURBATION EXPANSION

We assume that our wave functions and energies can be expanded in terms of the parameter

$$\lambda = c^{-1} \tag{17}$$

It is convenient to introduce scaled energies defined by

$$\varepsilon_m = \lambda^2 E_m ; \tag{18}$$

we then put forward the perturbation expansions

$$\begin{aligned} \Psi_{m\alpha} &= \sum_{p=0}^{\infty} \Psi_{m\alpha,p} \lambda^p , \\ \varepsilon_m &= \sum_{p=0}^{\infty} \varepsilon_{m,p} \lambda^{2p} . \end{aligned} \tag{19}$$

Note that only even powers of λ appear in the expansion of ε_m; as shown in detail elsewhere[2], the odd powers are expected to vanish. Since Eq. (19) obviously constitutes a generalization of the well-known Pauli approximation, we call this scheme Pauli Perturbation Expansion.

It is convenient to introduce the operators given by

$$M = m_e \Sigma_\mu \beta_\mu , \tag{20}$$

$$P = \Sigma_\mu \vec{\alpha}_\mu \cdot \vec{p}_\mu , \tag{21}$$

$$K = (2m_e)^{-1} \Sigma_\mu \vec{\alpha}_\mu \cdot \vec{p}_\mu \beta_\mu , \tag{22}$$

$$T = (2m_e)^{-1} \Sigma_\mu \beta_\mu \vec{p}_\mu \cdot \vec{p}_\mu , \tag{23}$$

$$U = \Sigma_\mu U(\vec{r}_\mu) . \tag{24}$$

The operators M, P and U emerge from H_D as given by Eq. (8). The remaining operators satisfy the commutation relations

$$P = [K,M] \, , \tag{25}$$

$$T = -\frac{1}{2}[K,P] = -\frac{1}{2}[K,[K,M]] \, . \tag{26}$$

where in general [A,B] designates the commutator of the operators A and B. In addition, we have

$$[M,T] = 0 \, , \tag{27}$$
$$[M,U] = 0 \, . \tag{28}$$

Another operator of interest is given by the product

$$\beta = \beta_1\beta_2..\beta_N \, ; \tag{29}$$

we have the commutation and anticommutation relations

$$[\beta,M] = [\beta,T] = [\beta,U] = 0 \, , \tag{30}$$
$$\{\beta,P\} = \{\beta,K\} = 0 \, , \tag{31}$$
$$[\beta,\rho_\mu(\vec{r}\,')] = 0 \, , \tag{32}$$
$$\{\beta,j_\mu(\vec{r}\,')\} = 0 \, , \tag{33}$$

where in general {A,B} designates the anticommutator of the operators A and B.

We insert the expressions given by Eq. (19) into Eq. (16) and develop the resulting energy expansion. This is not completely straightforward because of the factor

$$\cos[(E_m-E_n)R/\hbar c] = \cos[(\varepsilon_m-\varepsilon_n)R\hbar^{-1}\lambda^{-1}] \, . \tag{34}$$

In view of the second Eq. (19), $(\varepsilon_m-\varepsilon_n)\lambda^{-1}$ is of order λ if $\varepsilon_{m,o} = \varepsilon_{n,o}$ and of order λ^{-1} otherwise. For the purpose of evaluating an integral containing the factor (34) we need two different expansions. Accordingly, we partition the summation over n in Eq. (16) into two complementary sums, namely

$$\begin{aligned} &\textstyle\sum_n = \sum_{n\subset\Theta(m)} + \sum_{n\subset\Lambda(m)} \, , \\ &n\subset\Theta(m) \, , \; \varepsilon_{m,o} = \varepsilon_{n,o} \, , \\ &n\subset\Lambda(m) \, , \; \varepsilon_{m,o} \neq \varepsilon_{n,o} \end{aligned} \tag{35}$$

After some manipulation, as shown in detail elsewhere[2], we obtain for the scaled energy up to order λ

$$\begin{aligned}
\varepsilon_{m,o}\delta_{\alpha\beta} &= \langle\Psi_{m\alpha,o}|M|\Psi_{m\beta,o}\rangle \ , \\
\varepsilon_{m,1}\delta_{\alpha\beta} &= \Sigma_{q=0}^{2}\langle\Psi_{m\alpha,2-q}|M|\Psi_{m\beta,q}\rangle \\
&\quad + \Sigma_{q=0}^{1}\langle\Psi_{m\alpha,1-q}|P|\Psi_{m\beta,q}\rangle \\
&\quad + \langle\Psi_{m\alpha,o}|U|\Psi_{m\beta,o}\rangle \\
&\quad + \Sigma_{n\subset\theta(m)}\int d\vec{r}\,'\int d\vec{r}\,''R^{-1}F_{nm\alpha\beta,o}(\vec{r}\,',\vec{r}\,'') \ ; \\
\varepsilon_{m,2}\delta_{\alpha\beta} &= \Sigma_{q=0}^{4}\langle\Psi_{m\alpha,4-q}|M|\Psi_{m\beta,q}\rangle \\
&\quad + \Sigma_{q=0}^{3}\langle\Psi_{m\alpha,3-q}|P|\Psi_{m\beta,q}\rangle \\
&\quad + \Sigma_{q=0}^{2}\langle\Psi_{m\alpha,2-q}|U|\Psi_{m,\beta,q}\rangle \\
&\quad + \Sigma_{n\subset\theta(m)}\int d\vec{r}\,'\int d\vec{r}\,''R^{-1}F_{nm\alpha\beta,1}(\vec{r}\,',\vec{r}\,'') \\
&\quad - \frac{1}{2}\hbar^{-2}\Sigma_{n\subset\theta(m)}(\varepsilon_{m,1}-\varepsilon_{n,1})^{2}\int d\vec{r}\,'\int d\vec{r}\,''RF_{nm\alpha\beta,o}(\vec{r}\,',\vec{r}\,'') \\
&\quad - 4\pi\hbar^{2}\Sigma_{n\subset\Lambda(m)}(\varepsilon_{m,o}-\varepsilon_{n,o})^{-2}\int d\vec{r}\,'\int d\vec{r}\,''\delta(\vec{r}\,'-\vec{r}\,'') \\
&\qquad\qquad F_{nm\alpha\beta,o}(\vec{r}\,',\vec{r}\,'') \ ,
\end{aligned} \tag{36}$$

where we have used

$$\begin{aligned}
F_{nm\alpha\beta,p}(\vec{r}\,',\vec{r}\,'') = \frac{1}{2}\sum_{q=0}^{2p}\sum_{\mu}\sum_{\nu\neq\mu}\sum_{j}[&\rho_{\mu,m\alpha,n\gamma,2p-q}(\vec{r}\,')\rho_{\nu,n\gamma,m\beta,q}(\vec{r}\,'') \\
&-\vec{j}_{\mu,m\alpha,n\gamma,2p-q}(\vec{r}\,')\cdot\vec{j}_{\nu,n\gamma,m\beta,q}(\vec{r}\,'')] \ ,
\end{aligned} \tag{37}$$

$$\left.\begin{aligned}
\rho_{\mu,m\alpha,n\beta,p}(\vec{r}\,') &= \Sigma_{q=0}^{p}\langle\Psi_{m\alpha,p-q}|\rho_{\mu}(\vec{r}\,')|\Psi_{n\beta,q}\rangle \ , \\
\vec{j}_{\mu,m\alpha,n\beta,p}(r') &= \Sigma_{q=0}^{p}\langle\Psi_{m\alpha,p-q}|\vec{j}_{\mu}(\vec{r}\,')|\Psi_{n\beta,q}\rangle \ .
\end{aligned}\right\} \tag{38}$$

We apply the variation principle piecemeal to these expressions. In general, we expect to obtain an equation determining $\Psi_{m\alpha,p}$ from the variation of $\varepsilon_{m,p}$. In the case p=0 we immediately find

$$M\Psi_{m\alpha,o} = \varepsilon_{m,o}\Psi_{m\alpha,o} \tag{39}$$

In view of Eq. (20), $\varepsilon_{m,o}$ can be written in the form

$$\varepsilon_{m,o} = m_e(N-2h) \ , \ h = 0, 1, \ldots, N \ , \tag{40}$$

since the eigenvalues of β_μ are ±1. Here, h is the number of negative rest mass electrons, or "holes" in our usage. Note that the number of holes has nothing to do with space and spin behavior; the solutions of Eq. (39) have a degeneracy that encompasses all space or spin behavior.

Since β commutes with M, we can write

$$\beta\Psi_{m\alpha,o} = \tau_m\Psi_{m\alpha,o} \tag{41}$$

$$\tau_m = (-1)^h \ . \tag{42}$$

We call β the hole parity operator since its eigenvalues are +1 or -1 according to whether the number of holes is even or odd, respectively. We see from Eqs. (30) and (31) that the operator P flips hole parity while M and U preserve it. Arguments of this type can be used to establish the relation

$$\beta\Psi_{m\alpha,p} = \tau_m(-1)^p\Psi_{m\alpha,p} \ , \tag{43}$$

which is very useful in the systematic reduction of Eq. (36).

When this reduction is carried out, application of the variation principle to $\varepsilon_{m,1}$ yields

$$\Psi_{m\alpha,1} = K\Psi_{m\alpha,o} \ , \tag{44}$$

and

$$(T+U+V)\Psi_{m\alpha,o} = \varepsilon_{m,1}\Psi_{m\alpha,o} \ , \tag{45}$$

where

$$V = \frac{1}{2}\Sigma_\mu\Sigma_{\nu\neq\mu}e^2r_{\mu\nu}^{-1} \ . \tag{46}$$

Since T, U and V all commute with M, Eq. (45) does not conflict with Eq. (39); the effect of Eq. (45) is to remove the degeneracy due to space and spin in Eq. (39). For no-hole solutions, Eq. (45) is equivalent to the non-relativistic Schrödinger equation for N electrons: the non-vanishing components of the spinor $\Psi_{m\alpha,o}$ are proportional to the Schrödinger wave function, while their ratios represent the spin of the system.

We have to use Eq. (45) in order to reduce our expression for $\varepsilon_{m,2}$ in Eq. (36). After a considerable amount of manipulation, we find that application of the variation principle yields

$$\Psi_{m\alpha,2} = \frac{1}{2} K^2 \Psi_{m\alpha,o} + \Psi''_{m\alpha,o} , \tag{47}$$

where $\Psi''_{m\alpha,o}$ is any function with the same rest mass $\varepsilon_{m,o}$ as $\Psi_{m\alpha,o}$. One then obtains an expression for $\varepsilon_{m,2}$ entirely in terms of $\Psi_{m\alpha,o}$, namely

$$\varepsilon_{m,2}\,\delta_{\alpha\beta} = \langle \Psi_{m\alpha,o} | T' + U_{LS} + V_{LS} + B_{LL} + B_{SS} + B' | \Psi_{m\beta,o} \rangle , \tag{48}$$

where

$$T' = -(2m_e)^{-3} \Sigma_\mu \beta_\mu (\vec{p}_\mu \cdot \vec{p}_\mu) , \tag{49}$$

$$U_{LS} = \frac{1}{2} m_e^{-2} \sum_\mu \{ [\vec{\nabla}_\mu U(\vec{r}_\mu) \times \vec{P}_\mu] \cdot \vec{s}_\mu + \frac{1}{4} \hbar^2 (\vec{\nabla}_\mu \cdot \vec{\nabla}_\mu) U(\vec{r}_\mu) \} , \tag{50}$$

$$V_{LS} = -\frac{1}{2} e^2 m_e^{-2} \sum_\mu \sum_{\nu\neq\mu} [r_{\mu\nu}^{-3} \vec{\ell}_{\mu\nu} \cdot \vec{s}_\mu + \pi\hbar^2 \delta(\vec{r}_\mu - \vec{r}_\nu)] , \tag{51}$$

$$B_{LL} = -\frac{1}{2} e^2 m_e^{-2} \sum_\mu \sum_{\nu\neq\mu} \beta_\mu \beta_\nu [\vec{p}_\mu \cdot r_{\mu\nu}^{-1} \vec{p}_\nu + \frac{1}{2} r_{\mu\nu} \vec{\ell}_{\mu\nu} \cdot \vec{\ell}_{\nu\mu}] , \tag{52}$$

$$B_{LS} = -e^2 m_e^{-2} \sum_\mu \sum_{\nu\neq\mu} \beta_\mu \beta_\nu r_{\mu\nu}^{-3} \vec{\ell}_{\mu\nu} \cdot \vec{s}_\nu , \tag{53}$$

$$B_{SS} = -\frac{1}{2} e^2 m_e^{-2} \sum_\mu \sum_{\nu\neq\mu} \beta_\mu \beta_\nu [3 r_{\mu\nu}^{-5} (\vec{r}_{\mu\nu} \cdot \vec{s}_\mu)(\vec{r}_{\mu\nu} \cdot \vec{s}_\nu) - r_{\mu\nu}^{-3} (\vec{s}_\mu \cdot \vec{s}_\nu) + \frac{8\pi}{3} \delta(\vec{r}_\mu - \vec{r}_\nu)(\vec{s}_\mu \cdot \vec{s}_\nu)] , \tag{54}$$

$$B' = -\frac{1}{2} \pi\hbar e^2 m_e^{-2} \sum_\mu \sum_{\nu\neq\mu} \delta(\vec{r}_\mu - \vec{r}_\nu) \beta_\mu \beta_\nu \vec{\alpha}_\mu \cdot \vec{\alpha}_\nu , \tag{55}$$

with the abbreviation

$$\vec{\ell}_{\mu\nu} = (\vec{r}_\mu - \vec{r}_\nu) \times \vec{p}_\mu . \tag{56}$$

Eqs. (49) to (55) are the generalizations of the terms found, for example, by Bethe and Salpeter[3]. T' gives the relativistic mass correction, U_{LS} is the nuclear spin-orbit coupling, and we have written V_{LS} in a form that clearly shows it to be the analogue of U_{LS}, but with electrons replacing the nuclei as magnetic field sources. The next three terms, B_{LL}, B_{LS} and B_{SS} are usually called orbit-orbit, spin-other-orbit and spin-spin interactions, respectively. Finally, B' is a term that will not contribute in case $\Psi_{m\alpha,o}$ contains only positive rest masses, but may be of interest in other contexts.

IV. CONCLUSIONS

Apparently, our formalism provides the vehicle for a straightforward derivation of the many-electron Pauli approximation. The validity of our approach hinges on the fact that the interaction of electromagnetic particles, in classical as well as in quantum mechanics, must always be consistent with the classical Maxwell-Lorentz formalism for such a system. We can easily extend our theoretical framework in several directions. First, the calculation of relativistic effects of order (v/c), and even higher, is also straight-forward. Second, we can choose more general charge densities $\rho_\mu(\vec{r}')$ than Dirac delta functions; hence, we can accommodate particles with a more complicated internal structure. In a similar vein, the problems of particle self-energy can be re-examined by including the electromagnetic self-interaction terms, and choosing $\rho_\mu(\vec{r}')$ appropriately. We intend to explore these and similar questions in the future.

Another significant feature of our formalism is that all relativistic effects are calculated as expectation values of appropriate operators over Schrödinger wave functions. Thus, an accurate calculation of a relativistic effect requires no more and no less than a high-quality non-relativistic Schrödinger wave function. In many instances, this will of course entail a correlated wave function. We have thus clearly separated the difficulties associated with relativistic effects from those associated with correlation.

We also want to emphasize the practical importance of solving formally for the higher order wavefunctions $\Psi_{m\alpha,p}$, p=1,2,... in terms of the zero-order (Schrödinger) function $\Psi_{m\alpha,o}$. In this context it should be noted that in the commonly used Hartree-Fock-Dirac scheme the large and small components of the individual orbitals, which in essence constitute $\Psi_{m\alpha,o}$ and $\Psi_{m\alpha,1}$, are co-determined numerically in coupled eigenvalue problems. The calculation of energies is extremely sensitive to the balance between these large and small components; this puts a heavy burden on the numerical calculation. Our formal solution of the small components in terms of the large components guarantees that this balance is always correct.

Another practical concern is the feasibility of molecular as well as atomic calculations. In a many-electron molecule the use of orbitals expanded in a basis set is virtually mandatory. Now, in a Hartree-Fock-Dirac type calculation, using expanded orbitals, the correct balance between large and small components can be achieved only by choosing separate but properly "matched" basis sets for the large and small components. This obviously renders the calculation complicated and expensive. In contrast, our present approach needs no other basis set than what is required for a good non-relativistic calculation.

REFERENCES

1. J.A. Wheeler and R.P. Feynman, Revs. Mod. Phys. 17, 157 (1945); ibid. 21, 245 (1949).
2. J.H. Detrich and C.C.J. Roothaan, Chapter 20, The Uncertainty Principle and Foundations of Quantum Mechanics, W.C. Price and S.S. Chissick, Eds. (Wiley-Interscience, London, 1977).
3. H.A. Bethe and E.E. Salpeter, Quantum Mechanics of One- and Two-Electron Atoms (Springer-Verlag, Berlin, 1957), P. 181.

RELATIVISTIC SELF-CONSISTENT-FIELD THEORY FOR MOLECULES*

G. L. Malli

Department of Chemistry and Theoretical Sciences
Institute, Simon Fraser University,
Burnaby, British Columbia, Canada V5A 1S6

INTRODUCTION

The well-known Hartree-Fock[1] scheme has been used for decades to understand the electronic structure of atoms and molecules. However, this scheme is based upon the Schrödinger equation which is not Lorentz invariant; i.e., it does not obey the principle of relativity and, therefore, it is called the non-relativistic Hartree-Fock (NRHF) scheme. Furthermore, the use of spin functions is purely formal in this scheme. More than half a century ago, the relativistic quantum mechanics for an electron was developed by Dirac[2] and it differs from the Schrödinger theory by the fact that the wavefunction for an electron is a four-component spinor and that the spin of the electron is built into the theory from the beginning.

During the last decade, there has been tremendous activity in the relativistic treatment of many-electron systems and, in particular, the effect of relativity on the electronic structure of atoms, molecules and solids has been investigated by many workers. In these lectures, I shall present the work of my co-workers and myself on the relativistic Hartree-Fock-Roothaan (RHFR) or the Dirac-Hartree-Fock-Roothaan (DHFR) method for molecules.

* This work was supported, in part, through a grant (No. A3598) by the Natural Science and Engineering Research Council (NSERC) of Canada.

In Section I we shall briefly discuss Dirac's equation[2] for an electron and its extension to N-electron molecular systems. In addition, we shall also present a notation to be used later on in these lectures. Sections II and III will describe the DHFR SCF theory for closed-shell and open-shell molecules, respectively, while, in Section IV, we shall present the multi-configuration (MC) DHFR SCF theory for molecular systems and Section V will present a critique of some recent DHFR SCF calculations on molecules. I should mention that Sucher[3], Grant[4] and Desclaux[5] have discussed the relativistic theory for many-electron systems - especially for atoms - in these proceedings and their lectures should be consulted for further details on this topic.

I. RELATIVISTIC THEORY

A. Dirac's Equation

In 1928, Dirac[2] proposed a relativistic (Lorentz covariant) equation for an electron which, in the central field of a nucleus with Charge Z, can be written as follows (in atomic units (a.u.)):

$$H_D\psi = E\psi, \tag{1}$$

where the Dirac Hamiltonian, H_D, is of the form:

$$H_D = c\alpha\cdot p + c^2\beta - Z/r, \tag{2}$$

$$\text{where } \alpha = \begin{pmatrix} 0 & \sigma^P \\ \sigma^P & 0 \end{pmatrix}, \ \beta = \begin{pmatrix} I & 0 \\ 0 & -I \end{pmatrix}. \tag{3}$$

In Eq. (2) c is the velocity of light, r is the electron-nucleus distance (point nucleus is used, although the finite size of 'nucleus' presents no difficulties) and p is the linear momentum operator of the electron. The α and β are the Dirac 4×4 matrices, which can be written in the standard representation in terms of 2×2 Pauli matrices, σ^P, and 2×2 unit matrix, I, as in Eq. (3).

The eigensolutions, ψ, of Eq. (1) are four-component spinors and these can be chosen as eigenfunctions of the set of commuting operators: H_D, J^2, the total angular momentum squared, J_z, the projection of J on the z-axis and $K=\beta(1+\sigma^D\cdot L)$, where L is the orbital angular momentum operator and σ^D is the Dirac spin-matrix operator of the form:

$$\sigma^D = \begin{pmatrix} \sigma^P & 0 \\ 0 & \sigma^P \end{pmatrix} \tag{4}$$

The eigenvalues of these commuting operators are E, j(j+1), m and κ, respectively, where $j=|\kappa|-1/2$, $-j\leq m\leq j$ and κ is an integer (nonzero), while j and m are half-integers. The four-component eigenspinors, $\psi_{n\kappa m}$, can be written as:

$$\psi_{n\kappa m}(r,\theta,\phi) = \begin{pmatrix} r^{-1} P_{n\kappa}(r)\ \chi_{\kappa m}(\theta,\phi) \\ \\ ir^{-1} Q_{n\kappa}(r)\ \chi_{-\kappa m}(\theta,\phi) \end{pmatrix} \tag{5}$$

The $r^{-1}P_{n\kappa}$ and $r^{-1}Q_{n\kappa}$ are the large and small component radial functions and the $\chi_{\kappa m}(\theta,\phi)$, the two-component spinors, have the form:

$$\chi_{\kappa m} = \sum_{\sigma=\pm 1/2} C(\lambda\, 1/2\, j; m-\sigma, \sigma) Y_{\lambda, m-\sigma}(\theta,\phi)\phi_{\sigma}, \tag{6}$$

where, in Eq. (6), $C(\lambda 1/2 j;m-\sigma,\sigma)$ are the Clebsh-Gordan (CG) coefficients, the $Y_{\lambda,m-\sigma}(\theta,\phi)$ are the normalized spherical harmonics,

$$\phi_{1/2} = \begin{pmatrix} 1 \\ 0 \end{pmatrix}, \quad \phi_{-1/2} = \begin{pmatrix} 0 \\ 1 \end{pmatrix}$$

and $\lambda=|\kappa+1/2|-1/2$. We shall use $\lambda=j+1/2\ \beta\ \mathrm{sgn}\ \kappa$, where $\beta=+1$ for the upper (large) component and $\beta=-1$ for the lower (small) component, and $\mathrm{sgn}\ \kappa=+1$, if κ is positive, and $\mathrm{sgn}\ \kappa=-1$, if κ is negative. Thus, $\chi_{\pm\kappa,m}$ will be denoted by $\chi_{\kappa,m,\pm 1}$ or, in general, by $\chi_{\kappa,m,\beta}$. Similarly, $R_{n\kappa\beta}\equiv R_{n,\kappa,\beta}$ will denote both the large and small component radial functions; i.e., $R_{n,\kappa,+1}=r^{-1}P_{n\kappa}$ and $R_{n,\kappa,-1}=r^{-1}Q_{n\kappa}$ and, therefore, one can write the spinor, $\psi_{n,\kappa,m}$, in this notation as follows:

$$\psi_{n\kappa m} = \begin{pmatrix} R_{n,\kappa,+1}\ \chi_{\kappa,m,+1} \\ \\ iR_{n,\kappa,-1}\ \chi_{\kappa,m,-1} \end{pmatrix} \tag{7}$$

This presents a very brief treatment of Dirac's equation. The lectures by Sucher[3], Grant[4] and Desclaux[5], <u>in these proceedings</u>, should be consulted for further details.

B. Relativistic Many-Electron Systems

It is well known that Dirac's theory accounts for many experimental results for one-electron atomic systems, apart from quantum electrodynamical (QED) corrections, and that the spin of the electron is built into the theory from the beginning. One may, therefore, believe that the Dirac theory is the starting point for a relativistic treatment of many-electron systems. However, it is not possible to generalize the Dirac theory to many-electron systems. Thus, a fully Lorentz invariant Hamiltonian for many-electron systems cannot be obtained in a closed-form because the complete electron-electron interaction involves, in addition to the instantaneous Coulomb interaction (r_{ij}^{-1}, which is not Lorentz invariant), the exchange of virtual photons between the interacting electrons. However, Breit[6] proposed H_{Br} as a correction to the Coulomb interaction, r_{12}^{-1} for two electrons; viz.,

$$H_{Br}(12) = -\left[\frac{(\alpha_1\cdot\alpha_2)}{2r_{12}} + \frac{(\alpha_1\cdot r_{12})(\alpha_2\cdot r_{12})}{2r_{12}^3}\right] \quad (8)$$

H_{Br} takes into account the magnetic and retardation terms and is of order $(Z/c)^2$, relative to the Coulomb interaction r_{12}^{-1}. Thus, it should be used as the first-order perturbation to the zeroth order Hamiltonian, consisting of the Dirac Hamiltonians for the two electrons plus the inter-electron Coulomb repulsion term. This latter Hamiltonian is the so-called Dirac-Coulomb-Hamiltonian (H_{DC}) for two-electrons:

$$H_{DC} = H_D(1) + H_D(2) + r_{12}^{-1}, \text{ where}$$

$$H_D(\mu) = c\alpha_\mu\cdot p_\mu + c^2\beta_\mu - Z/r_\mu \quad (9)$$

It should be remarked that the rest mass energy of the electron is included in Eq. (9) and this is usually subtracted from H_D to obtain the binding energy of an electron. This is generally achieved by using β' instead of β in Eq. (9), where

$$\beta' = \begin{pmatrix} 0 & 0 \\ 0 & -2I \end{pmatrix}$$

In 1935, Swirles set up a relativistic self-consistent-field scheme (SCF) for closed-shell many-electron atoms and she derived Dirac-Fock (DF) equations, using the generalized H_{DC} as the Hamiltonian H for N-electrons:

$$H(1,\ldots N) = \sum_{\mu=1}^{N} H_D(\mu) + \frac{1}{2}\sum_{\mu\neq\nu} r_{\mu\nu}^{-1} \quad (10)$$

However, Brown and Ravenhal[8] were the first to point out that, as a result of the existence of negative energy states in the Dirac equation, bound-state eigenfunctions do not exist for H_{DC} (Eq. 9) or its generalization H of Eq. (10). Their argument was as follows:

for two electrons starting with the solutions of $H_o=H_D(1)+H_D(2)$, when the inter-electron interaction r_{12}^{-1} is 'switched on', slowly, the system can make transitions to states where one electron has a large negative energy and the other electron is in the positive-energy continuum and, thus, H_{DC} has no stationary solutions. Brown and Ravenhal used 'projection operators' to project out the negative energy solutions and derived an equation for two-electrons, involving projection operators, which led to the momentum transform of the Breit interaction after appropriate approximations. However, this paper was ignored for almost twenty years!

The relativistic SCF theory for atoms was reformulated by Grant[4,9] in 1961 using, extensively, the tensor operator algebra of Racah and has been used to calculate wavefunctions and energies of ground states of atoms via the numerical integration[5,10,11] methods. Mittleman[12] and Sucher[3,13], using quantum electrodynamics (QED), have derived the correct form of H such that the negative-energy states will be decoupled from the positive-energy states and their papers should be consulted for details. A brief outline of the method of Sucher[3,13] is given below. For N-electron atomic systems, the so-called 'Dirac-Fock' Hamiltonian, following Sucher, can be written as follows:

$$H_+ = \sum_{i=1}^{N} H_D(i) + L_+ \sum_{i<j} r_{ij}^{-1} L_+ \tag{11}$$

$$\text{where } L_+ = \prod_{i=1}^{N} L_+(i), \tag{12}$$

$$\text{and } L_+(i) = \sum_n U_n(i)\rangle\langle U_n(i) \tag{13}$$

In Eq. (11), L_+ is the projection operator spanned by the product of the postive-energy eigenfunctions $U_n(i)$ of H_o, where

$$H_o = \sum_i H_D(i)$$

Sucher[3,13] has shown that H_+ is free from the defects of H_{DC} and that the bound state eigenfunctions, ψ, are those which simultaneously satisfy

$$H_+\psi = E\psi \tag{14a}$$

$$\text{and } L_+(i)\psi = \psi (i=1,2,\ldots N) \tag{14b}$$

It is obvious that H_+ depends, through L_+, upon the choice made for H_o. Mittleman[12] has shown that, for a single Slater determinant (SD) wavefunction, H_o should be the (Dirac) Hartree-Fock operator. So far, no such calculations have been reported for many-electron atoms.

Sucher[3,13] has rederived relativistic (Dirac) Hartree-Fock (DHF) equations, starting with H_+, subject to the constraints given in Eq. (14), in addition to the orthonormal constraints of the positive-energy Dirac spinors; i.e.,

$$\langle\psi_i|\psi_j\rangle = \delta(i,j)$$
$$\text{and } L_+(1)\psi_i(1) = \psi_i(1) \tag{15}$$

In these DHF equations, there are projection operators, $L_+(1)$, standing on the left and right of all the inter-electron Coulomb terms, but normalizable solutions can exist for these DHF equations, even if the projectors $L_+(i)$ are omitted.

II. RELATIVISTIC SCF THEORY FOR CLOSED-SHELL MOLECULES

A. Dirac-Fock SCF Equations

The non-relativistic Hartree-Fock (NRHF) equations for molecules are well known[14]; however, these have not been solved exactly, as in the case of atomic systems, by the numerical integration methods. The expansion method of Roothaan[14] has been used almost exclusively for molecules without the inclusion of the relativistic effects. It should be remarked that the Dirac equation for the H_2^+ molecular ion cannot be solved exactly and approximate schemes for the relativistic treatment of molecules should be developed.

A relativistic (Dirac) Hartree-Fock-Roothaan (RHFR or DHFR) method was developed by Malli and Oreg[15] for closed-shell molecules for which the wavefunction consits of a single antisymmetrized product (AP) or a single Slater determinant (SD) of four-component molecular spinors (MS), where each MS is expressed as a linear combination of basis atomic spinors (LCAS/MS), analogous to the well known LCAO/MO scheme for molecules[14].

It was shown[15] that, in this scheme, the relativistic Hartree-Fock-Roothaan (RHFR) SCF calculations for any molecular system can be reduced essentially to the calculation of the non-relativistic-type integrals which can be evaluated by well-known techniques which have been developed over many years.

The 'relativistic' Hamiltonian H for a molecular system of n nuclei and N electrons in Born-Oppenheimer (BO) approximation (omitting the internuclear repulsion terms which are constant for a given molecular geometry) will be taken in atomic units (a.u.) as

$$H = \sum_{\mu=1}^{N} H(\mu) + \frac{1}{2} \sum_{\mu \neq \nu} r_{\mu\nu}^{-1} , \tag{16}$$

where $H(\mu)$ is the Dirac Hamiltonian for the μth electron in the potential of all the nuclei of the molecule; viz.,

$$H(\mu) = c\alpha_\mu \cdot p_\mu + c^2\beta_\mu + \sum_{a=1}^{n} \frac{Z_a}{r_{a\mu}} \tag{17}$$

The second term in Eq. (16) represents the inter-electron Coulomb repulsion terms. The Breit-Brown[15] magnetic interaction terms H' can be used as perturbation to H, if necessary, where

$$H' = \frac{1}{2} \sum_{\mu \neq \nu} \frac{\alpha_\mu \cdot \alpha_\nu}{r_{\mu\nu}} = \frac{1}{2} \sum_{\mu \neq \nu} M_{\mu\nu} . \tag{18}$$

In Eq. (17) Z_a is the nuclear charge on the ath nucleus, while $r_{a\mu}$ is the distance between the ath nucleus and the μth electron (point nuclei are assumed for simplicity) and α_μ and β_μ are the well-known Dirac matrices for the μth electron, as defined in Eq. (3), and the sum over 'a' runs over all nuclei. It should be pointed out that H is the generalized H_{DC} for molecules and does not involve the projection operators L_+; i.e., L_+ is set equal to unity, as discussed in Section I.

The N-electron wavefunction, Φ, for the closed-shell molecular system is taken as a single Slater determinant (SD) of one-electron, four-component molecular spinors, Ψ_k; i.e.,

$$\Phi = (N!)^{-1/2} \left| \Psi_1(1) \ldots \Psi_N(N) \right| \tag{19}$$

In. Eq. (19), each $\Psi_k(\mu)$ is taken to transform like the additional irreducible representations (AIR) of the double group G' of the molecular system under investigation and $\Psi_k(\mu)$ can be regarded as the relativistic analogue of the non-relativistic molecular spin-orbital. Each $\Psi_k(\mu)$ is of the form:

$$\Psi_k(\mu) = \begin{pmatrix} \Psi_{k1}(\mu) \\ \Psi_{k2}(\mu) \\ \Psi_{k3}(\mu) \\ \Psi_{k4}(\mu) \end{pmatrix} , \tag{20}$$

where each component, $\Psi_{k\rho}(\mu)$ (ρ = 1 to 4), is a function of the space coordinates of the μth electron and the Ψ_k are taken to form an orthonormal set; i.e.,

$$\langle\Psi_k|\Psi_\ell\rangle = \int \Psi_k{}^\dagger\Psi_\ell d\tau = \delta(k,\ell), \tag{21}$$

where $\Psi_k{}^\dagger$ is the Hermitian conjugate of Ψ_k.

It follows, from Eq. (21), that

$$\langle\Phi|\Phi\rangle = 1 \tag{22}$$

Further, using Eq. (22), the expectation value, E, of H can be written as

$$E = \sum_i H_i + \frac{1}{2}\sum_{ij}(J_{ij}-K_{ij}) \tag{23}$$

The summation in Eq. (23) runs over all occupied one-electron states; i.e., molecular spinors and i,j, etc., denote the symmetry species, the subspecies and an index, or set of indices, characterising molecular spinor, using the theory of double-groups as mentioned earlier.

The various matrix elements in Eq. (23) can be defined over molecular spinors Ψ_k, in analogous manner, to the corresponding matrix elements over molecular spinorbitals[14]; viz.,

$$H_i = \langle\Psi_i|H|\Psi_i\rangle = \int \Psi_i{}^\dagger(\mu)H(\mu)\Psi_i(\mu)d\tau_\mu \tag{24}$$

$$J_{ij} = \langle\Psi_i|J_j|\Psi_i\rangle = \langle\Psi_j|J_i|\Psi_j\rangle \tag{25}$$

$$K_{ij} = \langle\Psi_i|K_j|\Psi_i\rangle = \langle\Psi_j|K_i|\Psi_j\rangle \tag{26}$$

The Coulomb and exchange operators J_j and K_j are defined by the relations given below[14]:

$$J_j^\mu\Psi^\mu = \int\left[\Psi_j{}^{\nu\dagger}\Psi_j{}^\nu r_{\mu\nu}{}^{-1}\,d\tau_\nu\right]\Psi^\mu \tag{27}$$

$$K_j^\mu\Psi^\mu = \int\left[\Psi_j{}^{\nu\dagger}\Psi^\nu r_{\mu\nu}{}^{-1}\,d\tau_\nu\right]\Psi_j{}^\mu \tag{28}$$

A straightforward application of the variational principle to E in Eq. (23), subject to orthonormality constraints given by Eq. (21), leads to the following set of equations for the best molecular spinors which minimize the energy E:

$$F\Psi_k = \varepsilon_k \Psi_k, \; k = 1,2,\ldots N \;, \tag{29}$$

where the DHF operator F has the form:

$$F = H+G = H+\sum_i (J_i - K_i) \tag{30}$$

The equations given in Eq. (29) are the (relativistic) Dirac-Hartree-Fock (DHF) equations for closed-shell molecular systems and these are generally solved by a self-consistent-field (SCF) procedure. This technique has been applied with great succes to atoms, as described in detail by Grant[4,9] and Desclaux[5,10].

Due to the central symmetry of an atom, the solution of the Dirac-Hartree-Fock equations is greatly simplified for atoms. However, for molecules, even the non-relativistic Hartree-Fock equations have remained, so far, too difficult to solve and, as the DHF equations for molecules are a great deal more complicated, there is little hope, at present, to solve the DHF equations for molecules directly by the numerical integration method. However, the DHF equations, given in Eq. (29), can be solved for molecules by the application of the Roothaan's expansion method[14]. In this scheme, the molecular spinors, Ψ_k, are written as linear combination of basis atomic spinors, ψ_p, centred on various atoms of the molecular system; i.e.,

$$\Psi_k = \sum_{p=1}^{m} \psi_p C_{pk} \;, \tag{31}$$

where ψ_p is a four-component basis atomic spinor (BAS), the summation over p runs over all the m members of the basis set chosen and C_{pk} are the expansion coefficients. The orthonormal constraints of Eq. (21) can then be written as:

$$\sum_{pq} C^*_{pk} S_{pq} C_{q\ell} = \delta(k,\ell) \;, \tag{32}$$

where the overlap matrix element, $S_{pq} = \langle\psi_p|\psi_q\rangle$.

Then, by subsituting Eq. (31) in Eq. (23), the expectation value, E, can be written as follows:

$$E = \sum_{pq} H_{pq} D_{pq} + \frac{1}{2} \sum_{pq} \sum_{rs} D_{pq} P_{pqrs} D_{rs} \;, \tag{33}$$

where, in Eq. (33),

$$H_{pq} = \langle\psi_p|H(\mu)|\psi_q\rangle \tag{34}$$

$$D_{pq} = \sum_k C^*_{pk} C_{qk} \tag{35}$$

$$P_{pqrs} = J_{pqrs} - K_{pqrs} , \tag{36}$$

where, in Eq. (36),

$$J_{pqrs} = \langle\psi_p(1)\psi_r(2)| r_{12}^{-1} |\psi_q(1)\psi_s(2)\rangle \tag{37}$$

$$K_{pqrs} = \langle\psi_p(1)\psi_r(2)| r_{12}^{-1} |\psi_s(1)\psi_q(2)\rangle \tag{38}$$

The equations (32) and (33) can also be rewritten in the matrix notation of Roothaan; viz.,

$$E = H^+ D + \frac{1}{2} D^+ PD \tag{39}$$

$$C_k^+ SC_\ell = \delta(k,\ell) \tag{40}$$

The application of the variational treatment to E in Eq. (39), subject to the constraints given in Eq. (40) leads to the relativistic or Dirac Hartree-Fock-Roothaan (DHFR) equations for the best set of molecular spinors expressed as LCAS; viz.,

$$FC_k = \varepsilon_k SC_k , \tag{41}$$

$$F = H+PD , \tag{42}$$

where H, P and D are defined in equations (34) to (36). The set of DHFR equations are solved via a SCF procedure and these equations hold good for any closed-shell molecular system. But, in order to solve these equations, one must evaluate the various matrix elements H_{pq}, etc., with a given basis set. Once it can be shown that it is possible to do so, the DHFR equations can be solved for any closed-shell molecular system. Therefore, we next consider the evaluation of the various relativistic matrix elements.

B. Relativistic Integrals in Terms of Non-Relativistic Integrals for any Molecular System

There are two categories of the relativistic integrals for any molecular system; viz.,

a) One-Electron Integrals: The matrix elements of H_{pq} and S_{pq} give rise to such integrals which, in general, can be written as

$$\langle \psi_A{}^a | O(\mu) | \psi_B{}^b \rangle = \int \psi_A{}^{a\dagger}(\mu)\; O(\mu)\; \psi_B{}^b(\mu) d\tau_\mu \tag{43}$$

In Eq. (43) $O(\mu)$ is a one-electron operator, $H(\mu)$, for H_{pq} and unity for S_{pq}. $\psi_A{}^a(\mu)$ is the basis atomic spinor centred on nucleus 'a' with a set of quantum numbers, denoted by 'A'; i.e., $(n_A \kappa_A m_A)$, and is a function of the coordinates of the μth electron. As $H(\mu)$ involves the nucleus-attraction term, (Z/r), the most complex integral arising from the one-electron matrix elements will be the three-centre-integral; viz.,

$$\langle \psi_A{}^a \Big| \frac{1}{r_c} \Big| \psi_B{}^b \rangle = \int \psi_A{}^{a\dagger}(\mu)\; \frac{1}{r_{c\mu}}\; \psi_B{}^b(\mu) d\tau_\mu \;. \tag{44}$$

Of course, there will arise, in addition, one- and two-centre, one-electron integrals, in general, for any molecule.

b) Two-Electron Integrals: These are the most difficult integrals and occur in J_{pqrs} and K_{pqrs}. A general integral in this category is of the form:

$$\langle \psi_A^a(1) \psi_B^b(2) | r_{12}{}^{-1} | \psi_C^c\,(1) \psi_D^d(2) \rangle \tag{45}$$

This, at most, can be a four-centre, two-electron integral if a, b, c and d stand for four different atoms; its special cases are three-, two- and one-centre, two-electron integrals.

It is obvious that, if we consider only diatomics, the most complex integrals in both the categories (a) and (b) would be, at most, two-centre integrals.

Now, we shall consider the integrals given in equations (44) and (45) and show that these integrals can be expressed in terms of non-relativistic type integrals.

Let us first consider the three-centre integral of Eq. (44). If we substitute Eqs. (5) to (7) in Eq. (44), then it can be shown that

$$\langle \psi_A^a \Big| \frac{1}{r_c} \Big| \psi_B^b \rangle = \sum_{\sigma_A \sigma_B} \sum_{\beta_A \beta_B} \delta(\sigma_A, \sigma_B)\; \delta(\beta_A, \beta_B)$$

$$\times C(\lambda_A\, 1/2 j_A; m_A-\sigma_A, \sigma_A)\ C(\lambda_B\, 1/2 j_B; m_B-\sigma_B, \sigma_B)$$

$$\times \langle Y_{\lambda_A, m_A-\sigma_A}(\theta_a, \phi_a) R_A(r_a) \left| \frac{1}{r_c} \right| Y_{\lambda_B, m_B-\sigma_B}(\theta_b, \phi_b) R_B(r_b) \rangle \quad (46)$$

Thus, the relativistic three-centre nuclear attraction integral of Eq. (44) has been expressed in terms of the corresponding non-relativistic integral which has been treated by many workers.

Next, we consider the four-centre, two-electron integral, given in Eq. (45). However, we shall consider the Breit-Brown operator, M_{12}, defined in Eq. (18), instead of the Coulomb interaction operator (r_{12}^{-1}), as M_{12} reduces to r_{12}^{-1} if $\alpha_1 \cdot \alpha_1 = 1$.

Furthermore, integrals involving the operator M_{12} will be needed if H' is treated as perturbation to H. We shall denote the integrals involving r_{12}^{-1} and M_{12} as

$$C_{ABCD}^{abcd} \text{ and } M_{ABCD}^{abcd},$$

respectively; i.e.,

$$C_{ABCD}^{abcd} = \langle \psi_A^a(1) \psi_B^b(2) | r_{12}^{-1} | \psi_C^c(1) \psi_D^d(2) \rangle \quad (47)$$

$$M_{ABCD}^{abcd} = \langle \psi_A^a(1) \psi_B^b(2) | M_{12} | \psi_C^c(1) \psi_D^d(2) \rangle \quad (48)$$

Using Eqs. (3-7), and after some algebra, one can write:

$$M_{ABCD}^{abcd} = \sum_{\beta_A \beta_B} \sum_{\beta_C \beta_D} \sum_{\sigma_A \sigma_B} \sum_{\sigma_C \sigma_D} 2\delta(\sigma_A+\sigma_B, \sigma_C+\sigma_D)\delta(\beta_A, -\beta_C)\delta(\beta_B, -\beta_D)$$

$$\times (-1)^{(1/2)(\beta_A+\beta_B) + 1-\sigma_A-\sigma_B}$$

$$\times C(1/2\ 1/2\ 1;\ \sigma_C, -\sigma_A)\ C(1/2\ 1/2\ 1;\ \sigma_B, -\sigma_D)$$

$$\times \prod_{i=A,B,C,D} C(\lambda_i\ 1/2\ j_i; m_i-\sigma_i, \sigma_i)\ I_{ABCD}^{abcd}, \quad (49)$$

where I_{abcd}^{abcd} is the non-relativistic type integral:

$$I_{ABCD}^{abcd} = \left\langle R_A^a(1)Y^a_{\lambda_A, m_A-\sigma_A}(1)R_B^b(2)Y^b_{\lambda_B, m_B-\sigma_B}(2) \left|\frac{1}{r_{12}}\right| R_C^c(1)Y^c_{\lambda_C,m_C-\sigma_C}(1)R_D^d(2)Y^d_{\lambda_D,m_D-\sigma_D}(2)\right\rangle \tag{50}$$

Similarly, one can write for C_{ABCD}^{abcd} the following expression:

$$C_{ABCD}^{abcd} = \sum_{\beta_A\beta_B} \sum_{\beta_C\beta_D} \sum_{\sigma_A\sigma_B} \sum_{\sigma_C\sigma_D} \delta(\beta_C,\beta_A)\delta(\beta_D,\beta_B)\delta(\sigma_C,\sigma_A)\delta(\sigma_D,\sigma_B) \prod_{i=A,B,C,D} C(\lambda_i 1/2 j_i;m_i-\sigma_i,\sigma_i)\, I_{ABCD}^{abcd} \tag{51}$$

Thus, we have shown that it is possible to express all the relativistic matrix elements occurring in our DHFR (LCAS/MS HF SCF) scheme for any molecular system in terms of the corresponding non-relativistic type multi-centre integrals. We have used a coordinate system in which all the corresponding axes of all the centres point in the same direction; such a system is known as the parallel system of coordinates[15]. Furthermore, it should be pointed that the results in Eqs. (46-51) are quite general and, for the evaluation of I_{ABCD}^{abcd}, any convenient basis set for the radial components, R_A^a, etc., can be used.

C. Choice of the Basis Set for Relativistic Molecular Calculations

In Section IIB we have expressed the molecular relativistic matrix elements in terms of the non-relativistic type matrix elements and, in order to evaluate the relativistic integrals, one must consider the question of the choice of the basis set to be used for the radial functions $R_{n\kappa\beta}$ because the method and feasibility of the evaluation of relativistic integrals will depend upon the basis set chosen.

One may, of course, use the radial functions of the exact solutions of the Dirac equation for an electron in a central field; i.e., the confluent hypergeometric functions (CHGF) which can be expressed in terms of Laguerre polynomials. However, the multi-centre integrals involving CHGF's are expected to be extremely complicated and not much work has been done on this subject. Since multi-centre integrals involving Slater-type-functions (STF) and Gaussian-type-functions (GTF) have been investigated extensively for many years in non-relativistic molecular calculations, it is reasonable to use STF's or GTF's for the expansion of $R_{n\kappa\beta}$. It should be remarked that Kim[16] and Kagawa[17] used STF's with non-integer principal quantum number n for expanding $R_{n\kappa\beta}$ in their DHFR schemes for atoms; however, Malli[18] used GTF's with integer n for atoms and proposed their use for relativistic molecular calculations. It should be pointed out that Kim[16] and Kagawa[17] used non-integer n STF's for atoms so as to satisfy the cusp at the point nucleus. However, for a finite nucleus (and all nuclei are finite; in particular, those of heavy atoms in which relativistic effects are very significant) it is not necessary to use non-integer n. Furthermore, for molecules, the cusp conditions are not crucial and, thus, one may use either integer n STF's[15] or GTF's[18]. For diatomics and simple linear molecules STF's have been used but GTF's are usually preferred for a molecule in general for non-relativistic calculations.

The expansion of $R_{n\kappa\beta}$ in terms of STF's can be written as follows:

$$R_{n\kappa\beta} = \frac{1}{r} \sum_p \xi_{pn\kappa\beta} f_{\kappa p} , \qquad (52a)$$

where $\xi_{pn\kappa\beta}$ are the coefficients of expansion and $f_{\kappa i}$ are the Slater-type radial functions:

$$f_{\kappa i} = (2\zeta_{\kappa i})^{n_{\kappa i}+1/2} [2n_{\kappa i}!]^{-1/2} r^{n_{\kappa i}} \exp(-\zeta_{\kappa i} r) , \qquad (52b)$$

where $n_{\kappa i}$ is taken to be integer.

The evaluation of all the relativistic matrix elements for diatomics using STF's as basis set has been treated in Malli and Oreg[15], which should be consulted for details.

Multi-centre integrals involving spherical gaussians have been treated in the literature[19-22]; however, most NRHF calculations on molecules have been carried out using Cartesian gaussians. Aoyama[23] et al. have developed a computer program for relativistic calculations on molecules based upon the integral analysis of Taketa[24] et al. for cartesian gaussians. They computed the non-redundant integrals involving cartesian gaussians and transformed them directly to integrals over basis spinors, thus bypassing the transformation from the cartesian to spherical gaussians.

III. RELATIVISTIC THEORY FOR OPEN-SHELL MOLECULES

A. Open-Shell Formalism

We shall consider open-shell formalism for simple rotation double groups (SRDG), including the finite double groups of the linear molecules; i.e., for C_n, C_{nv}, D_n, D_{nh}, D_{nd}, C_{nh}, S_n, $D_{\infty h}$, and $C_{\infty v}$ double groups for any value of n which cover a large number of molecules. It has been shown[25,26] that the additional irreducible representations (IR) of SRDG's are, at most, two-dimensional and thus their open-shells will be singly occupied. In addition, there are very useful relations[27] between the electron-repulsion interaction integrals over the symmetry spinors of the SRDG's arising in the DHFR theory. We shall give here a summary of the relativistic open-shell theory presented by us[28]. We shall consider open-shells which satisfy the following conditions:

(i) The total wavefunction Φ will, in general, be taken as a sum of several Slater determinants (SD) where each SD will consist of a core of doubly occupied closed-shell $\{\Psi_c\}$ and singly occupied open-shells $\{\Psi_o\}$ and the set of occupied molecular spinors is assumed to be orthonormal.

(ii) Only one open-shell for a given symmetry is allowed.

(iii) In the direct product of the additional IR's of two or more open-shells, a given IR will not occur more than once.

It should be pointed out that this formalism[28] is also applicable to singly occupied two-dimensional open-shells of the non-SRDG's; e.g., tetrahedral, octahedral and icosahedral double groups as long as the conditions (i) to (iii) are satisfied.

The total energy E will be written in the same form as given in Roothaan and Bagus[29] for the non-relativistic open-shell theory; viz.

$$E = H^{\dagger}D_T + \frac{1}{2}(D_T{}^{\dagger}PD_T - D_O{}^{\dagger}QD_O) \tag{53}$$

The scalar products between the supervectors and supermatrices in Eq. (53) are defined as in Wahl et al.[30], except that H(u) given in Eq. (17) should replace its non-relativistic counterpart given in Wahl et al.

The various matrix elements in Eq. (53) are defined[28] elsewhere, where the coefficients for the states arising from two open-shells for linear molecules are also given.

The application of the variational principle to E of Eq. (53) subject to orthonormality constraints; viz.

$$C^{\dagger}_{i\lambda}\, S_{\lambda}\, C_{j\lambda} = \delta(i,j)$$

leads to the following pseudo-eigenvalue equations for closed and open-shells as in the non-relativistic treatment[29]:

$$F_c C = \varepsilon \, SC \; , \tag{54}$$

$$F_o C_o = \varepsilon \, SC_o \; , \tag{55}$$

where the Fock operators for closed-shell F_c and F_o are defined below:

$$F_c = G + P + R_o \; , \tag{56}$$

$$F_o = H + P - Q + R_c \; , \tag{57}$$

The coupling operators R_c and R_o are defined in Roothaan and Bagus[29].

IV. RELATIVISTIC MULTICONFIGURATION SCF THEORY FOR MOLECULES

A. Introductory Remarks

The relativistic treatment for molecules presented so far is limited to a 'single configuration (SC)' which suffers from two major defects; viz., (i) electron-correlation effects are not included, (ii) a SC wavefunction will not lead to proper dissociation limits in all cases. In addition, there is a large amount of experimental data available on the binding energies and excitation spectra of molecular complexes and clusters of heavy atoms, etc., which may be explained using the relativistic MCSCF treatment for these molecular systems.

A Dirac-Fock (DF) multiconfiguration (MC) self-consistent-field (SCF) formalism for molecules has recently been presented[31]; we give here its salient features and the reader is referred to the original paper[31] for further details.

B. Wavefunction and the Hamiltonian

The wavefunction Φ for the system in the multi-configuration Dirac-Fock (MCDF) theory is expressed in terms of configuration state functions (CSF) Φ_I with expansion coefficients $\{C_I\}$ as follows:

$$\Phi = \sum_I \Phi_I C_I \tag{58}$$

The various CSF's Φ_I are formed with the minimum number of Slater determinants of one-electron molecular spinors $\{\phi_i\}$ so as to satisfy the symmetry requirements of the wavefunction Φ, where the four-component MS's are assumed to form an orthonormal set; i.e.,

$$\langle \phi_i \mid \phi_j \rangle = \delta(i,j).$$

The Hamiltonian H for the molecular system will be taken, for simplicity, as given in Eq. (16).

C. Dirac-Fock MCSCF Equations

The derivation of the DF MCSCF equations is carried out by following the well-known procedures[32-34] used for the non-relativistic MC SCF equations. When the expectation value of energy E, where

$$E = \langle \Phi \mid H \mid \Phi \rangle / \langle \Phi \mid \Phi \rangle \tag{59}$$

is varied with respect to the expansion coefficients $\{C_I\}$ with fixed CSF's, one obtains the set of equations:

$$\sum_J H_{IJ} C_J = E C_I \,, \tag{60}$$

where

$$H_{IJ} = \langle \Phi_I \mid H \mid \Phi_J \rangle \tag{61}$$

The E as well as the $\{C_I\}$ are determined by solving Eq. (60). It is found convenient to express E in terms of the reduced density matrices γ and Γ of the first and second orders[35], respectively, which occur in the expressions of reduced density operators in terms of the $\{\phi_i\}$:

$$\gamma(1 \mid 1') = \sum_{i,j=1}^{m} \gamma_{ij} \phi_j(1) \phi_i^{\dagger}(1') \,, \tag{62}$$

$$\Gamma(1,2 \mid 1',2') = \sum_{i,j,k,\ell=1}^{m} \Gamma_{ij,k\ell} \phi_\ell(2) \phi_j(1) \phi_i^{\dagger}(1') \phi_k^{\dagger}(2'), \tag{63}$$

where γ_{ij} and $\Gamma_{ij,k\ell}$ are the reduced density matrices of first and second order which obey the following symmetry relations:

$$\gamma_{ij} = \gamma_{ij}^{*}, \text{ and } \Gamma_{ij,k\ell} = \Gamma_{k\ell,ij} = \Gamma_{ji,\ell k}^{*} \tag{64}$$

The E can be written in terms of these reduced density matrices as follows:

$$E = \sum_{ij}^{m} \gamma_{ij} \langle\phi_i|H(\mu)|\phi_j\rangle + \frac{1}{2} \sum_{ijk\ell}^{m} \Gamma_{ij,k\ell} \times \langle\phi_i\langle\phi_k|r_{\mu\nu}^{-1}|\phi_\ell\rangle\phi_j\rangle , \tag{65}$$

where

$$\langle\phi_i|H(\mu)|\phi_j\rangle = \int d\tau_\mu \phi_i^\dagger(\mu)H(\mu)\phi_j(\mu) , \tag{66}$$

$$\langle\phi_i\langle\phi_k|r_{\mu\nu}^{-1}|\phi_\ell\rangle\phi_j\rangle = \iint d\tau_\mu d\tau_\nu \phi_i^\dagger(\mu)\phi_k^\dagger(\nu)r_{\mu\nu}^{-1}\phi_\ell(\nu)\phi_j(\mu) . \tag{67}$$

If the molecular spinors $\{\phi_i\}$ are varied in Eq. (65) subject to the orthonormal conditions $\langle\phi_i|\phi_j\rangle = \delta(i,j)$, one can obtain, using the relations given in Eq. (64), the Dirac-Fock MC SCF equations; viz.

$$\sum_{j=1}^{m} F_{ij}\phi_j = \sum_{j=1}^{m} \phi_j\varepsilon_{ji} , \tag{68}$$

where the matrix of the Langrangian multipliers can be shown to be Hermitian; viz.

$$\varepsilon_{ij} = \varepsilon_{ji}^* \tag{69}$$

The F_{ij} in Eq. (68) can be expressed in terms of γ_{ij} and $\Gamma_{ij,k\ell}$ as follows:

$$F_{ij} = \gamma_{ij}H(\mu) + \sum_{k,\ell=1}^{m} \Gamma_{ij,k\ell}\langle\phi_k|r_{\mu\nu}^{-1}|\phi_\ell\rangle \tag{70}$$

The $\{\phi_i\}$ can be determined from the set of Eqs. (68-69) which can be reduced to pseudo-eigenvalue equations by means of the generalized coupling operator techniques[36].

D. Matrix MCDF SCF Equations

If the $\{\phi_i\}$ are expanded in terms of four-component basis spinors $\{\chi_p\}$ as

$$\phi_i = \sum_{p=1}^{m} C_{ip}\,\chi_p , \tag{71}$$

where the expansion coefficients are assumed to be scalar for simplicity; then E of Eq. (65) can be rewritten as follows:

$$E = \sum_{ij}^{m} \gamma_{ij} \sum_{pq}^{n} C_{ip}^{*} \langle \chi_p | H(\mu) | \chi_q \rangle C_{jq}$$

$$+ \frac{1}{2} \sum_{ijk\ell}^{m} \Gamma_{ij,k\ell} \sum_{pqrs}^{n} C_{ip}^{*} C_{kr}^{*} \langle \chi_p \langle \chi_r | r_{\mu\nu}^{-1} | \chi_s \rangle \chi_q \rangle C_{\ell s} C_{jq} \quad (72)$$

The variation of E with respect to the expansion coefficients $\{C_{ip}\}$ of the molecular spinors $\{\phi_i\}$ finally leads to the matrix MCDF SCF equations:

$$\sum_{j=1}^{m} F_{ij} C_j = S \sum_{j=1}^{m} C_j \, \varepsilon_{ji} \, , \quad (73)$$

and the hermiticity of the matrix of Langrangian multipliers,

$$\varepsilon_{ji}^{*} = \varepsilon_{ij} \, , \quad (74)$$

where

$$\left(F_{ij}\right)_{pq} = \gamma_{ij} \langle \chi_p | H | \chi_q \rangle + \sum_{k\ell}^{m} \Gamma_{ij,k\ell} \sum_{rs}^{m} C_{kr}^{*} \langle \chi_p \langle \chi_r | r_{\mu\nu}^{-1} | \chi_s \rangle \chi_q \rangle C_{\ell s} \, , \quad (75)$$

$$S_{pq} = \langle \chi_p | \chi_q \rangle \text{ and } (C_j)_p = C_{jp} \quad (76)$$

The expansion coefficients $\{C_{ip}\}$ are determined from Eqs. (73-74). In order to test the validity of the MCDF SCF equations, we shall apply them to a closed-shell molecular system whose wavefunction Φ consists of a single SD of N molecular spinors $\{\phi_i\}$, where $i=1,2...N$. The E for such a system can be written using Eq. (65) as follows:

$$E = H_0 + \sum_{i=1}^{N} \langle \phi_i | H(\mu) | \phi_i \rangle$$

$$+ \frac{1}{2} \sum_{i,j=1}^{N} \left(\langle \phi_i \langle \phi_j | r_{\mu\nu}^{-1} | \phi_j \rangle \phi_i \rangle - \langle \phi_i \langle \phi_j | r_{\mu\nu}^{-1} | \phi_i \rangle \phi_j \rangle \right) \, , \quad (77)$$

In the derivation of Eq. (77), the following non-zero reduced density matrices were used:

$$\gamma_{ii} = 1, \ \Gamma_{ii,jj} = 1, \text{ and } \Gamma_{ij,ji} = -1 \tag{78}$$

Then it can be shown[31] that the matrix MCDF SCF equations given in Eq. (73) reduce for a single configuration to the form:

$$FC_i = \varepsilon_i SC_i \ , \tag{79}$$

where

$$F_{pq} = H_{pq} + \sum_{rs} (J_{pq,rs} - K_{pq,rs}) D_{rs} \ , \tag{80}$$

$$H_{pq} = \langle \chi_p | H(\mu) | \chi_q \rangle \ , \tag{81}$$

$$J_{pqrs} = \langle \chi_p \langle \chi_r | r_{\mu\nu}^{-1} | \chi_s \rangle \chi_q \rangle \ , \tag{82}$$

$$K_{pqrs} = \langle \chi_p \langle \chi_r | r_{\mu\nu}^{-1} | \chi_q \rangle \chi_s \rangle \ , \tag{83}$$

$$D_{rs} = \sum_{j=1}^{N} C_{jr}^{*} C_{js} \ , \tag{84}$$

$$S_{pq} = \langle \chi_p | \chi_q \rangle \ . \tag{85}$$

It should be pointed out that exactly these DF SCF equations (given in Eq. (79)) were derived by Malli and Oreg[15] for a closed-shell molecular system which were presented in Section II; however, Matsuoka et al.[31] were able to derive these DF SCF equations as special case of the MCDF SCF equations (68-69) and (73).

Finally, the total energy E can be written as

$$E = \sum_{pq} H_{pq} D_{pq} + \frac{1}{2} \sum_{pqrs} D_{pq} (J_{pqrs} - K_{pqrs}) D_{rs} \ . \tag{86}$$

$$= H^{\dagger} D + \frac{1}{2} D^{\dagger} P D \tag{87}$$

where

$$P_{pqrs} = J_{pqrs} - K_{pqrs} \tag{88}$$

Equations (86) to (88) agree exactly with the corresponding equations of Malli and Oreg[15] given in Section II.

The MCDF SCF formalism is necessarily much more complicated than the SC DF SCF formalism and its application to a molecular system is a possibility for the future.

V. DIRAC-FOCK SCF CALCULATIONS ON MOLECULES

A. Dirac-Hartree-Fock-Roothaan Calculations (DHFR)

In Sections II-IV we have presented the DF SCF formalisms within a single configuration (SC) as well as mutli-configuration schemes for molecular systems.

We shall now report on the application of these formalisms to molecular systems. The first ab initio DHFR calculation on the closed-shell light atom diatomics Li_2 and Be_2 was carried out by Malli and Oreg[37] with a minimum basis set of atomic spinors on Li and Be using integer n STF's to expand the radial functions $R_{n\kappa\beta}$. The 1p function occurring in the small components of the 1s atomic spinors of Li and Be atoms was approximated by a 2p function and orbital exponent optimization was not performed. This calculation was carried out to test whether DHFR calculations can be carried out on molecules. It was found that Be_2 was unbound, as expected, but even Li_2, which has a very small D_e, was also unbound. Unfortunately, relativistic effects for the spinor energies of Li_2 and Be_2 are overestimated; it was due to the fact that different basis sets were used in the relativistic and non-relativistic calculations.

DHFR calculations on H_2 and LiH molecules have also been reported[31] using STF's and including the 1p functions on H as well as Li. The relativistic contribution to the total molecular energies was calculated to be -1.3×10^{-5} a.u. for H_2 and -1.1×10^{-3} a.u. for LiH which are of the right order of magnitude considering the relativistic contributions to the energy of H and Li atoms. In both the calculations, the single-zeta basis set yielded about half of the relativistic energies obtained by using the double-zeta basis set.

The single-determinant DHFR SCF calculations have been performed[23] on the ground states of the diatomics H_2, LiH, Li_2, HF, F_2 and the polyatomic formaldehyde H_2CO using cartesian GTF's as basis set. In this paper[23], the non-relativistic calculations were made by solving the non-relativistic limit of the DHFR equations, which can be written as follows[23]:

$$\left[F_{\lambda}^{LL} + X_{\lambda}\right] C_{i\lambda}^{L} = \varepsilon_{i\lambda} S_{\lambda}^{L} C_{i\lambda}^{L} \tag{89}$$

where

$$X_\lambda = -T_\lambda^{LS} (S_\lambda^S)^{-1} T_\lambda^{SL}/2. \tag{90}$$

The various matrix elements occurring in Eqs. (89-90) are defined in the Equations (3.8)-(3.19) of Matsuoka et al.[31]. The expression for the total energy; i.e., the non-relativistic energy in this limit can be written as

$$E = H_o + \sum_{pq} (U_{pq}^L + X_{pq}^{SL}) D_{pq}^{LL} \tag{92}$$

$$+ \frac{1}{2} \sum_{pq,rs} D_{pq}^{LL} (J_{pq,rs}^{LL} - K_{pq,rs}^{LL}) D_{rs}^{LL},$$

where the repulsion energy between nuclei H_o has the form

$$H_o = \frac{1}{2} \sum_{a,b=1}^{n} \frac{Z_a Z_b}{R_{ab}} \tag{93}$$

The density matrix D_{pq}^{LL} and the nuclear potential terms U_{pq}^L are as defined in Matsuoka et al.[31]. It should be pointed out that Eq. (92) is an extension of the result for atoms derived by Kim[16], who assumed that $n^L = n^S$, where n^L and n^S are the number of basis functions for the expansion of large and small radial components, respectively, of an atomic spinor.

If a complete or a near complete basis set is used for the small components of the molecular spinors $\{\phi_i\}$, the energy of Eq. (92) gives the non-relativistic[23] Hartree-Fock energy.

The DHFR SCF calculations were performed, however, by solving the Equation (79) using basis sets obtained by fitting the numerical Dirac-Fock[10] atomic wavefunctions in terms of cartesian gaussians by the method of least squares, except that 1p functions were approximated by 2p.

It should be pointed out that non-relativistic Hartree-Fock-Roothaan (NRHFR) calculations are also available on these molecules using fairly extended basis sets of STF's for the diatomics[38,39], and CGTF's for H_2CO[40]. The calculations on LiH were performed using three different gaussian basis sets. The smallest basis set consists of five 1s-type GTF's for the expansion of the large components but of three 2p-type GTF's for the expansion of the small components of the spinors $1s_{1/2}$ and $2s_{1/2}$ on Li atom; however, three 1s-type GTF's and only two 2p-type GTF's were used for the small components of the spinor $1s_{1/2}$ on H. This basis set gave the total relativistic energy of -8.181743 a.u., and the non-relativistic limit for the total energy as -8.181141 a.u. The energies of the $1e_{1/2}$ and $2e_{1/2}$ spinors were calculated to be -2.56049 a.u. and -0.30225 a.u., respectively.

It should be noted that the non-relativistic limit of this calculation was about 0.2 a.u. lower than the total energy reported by Cade and Huo[38] in the NRHFR calculation, which is alleged to be of true Hartree-Fock accuracy. In addition, the non-relativistic limit of orbital energy of the spinor ($1e_{1/2}$) was 0.12 a.u. lower than the value reported by Cade and Huo[38]. There is, however, a more extended basis set calculation which gave non-relativistic limits for the total, as well as the orbital energies, in excellent agreement with the NRHFR results[38] and the DHFR results of Matsuoka et al.[31], using STF's as basis sets. It seems that even for light atom systems, extended basis sets are necessary for accurate DHFR calculations.

In Table 1 we present the results of the DHFR, its non-relativisitic limit (NRL) and the non-relativistic Hartree-Fock-Roothaan (HFR) calculations on Li_2 and F_2. It can be seen from Table I that there must be serious error in the Li_2 calculation because the calculated NRL total energy is too low. Furthermore, using the fact that the DF total energy for two Li atoms at infinite distance apart is -14.8672 a.u., the binding energy of Li_2 from this calculation would be predicted to be 0.65 a.u. or about 17 e.v. Moreover, the orbital energies of the molecular spinors $2e_{1/2}g$ and $1e_{1/2}u$ are definitely wrong, as these are quite different from the orbital energies of the atomic spinors $2s_{1/2}$ and $1s_{1/2}$ of Li, respectively. However, in the calculation of Malli and Oreg[37], Li_2 was unbound but the orbital energies of the molecular spinors were very close to the non-relativistic HFR values of Cade and Wahl[39] and to the atomic spinor values as expected. As far as our calculation on Li_2 is concerned, basis set must be extended in order to get a bound state for Li_2; however, the dissociation energy of Li_2 is very small (1.3 e.v.) and it may be found that this molecule is not bound within a single-configuration (SC) DHFR theory. There is already such an example in the case of F_2 in the NRHFR theory[40].

As far as F_2 is concerned, the non-relativistic limit value of -199.2054 a.u. for the total energy of F_2 calculated by Aoyama et al.[23] is about 0.5 a.u. lower than that obtained by Cade and Wahl[39].

The dissociation energy of F_2 is 1.68 e.v. and the HFR calculation[40] did not lead to any binding, as mentioned earlier. Furthermore, the total DF energy of two atoms infinitely apart is -199.023 a.u. and the SC DHFR calculation of Aoyama et al. gave, at Re, a total energy of -199.3782 and, thus, F_2 was found to be bound by about 10 e.v.!

Table 1. DHFR, Non-relativistic Limit (NRL) and Hartree-Fock-Roothaan (HFR) Total Energy (-E) and Orbital Energies (-ε) for the Ground States of Li_2 and F_2 in a.u.

ϕ_i	DHFR (-ε)		NRL (-ε)	HFR (-ε)
		Li_2		
$1e_{1/2g}$	2.4705	2.6074	2.6072	2.4523
$2e_{1/2g}$	0.1871	0.7682	0.7679	0.1816
$1e_{1/2u}$	2.4694	0.4924	0.4921	2.4520
-E	14.8582	15.5184	15.5160	14.8715
Reference	37	23	23	39
		F_2		
$1e_{1/2g}$		26.4996	26.2054	26.4289
$1e_{1/2u}$		26.4908	26.4552	26.4286
$2e_{1/2g}$		1.7768	1.7731	1.7620
$2e_{1/2u}$		1.5170	1.5124	1.4997
$1e_{3/2u}$		0.8833	0.8832	0.8097
$3e_{1/2u}$		0.8471	0.8460	0.8097
$3e_{1/2g}$		0.8121	0.8125	0.7504
$1e_{3/2g}$		0.7528	0.7526	0.6682
$4e_{1/2g}$		0.6811	0.6780	0.6682
-E		199.3782	199.2054	198.7701
Reference		23	23	39

Thus, we have seen that the DF total energies as well as the non-relativistic limits calculated by Aoyama et al. for LiH, Li_2 and F_2 are too low; in addition, in some cases, the molecular spinor energies are grossly in error. This is partly due to the variational instability problem mentioned earlier and partly due to the 'basis set' problem in DHFR calculations.

However, their calculations on HF and H_2CO are extremely good and the NRL results are in very good agreement with the non-relativistic HFR calculations on HF[38] and H_2CO[41]. Table 2 collects the results of H_2CO.

So far, we have discussed DHFR calculations on light atom molecules in which relativistic effects are not expected to be very significant. Unfortunately, with the exception of a preliminary calculation[42] on AgH and AuH, ab-initio, all electron DHFR calculations are not available at present for systems containing heavy atoms in which relativistic effects would be very important. However, relativistic calculations on molecules containing heavy atoms have been carried out using the discrete variational[43] and scattered wave[44] methods; the latter method and its applications is described by Yang[44] in these proceedings which should be consulted for further details.

Table 2. DHFR and Non-relativistic Total (-E) and Orbital Energies (-ε) for the Ground States of Formaldehyde (H_2CO) in a.u.

	DHFR	Non-relativistic Limit	Hartree-Fock
-E	113.8641	113.7970	113.8917
$-\varepsilon(1e_{1/2})$	20.6128	20.593	20.5738
$-\varepsilon(2e_{1/2})$	11.5876	11.580	11.3431
$-\varepsilon(3e_{1/2})$	1.4708	1.470	1.4038
$-\varepsilon(4e_{1/2})$	0.9145	0.914	0.8646
$-\varepsilon(5e_{1\ 2})$	0.7836	0.7843	0.6893
$-\varepsilon(6e_{1/2})$	0.7042	0.7049	0.6506
$-\varepsilon(7e_{1/2})$	0.6328	0.6335	0.5341
$-\varepsilon(8e_{1/2})$	0.5114	0.5110	0.4402
Reference	23	23	40

Lastly, we should like to mention that calculations on heavy atom hydrides have been performed by Desclaux et al.[45], using a modified atomic numerical Dirac-Hartree-Fock program - the so-called DF one-centre-expansion (OCE) method, developed originally by

Mackrodt[46]. Desclaux[45] has discussed, in detail, the DF OCE method and its applications in these proceedings, and we shall only remark that the relativistic CuH, AgH and AuH were found to be unbound[45] by 0.156, 0.171 and 0.142 a.u.'s, respectively.

The Re's, calculated relativistically, were larger than the experimental values, although a minor bond contraction was found, due to relativity in each molecule.

The results of the DHFR[42] and DF OCE[45] calculations for AgH and AuH are collected in Table 3. It can be seen from this table that the total energies calculated for AgH and AuH by Lee and McLean[42] are even higher by 0.60 a.u. and 17 a.u., respectively, than those calculated by Desclaux et al.[45]; therefore, our remark about the unboundness of these two molecules would apply also to the calculation of Lee and McLean[42]. The calculation of Lee and McLean on AuH is

Table 3. Relativistic (REL) and Non-relativistic Limit (NRL) Total (-E) Energies for AgH and AuH in a.u.

	REL	NRL
	AgH	
-E	5315.070^a	5198.016
	5314.475^b	5198.320
	AuH	
-E	19040.067^a	17865.593
	19040.176^a	17865.716
	19023.065^b	17866.259

a Ref. 45
b Ref. 42

seriously in error, since the total relativistic energy calculated for AuH was found to be 17 a.u. higher than the total DF energy of a point nucleus Au atom assumed in their calculation. It should be remarked that these relativistic calculations were 20 times as expensive as the corresponding non-relativistic calculations.

In conclusion, we would like to remark that it is gratifying that the computational machinery is at hand for relativistic calculations on diatomics as well as polyatomics containing first row atoms. However, the calculations on heavy atom molecular systems would be prohibitively expensive, at present, and may sometimes lead to improper results. Since very large basis sets would be needed for calculations on heavy-atom molecular systems, much faster and more efficient computer programs would be needed to evaluate and handle millions of molecular integrals in future relativistic molecular calculations.

Finally, I would like to quote Dirac[47], the father of relativistic quantum mechanics, who stated more than half a century ago:

> "The underlying physical laws necessary for the mathematical theory of a large part of physics and the whole of chemistry are thus completely known, and the difficulty is only that the exact application of these laws leads to equations much too complicated to be soluble. It therefore becomes desirable that approximate practical methods of applying quantum mechanics should be developed, which can lead to an explanation of the main features of complex atomic systems without too much computation.".

I would only remove the word 'atomic' from the last line, and agree with Dirac wholeheartedly!

ACKNOWLEDGEMENTS

This work would not have been possible without the cooperation and the enthusiasm of my co-workers over many years. In particular, I acknowledge my debt to Joseph Oreg, Taskashi Kagawa, Osamu Matsuoka and Yasuyuki Ishikawa, who played very important roles in the development of the theory and the various computer programs for relativistic atomic and molecular calculations.

REFERENCES

1. (a) D.R. Hartree, Proc. Camb. Phil. Soc. 24, 89 (1928),
   (b) V.A. Fock, Z. Physik 61, 126 (1930).
2. P.M. Dirac, Proc. Roy. Soc. Lond. A117, 610 (1928).
3. J. Sucher, these proceedings.

4. I.P. Grant, these proceedings.
5. J.P. Desclaux, these proceedings.
6. G. Breit, Phys. Rev. 34, 553 (1929).
7. B. Swirles, Proc. Roy. Soc. Lond. A152, 625 (1935).
8. G.E. Brown and D.G. Ravenhall, Proc. Roy. Soc. Lond. A208, 552 (1951).
9. I.P. Grant, Proc. Roy. Soc. Lond. A262, 555 (1961); Adv. Phys. 17, 747 (1971).
10. J.P. Desclaux, At. Data. Nucl. Data. Tables 12, 311 (1973).
11. J. Maly and M. Hussonnois, Theoret. Chim. Acta 28, 363 (1973).
12. M.H. Mittleman, Phys. Rev. 4A, 893 (1971); ibid 5A, 2395 (1972).
13. J. Sucher, Phys. Rev. A22, 348 (1980).
14. C.C.J. Roothaan, Revs. Mod. Physics 23, 69 (1951).
15. G. Malli and J. Oreg, J. Chem. Phys. 63, 830 (1975).
16. Y.K. Kim, Phys. Rev. 17, 154 (1967).
17. T. Kagawa, Phys. Rev. A12, 2245 (1975).
18. G. Malli, Chem. Phys. Letts. 68, 529 (1979).
19. F.E. Harris, Rev. Mod. Phys. 35. 559 (1963).
20. M. Krauss, J. Res. Natl. Bur. Stand. Sect. B68, 35 (1964).
21. K. O-Ohata, Mem. Fac. Sci. Kyushu Univ. Ser. B5, 7 (1974).
22. G. Fieck, J. Phys. B12, 1063 (1979).
23. T. Aoyama, H. Yamakawa and O. Matsuoka, J. Chem. Phys. 73, 1329 (1980).
24. H. Taketa, S. Huzinaga and K. O-Ohata, J. Phys. Soc. Japan 21, 2313 (1966).
25. J. Oreg and G. Malli, J. Chem. Phys. 61. 4349 (1974).
26. J. Oreg and G. Malli, J. Chem. Phys. 65, 1746 (1976).
27. J. Oreg and G. Malli, Mol. Phys. 37, 265 (1979).
28. G. Malli, Chem. Phys. Letts. 73. 510 (1980).
29. C.C.J. Roothaan and P.S. Bagus, Methods Comput. Phys. 2, 47 (1963).
30. A.C. Wahl, P.J. Bertonicini, K. Kaiser and R.H. Land, Argonne National Laboratory Report ANL-7271, Argonne, U.S.A. (1968).
31. O. Matsuoka, N. Suzuki, T. Aoyama and G. Malli, J. Chem. Phys. 73, 1320 (1980).
32. J. Hinze and C.C.J. Roothaan, Progr. Theor. Phys. Suppl. 40, 37 (1967).
33. J. Hinze, J. Chem. Phys. 59, 6424 (1973).
34. J. Hinze, Adv. Chem. Phys. 26, 213 (1975).
35. P.O. Löwdin, Phys. Rev. 97, 1474 (1955).
36. K. Hirao, J. Chem. Phys. 60, 3214 (1974).
37. G. Malli and J. Oreg, Chem. Phys. Letts. 69, 313 (1980).
38. P.E. Cade and W. Huo, At. Data Nucl. Data Tables 12, 415 (1973).
39. P.E. Cade and A.C. Wahl, At. Data Nucl. Data Tables 13, 339 (1974).
40. A.C. Wahl, J. Chem. Phys. 41, 2600 (1964).
41. D.B. Neumann and J.W. Moskowitz, J. Chem. Phys. 50, 2216 (1969).

42. Y.S. Lee and A.D. McLean, J. Chem. Phys. 76, 735 (1982).
43. A. Rosen and D.E. Ellis, J. Chem. Phys. 62, 3039 (1975).
44. (a) C.Y. Yang, J. Chem. Phys. 68, 2626 (1978),
(b) C.Y. Yang, these proceedings.
45. (a) J.P. Desclaux and P. Pyykkö, Chem. Phys. Letts. 39, 300 (1976),
(b) J.P. Desclaux, these proceedings.
46. W.C. Mackrodt, Mol. Phys. 18, 697 (1970).
47. P.A.M. Dirac, Proc. Roy. Soc. A123, 714 (1929).

DIRAC-FOCK ONE-CENTRE EXPANSION METHOD

J.P. Desclaux

Centre d'Etude Nucléaires de Grenoble
Département de Recherche Fondamentale
LIH 85 X 38041 Grenoble, France

INTRODUCTION

It is by now well recognized that relativistic corrections are important not only for electrons which have a mean velocity not too small compared with the speed of light, i.e. the inner electrons of heavy atoms, but also for the valence electrons of medium and high Z atoms. The underlying reasons are first that the valence electrons themselves have an instantaneous velocity which becomes appreciable when they come close to the nucleus. Second we are dealing with a N-electron interacting system and modifications of the inner charge distribution will be reflected in the behaviour of all the other electrons. Futhermore the valence orbitals have to be orthogonal to the inner ones and this constraint will also induce changes when relativistic corrections are introduced. Carrying out full relativistic Hartree-Fock calculations for molecules with very heavy atoms is still a formidable task as will be illustrated by other lectures during this school. For this reason, progresses have been achieved either by reducing the complexity of the problem in the framework of local exchange approximations as used in the multiple scattering[1] or discrete variational[2] methods or by carrying out model calculations with the one-centre expansion method we shall describe here. In this method the exchange interaction is treated exactly and the one-electron wave functions are four component Dirac spinors. On the other hand both potential and wave functions are expanded about a given centre generally taken to be the heaviest nucleus of the system under consideration. Because of this expansion, the method is practically applicable only to hydrides XH_n and even for them the expansion is only slowly converging if accurate results are required. Neverthless experience obtained in the non-relativistic case, even at the very crude spherically symmetric level, has shown surprisingly good results for the equilibrium distances of the X-H bonds and force constants. As many elements form chemical bonds with hydrogen, this was a logical step to start the investigation of the influence of relativistic corrections on the calculation of chemical properties. In fact it appears now that the qualitative insight obtained from these calculations on model systems is just confirmed by present day calculations using other approaches.

I. THE DIRAC-FOCK ONE-CENTRE EXPANSION METHOD

I.1 Nuclear potential expansion

Let the off centre nuclei be protons and choose for each of them a coordinate system with its origin at the heavy nucleus and having the z_i axis passing through the ith proton. Then in this coordinate system, the potential due to this proton is given by :

$$V_i(r) = \sum_{\ell=0}^{\infty} \frac{r_<^{\ell}}{r_>^{\ell+1}} (4\pi/2\ell+1)^{1/2}\, Y_{\ell,0}(\theta_i,\varphi_i) \tag{1}$$

with

$$r_< = \mathrm{Min}(r,R_i) \quad \text{and} \quad r_> = \mathrm{Max}\,(r,R_i) \tag{2}$$

r_i being the distance between origin and the ith proton. We can now rotate all these proton potentials to a common coordinate system using the Wigner rotation matrices $D^{\ell}_{m,m}(\alpha_i,\beta_i,\gamma_i)$ defined with respect to the Euler angles α, β and γ of each proton. We obtain :

$$Y_{\ell,0}(\theta_i,\varphi_i) = (4\pi/2\ell+1)^{1/2}\, D^{\ell}_{m,0}(\alpha_i,\beta_i,\gamma_i)\, Y_{\ell,m}(\theta,\varphi) \tag{3}$$

Then the total potential of the N protons becomes :

$$V_p(r) = \sum_{\ell=0}^{\infty} \frac{r_<^{\ell}}{r_>^{\ell+1}} (4\pi/2\ell+1)^{1/2} \sum_{m=-\ell}^{+\ell} Y_{\ell,m}(\theta,\varphi) \sum_{i=1}^{N} D^{\ell}_{m,0}(\alpha_i,\beta_i,\gamma_i) \tag{4}$$

As an illustration we consider the case of tetrahedral symmetry with the four protons located at $(R,R,R)/\sqrt{3}$, $(-R,-R,R)/\sqrt{3}$, etc..., and obtain the expression :

$$\begin{aligned} \frac{1}{4} V_p(r) = {} & \frac{-1}{r_>} - \frac{i}{3}(10\pi/7)^{1/2} \frac{r_<^3}{r_>^4} [Y_{3,-2} - Y_{3,2}] \\ & + (\pi)^{1/2}/27 \frac{r_<^4}{r_>^5} [(35/2)^{1/2}[(Y_{4,-4} + Y_{4,4}) + 7Y_{4,0}] \\ & + \frac{4}{9}(\pi/13)^{1/2} \frac{r_<^6}{r_>^7} [(7/2)^{1/2}[(Y_{6,-4} + Y_{6,4}) - Y_{6,0}] \end{aligned} \tag{5}$$

I.2 Symmetry orbitals

The next problem to be considered in relativistic molecular calculations is the construction of symmetry adapted orbitals. We may note as a parenthesis that some methods (like the relativistically parameterized extended Huckel[3] one) do not require such information since the relativistic orbitals are produced by a direct diagonalization of the Hamiltontian matrix. The purpose of this section is not to consider in detail the theory of double point groups which are required in the relativistic case but only to illustrate how to use them for building the symmetrized molecular orbitals.

For a rotation ω, a Dirac spinor $\psi(x)$ will transform like :

$$\psi'(x') = e^{i\omega j}\ \psi(x) \tag{6}$$

As :

$$j = (2k+1)/2 \qquad k = 0,\ 1,\ 2,\ldots \tag{7}$$

is a half integer, we get for a rotation through 2π :

$$\psi'(x') = e^{i\pi}\ \psi(x) = -\psi(x) \tag{8}$$

Thus only a rotation through 4π will leave the wave function unchanged. As this operation is not included among the usual group operations, Bethe proposed in 1929 to add to the non-relativistic point group G a new element $\overline{E}$ for this rotation through 2π and having the property :

$$\overline{E}^2 = E \tag{9}$$

where E is the identity. The new elements $\overline{A} = \overline{E}A$ associated to each element A of G double the order of the group G to give the double group G^*. Note that this double group is different from the non-relativistic full group of the system :

$$G_{Full} = G \times SU(2) \tag{10}$$

given, because the spin and coordinate spaces are decoupled, by the direct product of G and SU(2). Taking the proper rotations as an example, we have to consider in the relativistic case only the particular rotations which are the same in spin and coordinate spaces since these two spaces are now coupled. Because the order of G^* is doubled compared to the order g of G, we have additional irreducible representations due to the fact that the number r and dimensions n_i of the irreducible representations of G^* must verify the relation :

$$\sum_{i=1}^{r} n_i^2 = 2g \tag{11}$$

The number of additional irreducible representations is, in many cases, smaller than that of the single group. For example the group C_{2v} of order g=4 has four one-dimensional irreducible representations and the corresponding double group has only one additional two-dimensional irreducible representation. As the relativistic symmetry orbitals belong to this unique additional representation, we see that several non-relativistic symmetries will be mapped to a single relativistic one. This result is called "spin-orbit induced hybridization". A counter example is provided by the single group C_{2h} also of order four with four one-dimensional representations like C_{2v} but for which the associated double group has four one-dimensional additional irreducible representations. Two methods have been used to construct the symmetry adapted relativistic molecular orbitals.

I.2.1 Projection operator method

Assume that the representation matrices $\Gamma^k(R)$ of the kth irreducible representation are known for all the elements R of the group. Then consider the projection operator :

$$P^k_{\lambda\nu} = \frac{n_k}{g} \sum_R \Gamma^k_{\lambda\nu}(R)^* O_R \tag{12}$$

where n_k is the dimension of the kth irreducible representation, g the order of the double group, λ the row index and ν the column index and O_R the symmetry operation. The sum is taken over all the elements of the group. Operating on an arbitrary function ψ, the operator $P^k_{\lambda\lambda}$ will project out the part transforming as the λ row of k if any, i.e.

$$\phi^k_\lambda = P^k_{\lambda\lambda}\psi \tag{13}$$

Then the partners of ϕ^k_λ are obtained by :

$$\phi^k_\nu = P^k_{\nu\lambda}\ \phi^k_\lambda \tag{14}$$

For more details concerning this method see Rosén and Ellis[4].

I.2.2 Coupling constant method

In the one-centre calculations we have used the following method to obtain the symmetry orbitals. First we construct the non-relativistic u^ℓ_ν symmetry orbitals in a $|\ell m>$ basis for a given ℓ value of the angular momentum. We then combine these functions with the spin functions which belong to an additional irreducible representation j. The general expression is :

$$\phi^k_\lambda = \sum_{\nu,\mu} C^{ij,k}_{\nu\mu,\lambda}\ u^i_\nu\ V^j_\mu \tag{15}$$

where the coupling constants $C^{ij,k}_{\nu\mu,\lambda}$ are obtained from the tables of Koster at al[5]. This yields every ϕ^k_λ as a linear combination of

$|\ell m_\ell \tfrac{1}{2} m_s\rangle$ orbitals. Finally we use Clebsch-Gordan coefficients to express them in term of $|jm\rangle$ orbitals. As the last step we read out from the expressions thus obtained the coefficients of the $|jm\rangle$ orbitals and use them with fourth order Dirac spinors. This method has the advantage of giving the correct relative phase between the two $\ell = j \pm 1/2$ orbitals. Taking again the example of tetrahedral symmetry, we obtain for the two two-dimensional irreducible representations Γ_6 and Γ_7 and for the one four-dimensional one Γ_8, the following symmetry orbitals:

Γ_6

s : $|1/2\rangle$; $|-1/2\rangle$

$f_{5/2}$: $i\{|5/2\rangle - (5)^{\frac{1}{2}}|-3/2\rangle\}/(6)^{\frac{1}{2}}$; $i\{|-5/2\rangle - (5)^{\frac{1}{2}}|3/2\rangle\}/(6)^{\frac{1}{2}}$

$f_{7/2}$: $i\{|-3/2\rangle - (3)^{\frac{1}{2}}|5/2\rangle\}/2$; $i\{(3)^{\frac{1}{2}}|-5/2\rangle - |3/2\rangle\}/2$

Γ_7

$p_{1/2}$: $i|1/2\rangle$; $i|-1/2\rangle$

$d_{5/2}$: $\{(5)^{\frac{1}{2}}|-3/2\rangle - |5/2\rangle\}/(6)^{\frac{1}{2}}$; $\{(5)^{\frac{1}{2}}|3/2\rangle - |-5/2\rangle\}/(6)^{\frac{1}{2}}$

$f_{7/2}$: $-i\{(7)^{\frac{1}{2}}|1/2\rangle + (5)^{\frac{1}{2}}|-7/2\rangle\}/(12)^{\frac{1}{2}}$; $i\{(7)^{\frac{1}{2}}|-1/2\rangle + (5)^{\frac{1}{2}}|7/2\rangle\}/(12)^{\frac{1}{2}}$

Γ_8

$p_{3/2}$: $i|1/2\rangle$; $-i|3/2\rangle$

$-i|-3/2\rangle$; $i|-1/2\rangle$

$d_{3/2}$: $|-3/2\rangle$; $|-1/2\rangle$

$|1/2\rangle$; $|3/2\rangle$

$d_{5/2}$: $\{|-3/2\rangle + (5)^{\frac{1}{2}}|5/2\rangle\}/(6)^{\frac{1}{2}}$; $-|-1/2\rangle$

$|1/2\rangle$; $-\{|3/2\rangle + (5)^{\frac{1}{2}}|-5/2\rangle\}/(6)^{\frac{1}{2}}$

$f_{5/2}$: $-i|1/2\rangle$; $-i\{(5)^{\frac{1}{2}}|-5/2\rangle + |3/2\rangle\}/(6)^{\frac{1}{2}}$

$i\{(5)^{\frac{1}{2}}|5/2\rangle + |-3/2\rangle\}/(6)^{\frac{1}{2}}$; $i|-1/2\rangle$

$f_{7/2}$: $i\{(5)^{\frac{1}{2}}|1/2\rangle - (7)^{\frac{1}{2}}|-7/2\rangle\}/(12)^{\frac{1}{2}}$; $i\{|-5/2\rangle + (3)^{\frac{1}{2}}|3/2\rangle\}/2$

$i\{|5/2\rangle + (3)^{\frac{1}{2}}|-3/2\rangle\}/2$; $i\{(5)^{\frac{1}{2}}|-1/2\rangle - (7)^{\frac{1}{2}}|7/2\rangle\}/(12)^{\frac{1}{2}}$

where only the values of m of the $|jm\rangle$ orbitals are listed.

I.3 Radial equations

In practice the one-electron orbitals are divided in two groups. The core orbitals, each of them completly filled with 2j+1 electrons and treated as spherically symmetric atomic orbitals. The valence molecular orbitals constructed as outlined above and which describe the non spherical charge distribution of the outermost electrons. For example in the case of AuH all the electrons up to and including the 5d ones have been taken as core electrons while the two remaining electrons are described by the valence σ molecular orbitals

$$\begin{aligned} \phi_\sigma(1/2) &= |6s,1/2> - |6p^*,1/2> + |6p,1/2> \\ \phi_\sigma(-1/2) &= |6s,-1/2> + |6p^*,-1/2> + |6p,-1/2> \end{aligned} \tag{17}$$

where $6p^*$ stands for $6p_{1/2}$ and 6p for $6p_{3/2}$. The above orbitals are normalized so that :

$$\begin{aligned} <\phi|\phi> &= <6s|6s> + <6p^*|6p^*> + <6p|6p> \\ &= N(6s) + N(6p^*) + N(6p) \end{aligned} \tag{18}$$

During the iterative process both the radial wave function and their norms N are optimized. Requiring that the total energy be stationary, the Dirac-Fock equations are obtained in a form quite similar to their pure atomic conterpart, i.e. :

$$\begin{aligned} \frac{dP_A}{dr} + \frac{k_A}{r} P_A &= [2c + \frac{1}{c}\{\varepsilon_A - V_A(r)\}]Q_A + X_A^Q(r) + \sum_{B\neq A} \varepsilon_{AB}\, Q_B \\ \frac{dQ_A}{dr} - \frac{k_A}{r} Q_A &= -\frac{1}{c}\{\varepsilon_A - V_A(r)\}\, P_A - X_A^P(r) - \sum_{B\neq A} \varepsilon_{AB}\, P_B \end{aligned} \tag{19}$$

However the direct and exchange potentials (V and X) include additional terms which arise from the proton potential as well as from the core-valence and valence-valence Coulomb interaction. Furthermore extra off-diagonal Lagrange multipliers ε_{AB} are required to insure orthogonality between the core orbitals and the components of the valence molecular orbitals having the same symmetry. These extra multipliers are needed even in the case of closed shells and arise from the fact that the potential terms for a core orbital are different from those of the valence orbitals.

I.3.1 Numerical solution of the radial equations

The main difference between the system of radial equations (19) and its atomic Dirac-Fock equivalent is that we have now an eigenvalue ε_A which must be the same for all the atomic components belonging to the same molecular orbital. Instead of trying to solve a system of 2N coupled equations (N being the number of atomic components) we consider, in the spirit of the self-consistent method, all the potentials as fixed and calculate them with the help of the estimates available at each step. The program uses the predictor corrector algorithm dexcribed in the first lecture and contains four different methods to solve the system 19 :

1) solve the inhomogeneous system both outward and inward after selecting a matching point R. Then solve the associated homogeneous system outside the matching point. Form a linear combination of inhomogeneous and homogeneous solutions such that the large component is continuous everywhere. Change the eigenvalue ε until the small component is also continuous everywhere.

2) Same as 1 but inforce normalization by calculating the derivatives $p = \partial P/\partial\varepsilon$ and $q = \partial Q/\partial\varepsilon$.

3) Solve both inhomogeneous and homogeneous systems outward and inward and form linear combinations such that both the large and small components are continuous everywhere. Calculate the normalization derivatives p and q. Check the number of nodes and if incorrect use method 4.

4) Same as 3 but modify the eigenvalue ε until the correct number of nodes is obtained.

The following process is used : for a core orbital method 1 is generally used. If it fails the other methods are tried sequentially. For a valence molecular orbital method 1 is used for the dominant atomic component and the normalization derivatives p and q are calculated. The eigenvalue ε is thus determined and used in method 3 to calculate the other atomic components. If necessary go from method 3 to method 4 and calculate a new orbital energy as :

$$\varepsilon = \sum_i \varepsilon_i N_i \qquad (20)$$

Then the molecular orbital energy is corrected by adding to it :

$$\Delta\varepsilon = (1 - \sum_i N_i) / \left[2 \sum_i \int_0^\infty (P_i p_i + Q_i q_i) dr\right] \qquad (21)$$

and the molecular orbital is renormalized by multiplying all the atomic components by $(\sum_i N_i)^{-1/2}$. This method of solution has been used in all our one-centre molecular calculations and in the atomic polarizability calculations we shall discuss in the last section.

II. SURVEY OF DF-OCE RESULTS

The Dirac-Fock One-Centre Expansion (DF-OCE) method has now been applied to a large number of hydrides or model systems involving

atoms of almost any column and row of the periodic table. The results are summarized in Table I which provides thus an overview of the importance of relativistic corrections on the calculation of some properties of chemical interest. More systematic analysises can be found in the review articles of Pyykkö[15] and Pyykkö and Desclaux[16]. Only few points will be considered here in some details.
As indicated in the description of the method, the norms N of the individual atomic components of a given molecular orbital are optimized during the self-consistent process. For a p_σ orbital, the ratio of the norms of the p^* and p atomic orbitals, is at the non-relativistic limit :

$$N(p)/N(p^*) = 2 \qquad (22)$$

Thus any deviation from this limit will reflect the importance of spin-orbit interaction and consequently the transition towards "j-j" coupling. As an example the value of this ratio was found to be 1.46 for InH and only 0.65 for TlH. This latter result indicates a transition to predominant $p_{1/2}$ bonding instead of p_σ bonding providing thus a partial explanation for the dominant monovalency of Tl while the elements B to In are trivalent. This transition makes also the bond weaker as first explained by Pitzer[17].
In the OCE method the charge density of the valence orbitals serves two purposes : to describe the density along the bonds and also to represent the 1s density peaked at each proton site. Analyzing the partial densities (s,p,..) for UH_6[11] we found that the f density has two maxima, one at the proton sites as expected and the second one in the middle of the bonds. The existence of this latter inner maximum suggests a true 5f participation in the bonding of this UH_6 model system. For ThH_4, which is known to have no 5f electrons, the charge distribution was found to be qualitatively different from that of UH_6 with almost no 5f charge in the bond region, showing that for ThH_4 the 5f atomic component describes only the 1s density around the protons.

The last point we want to discuss is the interpretation of changes in the equilibrium distances induced by relativistic effects. The first proposed explanation was that this was a consequence of the modification in the charge density of the atomic orbitals with the simple minded point of view that to form a stable bond it is necessary to achieve a correct overlap between the wave functions of the individual atoms. Thus if relativistic corrections result in a contraction of the valence atomic orbitals, the atoms have to come closer to achieve the necessary overlap. This explanation seemed also be supported by the fact that when d electrons were involved in the bonding the DF-OCE method predicts either no significant decrease of the equilibrium distances or on the contrary a small increase. Not a too surprising result since we know from atomic calculations that the behaviour of the d electrons is much less

Table I

Molecules treated by the DF-OCE method and main results.

- CH_4 to SnH_4 and HF to HI (reference 6)
  Relativistic contribution to total energies. Unfortunately the conclusions were invalidated because of numerical inaccuracies.

- CH_4 to PbH_4 (reference 7)
  Bond length contraction and increase of the force constant due to relativistic corrections.

- CuH, AgH and AuH (reference 8)
  Relativistic increase of the dissociation energy. Chemical difference between Ag and Au "mainly a relativistic effect".

- BH to TlH (reference 9)
  Relativistic decrease of the dissociation energy for TlH. Transition from $p\sigma$ to $p_{1/2}$ bonding. Monovalency of TlH partially due to 6p spin-orbit splitting.

- TiH_4 to $(104)H_4$ (reference 10)
  Chemical similarity of Zr and Hf attributed to cancellation of relativistic effects. Small relativistic bond length expansion for TiH_4 and ZrH_4.

- CeH_4, ThH_4, UH_6, CrH_6 to $(106)H_6$ (reference 11)
  Relativistic effects explain why W-H bonds are stronger but not longer than Mo-H ones. 5d density moves to bonding region due to relativistic corrections and explains higher valency of W. Lanthanoid contraction is a shell effect. Actinoid contraction found to be 30 pm. Further evidence for 5f participation in U-H bonds.

- MH^+ and MH_2 with M=Be to Ra, Zn to Hg, Yb and No (reference 12)
  Strong d contribution to the bonds of Ca and Ra, relativistic effects make them smaller for Ra than for Ba. Small bond length contractions or even expansions. Ra-H bonds longer than Ba-H ones. Yb-H and No-H bonds lengths comparable. Linear two-coordination of Hg attributed to relativistic effects.

- $^1\Sigma$ states of ScH to AcH, TmH, LuH and LrH (reference 13)
  SCF lanthanoid expansion of the 5d shell from La to Lu. Trends in group IIIa. Lu-H and Lr-H bond lengths comparable.

- AuH and TlH (reference 14)
  Confirms that bond length contractions are not a direct consequence of atomic orbital contractions.

affected by relativistic corrections than that of s or p electrons. Furthermore the indirect effect may even dominate the direct one and the net result is an expansion of the charge density. Subsequent calculations[18] performed with the perturbative Hartree-Slater method have shown that most of the contraction of the bond lengths can be obtained with uncontracted non-relativistic orbitals used to calculate the first order relativistic corrections of order α^2. This result has been confirmed by DF-OCE calculations in which the wave functions were obtained at the non-relativistic limit by increasing artificially the velocity of light. Then these wave functions were unchanged, excepted for renormalization, to calculate the total energy with in the Hamiltonian the actual velocity of light restored. Furthermore, new results[19] seem to indicate that a perturbative treatment is also able to reproduce the small bond length expansion found with the DF-OCE method when d electrons are involved. The following interpretation may be given. Write the molecular orbital as a linear combination of atomic orbitals :

$$\psi = \sum_i c_i |v_i\rangle + a|H,1s\rangle + b|core\rangle \tag{23}$$

where the sum runs over the valence oribtals v_i, $|H,1s\rangle$ are the hydrogen 1s orbitals and the core contribution results from the orthogonality constraint. Neglecting the contribution of hydrogen orbitals, we can write the first order relativistic correction as the sum of three terms : core-core, core-valence and valence-valence, i.e. :

$$E^{(1)} = E_{cc}^{(1)} + E_{cv}^{(1)} + E_{vv}^{(1)} \tag{24}$$

Substracting from the above equation the first order correction for the atoms, we obtain again a sum of three terms for the variation of the relativistic corrections with the internuclear distance R.

$$\begin{aligned} \Delta E^{(1)}(R) &= E^{(1)}(R) - E^{(1)}(\text{atoms}) \\ &= \Delta E_{cc}^{(1)}(R) + \Delta E_{cv}^{(1)}(R) + \Delta E_{vv}^{(1)}(R) \end{aligned} \tag{25}$$

It has been found[19] that $\Delta E_{cc}^{(1)}$ is the most important term and, as the overlap with the core (given by the coefficient b of Eq. 23) increases when R decreases, we see that the mixing with the core orbitals may also result in a contraction of the equilibrium distances. Obviously the dominant contribution to ΔE_{cc} arises from s and $p_{1/2}$ orbitals. As it was also found that contribution of d orbitals in the bonding diminishes the 1s one, the above interpretation is coherent for both cases : contraction and expansion. In the DF-OCE method the contribution of the core is less obvious but manifests itself through the contribution of the off-diagonal multipliers introduced to insure orthogonality between valence and core orbitals. A more exhaustive discussion will be found in the reference 19. Note neverthless that if the contraction of atomic orbitals and that of the equilibrium distances can be viewed as distinct, they have both the same origin i.e. the relativistic decrease of the kinetic energy due to the dominant mass-velocity correction.

III. ELECTRIC DIPOLE POLARIZABILITIES

Having a program able to handle non spherical potentials it is possible to consider other applications than the one-centre expansion method. Recently we have used a sightly modified version to study the electric dipole polarizability of atoms with the ns and ns^2 ground state configurations[20]. In that case, the Dirac-Fock Hamiltonian is the sum of one-electron operators plus the coulomb repulsion between the electrons :

$$H = \sum_i h_i + \sum_{i>j} 1/r_{ij} \tag{26}$$

where h_i contains the external electric field F, i.e. :

$$h_i = h_D(i) + Fz_i \tag{27}$$

with $h_D(i)$ the usual Coulomb Dirac one-electron operator. For the atoms considered, the polarized valence one-electron orbitals were restricted to be linear combination of ns and np orbitals only :

$$\begin{aligned} \phi(\tfrac{1}{2}) &= |ns,1/2\rangle - |np^*,1/2\rangle + |np,1/2\rangle \\ \phi(-\tfrac{1}{2}) &= |ns,-1/2\rangle + |np^*,-1/2\rangle + |np,-1/2\rangle \end{aligned} \tag{28}$$

The method is thus equivalent to a coupled Hartree-Fock calculation. Several values of the electric field were used for each atom and the change in energy was fitted by :

$$\Delta E = \frac{-1}{2}\alpha F^2 - \frac{1}{24}\gamma F^4 \tag{29}$$

giving thus the polarizabilities α. The main objective was to obtain a feeling for relativistic corrections on this property rather than to obtain very accurate results. We give in Table II the ratio of the relativistic to the non-relativistic calculated polarizabilities for the heaviest atoms considered.

As can be seen from these results, the inclusion of relativistic corrections decreases the calculated polarizabilities. Furthermore while the non-relativistic α value increases from Ba to Ra, the relativistic one decreases. The same trend was observed when going from Cd to Hg. A result which parallels the decrease of p character

Table II

Ratio of relativistic to non-relativistic polarizabilities

Atoms	Sr	Ba	Ra	Cd	Hg	Cs
α^{Rel}/α^{NR}	0.94	0.89	0.67	0.83	0.54	0.83

in the valence σ molecular orbital from BaH^+ to RaH^+ and from CdH^+ to HgH^+[12]. The largest decrease happens for Hg a not too unexpected result knowing the "pathological" contraction of the 6s orbital near gold. As a final remark concerning these results, let us point out that they certainly give an upper limit for relativistic corrections because of the neglected d contribution.

CONCLUSION

From a systematic comparison between relativistic and non-relativistic results the one-centre expansion method has proved to be more than useful to investigate the importance of relativistic corrections on the calculation of chemical properties across the periodic table. We have always been careful to avoid spurious effects by performing the calculations at exactly the same level of approximation and, for this reason, the non-relativistic calculations where obtained with the same program by simply increasing the velocity of ligth. From this comparison at least qualitative explanations, sometimes only partial, have been proposed to explain "anomalies" in the second part of the periodic table. Examples are : why is gold so different from silver ? Why is mercury a liquid and what causes its strong tendancy for two coordination ? Why is thallium mainly monovalent while elements B to In are trivalent ? Despite the approximations and the slow convergence (we have verified that increasing the maximum ℓ value in the expansion does not change the qualitative feature of the results) of the method, we believe that the results will not be qualitatively invalidated when more rigorous methods will become available to handle sophisticated molecular systems with very heavy atoms. But today we are still far from being able to do routine full Dirac-Fock calculations for heavy hydrides not saying anything about molecules like Au_2. Nevertheless my own feeling is that the accuracy of the DF-OCE method is not good enough to reach the quantitative level required for chemical purpose and that we have now to concentrate our efforts toward other models or find new fields of application for the one-centre expansion method.

ACKNOWLEDGEMENTS

It is a pleasure to thank professor P. Pyykkö for more than seven years of uninterrupted collaboration.

REFERENCES

1) For a review of the relativistic multiple scattering method see C.Y. Yang, these proceedings.

2) A. Rosén, D.E. Ellis, H. Adachi and F.W. Averill J. Chem. Phys. 65, 3629 (1976)

3) L.L. Lohr,Jr. and P. Pyykkö Chem. Phys. Lett. 62, 333 (1979)

4) A. Rosén and D.E. Ellis J. Chem. Phys. 62, 3039 (1975)

5) G.F. Koster, J.O. Dimmock, R.G. Wheeler and H. Statz *Properties of the thirty-two point groups*, MIT Press, Cambridge, Massachussets (1963)

6) W.C. Mackrodt Mol. Phys. 18, 697 (1970)

7) J.P. Desclaux and P. Pyykkö Chem. Phys. Lett. 29, 534 (1974)

8) J.P. Desclaux and P. Pyykkö Chem. Phys. Lett. 39, 300 (1976)

9) P. Pyykkö and J.P. Desclaux Chem. Phys. Lett. 42, 545 (1976)

10) P. Pyykkö and J.P. Desclaux Chem. Phys. Lett. 50, 503 (1977)

11) P. Pyykkö and J.P. Desclaux Chem. Phys. 34, 261 (1978)

12) P. Pyykkö J. Chem. Soc. Faraday II 75, 1256 (1979)

13) P. Pyykkö Phys. Scripta 20, 647 (1979)

14) J.G. Snijders and P. Pyykkö Chem. Phys. Lett. 75, 5 (1980)

15) P. Pyykkö Adv. Quantum Chem. 11, 363 (1978)

16) P. Pyykkö and J.P. Desclaux Accounts Chem. Res. 12, 276 (1979)

17) K.S. Pitzer J. Chem. Phys. 63, 1032 (1975)

18) T. Ziegler, J.G. Snijders and E.J. Baerends Chem. Phys. Lett. 75, 1 (1980)

19) P. Pyykkö, J.G. Snijders and E.J. Baerends to be published

20) J.P. Desclaux, L. Laaksonen and P. Pyykkö J. phys. B 14, 419 (1981)

RELATIVISTIC EFFECTS IN SOLIDS*

D. D. Koelling and A. H. MacDonald*

Argonne National Laboratory
Argonne, Illinois 60439, USA
*National Research Council
Ottawa, Canada K1A 0R6

Because a solid can be viewed as a collection of interacting atoms, it will exhibit all the relativistic effects known to occur in atoms. In fact, the relativistic contributions in solids arise exclusively from the atomic-like regions of space near the nucleii with the interstitial region being quite well described in a non-relativistic approximation. Thus, for those band structure techniques such as the APW, KKR and related methods which partition space into atomic like spheres and an interstitial region, it is actually possible to "switch kinematics" at the boundary as well. Such a procedural step not only tells one how to include the relativistic contributions but also how to predict their effects: they will be the atomic results only slightly modified by the differing boundary conditions. We will discuss this using simplified views of the APW and LCAO methods as examples.

It is often necessary to make additional simplifying assumptions to deal with solids that are not essential in atomic structure calculations. For example, the spin orbit coupling is often removed from the kinematics. In addition, the relativistic features of the electron-electron interaction are much more easily included in a local density approximation in solids. We will consider the approximations to the kinematics briefly but more carefully examine the inclusion of the relativistic interaction terms.

---

*Work supported by the U.S. Department of Energy and National Research Council of Canada.

The effects of the relativistic contributions can be divided according to whether or not they break the symmetry of the problem. The "scalar" effects such as the mass velocity and Darwin terms primarily affect the relative position of orbitals with differing ℓ- (i.e., j-) character. For example, they produce s-d shifts in transition metals. These shifts are closely related to the solid state chemistry of the heavy transition metlas, rare-earths, and actinides. The spin-orbit coupling breaks the symmetry of the system and will thus lift degeneracies It can thus have dramatic effects but is often less significant than generally thought. However, it does affect the magnetic properties of solids and this is really very poorly understood in metals.

In these lectures, we are going to focus on the relativistic effects occurring in metals and predominantly in transition metals. The semiconductors and insulators will be mentioned briefly (in the figures) but the language we will adopt is not always the most appropriate. For example, in the covalent semiconductors, it would be better to discuss relativistic effects on bond lengths and strengths[1-4] (bonds are shortened but can be either weakened or strengthened, rather than the metals-oriented concepts at the root of our discussion. Nevertheless, much of what follows applies with equal force to the valence and conduction levels of all solid state systems.

In the classic Sommerfeld model of a metal, the valence and conduction electrons would be quite well described in the non-relativistic approximation. Only electrons externally injected with a large kinetic energy would need any relativistic corrections applied. Within the Sommerfeld model, we can define the Fermi velocity in terms of the Fermi wave vector

$$V_F = k_F/m = \frac{1}{m}\left(\frac{3}{8\pi}N\right)^{1/3} \qquad (1)$$

which in turn relates to the number density of electrons. Then for N electrons per unit cell whose volume is a^3/f ($f = 2$ for bcc and $f = 4$ for fcc), one finds

$$V_F/c = 0.0019\,\frac{[Nf]^{1/3}}{\text{Å}} \qquad (2)$$

So for an fcc metal with 2 electrons per unit cell with a = 4 Å, $(V_F/c) \sim 0.001$ which is clearly in the non-relativistic regime. However, the Sommerfeld model ignores the cores of the atoms and the strong attractive potentials found there. The picture to be evoked then is that the conduction and valence electrons course through the solid in a manner quite adequately described by non-relativistic kinematics (and instantaneous Coulomb interactions) except when they approach the core regions of the atomic constituents. The success of the old Sommerfeld model for simple metals is normally explained within pseudopotential theory by noting that the atomic cores occupy only a small fraction of the volume of the solid and that core orthogonalization produces an effective repulsive potential which cancels much of the strong core Coulomb potential. One could conceive of utilizing the same approach for the relativistic effects where one treated the cores and core region with relativistic kinematics but the full solid in a more non-relativistic fashion. This is certainly possible in pseudopotential theory but the actual details of "switching kinematics" can get complicated.

It is possible to accomplish such a switch in kinematics utilizing approaches that partition space such as the Augmented Plane Wave (APW) method and Greens Function or Korringa-Kohn Rostocker (KKR) method as well as more approximate versions of these. That partitioning occurs quite naturally in the development of these computational techniques. Here we will mention only a few relevant features of these techniques. The reader interested in computational techniques is referred to the more detailed discussion and references in Appendix A. In discussing the nature of relativistic effects in solids, we will primarily utilize simpler illustrative schemes such as the ASA, discussed in Appendix B.

The electronic potential in a solid is a complicated three dimensional function. Much of its essence, however, can be embodied in a very simple shape approximation. Around each atomic site, one draws non-interpenetrating spheres. Within the spheres, the potential is approximated by its spherically symmetric component. In the remaining interstitial region, the potential is approximated by its average (i.e., constant) value. This is the famous muffin-tin approximation[5] named for its appearance in two-dimensional drawings. It is actually the first term of a dual expansion in spherical harmonics inside the muffin-tin spheres and in plane waves in the interstitial region. It is also strongly physically motivated. Within the muffin-tin spheres is an atomic-like region which should be treated with a central field. On the other hand, the interstitial region is the region of the Sommerfeld model which was based on a constant potential. The approximation turns out to be astonishingly good even in cases where one might expect it to break down. One can appreciate that it should be very good in cubic systems since symmetry dictates that the L = 1,2 and 3 terms of the in-sphere spherical harmonic expansion should be zero. Thus, it is the L = 4 term which is the first term neglected. Its major effect is to couple $\ell = 2$ (i.e., d) states with each other to provide the t_{2g}-e_g crystal field splitting. The remaining components of the potential in this decomposition always have far less effect than the muffin-tin components. Thus, one's basis functions can be constructed using the muffin-tin shape approximation and then the final calculation performed utilizing either only the muffin-tin component or the full potential (the choice basically depending on a cost-benefit analysis). These basis functions always consist of atomic like solution inside the muffin-tin spheres. If one uses a plane wave basis in the interstitial region, one has an augmented plane wave or APW method. If one uses a solution obtained from a Green's funtion, one has the Green's function or KKR method. (See Appendix A.)

At the muffin-tin surface, one must match the wave (or basis) function and its derivative. For the KKR method applied to a muffin-tin potential, it is this matching condition which yields the secular equation. For the APW method, the standard procedure is to match the functional value in the basis function and include the matching of derivatives as a part of the variational procedure which generates the secular equation. However, more modern approaches, which linearize the problem by incorporating the energy derivative of the central field solutions (inside the muffin-tin), actually match function and derivative at the boundary in forming the basis functions. The variational quantity yielding the secular equation then need not be concerned with insuring the continuity of the first derivative.

The essential feature of the APW and KKR techniques is that space has been partitioned and different functional forms have been used in the two regions of space. Clearly, it is just as possible to switch kinematics at the boundary. Because the boundary is in a region of space where the relativistic effects are quite small, this transformation can be done easily. This view is summarized pictorially in Fig. 1. We can set up a formal scheme based on the Foldy Wouthuysen transformation. This transformation is a series of point transformations successively reducing the magnitude of odd terms (i.e., those coupling large and small components) of the Hamiltonian. The second order Foldy Wouthuysen transformation yields the familiar Pauli Hamiltonian with spin-orbit coupling. The transformed wave function and Hamiltonian thus appear as

$$\tilde{\psi} = T\,\psi_D$$

$$\tilde{H} = T\,H\,T^{-1}$$

$$T = \exp\left[-i\left(\frac{1}{4m^2}\,\alpha\cdot\nabla V\right)\right] \qquad (3)$$

$$\exp\left[-\frac{i}{6m^3}\,(\underline{\alpha}\cdot\underline{\rho})^3\right]\,\exp\left[\frac{-i}{2m}\,(\beta\,\alpha\cdot\rho)\right]$$

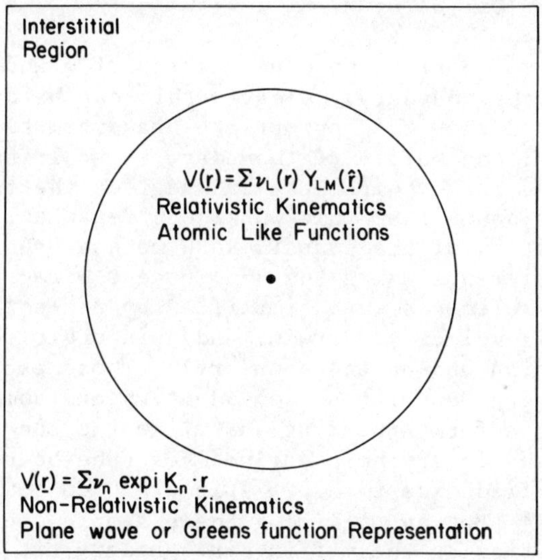

Fig. 1. Schematic representation of the unit cell.

The standard use of this transformation is to evaluate $\tilde{H}$ as a commutator expansion and then solve the wave equation involving $\tilde{H}$ to obtain $\tilde{\psi}$. However, because it assumes the dominance of the rest mass energy term, this expansion is actually slowly convergent in the extreme relativistic limit (near the nucleus) so we will take a different approach. In our case, we are dealing with matrix mechanics involving basis functions.

$$H_{ij} = \langle \phi_{Di} | H_D | \phi_{Dj} \rangle = \langle \tilde{\phi}_i | \tilde{H} | \tilde{\phi}_j \rangle \tag{4}$$

Because the Foldy Wouthuysen transformation is a point transformation and we are dealing with a local potential, the matrix element can be broken up into the contributions of the two regions of space.

$$H_{ij} = \langle\tilde{\phi}_i|\tilde{H}|\tilde{\phi}_j\rangle_{MT} + \langle\tilde{\phi}_i|\tilde{H}|\tilde{\phi}_j\rangle_I$$

$$H_{ij} = \langle\phi_{Di}|H_D|\phi_{Dj}\rangle_{MT} + \langle\tilde{\phi}_i|\tilde{H}|\tilde{\phi}_j\rangle_I \tag{5}$$

It remains only to specify the construction of ϕ and in particular the matching at the boundary. Clearly this can be done in either representation and then the appropriate transformations performed. But, in practice, the muffin-tin boundary is so located that one can just join the large components and neglect the further corrections. As evidence that this works out reasonably, we may consider the history of the classic RAPW method. Loucks[6] worked out the matrix elements using the 4-component Dirac representation. His final matrix elements were simplified by neglecting some very small terms (mass-velocity, Darwin, and spin orbit coupling in the interstitial region and on the boundary). Those expressions were transformed using a Bessel function identity and some reorganization of terms to a form appearing just like the non-relativistic form only with the logarithmic derivatives (the boundary value parameters) modified except that a spin orbit coupling term is also present.[7-10] But by utilizing these simple concepts of switching kinematics at the muffin-tin boundary, it is possible to rederive the same results with far less effort and a clearer statement of the approximations involved.[11]

Now we are ready to promulgate a simple view of the relativistic effects in solids as being basically the same as those known in atomic structure but modified by the differing boundary conditions. Let us consider atomic calculations for gold as a prelude to our discussions of the effects in solids. In Fig. 2, we illustrate the widely known relativistic core contraction by comparing the density obtained from a relativistic and a non-relativistic calculation. The core contracton can be understood in a Thomas-Fermi model as follows. The nuclear potential is very attractive and it is energetically favorable for the electrons to move towards it. However, this is counterbalanced by their own Coulomb repulsion (which we are currently treating non-relativistically) and their increase in kinetic energy to satisfy the requirements of the exclusion principal. One way to estimate the kinetic energy as a function of density is to use the value of a non-interacting electron gas. This we show in Fig. 3. Because the kinetic energy rises more slowly with increasing density when treated relativistically than non-relativistically, there will be less kinetic energy repulsion in the relativistic case and the density will contract towards the nucleus.

The contraction of the cores can also be seen quite dramatically by examining the difference in potentials (which is predomi-

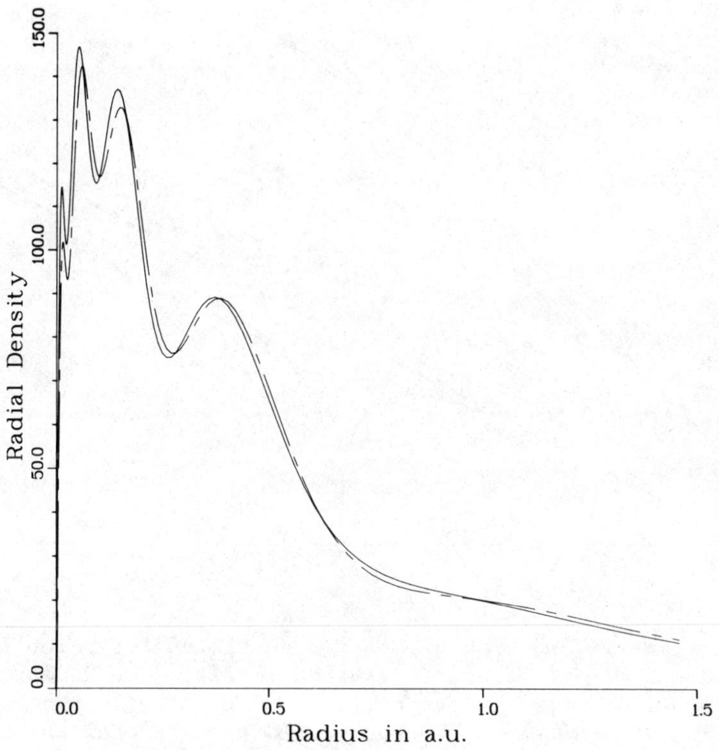

Fig. 2. Relativistic (full curve) and non-relativistic densities for gold. Note that the relativistic charge density is contracted relative to the non-relativistic. To expand the core region, only the results out to 1.5 a.u. are shown. The muffin-tin radius would typically occur at roughly 2.5 a.u.

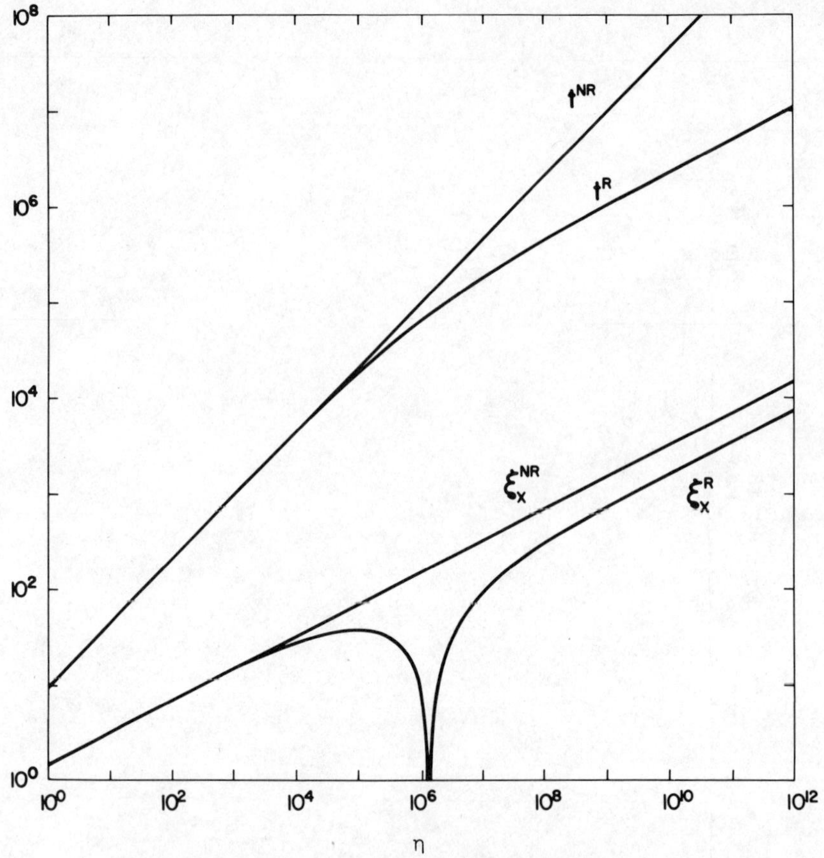

Fig. 3. Kinetic and exchange energy densities for relativistic and non-relativistic electron gases. The kinetic energy density t(n) is proportional to $n^{2/3}$ while the exchange energy density is negative and proportional to $n^{1/3}$ for the non-relativistic gas. For the relativistic gas t(n) is reduced due to mass-velocity effects and is proportional to $n^{1/3}$ in the ultrarelativistic limit while the exchange-energy density changes sign because of magnetic interactions and is reduced in magnitude by half in the ultrarelativistic limit.

nantly due to the change in Coulomb potential). This we do in Fig. 4 where we present (V^{REL} - $V^{N.R.}$) for gold. Note that near the nucleus V^{REL} is a great deal less attractive than $V^{N.R.}$ due to the increased electronic repulsion due to the core contraction. But also note how quickly it dies off with increasing radius. As a typical muffin-tin radius would be 2.5 - 3. a.u., this is well inside the muffin-tin boundary. We can see the lowering of the

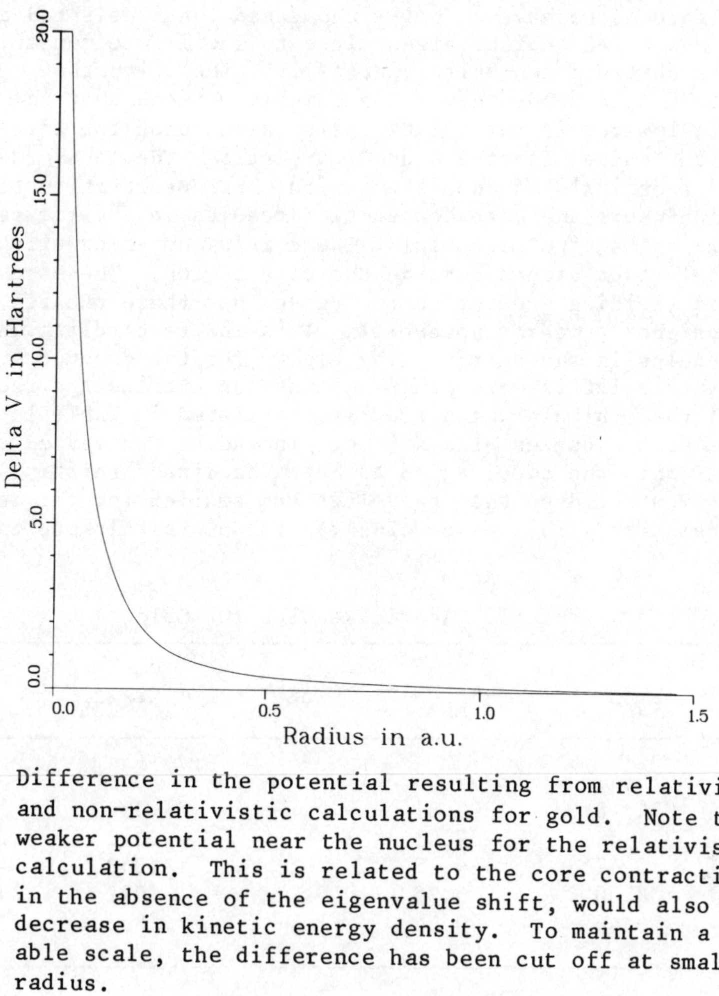

Fig. 4. Difference in the potential resulting from relativistic and non-relativistic calculations for gold. Note the weaker potential near the nucleus for the relativistic calculation. This is related to the core contraction but, in the absence of the eigenvalue shift, would also be the decrease in kinetic energy density. To maintain a reasonable scale, the difference has been cut off at small radius.

kinetic energy in the core region by writing a kinetic energy density as

$$t_n(r) \equiv \psi_n^+ (\varepsilon_n - V)\, \psi_n \tag{6}$$

Although the change in eigenvalue (Table 1) and in wave function will act to increase the relativistic kinetic energy, the effect of the potential change in this expression dominates and results in a net lowering of the kinetic energy density.

In Table 1, we have actually tabulated the j weighted average of the $j = \ell \pm 1/2$ orbital eigenvalues to facilitate comparison with those obtained non-relativistically. The spin-orbit splitting is then tabulated separately. One quickly notices that the eigenvalues are lowered in the relativistic calculation for the s and p orbitals but raised for the d and f orbitals. The lower energy for the s and p orbitals is consistent with their penetration to the core region where the core contraction results in their experiencing a more attractive potential. There is a competing effect for the orbitals that extend outside the core region. The core has contracted yielding greater screening so that these outside orbitals experience a weaker potential. This indirect relativistic effect results in the energy being higher for the d and f orbitals. To make this a little more concrete, one can examine the radial extent of the orbitals which are also tabulated in Table 1. One finds that the s states always (i.e., including the 6s) extend adequately into the core region to have the direct relativistic effect dominate and so the $\langle r \rangle$ values are smaller for the relativistic kinematics. This also holds for the p orbitals but there

Table 1. Atomic Results for Gold

	$-\bar{E}_{REL}$	$-E_{NR}$	ΔE^{SO}	$\overline{\langle r \rangle}_{REL}$	$\langle r \rangle_{NR}$
1s	5923.41	5367.05	-	.0017	.0019
2s	1044.40	895.81	-	.0070	.0080
3s	246.47	209.61	-	.018	.020
4s	53.47	44.18	-	.040	.045
5s	8.00	6.25	-	.092	1.01
6s	0.46	0.34	-	2.85	3.38
2p	911.84	861.48	134.79	.0063	.0068
3p	207.43	193.44	29.27	.018	.019
4p	40.51	37.18	7.16	.043	.045
5p	4.52	4.03	1.22	1.05	1.10
3d	161.72	163.05	6.39	.017	.017
4d	23.91	24.29	1.33	0.45	0.46
5d	0.55	0.63	0.12	1.53	1.49
4f	6.00	7.00	0.28	0.048	0.048

is a sizable cancellation. As for the d orbitals, the cancellation appears to be nearly complete for the 3d and 4d orbitals but the indirect relativistic effects dominate in the radial extent of the outer 5d orbital. (Especially the $5d_{5/2}$ orbital. The $5d_{3/2}$ orbital is slightly contracted.) The difference between the results for the energy and the radial extent is, of course, due to the fact that they apply different weights to the different regions of space. The 4f has roughly the same radial extent as the 4p and 4d. Thus, although it is excluded from the really small radius region, it still occurs far enough into the core to have the two effects nearly cancel: the orbital is expanded but kinetic energy lowered. (The contraction or expansion of the 4f orbital is actually dependent on the configuration used.)

It is also noteworthy that the contraction or expansion is much more dependent on the size of j than ℓ: The difference found for the $4p_{1/2}$ is about the size of the $4s_{1/2}$ while the $4p_{3/2}$ shows a smaller difference. The $4p_{3/2}$ and $4d_{3/2}$ show contractions of roughly the same size while the $4d_{5/2}$ is much closer to the non-relativistic orbital. The 4f's are much more sensitive so the difference of the $4f_{5/2}$ is much larger than the $4d_{5/2}$ but again you do have the ordering that the higher j states has a smaller difference. Even in light of this approximate j dependence, we will persist in our discussion to use a forced ℓ, s representation just for familiarity.

When examining the spin orbit splittings for the orbitals, one sees that they are rapidlly decreasing as a function of principal quantum number. This is easily understood from simple considerations of the Foldy-Wouthuysen transformed spin orbit operator for the central field:

$$H_{so} = \frac{\alpha^2}{2} \underline{L} \cdot \underline{\sigma} \left(\frac{dV}{dr}\right) \qquad (7)$$

As dV/dr increases dramatically going to small radii, the orbitals with the lower principal quantum number have the larger spin orbit coupling. The spin orbit is quite sensitive to the small r region as can be seen by examining the orbitals with the same principal quantum number. These orbitals all have very nearly identical radial extents as measured by the radial expectation values. However, the spin orbit coupling energy drops dramatically in spite of the fact that the magnitude of L is increasing. This sensitivity should be remembered when we discuss the solid state modifications to the relativistic effects.

In order to easily characterize some of the relativistic effects in solids, we need to introduce a few simple indices. The approach we are going to use is best adapted to face centered cubic and body centered cubic materials so we are going to treat all materials as having one of these two structures. Of course, many materials do not occur in these structures so we will make the following mapping to a fictitious structure. Those materials with the hexagonal close packed structure will be treated as fcc materials because they are closely related structures. Those with more unusual structures will be treated as bcc materials. The lattice constants of these fictitious solids will be chosen to maintain the same density as the actually occuring solid.

It is a reasonable approximation for the fcc and bcc structures to replace the actual unit cell with a spherical cell of the same volume. This was the approach used by Wigner and Seitz.[12] They then identified the bottom of the band as occuring when the radial function had zero derivative at the Wigner-Seitz sphere radius. The top of the band occurs where the radial solution goes to zero at this boundary. Using a formalism known as the linear combination of muffin-tin orbitals (LCMTO) theory,[13] one can make an approximation known as the atomic sphere approximation (ASA) discussed in Appendix B. It is basically the generalization of the Wigner-Seitz spherical cell method to the optimal inclusion of the boundary conditions which represent the periodicity requirement of the wave function. From the ASA treatment of these systems, one can identify one additional index which is the center of the band. This occurs at the energy where the logarithmic derivative of the radial soluton equals that of a spherical Neuman function in the limit of zero energy i.e., $-(\ell + 1)/R_{WS}$. This is the energy where the muffin-tin orbital most resembles an atomic like function. It should be noted that in this simple analysis we are assuming states of different angular momentum do not mix. This is a fairly crude approximation (and one which we will repair only much later) but it does allow us to extract the sense of the matter fairly effectively.

If we are to examine the trends across many systems, we need to produce potentials with great efficiency yet with reasonable realism. This we do using a time honored prescription for a model potential known as the overlapping charge density model. To set up this model, we first perform a self-consistent atomic calculation thereby properly setting up the core structure. We then place this charge density on each atomic site-without modification - and sum to obtain a model charge density. One then needs only the prescription to turn a charge density into a potential. The Coulomb contribution is straightforward and, for historical reasons, we choose the Slater exchange approximation.[14] (In many cases, this stronger exchange functional mocks up the effects of self-consistency.) The muffin-tin component of the resultant potential forms the basis for our model study. The model is admittedly not the most elegant but it is efficient, simple, and actually quite adequate to demonstrate the relativistic shifts we are interested in.

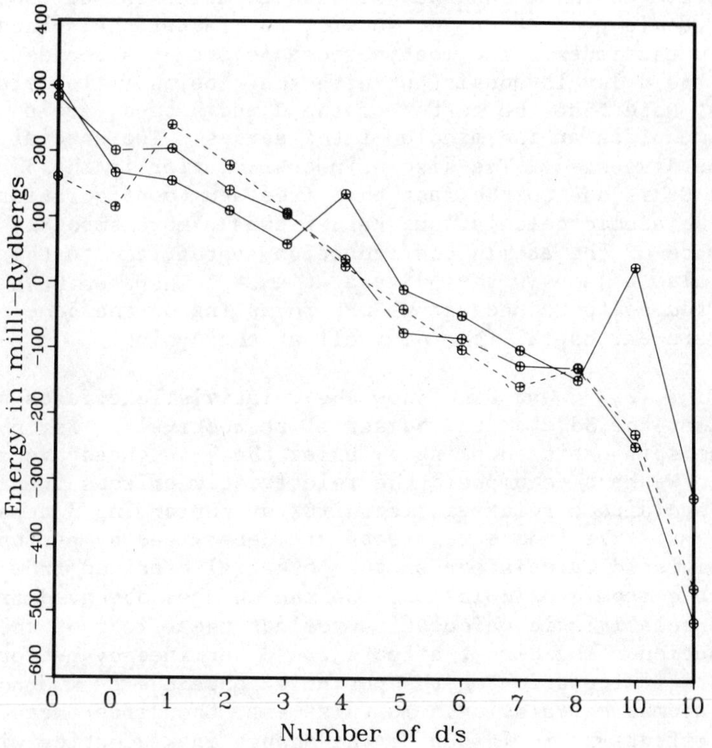

Fig. 5. Position of the d band center relative to that of the s band for the 3d (solid line), 4d (chain-dashed line), and 5d (dashed line) series. We have used a nominal number of d orbitals as the ordinate. This has no physical significance. The ordering is strictly that of increasing atomic number starting at the alkali metal in the appropriate row of the periodic table. The two anomalies are the 3d results at Cr and Cu due to the fact that the strict atomic configurations were used. These are non-relativistic results.

In the transition metals, the parameter of most interest is the relative position of the d bands ($\ell = 2$) relative to that of the s bands. In Fig. 5 we plot the variation of the position of the center of the d band with respect to that of the s band. The band structure expert might prefer to have the position relative to the bottom of the s-band but we wish to maintain our contact with the atomic properties and so will not factor the s-band-width into the index. The most noticeable trend is the downward shift of the d levels consistent with their being filled with electrons. Note that the center of the d and s bands are nearly at the same place at the middle of the series. Two special points need to be discussed. The discontinuous behavior in the 3d series at Cr and Cu is due to the fact that free atom configurations were used in the atomic calculations rather than a more smoothly varying estimate of the atomic configuration appropriate to the solid. There is also a jump in the 5d data at $n_d = 1$ where we have skipped from Ba to Lu because we are focussing on the 5d elements and the rare-earths fill the 4f shell at that point.

In Figs. 6, 7, and 8 we show the relativistic effects on this quantity for the 3d, 4d, and 5d series respectively. For now, we remove the spin orbit coupling by using the j-weighted average quantity. We have decomposed the relativistic effects into the indirect and direct relativistic shifts by performing 3 separate computations. The indirect effects are determined by performing a non-relativistic calculation on the potential obtained from a relativistic atomic calculation. So the changes observed from a fully non-relativistic calculation reflect the effect of the charge redistribution. The direct effects are determined by performing a relativistic calculation on the potential obtained from a non-relativistic atomic calculation. In this case, the differences represent the effects strictly due to the change in kinematics without a modification in charge. We also show the difference observed when comparing the fully relativistic and fully non-relativistic calculations and improperly refer to it as the total effect. It is seen that for the valence and conduction states the direct and indirect effects are additive to a pretty good approximation. There is a third effect present in metals which is the shift in occupation numbers due to these energy shifts. That feature will destroy much of the additivity. It will also diminish the size of the relativistic effects on this quantity since the shift of electron density from d- to s-orbitals will pull down the energy of the d-orbitals due to decreased intraatomic Coulomb interations. This we show for the 5d states in Fig. 8 using an extension of our model which we will discuss shortly.

Examining Fig. 6, 7, and 8 one sees immediately the increased size of the relativistic effects in going from the 3d to 4d to 5d series. The 3d shifts are small but not as small as many people would believe. The direct relativistic effects are dominated by the

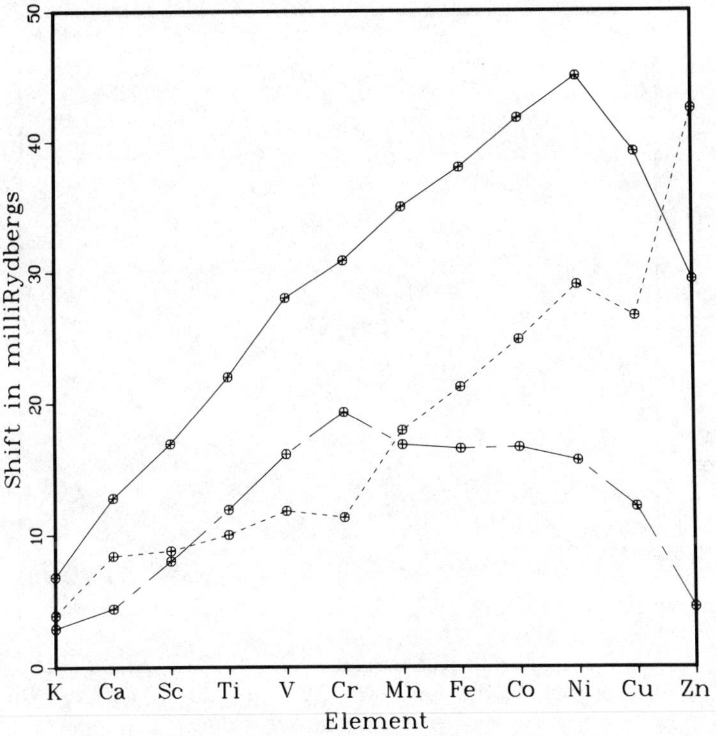

Fig. 6. Relativistic shift of the d-s position for the 3d series. The solid line represents the total shift. The chain-dashed line is the direct relativistic effect and the dashed line the indirect effect.

lowering of the s states thereby providing an upward shift in the d-s separation. The d states are also shifted downward-but by a much smaller amount. The indirect relativistic effects shift the d and s states up but the upward shift of the d states is dominant. However, both factors are such that they add rather than cancel.

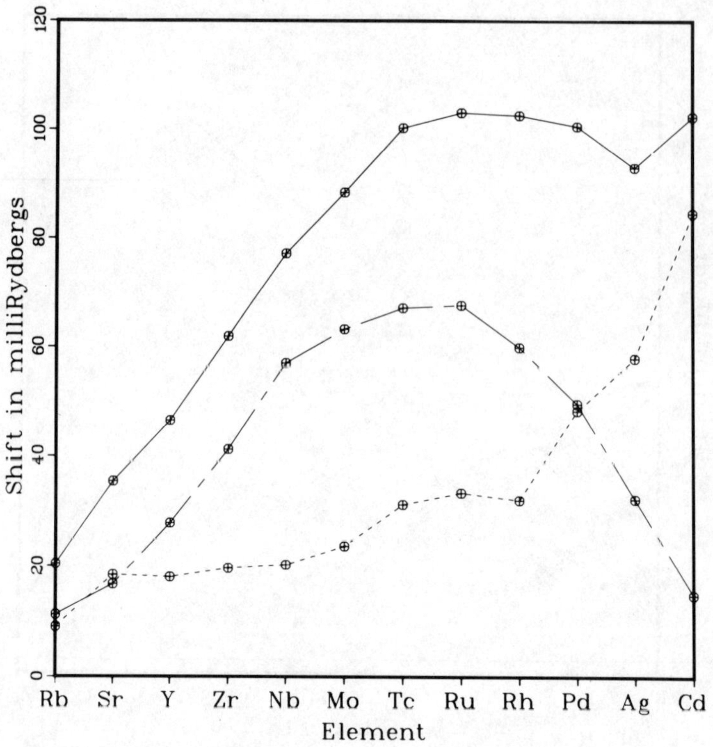

Fig. 7. Relativistic shift in the d-s position for the 4d series.

The indirect effects are monotonically increasing through the series reflecting the increasing contraction of the cores. The direct relativistic effects first rise, as expected, and then fall off. The direct effects first rise in the lower half of the series due to the s-states having an appreciable amplitude near the nucleus where the increased velocities due to the increased nuclear charge are acutely experienced. However, near the top of the series, the d-orbitals are far enough inside the atom to also have

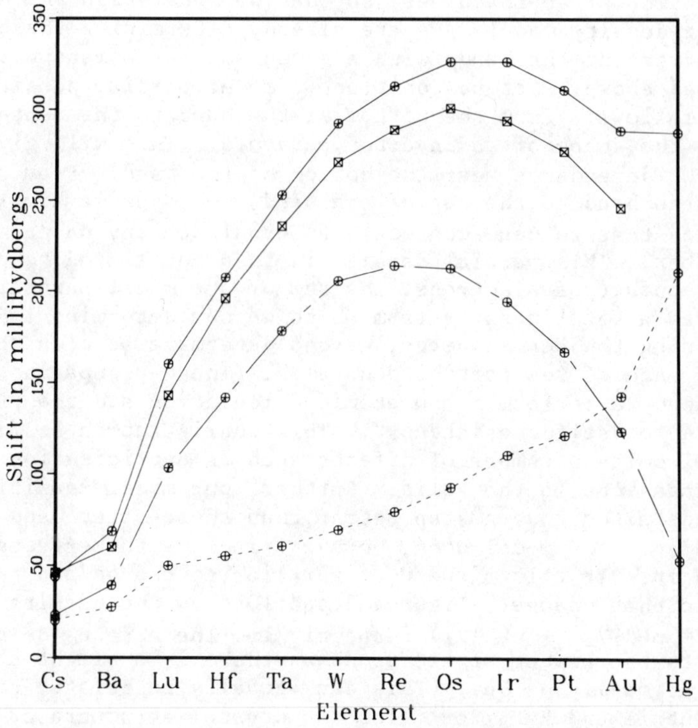

Fig. 8. Relativistic shift in the d-s position for the 5d series. The additional solid curve with the square symbols reflects the effect of configuration readjustment.

a significant relativistic shift and thus diminish the relative s-d shift. This is because the d-orbitals are being more dramatically pulled in with increasing atomic number and are going to become core states before the s states. (Otherwise you wouldn't have the simple metals and semiconductors on the right hand side of the periodic table. Remember also that it is the 3d-4s, 4d-5s,

5d-6s combination.) Thus they are contracting into the region where the direct relativistic effects are occuring.

To actually examine the repopulation effects, we must add that feature to our model. One could do this through the ASA but we will use a simpler approach here so that we can retain our overlapping charge density model. We are already determining the bottom, center, and top of the bands with a given angular momentum character as described above. Let us construct a model partial density of states as follows. From the bottom of the band to the center, it is to have the shape of an inverted parabola. This will give the correct $E-E_b$ dependence near the bottom of the band. From the center of the band to the top of the band, we adjust a single power $(E-E_c)^\gamma$ such that we can accomodate an equal density on each side of the center. This partial density is then normalized to contain the proper number of electrons. The sum of our model partial densities gives a total density from which we can determine a Fermi energy. Using the Fermi energy, we can determine an occupation number for each of the partial densities. These occupation numbers are then used to perform a new atomic calculation and the process is repeated to "self-consistency". This configuration self-consistency model omits a number of effects such as hybridization and orbital expansions in the solid. Further, our model densities are pretty crude although one step better than those often used in alloy studies. The model does, however, include the effects we are interested in. It allows the configuration to change and readjusts the core to that change. Figures 9 and 10 show the results for n_s and $n_d = n(d_{3/2}) + n(d_{5/2})$ respectively. The $n_s + n_d$ do not quite sum to the number of electrons as there is a slight n_p contribution primarily at Lu. (This can actually be related to the model of Duthie and Pettifor[15] for the crystal structure of the rare-earths.) The p-character actually lies quite high in the transition metal bands as can be seen from Fig. 11 where we show the non-relativistic p-s separations comparable to Fig. 5 for the d-s positions. Although these results come from basically atomic like considerations, they are consistent with combined (interpolation) scheme arguments which treats the d-bands via a tight-binding scheme and the s-p bands as orthogonalized plane waves.[16-17] The s-p or nearly free electron (NFE) bands have a roughly k^2 energy dependence and are predominately s-like near $k = 0$ and much more p-like at the edge of the Brillouin zone (large k). This is easily seen from the Rayleigh expansion of a plane wave

$$e^{i\underline{k}\cdot\underline{r}} = \sum (2\ell + 1)\, i \quad j_\ell(kr)\, P_\ell \quad (\hat{k}\cdot\hat{r})$$

and the functional dependence of the spherical Bessel functions. The d-bands occur in the middle of the NFE or s-p bands and so the p character is predominantly above the d-character. This is even more enhanced by the hybridization or Fano antiresonance which

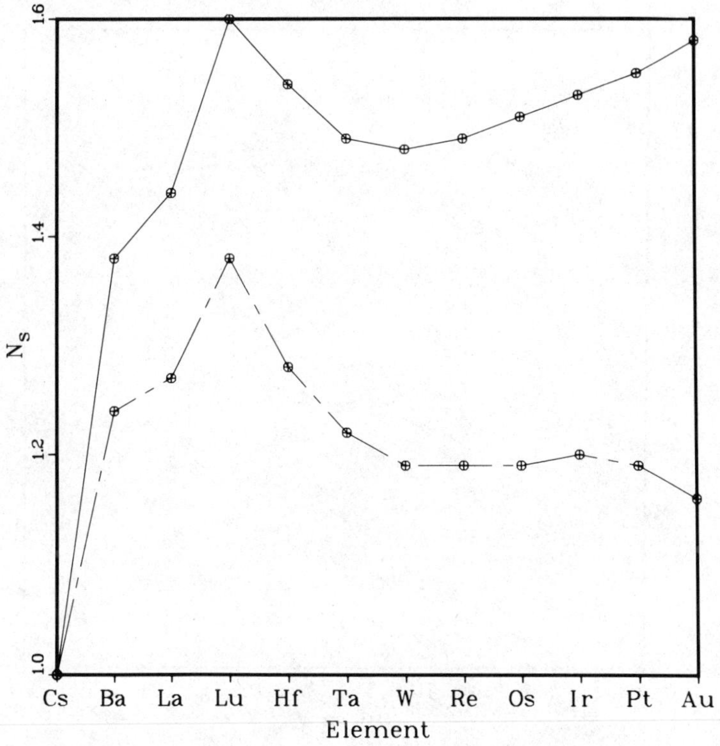

Fig. 9. s-Character found for the 5d series. The solid line is the relativistic result and the chain dashed line is the non-relativistic result. Note the dramatically reduced s-character found non-relativistically.

forces the NFE density out of the body of the d-density. In our model, this effect can be seen to be caused by the d-density occurring in a region just inside the principal maximum of the p-states. Then as the d contract to become more core like, this large Coulomb repulsion is reduced and the p states are lowered. For complete-

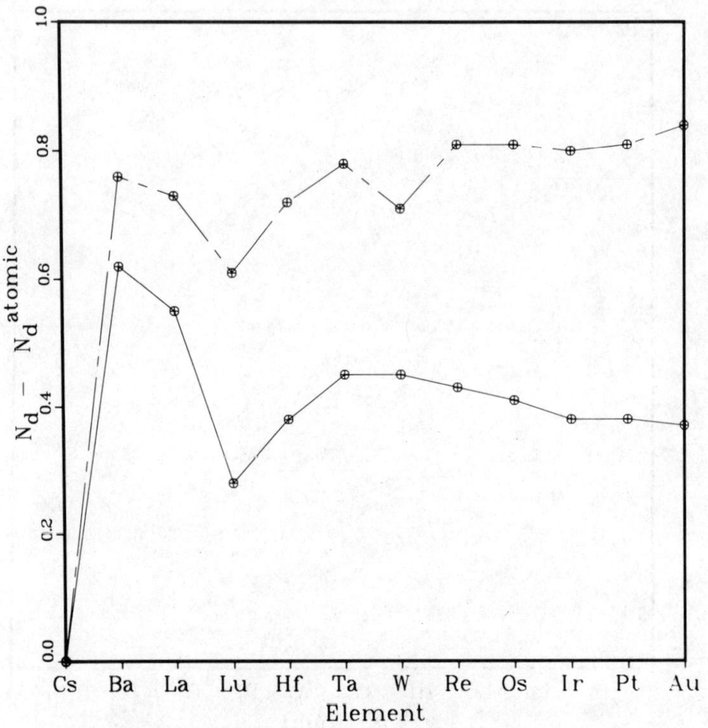

Fig. 10. d-Character found for the 5d series. Here we have subtracted an atomic-like number of d's to remove the simple linear dependence on Z due to the filling of the d shell. The non-relativistic result (chain dashed line) lies above the relativistic result (solid line) consistent with the opposite dependence in n_s.

ness we show in Figs. 12-14 the effect of relativistic kinematics on the 4(p-s), 5(p-s), and 6(p-s) separation through the transition element series and nearly free electron simple metals and semicon-

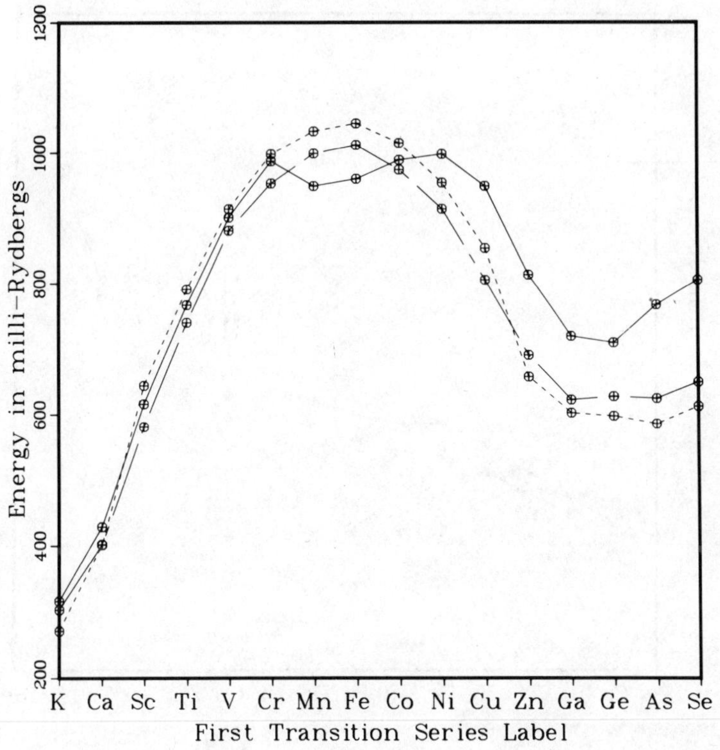

Fig. 11. Position of the p-band center relative to that of the s-band for the first (solid line), second (chain dashed line) and third (dashed line) transition rows.

ductors. In this case, the direct relativistic effect is a monotomic dependence increasing with atomic number representing the fact that the s states penetrate to smaller r.

The sudden rise in the direct relativistic contribution start-

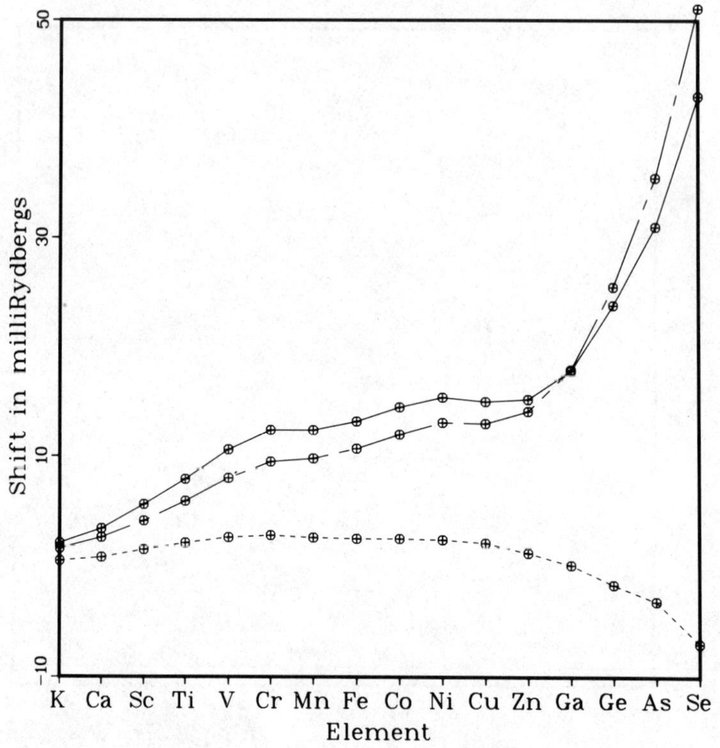

Fig. 12. Relativistic p-shift for the first transition series row. Solid line is the total relativistic shift, chain dashed is the direct relativistic effect and dashed line is the indirect effect.

ing at the noble metals occurs because the s-electrons are being occupied and pulled into the core region with the d-shell no longer holding them out as effectively. The energy of the p-states is also starting to fall as can be seen from Fig. 11 but they are still further out and higher in energy. The indirect relativistic

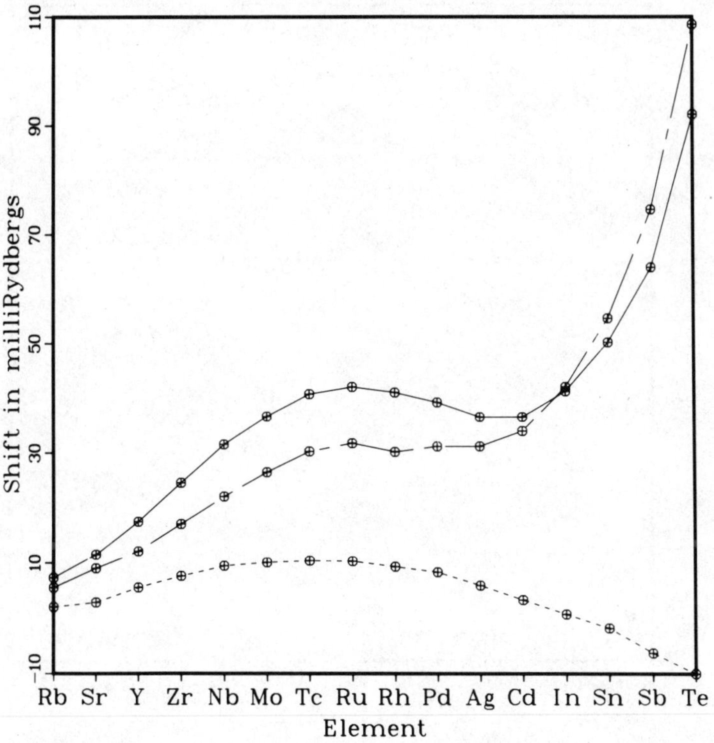

Fig. 13. Relativistic p-shift for the second transition series row.

effect first increases through the transition metal series but then decreases as the d-shell is filled and changes sign. To understand this one must note that the s and p orbitals are both experiencing rapidly increasing indirect relativistic effects but that the s is increasing more rapidly. This is because the charge redistribution yields its major potential energy changes in the outer regions of the core and this is precisely the region that the s- and p-orbitals are being collapsed into with the s-orbital "getting there first".

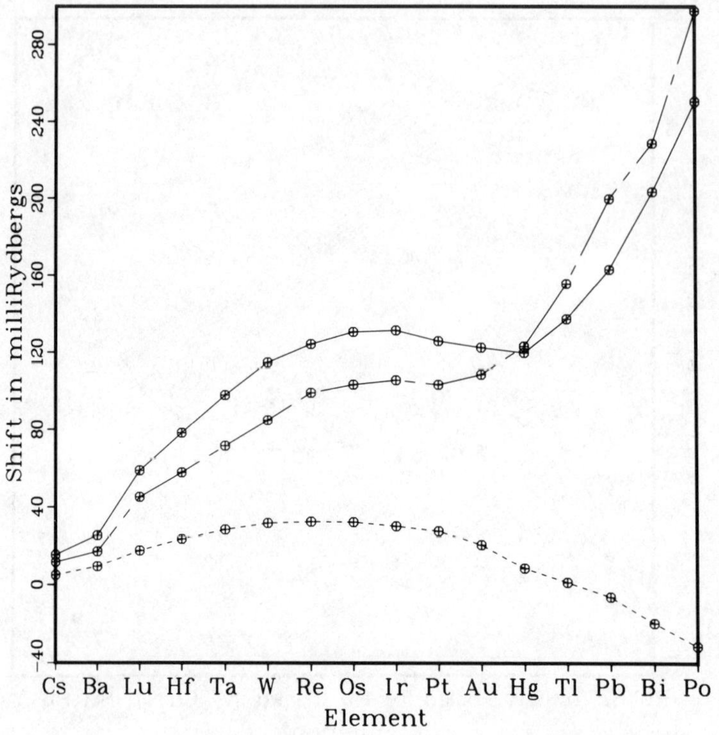

Fig. 14. Relativistic p-shift for the third transition series row.

In Figs. 15 and 16 we show the f-s shifts in the rare-earths and actinides. The actual positions of the 4f levels often have very little significance in a band calculation but one can see the sensitivity of the 4f levels to the charge density shifts which we are calling the indirect relativistic effects. The 5f states in the actinides are more often reasonably described by a band description. Again we see that the dominant effect on the 5f levels is through the indirect or charge redistribution effects. Note the

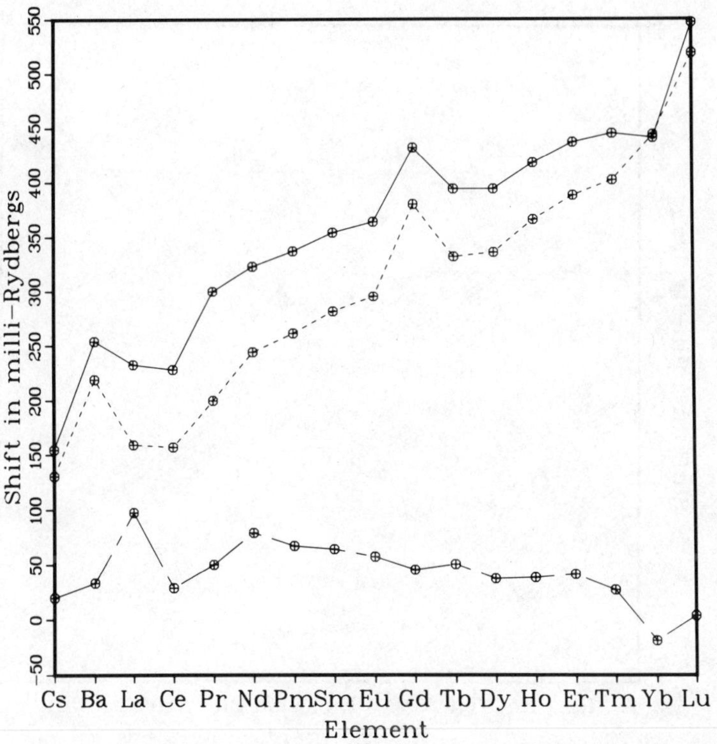

Fig. 15. f-s Shifts occurring in the rare-earths. Although the 4f position in a band structure often has very little significance, one can see the extreme sensitivity of the 4f's and the dominance of the indirect (dashed line) relativistic effects for these orbitals.

large upward shift of the 5f levels. There is a large upward shift of the 6d levels as well. In other words, a non-relativistic approximation will cause the s-states to be shifted upward relative to the d and f levels. This resulted in the early suggestion that

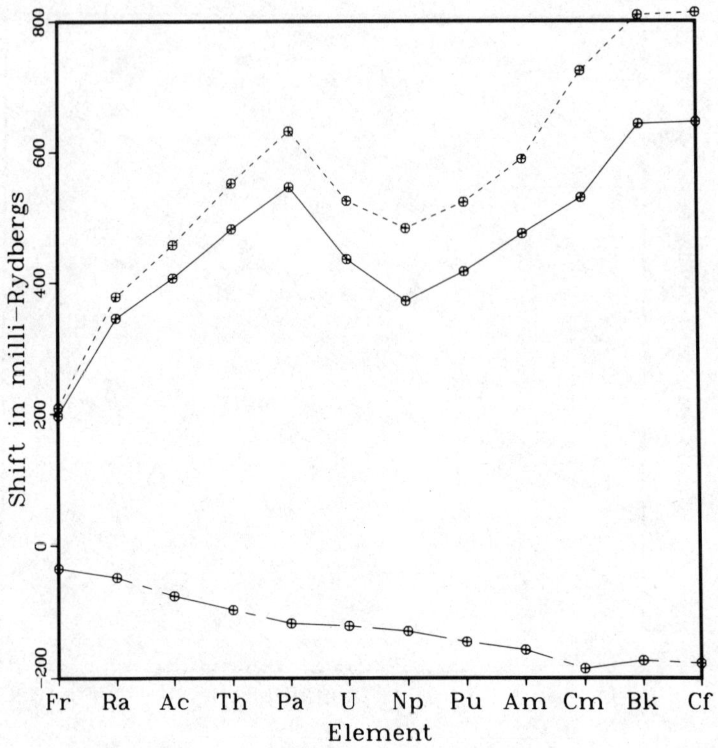

Fig. 16. f-s Shifts occurring in the actinides. Again note the dominance of the indirect effects (dashed line).

the actinides were like a transition metal where the d-states played the role of the s states and the f states the role of the d-states.[18,19] This, however, is incorrect. The s-p states are present in the metal and one has a situation of electrons with 3 different characteristics interacting. This we show in Fig. 17.

The next parameter of interest is the width of the bands. In

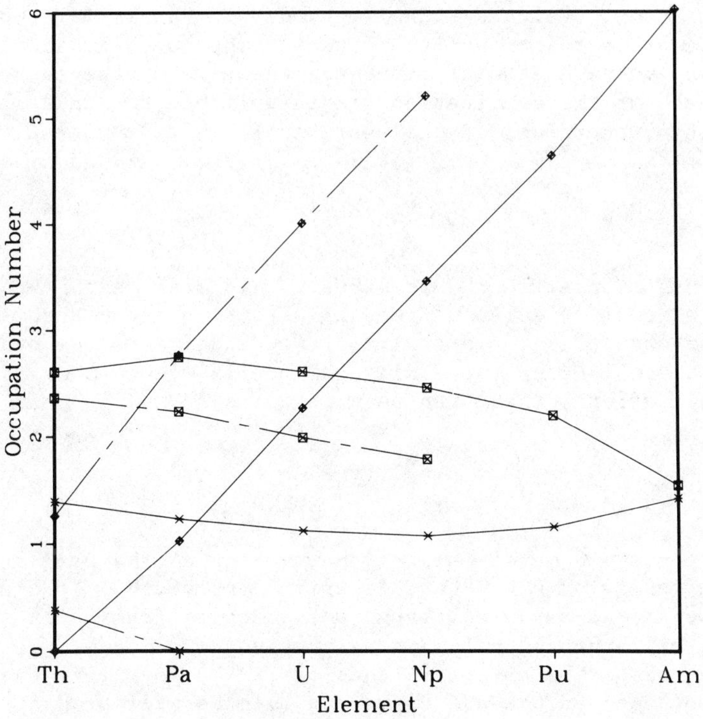

Fig. 17. Configurations obtained from our simple model for the actinides using relativistic and non-relativistic kinematics. The solid line connects the results obtained relativistically and the chain-dashed line connects those obtained non-relativistically. The s-orbital occupation is represented by an asterisk. It is non-zero only for thorium when the calculations are done non-relativistically. The d-orbital occupations are denoted by squares. They differ much less than that of the s- and f-orbitals being just above 2 in almost all cases. The f-orbital occupations are denoted by the diamonds. Note that the non-relativistic approximation greatly enhances the f-character. This together with the great sensitivity it induces forced us to stop the non-relativistic calculations at Np as Pu was just to sensitive to obtain a reasonable result for a reasonable effort. These problems did not occur in the relativistic calculations.

the case of the transition metals, the relative position of the d bands and their widths are really the only free parameters to be determined. Again, we will want to use a relatively simple model to focus on the relativistic effects. We can, in fact, get the required insight from a dimensional analysis of the linear combination of atomic orbitals (LCAO) method. In the LCAO method, one writes the basis functions as a linear combination of atomic orbitals placed on each atomic site-hence the name of the method. The coefficients of the combination are established to insure the proper periodicity properties. The classic choice (it is not unique)[20] is

$$\phi_n(k,r) = \frac{1}{\sqrt{N}} \sum_{\underline{R}} e^{-ik\cdot\underline{R}} \phi_n^{at}(\underline{r}-\underline{R}) \tag{8}$$

where $\underline{R}$ runs over all N atomic sites. This basis set is then used in a variational calculation resulting in a standard secular equation. For our very simple analysis, we will ignore the overlap matrix and consider only the diagonal matrix elements of H. With very little effort, these can be written as

$$H_{nn} = \sum_{\underline{R}} e^{ik\cdot\underline{R}} \int d^3r\, \phi_n^{at}(\underline{r}-\underline{R})\, H\, \phi_n^{at}(\underline{r}) \tag{9}$$

The term for $\underline{R} = \underline{0}$ merely places the average of the band energy. It is the terms for $\underline{R} \neq 0$ that determine its width. To crudely evaluate them, we may use a large r asymptotic form as it is only near $r \sim |R|/2$ that both functions have any amplitude. In that region, a reasonable ansatz is that $\phi \propto r^{-\ell-1}$ much like the muffin-tin orbital used in the ASA (Appendix B). We will need to know something about the extent or amplitude (i.e., normalization) of ϕ^{at} and this can be fixed by dimensional arguments. Since a wave function must have units of $L^{-3/2}$, an atomic orbital with extent $\langle r\rangle$ should have the form $\phi^{at} \sim \langle r\rangle^{\ell-1/2}/r^{\ell+1}$. Because we are merely doing a dimensional analysis, we can use $H \rightarrow p^2$. If we now scale all lengths in the integral by the Wigner-Seitz spherical cell radius, we get

$$H_{nn} \propto \frac{\langle r\rangle^{2\ell-1}}{R_{ws}^{2\ell+1}} \sum e^{i\underline{k}\cdot\underline{R}}\, h\left(\frac{R}{R_{ws}}\right) \tag{10}$$

range of variation of $\underline{k}\cdot\underline{R}$ is unaffected by a scaling since, as $\underline{R}$ is decreased by a contraction of the lattice, $\underline{k}$ will be increased. The bandwidth will be determined by the prefactor. Thus we have

$$W \sim R_{ws}^{-2} \left(\frac{\langle r\rangle}{R_{ws}}\right)^{2\ell-1} \tag{11}$$

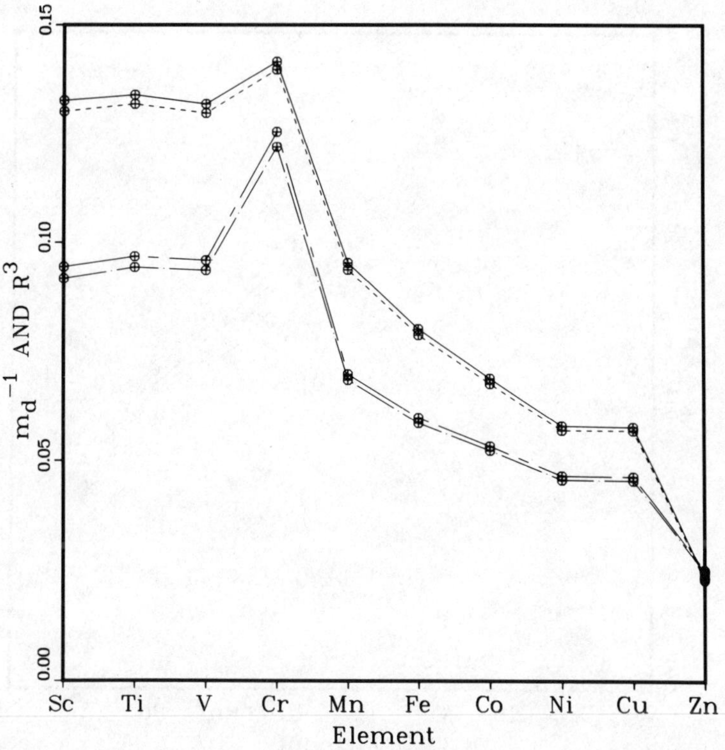

Fig. 18. Relativistic effects on d-band width for the 3d elements. The values of m_d^{-1} (see text) obtained relativistically are denoted by a solid line and that obtained non-relativistically by a dashed line. The values of $R = \langle r^{at} \rangle / R_{ws}$ obtained relativistically are denoted by a chain-dashed line while the non-relativistic values are denoted by a chain-dot line.

The R_{ws}^{-2} is the a^{-2} dependence which scales energies in a crystal. In the nearly free electron bands, it comes from the $E = k^2$ re-

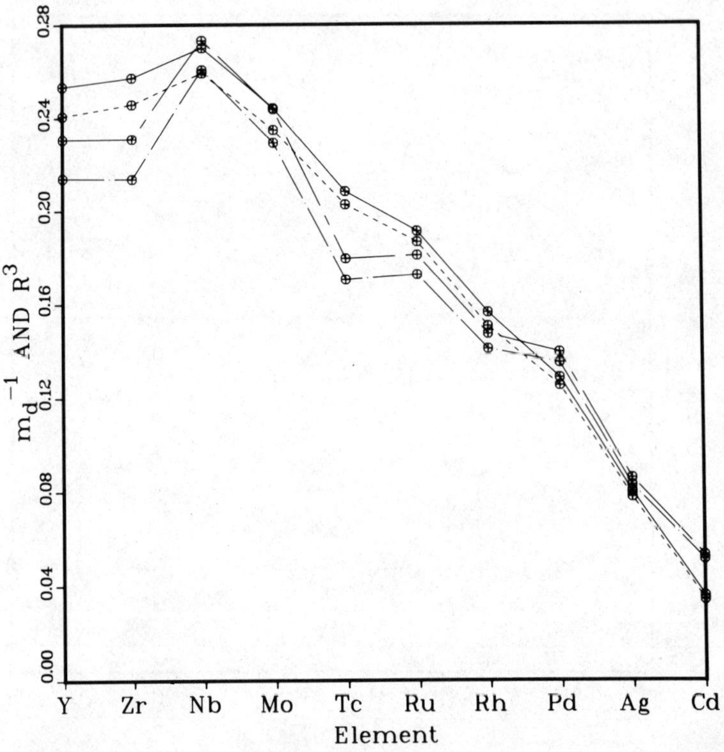

Fig. 19. Relativistic effects on d-band width for the 4d elements. Definition of connecting lines is the same as in Fig. 18. The effect is obviously larger for the 4d elements than for the 3d elements.

lation. The remaining factor yields the effect of expanding or contracting the orbital--sort of a comparison of the size of the peg to the size of the hole. This tells us that, if our simple

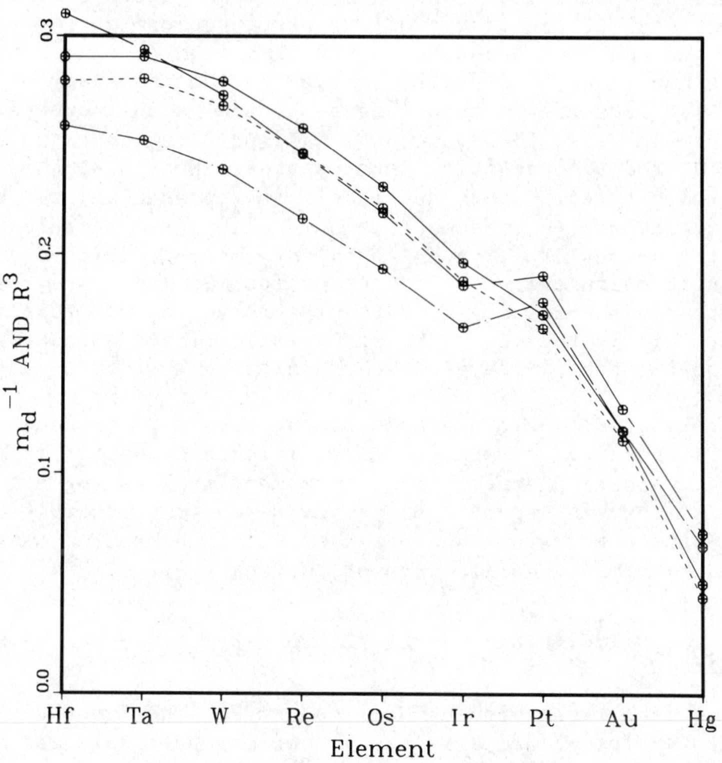

Fig. 20. Relativistic effects on d-band width for the 5d elements. Definition of connecting lines is the same as in Fig. 18.

ansatz is adequate, the effect of the relativistic kinematics on the bandwidth can be related to its effect on the size of the atomic orbital. This, of course, can only be applied to orbitals fairly well contained in a muffin-tin sphere (i.e., R/2 for the first nearest neighbors is already in the asymptotic region). What we will see is that it serves as a reasonable rule of thumb for the transition metal d states and actinide f states.

For the transition metals, we want to remove the R_{WS}^{-2} dependence. This we do by looking at the inverse mass parameter defined by Andersen.[13] This parameter relates to the bandwidth scaled by R_{WS}^{-2} as required and is defined so that the mass would be 1 in an empty lattice. Further, it has a nice simple interpretation as it is proportional to the square of the wavefunctions at the sphere boundary and thus the probability that the orbital tunnel out. Since the bandwidth is directly related to the itineracy of the electron, an increase in tunneling amplitude is an increase in bandwidth. In Figs. 18-20, we show m_d^{-1} and $R^3 = (\langle r_{at}\rangle/R_{WS})^3$ for the 3d, 4d, and 5d transition metal series. Note that this very simple analysis really does quite well in representing the qualitative situation. But, because we have only a simple rule of thumb, we plot the values of m_d^{-1} and R each for the relativistic and non-relativistic calculations. The curves for m_d^{-1} and R are quite close but with the relativistic results above the non-relativistic results. This is the indirect relativistic effect expansion of the conduction electron d-states which results in a wider d-band width.

We now begin to explore some of the physical consequences of spin-orbit coupling in metals. We will focus on metals in the 3d, 4d and 5d transition series, for which many interesting effects occur. As shown in Appendix B the influence of spin-orbit coupling on the d-band eigenvalues and eigenstates can be approximately described by a model Hamiltonian of the form

$$H_{is,js'} = H^{TB}_{i,j}(\underline{k})\,\delta_{s,s'} + (\xi_d/2)\,\underline{L}_{i,j}\cdot\underline{\sigma}_{s,s'} \qquad (12)$$

where $H^{TB}_{i,j}$ is a tight-binding Hamiltonian for the d-bands of the Slater-Koster form[21] and $\underline{s}$ is the vector of Pauli spin matrices. For the sake of definiteness we have listed the angular basis functions assumed in the following discussion in Table 2 and the matrix elements of L between these functions in Table 3. To our knowledge, this form of Hamiltonian was first used to explore spin-orbit effects in transition metals by Friedel et al.,[22] although its modern interpretation is somewhat different. An equation similar to Eq. (12) has recently been used by Rahman et al.[23] in discussing the g-factor of Pd.

The most important observation to make concerning the energy bands in the presence of spin-orbit coupling is that, provided the crystal has an inversion center[24] the 2 fold spin degeneracy existing at each point in $\underline{k}$ space is not lifted by the spin-orbit interaction. This property is a consequence of the time-reversal invariance of the Hamiltonian.[25] The eigenstates at wavevector $\underline{k}$ may be labelled by a band index, n, and a generalization of the nonrelativistic spin index, σ which we will allow to assume the values +1 and -1. The eigenstates at wavevector $\underline{k}$ can be chosen to obey

Table 2. Angular Basis Functions for $\ell = 2$

Index	Function	Spherical Harmonic Expansion
1	$\sqrt{\frac{5}{16\pi}}\frac{(3Z^2-r^2)}{r^2}$	$Y_{2,0}$
2	$\sqrt{\frac{15}{4\pi}}\frac{xz}{r^2}$	$\frac{1}{\sqrt{2}}(Y_{2,-1}-Y_{2,1})$
3	$\sqrt{\frac{15}{4\pi}}\frac{xz}{r^2}$	$\frac{i}{\sqrt{2}}(Y_{2,-1}+Y_{2,1})$
4	$\sqrt{\frac{15}{4\pi}}\frac{xy}{r^2}$	$\frac{i}{\sqrt{2}}(Y_{2,-2}-Y_{2,2})$
5	$\sqrt{\frac{15}{16\pi}}\frac{(x^2-y^2)}{r^2}$	$\frac{1}{\sqrt{2}}(Y_{2,-2}+Y_{2,2})$

$$C|n,\underline{k},+\rangle = -|n,\underline{k},-\rangle^* \tag{13a}$$

$$C|n,\underline{k},-\rangle = |n,\underline{k},+\rangle^* \tag{13b}$$

where the conjugation operator, C = JK, where J is the inversion operator and $K = i\sigma_y K_o$ where (K_o is the complex conjugation operator) is the time-reversal operator. Many useful relations for matrix elements of operators which transfer simply under conjugation follow from Eqs. (13a - 13b). For instance, for Hermitian operators O, which are invariant under conjugation, for example $\underline{\sigma}\cdot\underline{L}$, it is readily demonstrated that

$$\langle n,\underline{k},+|O|m,\underline{k},+\rangle = +\langle m,\underline{k},-|O|n,\underline{k},-\rangle^* \tag{14a}$$

Table 3. Matrix elements of the components of the angular momentum operator between the angular basis functions of Table 2

	i	j = 1	2	3	4	5
L^x_{ij}	1	0	0	$\sqrt{3}$	0	0
	2	0	0	0	1	0
	3	$-\sqrt{3}$	0	0	0	-1
	4	0	-1	0	0	0
	5	0	0	1	0	0
L^y_{ij}	1	0	$-\sqrt{3}$	0	0	0
	2	$\sqrt{3}$	0	0	0	-1
	3	0	0	0	-1	0
	4	0	0	1	0	0
	5	0	1	0	0	0
L^z_{ij}	1	0	0	0	0	0
	2	0	0	-1	0	0
	3	0	1	0	0	0
	4	0	0	0	0	2
	5	0	0	0	-2	0

and

$$\langle n,\underline{k},+|O|m,\underline{k},-\rangle = -\langle m,\underline{k},+|O|n,\underline{k},-\rangle^* \qquad (14b)$$

For operators which change sign under conjugation, for example $\underline{L}$, both of Eqs. (14) suffer a sign reversal. Note that these equations imply that matrix elements of $\underline{L}$ between the orbital eigenstates at $\xi^d = 0$, are pure imaginary (compare with Table 3) and hence that the expectation values are identically zero (i.e., angular momentum is "quenched" in the crystal).

The first question we wish to explore using Eq. (12) is the degree to which the band eigenvalues are altered by the inclusion of spin-orbit coupling. To consider this point we begin by treating the spin-orbit term via leading order perturbation theory. An important element in this development is the observation that angular momentum is "quenched" in the metal. Thus the eigenvalues at wavevector $\underline{k}$, when spin-orbit coupling is present, are given by

$$E_{n,\underline{k},\sigma} = \varepsilon_{n,\underline{k}} + \left(\frac{\xi_d}{2}\right)^2 \sum_{\substack{m\neq n \\ \sigma'}} \frac{\left|\langle\psi_{n,\underline{k}}|L|\psi_{m,\underline{k}}\rangle\cdot\underline{\sigma}_{\sigma\sigma'}\right|^2}{\varepsilon_{n,\underline{k}} - \varepsilon_{m,\underline{k}}} + \ldots \quad (15)$$

Equations (14) guarantee that Eq. (15) is consistent with the maintenance of a double degeneracy throughout the BZ. Note that if we assume the d-band energies are typically roughly evenly spaced we see from Eq. (15) that

$$|E_{n,\underline{k}}-\varepsilon_{n,\underline{k}}| \sim \xi_d \left(\frac{4\xi_d}{W_d}\right) \quad (16)$$

Equation (15) also shows that energies near the top of the band will be increased by spin-orbit coupling whereas energies near the bottom of the band will be decreased, i.e., spin-orbit coupling will produce a broadening of the d-band. Values for ξ_d in the transition series are presented in Fig. 21 and Fig. 22 shows the trends in ξ_d/W_d values for the same metals. In Fig. 21 the solid curve for the 5d electrons was calculated using overlapping atomic charge density potentials while the dash-dotted curve was calculated via a self-consistent configuration treatment As noted earlier the d-occupation is, in most cases, higher in the metal than in the atom. As a result in the self-consistent configuration treatment the d-electrons see a somewhat weaker potential and have a somewhat smaller spin-orbit splitting. Note that ξ_d/W_d increases through the transition series both because ξ_d increases and because W_d decreases sharply toward the top end of each series. This implies that the white line effects are to be seen only at the high end of the transition series.[26,27] We also note that ξ_d/W_d increases much more dramatically in going from 4d to 5d transition series than in going from 3d to 4d series. In going from the 3d series to the 4d series the increase in ξ_d is nearly compensated for by the increase in W_d. However even for Au(Z=79), which is located above the top end of the 5d series $(4\xi_d/W_d) < 1$. As a result the shifts in band energies induced by spin-orbit coupling are smaller in magnitude than the spin-orbit splitting parameters themselves. An exception to these remarks is the case in which a degeneracy in the band eigenvalues (apart from the spin degeneracy) exists. In this case we must use degenerate state

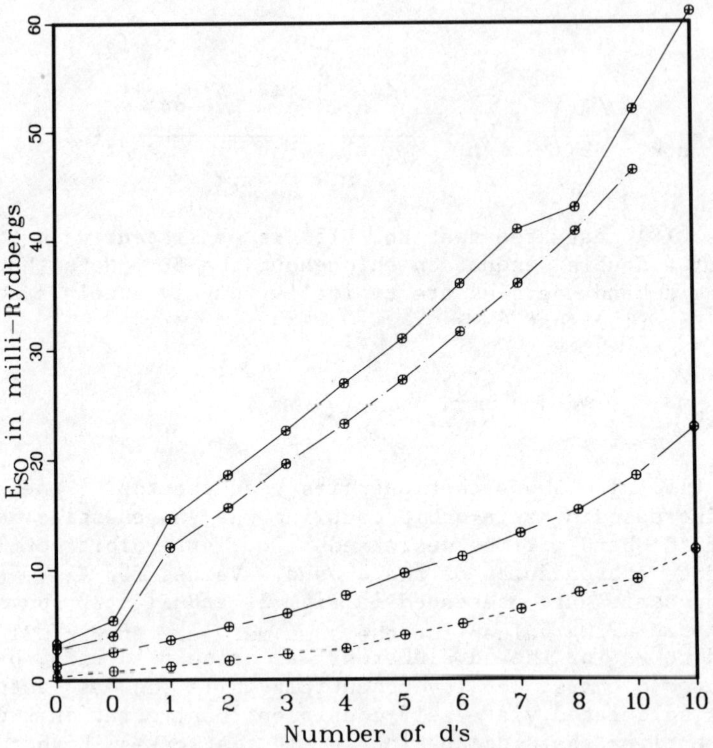

Fig. 21. Spin-orbit splitting parameters with potentials constructed from atomic configurations for the 3d (---), 4d (—— - ——) and 5d (——) transition series. For the 5d series the results obtained from a configurationally self-consistent calculation are also shown (—— • ——). The metallic d-occupancy is larger than the atomic d-occupancy so the configurational self-consistency weakens the crystal potential and reduces the spin-orbit splitting.

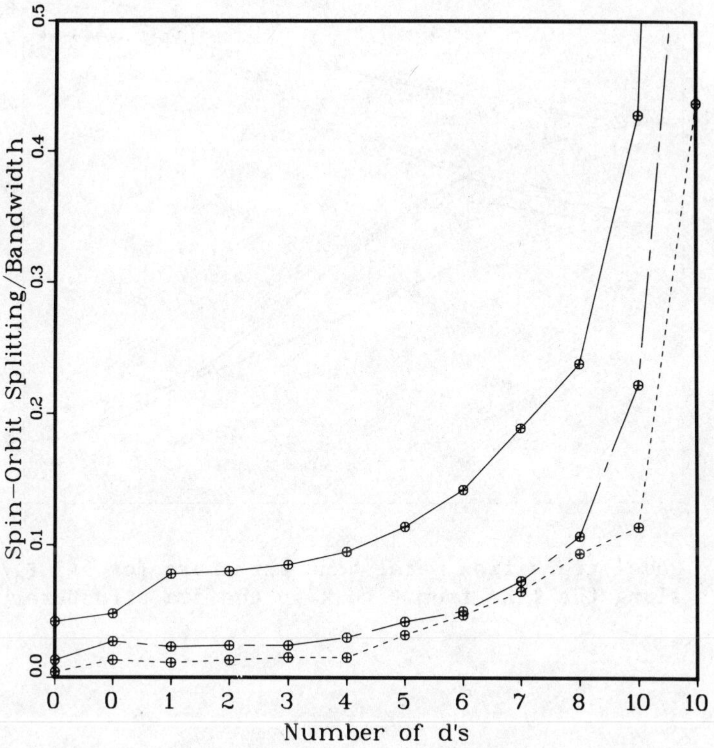

Fig. 22. Ratio of spin-orbit splitting to bandwidths calculated from atomic configurations for the 3d (---), 4d (—— – ——) and 5d (——) transition series.

perturbation theory and will generally find a reduction in the degeneracy at $\underline{k}$ and a leading order correction

$$|E_{n,\underline{k}}-\varepsilon_{n,\underline{k}}| \sim \xi_d \qquad (17)$$

A more quantitative picture of the influence of spin-orbit

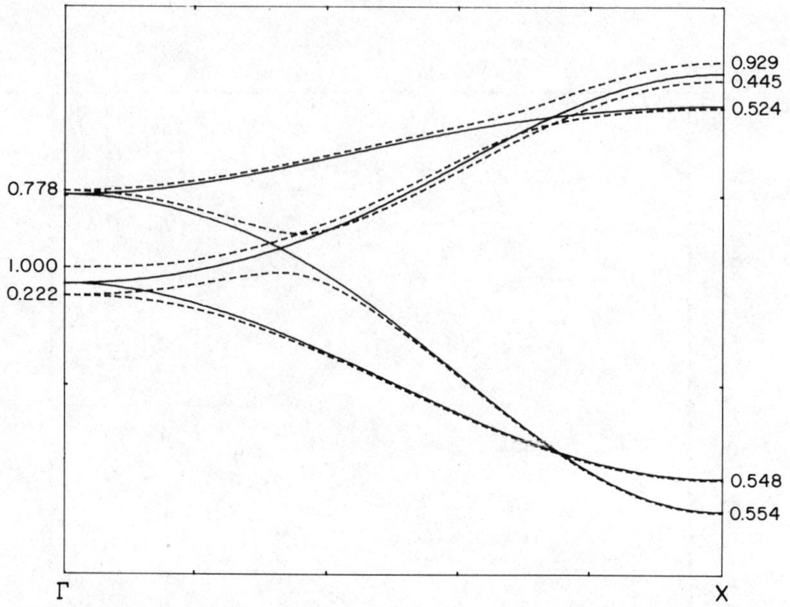

Fig. 23. Model transition metal band structure for $5/2\ \xi_d/W_d \sim 0.09$ along the line from Γ to X in the fcc structure.

coupling on the energy bands can be obtained by examining the eigenvalues of $H_{is,js'}$ [see Eq. (12)] as a function of ξ_d. In Figs. 23-26 we compare the band eigenvalues at $\xi_d = 0$ with those at finite ξ_d for the symmetry lines from Γ to X in the fcc structure and from Γ to H in the bcc structure. The value of $5\xi_d/2W_d$ was $\sim$0.09 in Figs. 23 and 25 and $\sim$0.18 in Figs. 24 and 26. Thus we may take the spin-orbit effects shown in Figs. 23 and 25 to be representative of the middle of the 5d transition series and the top of the 3d and 4d transition series while the comparisons shown in Figs. 24 and 26 are representative of spin-orbit effects near the top of the 5d transition series.

We consider the fcc case in detail. It follows from symmetry considerations that along the line from Γ to X = $2\pi/a$ (1,0,0), $H^{TB}_{4,4} = H^{TB}_{2,2}$ and all non-diagonal of $H^{TB}_{i,j}$ except $H^{TB}_{1,5}$ are zero. It follows that, in the absence of spin-orbit coupling the band eigenvalues are given by $H^{\pm} = (H_{11}+H_{55})/2 \pm \pi(H_{11}-H_{55})2/4+H_{1,5}{}^2$, H_{22} (twice) and H_{33}. The orbital eigenstates $|\pm\rangle$, corresponding to the eigenvalues $H^{\pm}$ are

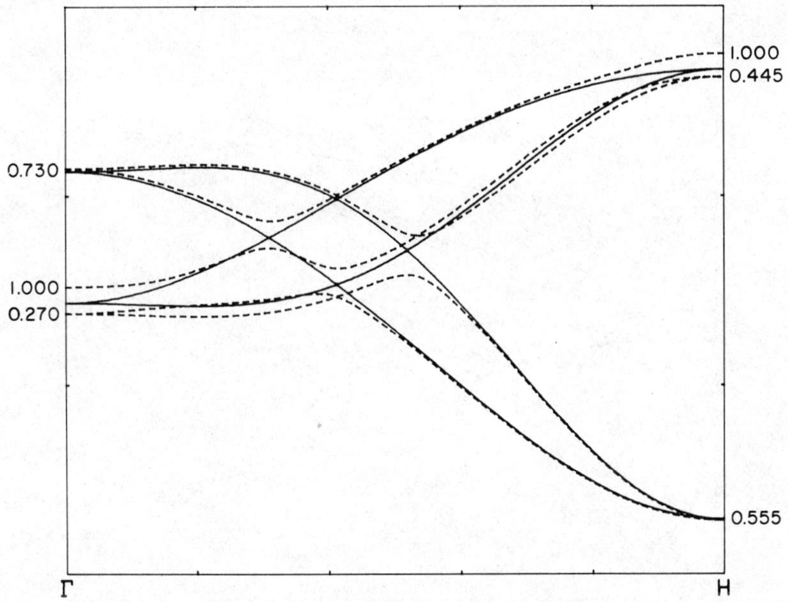

Fig. 24. Model transition metal band structure for $5/2\ \xi_d/W_d \sim 0.09$ along the line from Γ to H in the bcc structure.

$$|+\rangle = \frac{\sqrt{3}}{2}|1\rangle + \frac{1}{2}|5\rangle = \frac{15}{16\pi}\ \frac{z^2-y^2}{r^2} \qquad (18a)$$

and

$$|-\rangle - \frac{1}{2}|1\rangle + \frac{\sqrt{3}}{2}|5\rangle = \frac{15}{16\pi}\ \frac{3x^2-r^2}{r^2} \qquad (18b)$$

It is easily verified from Table 3 that $\langle +|\sigma\cdot L|-\rangle = 0$ and that the only non-zero matrix element between $\{|+\rangle, |-\rangle\}$ and $\{|2\rangle, |3\rangle, |4\rangle\}$ is that between $|-\rangle$ and $|3\rangle$. At Γ, $H_{22} = H_{33}$, $H_{11} = H_{55}$ and $H_{1,5} = 0$. The double degenerate level (H_{11}) corresponds to the e_g crystal field state and the triply degenerate level to the t_{2g} crystal

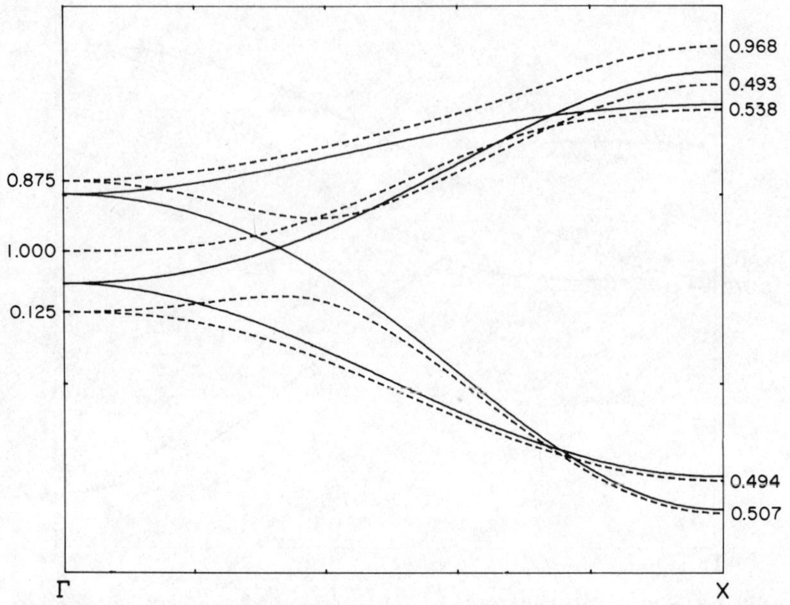

Fig. 25. Model transition metal band structure for $5/2\ \xi_d/W_d \sim 0.18$ along the line from Γ to X in the fcc structure.

field state. When the spin-orbit interaction is "turned on" the t_{2g} split is split by $\sim 3/2\xi_d$ and its center of gravity is lower because of coupling to the higher energy e_g levels. The e_g levels are not split because $\langle 1|\sigma\cdot\underline{L}|5\rangle = 0$ but they are raised in energy because of coupling to the nearby t_{2g} levels. As we move from Γ to X in the absence of spin-orbit coupling, H_{33} and H^{-} decrease while H_{22} and H_+ increase so that the band expands to its full width at X. The ordering of the energy levels at X is $H_{22} > H_+ > H_{33} > H_-$. Between Γ and X, H_{22} crosses H_- and H_+ while H_{33} crosses H_-. The orbital eigenstate $|3\rangle$ is coupled via $(\underline{\sigma}\cdot\underline{L})$ to $|+\rangle$, $|\bar{2}\rangle$ and $|4\rangle$ and therefore the amount by which H_{33} is lowered by spin-orbit coupling is reduced in accordance with the

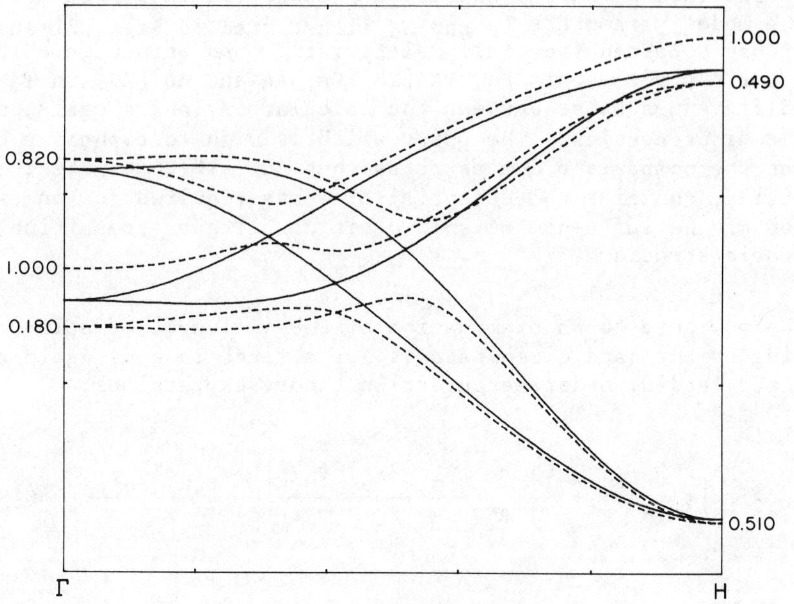

Fig. 26. Model transition metal band structure for $5/2\ \xi_d/W_d \sim 0.18$ along the line from Γ to H in the bcc structure.

increasing energy difference as we go from Γ to X. The crossings of H_{22} and H_- and H_{22} and H_+ are altered by the spin-orbit couplings within the $|2\rangle$, $|4\rangle$, $|-\rangle$ and $|2\rangle$, $|4\rangle$, $|+\rangle$ manifolds respectively. On the other hand the crossing between H_{33} and H_- is essentially unaltered since $\langle -|L|3\rangle = 0$. At X the H_-, and H_{33} levels are lowered only slightly by their coupling to the upper levels. At the top of the band the H_{22} levels are split and have their center of gravity shifted upward, principally due to coupling with the nearby H_+ level which is shifted downward.

The corresponding analysis of the results shown in Figs. 25 and 26 is left as an exercise for the reader. We should emphasize, of course, that the plots of the influence of spin-orbit coupling on the energy bands along symmetry directions tend to exaggerate the typical impact. This is because of the greater frequency of degeneracies and energy level bunching. Finally, we compare the spin-orbit induced energy shifts at Γ and X obtained from Eq. (12) and the model parameters ξ_d and W_d illustrated in Figs. 21 and 22 with those obtained from full relativistic band structure calculations for the Fcc metals Pd, Pt [28] Cu, Ag and Au [29] in Table 4. Many differences exist between the calculations which can account for the discrepancies. The point which we wish to emphasize here is that the comparison demonstrates that Eq. (12), despite its simplicity, contains the essential elements required for an examination of the influence of spin-orbit coupling on transition metal electronic structure.

We now turn to an examination of the influence of spin-orbit coupling on the band eigenstates. For a first look we again consider the leading order perturbation theory expressions.

$$|\Psi_{n,\underline{k},\sigma}\rangle = |\psi_{n,\underline{k}}\rangle\chi_\sigma + \frac{\xi_d}{2}\sum_{\substack{m\neq n\\ \sigma'}} \frac{|\psi_{m,k}\rangle\chi_{\sigma'}\langle\psi_{m,k}|L|\psi_{n,k}\rangle\cdot\underline{\sigma}_{\sigma'\sigma}}{\varepsilon_{n,\underline{k}} - \varepsilon_{m,\underline{k}}} \quad (19)$$

Once $\xi_d \neq 0$ the eigenstates are no longer products of orbital and spin eigenstates. Moreover, for a given value of ξ_d the eigenstates are more strongly altered than the band eigenvalues, the perturbation correction factor being $\sim(\xi_d/W_d)$. At a $\underline{k}$ point which possesses an orbital degeneracy, we must again use degenerate state perturbation theory, and so the band eigenstates will, in general be non-product states even for vanishingly small spin-orbit coupling. A convenient index of the amount by which the band eigenstates have been altered by the spin-orbit coupling can be obtained by evaluating their j = 5/2 and j = 3/2 projections.

$$F_j^{n,\underline{k}} = 1/2 \sum_{\sigma} \langle\Psi_{n,\underline{k},\sigma}|P_j|\Psi_{n,\underline{k},\sigma}\rangle \quad (20)$$

where, in Dirac bracket notation, the projection operators P_j are given by

$$P_j = \sum_{m_j=-j}^{j} |j\ m_j\rangle\langle j\ m_j| \quad (21)$$

and $|j\ m_j\rangle$ denotes the simultaneous eigenstates of J^2 and J_Z

Table 4. Comparison of spin-orbit coupling induced energy splittings obtained via the tight-binding model (M) with those obtained via a full band calculation (B). Energy levels at Γ and X are distinguished via their band indices.

Metal	Method of Calculation	Energy Level							
		Γ			X				
		1,2	3	4,5	1	2	3	4	5
Cu	M	-7	9	2	-1	-1	-1	-4	6
Cu	B	-6	8	1	0	0	-2	-5	6
Ag	M	-18	14	8	-2	-3	-3	-7	17
Ag	B	-13	15	5	-1	-1	-3	-9	15
Au	M	-59	52	34	-14	-16	-4	-12	46
Au	B	-43	44	21	-6	-8	-12	-24	48
Pd	M	-12	15	4	-1	-1	-2	-7	11
Pd	B	-9	13	3	-1	-1	-4	-8	13
Pt	M	-42	43	21	-6	-7	-6	-15	35
Pt	B	-32	38	13	-3	-6	-12	-24	42

($\underline{J} = \underline{L} + \underline{\sigma}/2$). It is easily established that in the limit $\xi_d/W_d \to 0$, $F_j^{n,\underline{k}} = (2j+1)/10$ for each band, and at each point in the zone, whereas for $\xi_d/W_d \to \infty$ $F_{3/2}^{n,\underline{k}} = 1$ for the four lowest eigenstates and $F_{5/2}^{n,\underline{k}} = 1$ for the six highest eigenstates. For intermediate values of ξ_d/W_d the atomic-like eigenstates are partially mixed by the crystal-symmetry imposed boundary conditions.

Using Eq. (19) and the expression

$$P_{\ell\pm1/2} = \pm(\underline{\sigma}\cdot\underline{L}+1/2\pm(\ell+1/2)/(2\ell+1) \tag{22}$$

it can be shown that to leading order in ξ_d

$$(2\ell+1)F_{\ell\pm1/2}^{n,\underline{k}} = \ell+1/2\pm1/2$$
$$\pm\xi_d \sum_{m\neq n} \frac{|\langle n,\underline{k}|L_x|m,\underline{k}\rangle|^2+|n,\underline{k}|L_y|m,\underline{k}\rangle|^2+|\langle n,\underline{k}|\underline{L_z}|m,\underline{k}\rangle|^2}{\varepsilon_{n,\underline{k}} - \varepsilon_{m,\underline{k}}} \tag{23}$$

(Once again we should emphasize that Eqs. (19) and (23) are not valid when an orbital degeneracy exists at $\underline{k}$ for band n.) Equation (23) shows that, as a rule, $F^{n,k}_{5/2}$ tends to increase with ξ_d toward the top of the band and decrease with ξ_d toward the bottom of the band while $F^{n,k}_{3/2}$ shows the opposite behaviour. In fact, it is easily demonstrated that, with Eq. (12) as the Hamiltonian, the centers of mass of the J = 5/2 and J = 3/2 partial densities of states are split by 5/2 ξ_d, the atomic spin-orbit splitting. The values of $F^{n,k}_{5/2}$ are listed for the high-symmetry points in Figs. 23 to 26. Note that, as long as we average over levels when an orbital denergacy exists, $F^{n,k}_{5/2}$ approaches 0.6 at all points as the spin orbit coupling parameter is reduced to zero.

To give a better indication of the overall degree of separation of the j = 5/2 and j = 3/2 character within the d-bands we have plotted in Fig. 27 $\eta^{5/2}$ versus η for the bcc structure. In these figures $\eta^{5/2}(\eta)$ is the number of j = 5/2 states occupied per atom when the Fermi level is chosen so that η states per atom are occupied. (The details of the calculation are presented elsewhere.[30]) Therefore, in the absence of spin-orbit coupling the plots would be straight lines from the lower-left to upper-right corners of the figures. On the other hand, in the limit of very strong spin-orbit coupling $(\xi_d/W_d) \to \infty$ the plots would again become straight lines to the upper right corner but beginning at the point $\eta = 4$ along the abcissa. Of the two values of (ξ_d/W_d) plotted the first corresponds to $5/2\xi_d/W_d \sim 0.18$ and is representative of the spin-orbit effects near the middle of the 5d transition series. From Fig. 27 we see that there is a spin-orbit induced drop of ∿40% in the 5/2 character among the lowest four d states per atom. Although the spin-orbit effects at a given point in the Brillouin-zone are obviously highly structure-dependent, once the averages are performed only a weak structural dependence persists and plots for the fcc and hcp structures are nearly indistinguishable from Fig. 27. The lines for stronger spin-orbit coupling are for $5/2\xi_d/W_d \sim 0.36$ and indicate the degree of j = 3/2, j = 5/2 separation which would be expected in Au and Zn. More detailed calculations of j = 3/2 and j = 5/2 partial densities of states for the noble metals have been presented by Christensen[26]. In closing this section we should note that indices of the eigenstates which are insensitive to their spin-composition are much more weakly dependent on spin-orbit coupling than the spin-sensitive index used here. This point has been illustrated in Fig. 28 by plotting the e_g partial occupation vs. η for two different values of ξ_d. Thus the total change density distribution in a metal will be relatively insensitive to spin-orbit coupling. It is this observation which can justify the omission of spin-orbit coupling in the self-consistency process which requires most of the computer time in a self-consistent band calculation.

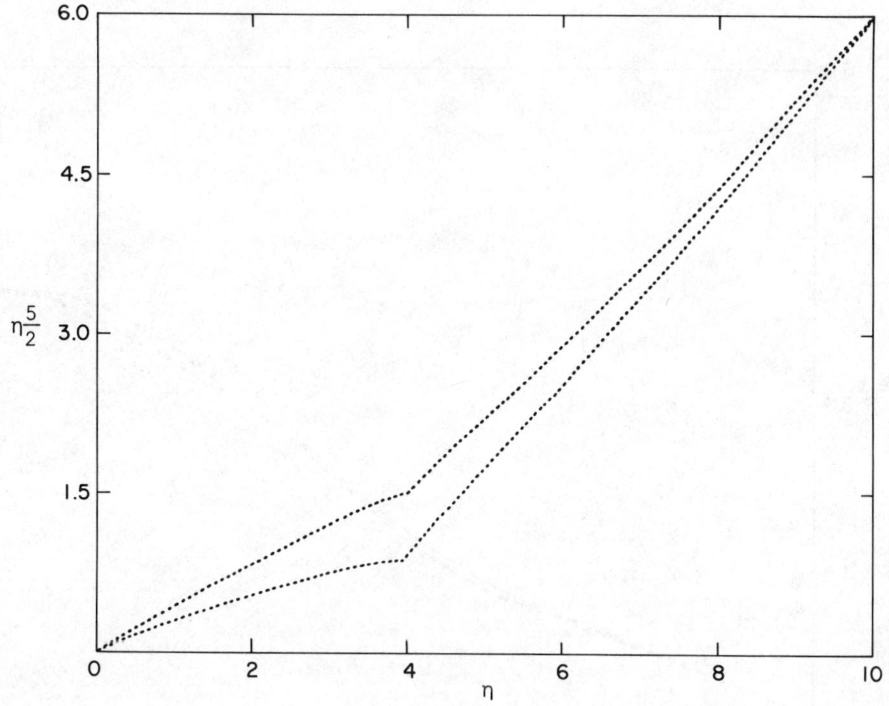

Fig. 27. The j = 5/2 fractional occupancy versus the total d-occupancy for $5/2\ \xi_d/W_d \sim 0.18$ for the model bcc transition metal band structure.

One area in which even moderate spin-orbit coupling can have a dramatic impact on transition metals, is that of magnetic properties. For example the connection between the Pauli exclusion principle and band magnetism, in terms of which many magnetic properties of transition metals can be understood, will obviously be altered by spin-orbit coupling. Even within a non-interacting electron-model qualitative changes occur as discussed below.

A quantity of fundamental importance in the discussing the magnetic response of paramagnetic transition metals in the local g-factor defined by

$$\mu_B H g^{\hat{\alpha}}_{n,\underline{k}} \equiv \Delta E_{n,\underline{k}}$$

where $\Delta E_{n,\underline{k}}$ is the splitting of the Kramers degeneracy at $\underline{k}$ for band n and a magnetic field H, applied in the α direction. For example, for metals with a high Fermi-level density of states the

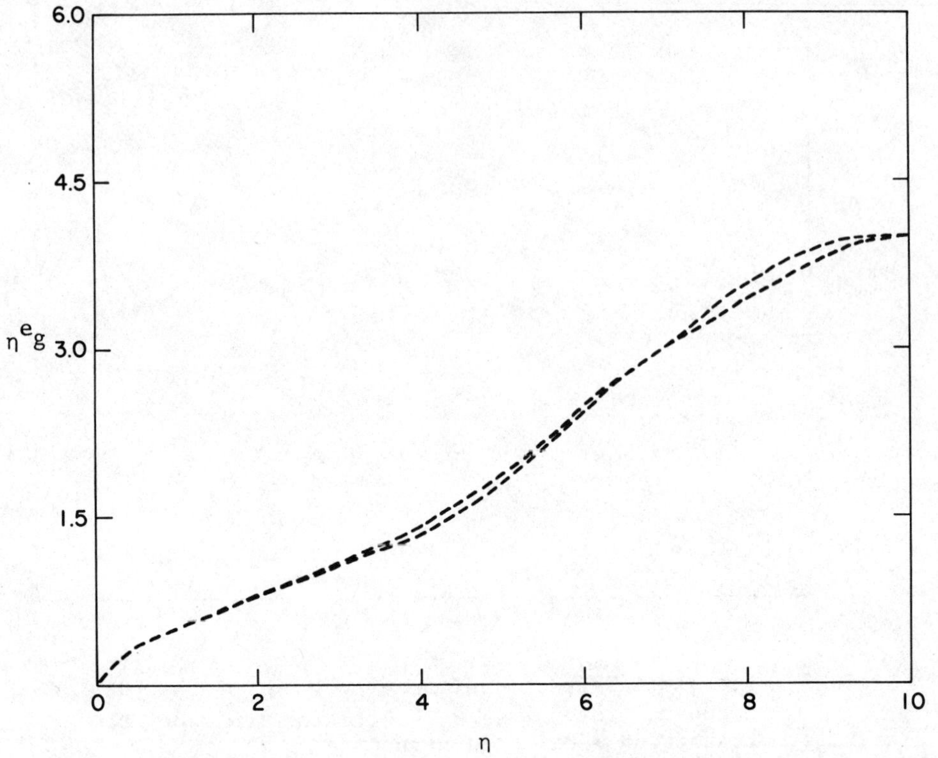

Fig. 28. The e_g fractional occupancy is plotted versus the total d-occupancy for a model transition metal band structure for two different values of ξ_d/W_d. Note the weak influence of spin-orbit coupling on this spin-independent property of the d-band eigenstates.

magnetic susceptibility will be proportional to the average of the square of this factor over the Fermi surface. Within a tight-binding model[31] this splitting can be determined by adding a Zeeman term, H_Z, to the Hamiltonian of Eq. (12).

$$H_Z = -\mu_B H\hat{\alpha}\cdot(\underline{\sigma}+\underline{L})$$

Since we can safely assume that $W_d >> \mu_B H$ the splitting can be determined by diagonalizing H_Z within the Kramers degeneracy. Since H_Z changes sign under conjugation it follows from Eqs. (14) that

$$\frac{g^{\hat{\alpha}\,2}_{n,\underline{k}}}{4} = \langle n,\underline{k},+|\alpha\cdot(\underline{\sigma}+\underline{L})|n,\underline{k},+\rangle^2 + |\langle n,\underline{k},+|\hat{\alpha}\cdot(\sigma+\underline{L})|n,k,\underline{-}\rangle|^2 \qquad (25)$$

In the absence of spin-orbit coupling, the eigenstates are product states, the matrix elements of $\underline{L}$ are quenched and we obtain

$$\frac{g^{\hat{\alpha}\,2}_{n,\underline{k}}}{4} = \frac{1}{2}\sum_{i,j} \alpha_i\alpha_j \,\mathrm{Tr}(\sigma^i\sigma^j) = 1 \qquad (26)$$

independent of the relationship of the applied field to the crystal symmetry. Equation (26) has recently been used to explain the dramatically anisotropic magnetic susceptibility observed in some metallic transition metal layer compounds.[32] It follows from

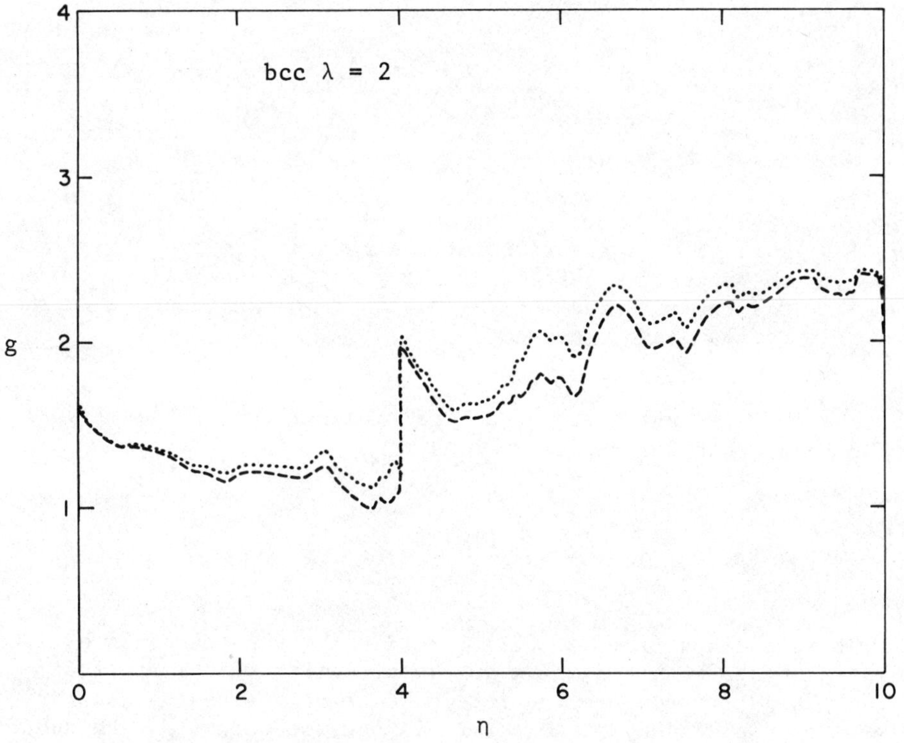

Fig. 29. Root mean square value of the g factor (dotted line) and the average magnitude of the g factor (dashed line) over the Fermi surface of the model bcc transition metal band-structure versus total d-band occupancy.

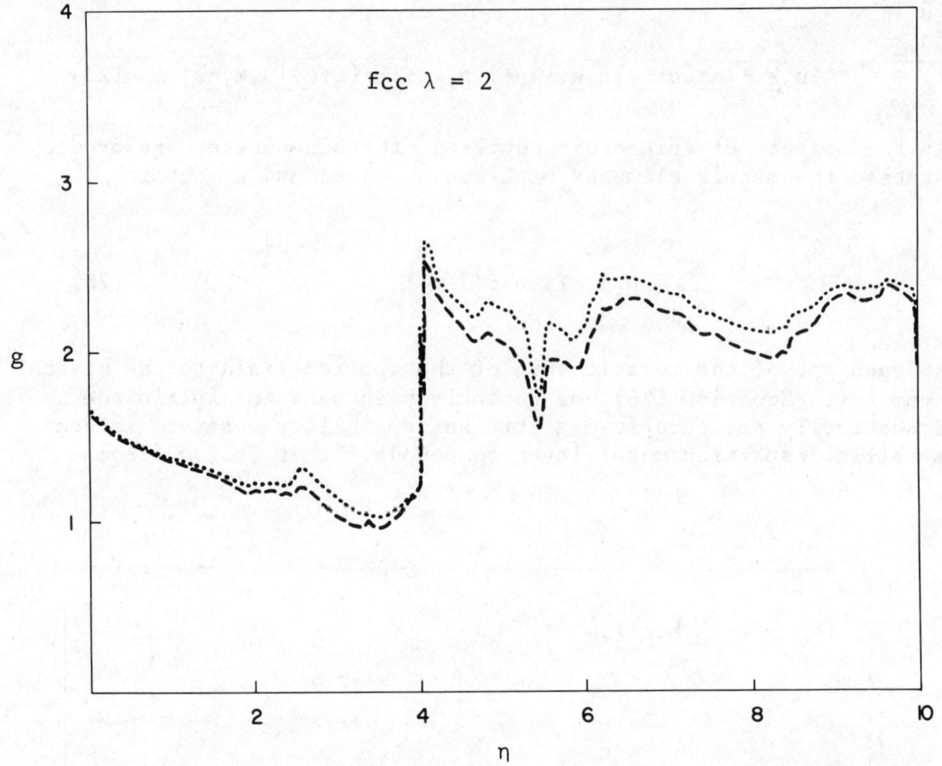

Fig. 30. Root mean square value of the g factor (dotted line) and the average magnitude of the g factor (dashed line) over the Fermi surface of the model fcc band structure versus total d-band occupancy.

Eq. (19) that, except where an orbital degeneracy has been lifted by the spin-orbit interaction

$$\frac{g^{\hat{\alpha}\,2}_{n,\underline{k}}}{4} = 1 + \xi_d \sum_{m\neq n} \frac{|\hat{\alpha}\cdot\langle n,\underline{k}|\underline{L}|m,\underline{k}\rangle|^2}{\varepsilon_n(\underline{k})-\varepsilon_m(\underline{k})} + \ldots \tag{27}$$

It is obvious from Eq. (27) that for a given state the g-factor will tend to be very sensitive to the direction of the applied field. This tendancy was the main conclusion reached by Mueller et al., in their study of g-factors in the fcc metals Ni, Pd and Pt and was also emphasized by Rahman et al.,[23] in their recent study of Pd. In especially sensitive cases such as the X pocket in Ni, even the very Fermi surface can be modified by the direction of the field.[34] Another observation which follows from Eq. (27)

is that the spin-orbit induced g-shifts will tend to be negative toward the bottom of the band and positive toward the top of the band. To give an indication of the average trends in g-factors which would be expected in transition metals we have plotted $\langle (g^{\hat{\alpha}}_{n,\underline{k}})^2 \rangle^{\omega/2}$ versus the number of d-electrons occupied in Figs. 29 to 31 for bcc, fcc and hcp structures respectively. (The average referred to above is over the Fermi surface; the details of these calculations are given elsewhere.) In a non-interacting electron model for metals with a high Fermi level density of states, the magnetic susceptibility is proportional to this quantity. For the cubic metals averaging over the Fermi surface removes the anisotropy and we plot $\langle g^{\hat{\alpha}}_{n,\underline{k}} \rangle$ as well as $\langle (g^{\hat{\alpha}}_{n,\underline{k}})^2 \rangle^{1/2}$. The degree to which these two averages differ gives a measure of the amount by which $g^{\hat{\alpha}}_{n,\underline{k}}$ varies over the Fermi surfaces. We note that the expected tendancy toward negative g-shifts toward the bottom of the band and

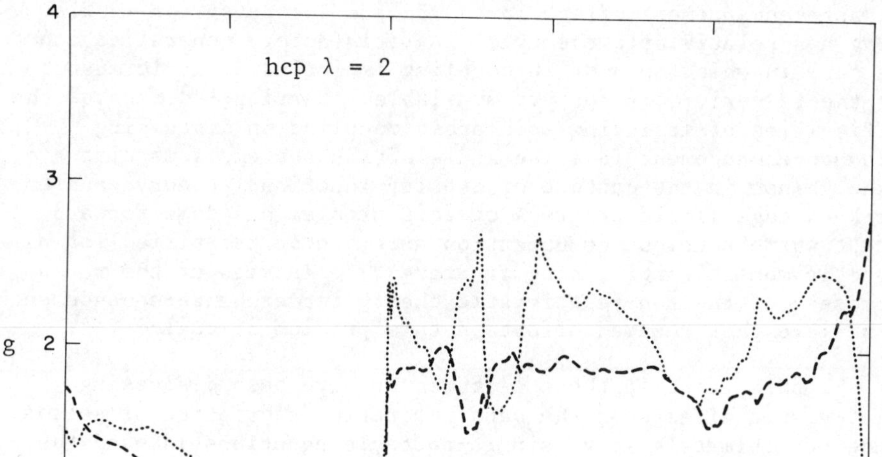

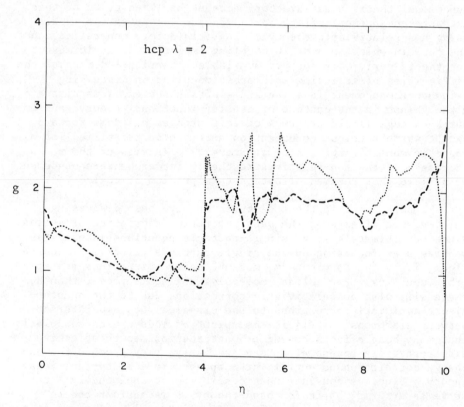

Fig. 31. Root mean square value of the g-factor averaged over the Fermi surface of the model hcp transition metal band structure for fields parallel (dashed line) perpendicular (dotted line) to the z-direction.

positive g-shifts toward the top of the band is seen in the figures. The fact that negative g-shifts tend to be larger than positive g-shifts results from the quenching of the spin-moment of the electrons. For the hcp structure some anisotropy remains after averaging over the Fermi surface and we have plotted results both for $\hat{\alpha} = (0,0,1)$ and $\hat{\alpha} = (1,0,0)$, using dashed lines and dotted lines respectively. This indicated degree of anisotropy corresponds to what might be expected to exist in the hexagonal 5d transition metals Re and Os.

In closing this section we mention a problem which we consider to be interesting and, as yet, essentially unsolved. As indicated by Figs. 29-31 the magnetic susceptibility, χ, of a metal may be strongly altered by spin-orbit coupling. We have mentioned that $\langle (g^{\hat{\alpha}}_{n,\underline{k}})^2 \rangle$ is proportional to χ for metals with a high density of states at the Fermi level. However, it is just such metals which tend to have a large exchange-enhancement of χ. Using spin-density functional theory a satisfactory account has been given of this exchange-enhancement effect in the 3d and 4d transition metals using non-relativistic theory.[34] A satisfactory generalization to the case in which spin-orbit coupling is included, as it must be for the 5d series, is not yet available. Herring[35] discussed the difficulties of including spin-orbit coupling in discussing exchange-enhancement in itinerant electron systems some time ago, although not in the context of density functional theory, and to our knowledge little progress on this problem has come forward except where a unique quantization axis can be identified for a magnetic moment partitioning of states.[35] In view of the recent successes of the non-relativistic theory for exchange-enhancement we believe that renewed effort on this problem is called for.

To this point in these lectures we have been discussing relativistic effects in the band theoretic description of solids. Since we ultimately solve single-particle equations in band-theory we have been focussing on the single-particle relativistic corrections, i.e., the Darwin term, mass-velocity corrections and spin-orbit coupling. For solids, molecules and atoms, other than H, there will also be relativistic corrections due to the quantum-electrodynamical corrections to the classical Coulomb interaction between electrons. It is in the spirit of modern band-calculations, that many body effects appear as contributions to an effective single-particle potential. More precisely, they produce the exchange-correlation potential which appears in density-functional theory. Since we consider that density-functional theory[38] can serve as a formal basis for band-theory it is appropriate to discuss relativistic interaction effects in solids in terms of the relativistic generalization[39,40] of this formalism. At the risk of going somewhat off the topic, we summarize the main results of that formalism in the following paragraph, and then return later to the question of relativistic interaction.

We consider a system of interacting Dirac and electromagnetic fields in the presence of a scalar non-quantised external field, $V(\underline{r})$. (We shall perform all calculations in the reference frame in which the momentum of the system is zero.) We can write the Hamiltonian as

$$\hat{H} = \hat{H}_{sys} + \int d\underline{r}\ \underline{n}(\underline{r})V(\underline{r}) \tag{28}$$

where $\hat{n}(\underline{r}) = \hat{\psi}^{+}(\underline{r})\hat{\psi}(\underline{r})$ is the Dirac number density operator and $\hat{H}_{sys}$ is the Hamiltonian in the absence of an external field. Then it can be shown that provided the ground state is non-degenerate:

i) the ground state $|\Psi\rangle$ can be expressed as a functional of the ground state density $n(\underline{r}) = \langle\Psi|\hat{n}(\underline{r})|\Psi\rangle$;

ii) the ground state energy functional

$$E_V[n] = \langle\Psi[n]|\hat{H}|\Psi[n]\rangle \tag{29}$$

is a minimum over densities corresponding to a given total charge at the correct density in the presence of the potential $V(\underline{r})$; and

iii) the ground state density can be determined by solving the following set of single particle equations self-consistently

$$(-i\hbar c\underline{\alpha}\cdot\underline{\nabla}+\beta mc^2+v_{eff}[n;\underline{r}])\psi_i(\underline{r}) = \varepsilon_i\psi_i(\underline{r}) \tag{30a}$$

$$\eta(\underline{r}) = \sum_i \psi_i^{+}(\underline{r})\psi_i(\underline{r})\theta(\mu-\varepsilon_i) \tag{30b}$$

$$N = \sum_i \theta(\mu-\varepsilon_i) \tag{30c}$$

$$v_{eff}[n;r] = v(r) + e^2\int \frac{d\underline{r}'\eta(\underline{r}')}{|\underline{r}-\underline{r}|} + \frac{\delta E_{xc}[n]}{\delta n(\underline{r})} \tag{30d}$$

where the total charge of the system is -eN and the sums over i in Eqs. (30b) and (30c) are over positive energy solutions to Eq. (30a) only. The exchange-correlation energy functional appearing in Eq. (30d) is defined by

$$E_{xc}[n] \equiv \langle\Psi[n]|\hat{H}_{sys}|\Psi[n]\rangle - T_s[n] - \frac{e^2}{2}\int d\underline{r}\int d\underline{r}'\ \frac{n(\underline{r})n(\underline{r}')}{|\underline{r}-\underline{r}'|} \tag{31}$$

where $T_s[n]$ is the internal energy functional for a system of non-interacting electrons ($e = 0$).

We see that, just as in the non-relativistic formalism, the many-body effects enter as a contribution to the effective potential However, in the relativistic case $E_{xc}[n]$ will contain effects in addition to the statistical and Coulomb correlations contained in the non-relativistic formalism. To be more explicit we consider a specific approximation, the local density approximation (LDA),

$$E_{xc}[n] \simeq d\underline{r} \int n(\underline{r})\, \epsilon_{xc}(n(\underline{r})) \tag{32}$$

where $\epsilon_{xc}(n) + t(n)$ is the energy density of a relativistic electron gas and $t(n)$ is the energy density of a non-interacting relativistic electron gas. It is customary to separate $\epsilon_{xc}(n)$ into the contribution that comes from treating interaction effects in leading order perturbation theory, $\epsilon_x(n)$, and the remaining contribution $\epsilon_c(n)$. These are properly regarded as the exchange and correlation contributions to $\epsilon_{xc}(n)$. Many calculations of $\epsilon_x(n)$ for a relativistic electron-gas have been presented in the literature[41-43] and the results can be expressed in the form

$$\epsilon_x(n) = -\frac{9}{4r_s}\left(\frac{2}{3\pi^2}\right)^{1/3}\left(1 - \frac{3}{2}\left(\frac{\beta\eta - \ell n\xi}{\beta^2}\right)^2\right)\ \mathrm{Ry.} \tag{33}$$

where $a_o r_s = (3/4\pi\eta)^{1/3}$ is the radius of a sphere containing one electron, $\beta = ƛ_c k_F(n) \simeq 1/(71.4r_s)$, $\eta = (1+\beta^2)^{1/2}$, and $u = b+h$. An approximate expression for $\epsilon_c(n)$, which reduces to the RPA in the non-relativistic limit has been given by Jancovici[42] and has recently been evaluated by Ramana and Rajagopal.[44]

Thus, in the local density approximation, the relativistic interaction effects appear as relativistic corrections to the functions $\epsilon_x(n)$, $\epsilon_c(n)$, $V_x(n) = d(n\epsilon_x(n))/dn$ and $V_c(n) = d(n\epsilon_c(n))/dn$. [See Eq. (30d).] From Eq. (33) it follows that

$$V_x(n) = \frac{-3}{r_s}\left(\frac{2}{3\pi^2}\right)^{1/3}\left(-\frac{1}{2} + \frac{3\ln\xi}{2\beta n}\right) \tag{34}$$

Thus, in the local density approximation, the relativistic interaction effects appear as relativistic corrections to the functions $\epsilon_x(n)$, $\epsilon_c(n)$, $V_x(n) = d(n\epsilon_x(n))/dn$ and $V_c(n) = d(n\epsilon_c(n))/dn$. [See Eq. (30d).] From Eq. (33) it follows that

$$V_x(n) = \frac{-3}{r_s}\left(\frac{2}{3\pi^2}\right)^{1/3}\left(-\frac{1}{2} + \frac{3\ln\xi}{2\beta n}\right) \tag{34}$$

The relativistic and non-relativistic forms for t(n) and $\varepsilon_x(n)$ are compared in Fig. 3. The leading relativistic interaction effects (i.e., those in $\varepsilon_x(n)$ correspond physically to the transverse magnetic current-current interaction and retardation corrections to the Coulomb interaction and to relativistic corrections to the local approximation to the exchange operator. In atoms, the exchange-operator can be included exactly while the transverse interactions may be treated perturlatively by the Breit interaction[45] or, for very high Z systems, generalizations thereof.[46] The LDA treatment of these effects certainly has the virtue of simplicity but we must compare with atomic calculations in order to access its accuracy.

Self-consistent Dirac-Slater type relativistic atomic calculations, both with and without the relativistic exchange correction, have been performed[47] for all atoms from Z = 19 to Z = 98. The shifts in the innermost core levels produced by the relativistic exchange-correction are illustrated in Fig. 32. A typical comparison between these calculations and those in which the transverse part of the interaction is treated perturbatively, that for the case of Hg, is illustrated in Table 5. We compare our energy shifts produced by the relativistic exchange corrections, with only the magnetic part of the perturbatively treated transverse interaction, since the retardation correction is cancelled, in our calculation, by the corrections to the local exchange approximation.[48] The magnetic interaction energies have been taken from Huang et al.[49] In studying this table several additional differences in the calculations must be kept in mind: i) The LDA binding energies come from a Koopman's theorem interpretation of the atomic energy levels (i.e., via a frozen orbital approximation) while the binding energies of Huang et al., include relaxation effects, ii) The LDA calculation includes the relativistic exchange correction self-consistently while Huange et al., include the transverse interaction energy only as a perturbation in both ion and atom states and iii) The LDA calculations used an $\alpha = 1$ non-relativistic exchange potential to simulate relaxation effects whereas Huange et al.,[49] used an $\alpha = 2/3$ exchange potential. The $\alpha = 1$ potential draws charge into the relativistic region and increases the transverse interaction by ∿20% relative to an $\alpha = 2/3$ calculation.

The first point to make concerning the results shown in Table 4 is that the LDA appears to overestimate the transverse interaction

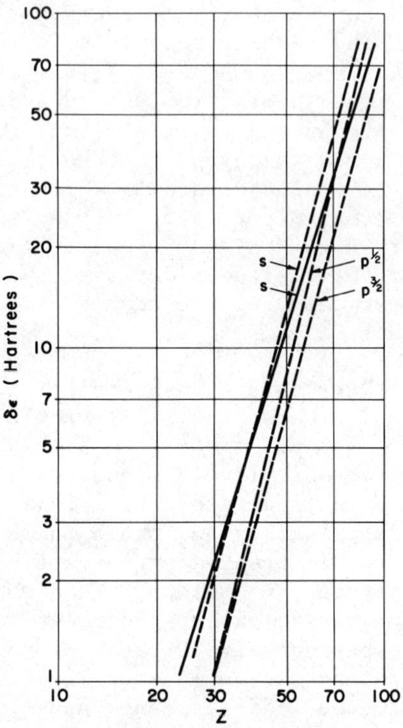

Fig. 32. Shifts in inner-core level eigenvalues produced by relativistic interaction effects in neutral atoms from Z=19 to Z=98. The shifts increase with Z approximately like $Z^{3.2}$ for 1s levels, like $Z^{3.7}$ for 2s levels, like $Z^{4.2}$ for $2p^{1/2}$ levels and like $Z^{4.1}$ for $2p^{3/2}$ levels.

energy by about a factor of 2 for s levels. This result is consistent with the comparisons of | s binding energies recently made by Ramana et al.[50] The overestimate of transverse interaction energies in s levels is probably responsible for the overestimate of the

Table 5. Relativistic interaction corrections to atomic energy levels for Hg. The column labelled LDA compares self-consistent Dirac-Slater type calculations with and without the relativistic exchange corrections while that labelled TI gives the perturbative value for the magnetic part of the transverse interaction energy as calculated by Huang et al.[49] The energies are in Ry units for $n < 3$ and mRy units for $n > 3$.

Orbital	LDA	TI
1s	49.8	24.1
2s	7.1	2.7
$2p^{1/2}$	5.7	4.7
$2p^{3/2}$	3.6	3.1
3s	1.49	0.50
$3p^{1/2}$	1.12	0.88
$3p^{3/2}$	0.68	0.56
$3d^{3/2}$	0.30	0.38
$3d^{5/2}$	0.26	0.26
4s	342	99
$4p^{1/2}$	232	183
$4p^{3/2}$	128	103
$4d^{3/2}$	27	49
$4d^{5/2}$	19	23
$4f^{5/2}$	-35	-14
$4f^{7/2}$	-36	-25
5s	59	14
$5p^{1/2}$	30	27
$5p^{3/2}$	12	12
$5d^{3/2}$	-43	0.7
$5d^{5/2}$	-4.8	-1.8
6s	4.4	1.0

total transverse interaction energy in the LDA discussed by MacDonald and Vosko.[40] What has not been noted earlier, however, is that other than for s levels, the transverse interaction energies seem to be well rendered by the LDA. This may indicate that the LDA breaks down badly only in the immediate vicinity of the nucleus. In the LDA calculation higher order effects of the transverse interaction energy, included by performing the calculation self-consistently, become important for the outer orbitals. The expansion of the inner oribtals results in a reduction of the screening of the nuclear charge and an attractive contribution to the effective potential which dominates the direct repulsive contribution in the

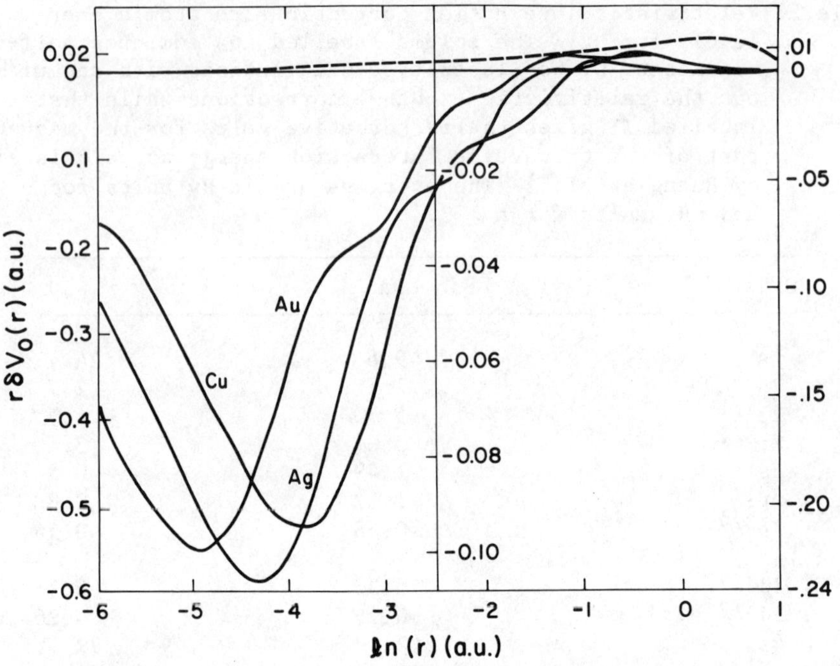

Fig. 33. Changes in self-consistent crystal potentials in the nobel metals due to relativistic interaction effects. Also illustrated, for the purpose of comparison, is the change in the self-consistent crystal potential of Au due to correlation treated in a local approximation (dashed line).

outer core region. This effect is illustrated in Fig. 33 by plotting the changes in the self-consistent crystal potentials in the noble metals which are produced by relativistic exchange corrections.[29] The desirability of a non-perturbative description of transverse interaction effects is evident. The relaxation effects in the Huang et al., calculation are also more important for outer orbitals and so the comparison of the two calculations becomes more muddled. Nevertheless, it certainly seems the transverse interaction effects are overestimated by the LDA, even for valence s-levels. This fact should be kept in mind when the influence of relativistic exchange on the valence levels of solids is discussed below.

The influence of relativistic interaction effects, treated

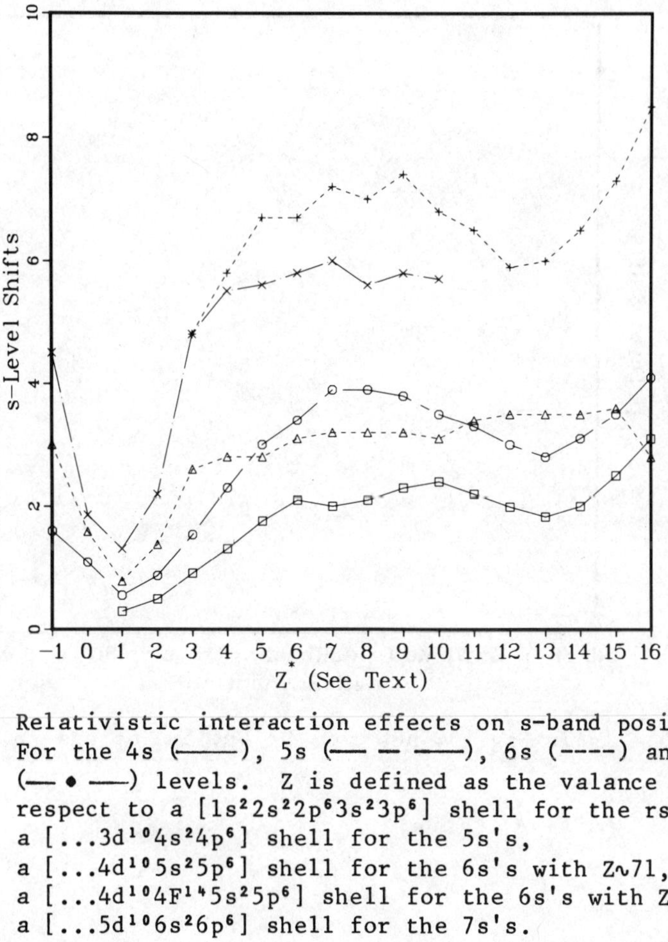

Fig. 34. Relativistic interaction effects on s-band positions. For the 4s (———), 5s (—— — ——), 6s (----) and 7s (— • —) levels. Z is defined as the valance with respect to a [$1s^2 2s^2 2p^6 3s^2 3p^6$] shell for the rs's, a [$...3d^{10} 4s^2 4p^6$] shell for the 5s's, a [$...4d^{10} 5s^2 5p^6$] shell for the 6s's with Z∿71, a [$...4d^{10} 4F^{14} 5s^2 5p^6$] shell for the 6s's with Z≥72 and a [$...5d^{10} 6s^2 6p^6$] shell for the 7s's.

in the local density approximation, on the energy bands in solids is illustrated in Figs. 34-37 for s-bands, p-bands, d-bands and

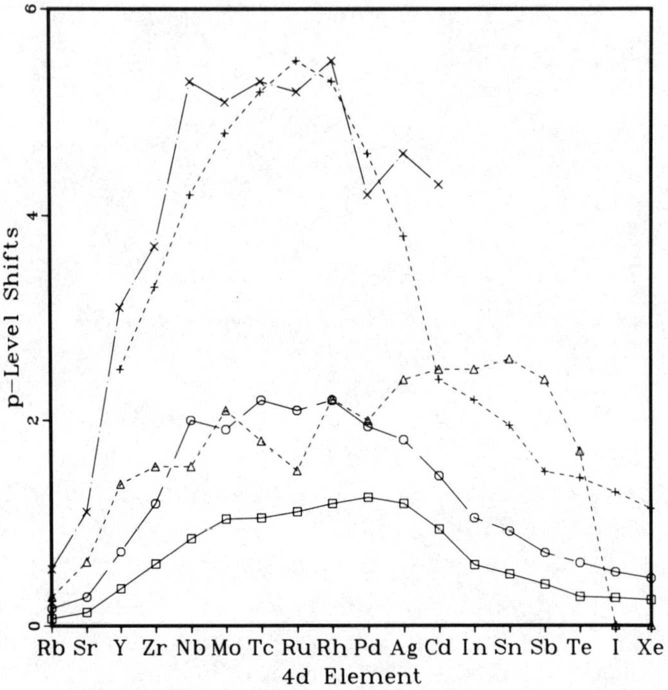

Fig. 35. As in Fig. 34 but for the 4p, 5p, 6p and 7p levels.

f-bands. The most important physical consequences occur in the transition series and the rare earths and actinides. In the transition series the s-bands are raised relative to the d-bands by up to ∿3.5 mRy, 5.0 mRy and 9.5 mRy for the 3d's, 4d's and 5d's respectively. The sizes of these shifts are comparable to those produced by correlation in these systems and they therefore should be included in first-principles band calculations.[28,29] Nevertheless it should be noted that such calculations as a rule place

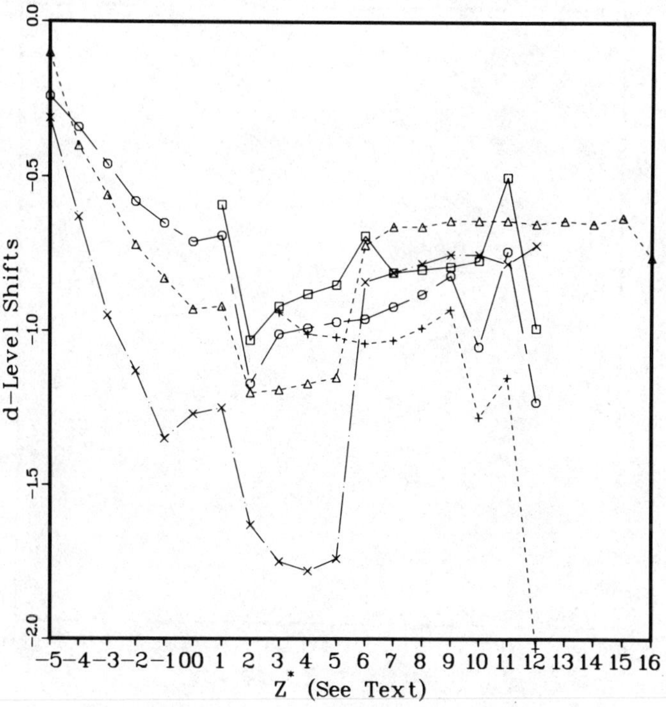

Fig. 36. As in Fig. 34 but for the 3d, 4d, 5d and 6d levels.

the d-bands too low relative to the s-bands so that no improvement in agreement with experiment can be expected to be produced. The s-f shifts for the rare earths and actinides are ∿16 mRy and ∿22 mRy respectively. Most of this shift is due to a downward shift of the localized f levels which see a stronger potential because of the expansion of the inner core levels. For the f-electron systems, first-principles band calculations tend to place the f-electrons too low in energy so that, again, it appears that the

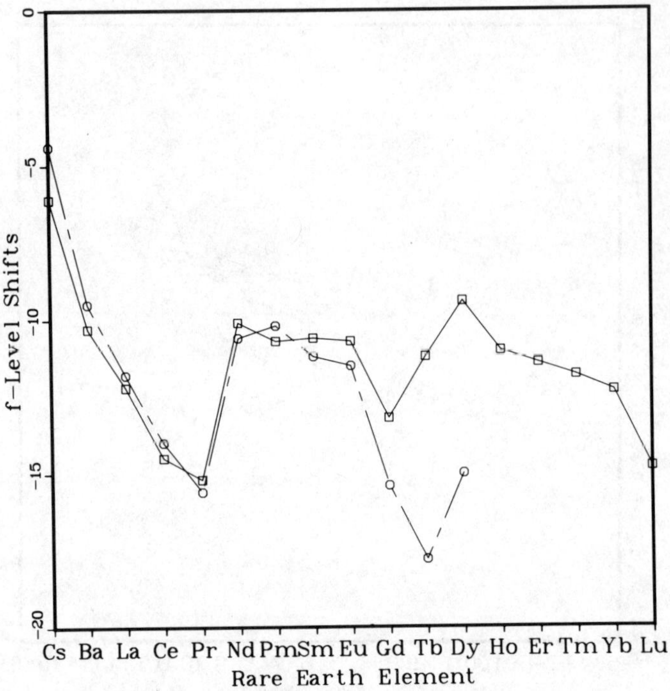

Fig. 37. As in Fig. 34 but for the 4f and 5f levels.

major source of error is not the ommission of relativistic exchange corrections.

Finally, we consider briefly the possibility of the development of a relativistic SPIN-density functional formalism. The most natural generalization is to consider a system of Dirac electrons in a magnetic field $\underline{B}(\underline{r})$, in addition to the potential $\nu(\underline{r})$, which leads to a Hamiltonian

$$\hat{H} = \hat{H}_{sys} + \int d\underline{r}\ \hat{\eta}(\underline{r})v(\underline{r}) - \frac{1}{c}\int d\underline{r}\ \underline{\hat{J}}(\underline{r})\cdot\underline{A}^{ext}(\underline{r}) \tag{35}$$

where $\underline{\nabla X A}^{ext}(\underline{r}) = \underline{B}(\underline{r})$ and $\hat{J}(\underline{r}) = ec\hat{\psi}^{+}(\underline{r})\underline{\alpha}\hat{\psi}(\underline{r})$ is the spatial part of the Dirac four-current density operator. As pointed out by Rajagopal and Callaway[51] this leads to a density functional formalism in which the fundamental density functions are the four-components of the ground-state expectation of the four-current density operator. However in metals this approach will lead to the severe complications associated with solving for Landau levels. An alternate approach, originally suggested by MacDonald and Vosko[40] and recently explored by Ramana and Rajagopal,[52,53] is to consider the Hamiltonian

$$\hat{H} = \hat{H}_{sys} + \int d\underline{r}[n(\underline{r})V(\underline{r}) - \hat{m}(\underline{r})\cdot\underline{B}(\underline{r})] \tag{36}$$

where $\underline{\hat{m}}(\underline{r}) = \mu_B\hat{\psi}+\beta\underline{\sigma}\hat{\psi}(\underline{r})$. The formalism emerging from Eq. (36) will be entirely analogous to the non-relativistic spin density functional formalism[54] and will provide a framework for describing internal magnetic effects in systems in which external magnetic fields are required only to break a ground-state degeneracy. Example of such systems are ferromagnets and non-closed-shell atoms.

The ingredient required to make this formalism useful is the local density approximation for $E_{xc}[n,\underline{m}]$,

$$E_{xc}[n,\underline{m}] \simeq \int d\underline{r}\ n(\underline{r})\ \varepsilon_{xc}(n(\underline{r}),\ |m(\underline{r})|) \tag{37}$$

$\varepsilon_{xc}(n,m)$ is the exchange-correlation energy density of a relativistic electron gas which has magnetization density m and is the lowest energy state of density n in the presence of some external potential

$$-\int d\underline{r}\ \underline{\hat{m}}(\underline{r})\cdot\underline{B} \tag{38}$$

Ramana and Rajagopal[52,53] have calculated the exchange-energy of a polarized electron gas but since the polarized states they consider are not ground-states in any external B, it is not strictly correct to use their results as the basis of a local density approximation for $E_{xc}[n,m]$. (There are obviously many states of the non-interacting electron gas with a given value of m. The one they choose makes the calculation of the exchange energy particularly straightforward.) In any case it is not clear that a local density approximation will capture the main relativistic effects on the $\underline{m}$ dependence[55] of $E_{xc}[n,\underline{m}]$. These questions should be explored in an attempt to understand exchange-enhanced susceptibility and ferromagnetism in relativistic metals.

APPENDIX A: RELATIVISTIC APW AND KKR TECHNIQUE

We here discuss APW techniques as background to the main discussion of relativistic effects in solids. We utilize the approach outlined in the main text where a relativistic treatment is applied to the muffin-tin regions and a non-relativistic treatment to the interstitial region. The basic procedure is a simple application of a Rayleigh-Ritz variational procedure for an expansion in a basis set. Thus we are writing

$$\psi(\underline{k},\underline{r}) = \sum_n a_n P_n(\underline{k},\underline{r}) \tag{A.1}$$

and varying $\langle H\rangle/\langle S\rangle$ with respect to a_n to get a secular equation

$$H\underline{a} = E\, S\underline{a} \tag{A.2}$$

where H is the Hamiltonian matrix and S is the overlap matrix. There are a number of possible choices for P_n and many of these have P_n depend on the eigenvalue. When P_n does depend on Eq. (A.2) becomes a non-linear problem which is typically handled by finding the zeroes of det (H-ES)[56] or by solving the secular equation for a series of choices of E.[57]

The constant feature of the different APW methods is the use of a plane wave expansion in the interstitial region. So in that region the basis function is

$$P_n(\underline{k},\underline{r}) = \begin{pmatrix} |s) \\ \sim 0 \end{pmatrix} \exp i(\underline{k}_n)\cdot\underline{r}$$

where $|s)$ is a two component spinor and the lower component is approximated as zero. K_n is a reciprocal lattice vector. In this region, we are using non-relativistic kinematics so the Hamiltonian is

$$H = \frac{1}{2m}\, p^2 + V(r) \tag{A.3}$$

and the contributions to the matrix elements are

$$\begin{aligned} H^I_{nm} &= \frac{k_m^2}{2m} U(\underline{K}_m - \underline{K}_n) + V(\underline{K}_m - \underline{K}_n)\,\delta_{ss'} \\ S^I_{nm} &= U(\underline{K}_m - \underline{K}_n)\,\delta_{ss'} \end{aligned} \tag{A.4}$$

Here $V(\underline{K})$ is the Fourier transform of $V(\underline{r})$ where it has been defined that $V(\underline{r})$ is zero inside the muffin-tin spheres. $U(\underline{k})$ is the Fourier transform of a step function which is zero inside the muffin-tin spheres and one in the interstial region. It has the form

$$U(\underline{K}) = \delta_{\underline{K},o} + \sum_i e^{iK\cdot\tau_i} \frac{4\pi R_i^3}{\Omega} \frac{j_1(KR_i)}{KR_i}$$

where τ_i is the position and R_i the radius of the i'th muffin-tin sphere The matching at the sphere boundary is facilitated by the Rayleigh expansion of the plane wave

$$P_n(k,r) = 4\pi \sum_{\ell m} i^\ell \begin{pmatrix} j_\ell(k_n)\, Y_{\ell m}(\hat{k}_n)\, Y_{\ell m}(\hat{r})\ |s) \\ \sim 0 \end{pmatrix} \exp\, i\underline{K}_n\cdot\underline{\tau}_i \quad (A.5)$$

However, this is to be joined to the relativistic solutions inside the muffin-tin spheres so it is useful to recombine terms to have an expansion in the $\chi_{k\mu}$ angular spinor harmonics (which are the relativistic generalizations of the spherical harmonics).

$$\chi_{\kappa\mu}^{(\hat{r})} = \sum_{s'} C(\kappa\mu s')\, Y_\ell^{\mu-s'}(r)\ |s') \quad (A.6)$$

The Clebsch-Gordon coefficeint C is indexed with κ,μ , and s because they determine the standard arguments and are a more compact notation.

$$j = \kappa - 1/2$$

$$\ell \begin{cases} = \kappa - 1 & \kappa < 0 \\ = \kappa & \kappa > 0 \end{cases} \quad (A.7)$$

$$\mu = m + s$$

Utilizing the orthonormality properties of the Clebsch-Gordon coefficients,

$$Y_\ell^m\ |s) = C(\kappa = \ell,\mu,s)\ \chi_{\ell\mu} + C(\kappa = -\ell-1,\mu,s)\ \chi_{-\ell-1\mu}$$

Clearly, one can transform between the ℓms and κ_μ representations only if the radial solutions for $\kappa = \ell (j = \ell - 1/2)$ and $\kappa = -\ell-1$ $(j = \ell + 1/2)$ are equal. This is the case of no spin orbit coupling and applies to the plane wave but not to the radial solutions inside the muffin-tin sphere. However, at any one radius they can be made equal since the normalization is unspecified. If we set them equal to the spherical Bessel function at the muffin tin radius, the basis function will be continuous but, in general, the derivative will not be. Although there are many possible variational quantities which incorporate a "penalty" for the discontinuity,[58] the one most often used in atomic calculations to estimate the eigenvalue correction when joining the inward and outward radial integrals. It is derived from a simple consideration of the kinetic energy in a thin shell of thickness ε at the boundary. The kinetic energy is $-1/_{\varepsilon}\left(\frac{\partial\psi^{I}}{\partial\theta} - \frac{\partial\psi^{MT}}{\partial\theta}\right)$. The ε will be cancelled by the radial integration so

$$\Delta H_{nm} = \int d^2\hat{r}\ \underline{P}_n\ [r\cdot\nabla\ \underline{P}_m^{MT} - \hat{r}\ \cdot\nabla\ \underline{P}_m^{I}] \qquad \text{(A.8)}$$

The component which involves the interstitial expressions alone is a simple plane wave integral and we can expand the P_n and P_m in their angular representations to get

$$\Delta H_{nn'} = 4\pi \sum_{i\ell m} e^{i(\underline{k}_{n'}-\underline{k}_n)\cdot\tau_i} (s_n|Y^*_{\ell m}(\hat{k}_n)\ j_\ell(k_nR_i)j_\ell(k_{n'}R_i)$$

$$\times \sum_{\ell(x)=\ell} C(\kappa\mu s)C(\kappa\mu s')\ \xi_\kappa\ Y^m_\ell(\hat{k})|s') \qquad \text{(A.9)}$$

$$+ (\underline{k}_n\cdot\underline{k}_{n'} - k^2_{n'})\ U\ (\underline{k}_{n'} - \underline{k}_n)\ (s_n|s_{n'})$$

where

$$\xi_\kappa \equiv \left.\frac{\partial g_x}{\partial r}\ g_\kappa^{-1}\right|_{r=R}$$

and g_κ is the radial solution from the interior. The basis function in the interior of the muffin-tin is to be made up of solutions of the Dirac radial equation multiplying the spinor harmonics with the coefficients determined by our matching coefficients. Even so, they are still undetermined because one must choose an energy to insert into the radial equations. The most accurate choice is to demand that this energy parameter match the eigenvalue. Then one has a non-linear problem and the historical solution has been to numerically tabulate the determinant of (H-ES) as a function of E seeking its zeroes (although it is not the only possible approach). The contribution of the interior region then cancels and after some rearranging[7-10] one gets

$$(H-ES)_{nn'} = (\underline{k}_n \cdot \underline{k}_{n'} - E)\, U(\underline{k}_{n'} - \underline{k}_n)\, (s_n | s_{n'})$$

$$+ (s_n | s_{n'}) \sum_i e^{i(\underline{k}_{n'} - \underline{k}_n) \cdot \tau_i} \sum_\ell (2\ell+1) P_\ell(\hat{k}_{n'} \cdot \hat{k}_n) j_\ell(n) j_\ell(n')\, \xi_\ell \tag{A.10}$$

$$+ \sum_i e^{i(\underline{k}_{n'} - \underline{k}_n) \cdot \tau_i} \sum_\ell j_\ell(k_n R_i) j_\ell(k_{n'} R_i)\, \eta_\ell$$

$$\times\, (s_n | \underline{L}_{\underline{k}_{n'}} \cdot \underline{\sigma}\, P_\ell\, (\hat{k}_n \cdot \hat{k}_{n'})\, | s_{n'})$$

$$\xi_\ell \equiv \frac{\ell}{2\ell+1}\, \xi_{k=\ell} = \frac{(\ell+1)}{2\ell+1}\, \xi_{k=-\ell-1}$$

$$\eta_\ell \equiv \xi_{k=-\ell-1} - \xi_{k=\ell}$$

$$L_{\underline{k}} \equiv -i\, (\underline{k} \times \frac{\partial}{\partial \underline{k}})$$

This is the standard relativistic APW method with the spin orbit splitting terms explicitly displayed. A significant feature to note is that the ξ_ℓ is just the j-weighted average of the ξ_k's Thus the terms not involving the spin orbit coupling have the same form as the non-relativistic expressions only with the logarithmic derivative for the ℓ partial wave replaced by the j-weighted sum of the two logarithmic derivatives for $j = \ell \pm 1/2$. In general, it is found that, when properly formulated, all APW and KKR prescriptions can be written in the form of the non-relativistic result plus a spin orbit term. The parameter of the formalism whether logarithmic derivative, phase shift, or normalization integral is replaced by a j-weighted sum of its relativistic analog. The spin-orbit term then involves the difference of the same parameter.

To normalize the wave function once the eigenvalue is found, it is necessary to calculate the contribution to the overlap integral from the interior of the MT spheres. Using the Green's identity relating the overlap integral to the values of the function at the sphere boundary, it is found that that contribution has the same functional form as the logarithm term in the Hamiltonian but with the logarithmic derivative replaced by the negative of the energy derivative of the logarithmic derivative. Again we see the form of the non-relativistic result with j-weighted parameter plus a spin-orbit term dependent on the difference of that parameter.

More recent developments in APW technique employ a different approach to the treatment of the interior of the muffin-tin spheres. The requirement that the basis functions be exact solutions inside the spheres is relinquished in favor of also satisfying the boundary condition of continuous derivative. This usually requires that one use two radial solutions but now the surface term no longer appears. The simplest choice to make is that one use two energies[59] "nearby". For example, one a quarter of the way up from the bottom of the bands and the other a quarter of the way down from the top. Or one may let the two energies approach each other and recombine to get the radial solution and its energy derivative which can then be used together as the necessary pair of functions.[13,60] A third alternative is to select an energy for each interstitial plane wave which gives a logarithmic derivative equal to that of the spherical Bessel function.[61] These so-called linearized versions all have the feature that they eliminate the dependence of H and S on E so that standard generalized eigenvalue problem techniques can be applied. The contribution of the interior must then be calculated and added into the Hamiltonian and overlap matrices. This is greatly facilitated by the radial functions being solutions of the Dirac equation so that the normalization and overlap integrals are related to the functional values at the surface. (This done by integration by parts which is the derivation of the Green's identity.) These linearized techniques usually require that one form linear combinations of radial solutions throughout the entire interior of the muffin-tin spheres to cast the problem in the form of the non-relativistic problem plus a spin orbit term. Thus, one gets more complicated average factors although they all turn out to be combinations of j-weighted sums of the various factors involved.

But what of the KKR method? In the Onodera and Okazaki formulation,[62] it is a bit difficult to see that it relates to our simple view. Nonetheless, Treusch and Sandrock[63] pointed out that the formalism as calculated could be transformed to the form they used when applying the Foldy-Wouthuyser transformed equations. It is much easier to see using the derivation of Takada.[64] In that case, he explcitly makes the appropriate non-relativistic approximations in the Green's function describing the interstitial region rather than as a mathematical simplification in the fully relativistic results. Briefly, the derivation proceeds as follows. One can express the wavefunction in the interstitial region in terms of its value inside the muffin-tin spheres by utilizing the Green's function relation

$$\psi = G\, V\, \psi$$

Then, following the Korringa form of the derivation of the method, one uses this to perform the matching at the sphere boundary. The result is that one only requires the Green's function in the limit where both arguments approach the muffin-tin sphere boundary--one from the inside and one from the outside. As we have stressed before, this boundary is well into the "non-relativistic" regime so a non-relativistic approximation can be made for the Green's function. Then the information that the relativistic effects have occurred is solely carried by the wavefunction from the interior and this is embodied by the phase shifts.

One final note should be made here. The difficulties with an orthogonalization catastrophe to the negative energy states that can occur when using a basis set expansion will not occur in these techniques. This is because they are actually using numerical solutions throughout the atomic core regions where the relativistic velocities occur. This is not the case for the relativistic orthogonalized plane wave method which extends the free particle spinor into the core region and then orthogonalizes it to the atomic core. The problem has not been observed by practitioners of the method but this is probably because their precision is 0.1 eV rather than the millielectron volt range discussed at this summer school.

APPENDIX B: THE ATOMIC SPHERE APPROXIMATION

Because we have utilized several concepts resulting from the atomic sphere approximation[13] (ASA), we here outline its development for the interested reader. As has been emphasized earlier, the electronic structure of solids may be thought of as emerging from the combination of atomic-like solutions within the muffin-tin sphere and (angular momentum destroying) boundary conditions imposed at the sphere boundary. The boundary conditions arise from the necessity of joining onto solutions in the interstitial region which satisfy the periodic boundary conditions. A particularly simple, though approximate, description of the electronic properties of solids which emphasises this point of view is obtained in the atomic sphere approximation (ASA). The ASA is the ultimate generalization of the Wigner-Seitz approximation which proceeds from the replacement of the actual Wigner-Seitz unit cell by a spherical shell of the same volume and nearly the same shape. Again, only the spherical component of the potential is retained. One can then treat the requirement of periodicity of the wavefunction as a boundary condition on the sphere surface. In practice, the only boundary condition Wigner and Seitz used was that the radial function should have zero derivative at the sphere boundary for the bottom of the band and zero value at the top. These boundary conditions ignore any coupling (hybridization) between solutions with different angular momentum. The ASA is derived using a different approach but will be reduced to a Wigner-Seitz like calculation in which the optimal boundary conditions are obtained.

The ASA is actually derived as an approximation to the linear combination of muffin-tin orbital (LCMTO) technique. We will follow here the derivation in which the muffin-tin orbitals are immediately approximated by ASA orbitals. This involves extending the matching condition from the muffin-tin sphere boundary to the Wigner-Seitz sphere boundary and setting the kinetic energy parameter equal to zero in the (zero volume) interstitial region.

Then, in the non-relativistic ASA, the eigenfunctions are expressed in terms of these atomic-sphere orbitals

$$X_{\ell,m}(E,r) = i^{\ell} Y_{\ell,m}(\hat{r}) \begin{cases} \phi_{\ell}(E,r) - \phi_{\ell}(E,S)\left\{\dfrac{D_{\ell}(E)+\ell+1}{2\ell+1}\right\}\left(\dfrac{r}{S}\right)^{\ell} & r \leq S \\ \phi_{\ell}(E,S)\left\{\dfrac{\ell-D_{\ell}(E)}{2\ell+1}\right\}\left(\dfrac{r}{S}\right)^{-(\ell+1)} & r \geq S \end{cases} \tag{B.1}$$

where S is the atomic sphere radius, $\phi_{\ell}(E,r)$ is a solution of the radial Schrodinger equation inside the atomic sphere centered at the origin and $D_{\ell}(E)$ is its logarithmic derivative at the sphere boundary. (We are following Andersen's notation here so it is worth noting that he has chosen to utilize the true logarithmic derivative

$$D_{\ell} \equiv d \ln(\phi)_{\ell}/d \ln (r)\big|_{r=S} \tag{B.2}$$

rather than the more conventional ϕ_{ℓ}'/ϕ_{ℓ}.) These atomic sphere orbitals are continuous, differentiable, and regular. They have the property that

$$\sum_{\vec{R}\neq 0} e^{i\vec{k}\cdot\vec{R}} X_{\ell m}(E, \vec{r}-\vec{R}) = - \sum_{\ell' m'} \left(\frac{r}{S}\right)^{\ell'} \frac{i^{\ell'} Y_{\ell' m'}(\hat{r})}{2(2\ell+1)} S_{\ell m;\ell' m'}(\vec{k})\, \phi_{\ell}(E,s)\left\{\frac{\ell-D_{\ell}(E)}{2\ell+1}\right\} \tag{B.3}$$

converges inside the atomic sphere. The structure factors S are the same standard KKR structure factors in the limit $E \to 0$. Expressions for them can be found in Andersen's original papers. If one ignores the fact that the Wigner-Seitz spheres overlap (and miss some of the interstitial volume), the condition for

$$\Psi(\vec{k},E,\vec{r}) = \sum_{\ell,m} B_{\ell,m}(\vec{k}) \sum_{\vec{R}} e^{i\vec{k}\cdot\vec{R}} X_{\ell,m}(E,\vec{r}-\vec{R}) \tag{B.4}$$

to be an eigenstate of the crystal Hamiltonian is that the terms proportional to $i^{\ell}Y_{\ell,m}(\hat{r})r^{\ell}$ must cancel inside each atomic sphere. This leads to the requirement that

$$\left| S_{\ell'm';\ell m}(\vec{k}) - \frac{2(2\ell+1)(D_{\ell}(E)+\ell+1)}{D_{\ell}(E)-\ell}\,\delta_{\ell',\ell'}\delta_{m,m'} \right| = 0 \tag{B.5}$$

This expression allows some very useful interpretations which we point out before proceeding to its relativistic generalization. First, note that the condition applied is the cancellation of the ASA orbital component inside the spheres which was not a solution of the radial equation. Thus the addition of that term may be viewed as a mathematical expediency to easily derive condition B.5. The reader need not be further concerned with its presence. Next observe that in Eq. (B.5) one has a complete separation of the structure factors S which embody the interstitial or boundary condition properties from the logarithmic derivatives D which embody the properties of the solution of the interior.

Although the structure factors are not diagonal in ℓ and the ASA can include some hybridization effects, one can consider the approximation which occurs when the $\ell \neq \ell'$ terms of S are neglected.

If one then diagonalizes $S_{\ell m,\ell m'}(k)$, one obtains a set of "canonical bands" which embody the boundary conditions for a band of pure ℓ character. These are a set of uncoupled equations which are solved by finding the energy for which $D_{\ell}(E)$ satisfies B.5 with S replaced by its diagonalized form. (The second term is a constant times a unit matrix so it is unaffected by the unitary transformation.)

A natural relativistic generalization of the ASA in the same spirit as Appendix A can be achieved by defining relativistic ASA Dirac spinors as

$$D_{k,mj} = i^{\ell} \begin{bmatrix} g_{\kappa}(r) - g_{\kappa}(S)\left(\dfrac{D_{\kappa}+\ell\ 1}{2\ell+1}\right)\left(\dfrac{r}{S}\right)^{\ell} \chi_{\kappa,m_j} \\ -if_{\kappa}(S)\vec{\alpha}\cdot\hat{r}\,\chi_{\kappa,mj} \end{bmatrix}^{r \leq S} \tag{B.6a}$$

$$D_{k,mj} = i^{\ell} \begin{bmatrix} g_{\kappa}(S) \dfrac{\ell - D_{\kappa}}{2\ell + 1} \chi_{\kappa,m_j} \\ 0 \end{bmatrix} \qquad r \geq S \tag{B.6b}$$

where g_{κ} and f_{κ} are the large and small component solutions of the radial Dirac equation and $\chi_{\kappa,mj}$ are the usual mixed angular and spin functions (generalized harmonics) associated with eigenstates of the Dirac equation in a central potential. To define notation we write the generalized harmonics is a slightly modified form

$$\chi_{k,m_j} = \sum_{\ell,m,s} U^{\ell,m,s}_{\kappa,m_j} Y_{\ell,m} \chi_s \tag{B.7}$$

and note that the matrix U which contains the Clebsch Gordon coefficients is unitary. Again, we assume that the small component of the atomic sphere spinor is negligible just inside the sphere boundary and we take these orbitals as being continuous and differentiable. It follows that the condition for

$$\psi(\vec{k},\vec{r}) = \sum_{k,mj} B_{k,m_j}(\vec{k}) \sum_{\vec{R}} e^{i\vec{k}\cdot\vec{R}} D_{\kappa,m_j}(E,\vec{r} - \vec{R}) \tag{B.8}$$

to be an eigenspinor of the Dirac equation in the crystal potential is that terms proportional to $r^{\ell}\chi_{\kappa,m_j}$ in Eq. (B.8) vanish inside each atomic sphere. This leads to the requirement that

$$0 = \left| \sum_{\substack{\ell,m,s \\ \ell'm's'}} U^{\ell,m,s}_{\kappa,m_j} S_{\ell m;\ell'm'} \delta_{s,s'} (U^{\ell'm's'}_{\kappa',m'm}) - 2(2\ell+1)\frac{D_{\kappa}+\ell+1}{D_{\kappa}-\ell} \delta_{\kappa'\kappa} \delta_{m_j' m_j} \right| \tag{B.9}$$

Which has the form of the non-relativistic result B.5 with a unitary transformation performed on S and κ replacing ℓ for the potential parameters. By taking advantage of the unitary nature of U Eq. (B.9) may be cast in a more revealing form

$$\left| S_{\ell'm';\ell m}(\vec{k})\, \delta_{s,s'} - \tilde{\Omega}_{\ell} \delta_{\ell,\ell'} \delta_{m,m'} \delta_{s,s'} - \Lambda_{\ell}\, \vec{S}\cdot\vec{L} = 0 \right| \tag{B.10a}$$

where

$$(2\ell+1)\tilde{\Omega}_{\ell} = \ell\, \Omega_{-\ell-1} + (\ell+1)\Omega_{\ell} \tag{B.10b}$$

$$(\ell+\frac{1}{2})\Lambda_\ell = \Omega_{-\ell-1} - \Omega_\ell \qquad \text{(B.10c)}$$

and

$$\Omega_\kappa = \frac{2(2\ell+1)(D_\kappa+\ell+1)}{D_\kappa - \ell} \qquad \text{(B.10d)}$$

Eq. (B.10a) emphasizes the fact that the boundary conditions at the atomic sphere are unchanged from the non-relativistic case and that the spin-orbit coupling, originating inside the atomic sphere, is refected by differing values for the potential parameters $\Omega_{-\ell-1}$ and Ω_ℓ. Further, we again see the situation where the non-relativistic form is retained with the relavant parameter replaced by a j-weighted average and then a spin-orbit term appearing containing the difference.

The same sorts of simplification to terms of canonical bands are obviously also possible in the relativistic case. Further progress in obtaining a qualitative understanding depends on a characterization of the energy dependence of the logarithmic derivative $D_{\ell_\kappa}(E)$. This can be done using as expansion in principal quantum number of the atomic like orbitals. To choose a starting point, it is observed that the center of the band occurs where $D_\kappa = -(\ell+1)$ which is where the ASA orbital becomes a pure atomic like orbital with a $r^{-(\ell+1)}$ tail attached. This energy C_κ is an excellent center for an expansion. We may view the following analysis as a simple Taylor series expansion (although the underlying analysis is somewhat more involved). Then

$$D_\kappa(E) \simeq -(\ell+1) - 1/2\ \mu_\kappa S^2(E-C_\kappa) + \ldots \qquad \text{(B.11)}$$

The derivative has been written in this peculiar form ($-1/2\ \mu_\kappa S^2$) because of the underlying analysis, of course, but one can understand much of it from simple considerations. First, it is negative because the normalization integral is proportional to the negative of the energy derivative of the logarithmic derivative. Since the norm must be positive, the logarithmic derivative must have a negative energy derivative. (One also gets from this relation that $\mu \propto \phi^{-1}_\kappa(S)$.) The band mass parameter is one for a constant potential. Now if D_κ varies more rapidly with E, the bands will be narrower and vice versa. So μ^{-1}_κ is roughly proportional to the bandwidth. The S^2 removes the standard a^{-2} dependence which is always one component of the band dependence on compression. Let us put the linear expansion B.11 in the diagonalized form of B.5 under the assumption that the expansion is valid over the entire band. (A rough but adequate approximation for our purposes. D_ℓ actually has a (-tangent) like behaviour). Then

$$\tilde{S}_{\ell\ell} = \frac{2(2\ell+1)(D+\ell+1)}{(D_\ell-\ell)} \approx \frac{1/2\, S^2(E-C_k)\,\mu_\kappa}{\left[1 + \frac{1}{2(2\ell+1)}\,\mu_\kappa S^2(E-C_\kappa)\right]} \tag{B.12}$$

so, if the denominator is approximately one, the bandwidth will be equal to the range of $\tilde{S}_{\ell\ell}$ (dependent only on crystal structure) divided by $(\mu_\kappa S^2)$. That is why we have used μ^{-1} as a measure of bandwidth in Figs. 17-18. The form of secular equation which emerges when the approximation of Eq. (B.11) is inserted into Eqs. (B.10) is what we take to be the basis of the model Hamiltonian used in Sec. 4 of these lectures.

APPENDIX C. APPROXIMATELY RELATIVISTIC TECHNIQUES

Given that the effects of relativity arise entirely within the muffin-tin spheres, the treatment of the relativistic equations can be taken from techniques used in atomic, electron scattering, or molecular physics. Furthermore, such considerations need only deal with the central field problem which occasionally provides some simplification. The main simplification sought in using approximate relativistic treatments is the elimination or delay in considering the spin orbit coupling. The reason is quite simple. If the spin orbit coupling can be neglected, the problem is the same as the non-relativistic treatment except that the relevant parameter is replaced by a j-weighted average value. In fact, there is no difference if those parameters are used as disposable fitting parameters as in phase shift analyses of Fermi surface data rather than calculated. However, the spin-orbit coupling is a symmetry breaking interaction which requires that both spins be considered at once thereby doubling the size of the secular equations involved and seriously complicating the analysis of magnetic interactions. It further complicates the computational aspects since it forces the secular equation to be complex. In the non-relativistic case, the secular equation can be made real whenever there is inversion symmetry and this is a great savings in effort.

The earliest attempts to include relativistic effects naturally enough used the Foldy-Wouthuysen transformed equations[53,65]

$$\begin{aligned} H = \frac{p^2}{2m} + V - \frac{h}{4mc^2}\,\sigma \cdot (\underline{\nabla} V \times \underline{p}) \\ - \frac{\hbar^2}{8m^2c^2}\,\nabla^2 V \end{aligned} \tag{C.1}$$

Here we have assumed zero magnetic field and removed the rest mass energy. Clearly the first two terms are the standard non-relativistic Hamiltonian, the third term is the spin orbit coupling as is more easily recognized when inserting $V = V(|r|)$, and the fourth term is the Darwin or Zitterbewegung term.

Difficulties obviously arise when one considers the $r \to 0$ limit where $V \to -Ze^2/r$. They can be handled, of course, but it is difficult to do when trying to apply a straightforward radial integration scheme. The best approach is probably to take a hint from the elimination schemes and replace m by $(m - 1/2c^2\ V)$ in the denominator of these two terms. Then the terms are naturally cut off as V and derivatives approach the size of mc^2 where the FW transformation is not valid anyway. Although not strictly a part of the FW transformed Hamiltonian at this level, the mass velocity term

$$H_{MV} = -\frac{1}{8m^3c^2}\ P^4 \tag{C.2}$$

is often included as a part of the Hamiltonian. It is, after all, primarily this term which lowers the kinetic energy such that the relativistic contraction occurs. The contribution of the Darwin term is mainly to lower the energy of the s states as its leading contribution is from the nuclear potential at $r \to 0$ since $-\nabla^2(-Ze^2/r) = 4\pi Ze^2\delta(r)$. Clearly this term will be modified by replacing a point nucleus approximation by a representation for a nucleus of finite extent. It will also be greatly affected by any imprecision in the radial integration technique. This is why relativistic calculations are usually done using a logarithmic grid where

$$r_n = r_o \exp(n\ \Delta) \tag{C.3}$$

so that the step is very fine near $r \to 0$. The FW approach is rarely used. Instead, a number of techniques <u>which are not</u> 1/c <u>expansions</u> are more commonly used.

As has been repeatedly noted above, the relativistic calculational techniques can be written in the form of their non-relativistic equivalent plus a spin orbit term. A reasonable approach is then to neglect the spin orbit term so the only reflection of the relativity is the formation of the appropriate j-weighted average parameter after having solved the radial Dirac equation inside the muffin-tin spheres. This has the advantage of avoiding the small r complications of the FW procedure while still retaining the non-relativistic form. It has two complications, however. First, because the parameters are j-weighted to form a particular spin combination, the associated wave function is not a pure spin basis function. In the case of the APW method--where this technique is most often used--one forms the j-weighted average of the logarithmic derivatives.[7-10] This corresponds to forming a pure spin state on the surface of the muffin-tin sphere and ignoring the spin dependent part of the derivatives while going into the interstitial region. This amounts to not allowing the boundary conditions to propagate a knowledge of the spin orbit coupling into the interstitial region. But the spin orbit does mix the spins within the muffin-tin spheres. Thus the resultant wave function is a pure spin state only in the interstitial region. Second, blindly ignoring the spin-orbit term can introduce fictitious results. In the case of the logarithmic derivative averaging, this occurs when one j component has a node at the MT sphere boundary. At that energy the logarithmic derivative goes through an asymptote like singularity. The result is a fictitious set of bands within the energy range between the asymptote for $j = \ell - 1/2$ and $j = \ell + 1/2$ resulting from the fact that the averaging procedure is poorly defined in this region. This difficulty is alleviated when one deals with the linearized technique which form continuous function and derivative because one effectively deals with the average of the derivative over the average of the function. Such a scheme could be introduced into the standard APW technique to deal with this dual asymptote problem. Nonetheless, one can see that the averaged parameter techniques require some care.

The remaining techniques all involve applying the Hamiltonian twice in one way or another. One is then reformulating the problem making use of properties of the solutions. This can lead to inconsistencies unless one is quite careful. An example has been discussed by Rose[66] for the Dirac technique of obtaining the Klein-Gordon equations. Although there are a number of choices in procedure which lead to different techniques, the first step is to write an equation for the large component alone.[67-73] The small component must then be obtained from the original Dirac equation to be consistent. One has (on subtracting out the rest mass energy)]

$$H_D = c\,\underline{\alpha} \cdot \underline{p} + V + (\beta - 1)\,mc^2 \tag{C.4a}$$

$$\alpha = \begin{matrix} 0 & \sigma \\ \sigma & 0 \end{matrix} \qquad \beta = \begin{matrix} 1 & 0 \\ 0 & -1 \end{matrix} \tag{C.4b}$$

$$\psi = \begin{matrix} u \\ \Lambda \end{matrix} \tag{C.4c}$$

yielding

$$c\,\sigma \cdot p\,\Lambda + V\,U = E\,U \tag{C.5a}$$

$$c\,\sigma \cdot p\,U + (V - 2mc^2)\,\Lambda = E \tag{C.5b}$$

with σ being the standard Pauli spin matrices. So one can readily solve C.5b (first application of the Hamiltonian) for Λ and insert into C.5a to get

$$c\ \underline{\sigma}\bullet\underline{p}\ \frac{c^2}{(2mc^2 + (E-V))}\ \sigma\bullet p + V \quad U = E\ U \qquad (C.6)$$

If the denominator in C.6 were to be $2mc^2$ alone, this would reduce to the non-relativistic equivalent as $(\sigma\bullet p)^2 = p^2$. Defining

$$M = m + 1/2c^2\ (E - V) \qquad (C.7a)$$

this becomes

$$\frac{1}{2M}\ p^2 + \frac{\hbar}{2M^2c^2}\ [\sigma \bullet (\underline{\nabla}\ V \times \underline{p}) - i\ \underline{\nabla}\ V\bullet p] + V \quad U = E\ U \qquad (C.7b)$$

For familiarity, we see that the mass velocity term can be obtained by expanding M^{-1} in the first term, the second term is the spin orbit coupling and the third term should relate to the Darwin term. One can see qualitatively that has the same effect as in C.1 from the interpretation of the Darwin term as the uncertainty in position due to the Zittenbewegung. If that uncertainty is wrapped up in a convolution integral, one can integrate by parts so that $-i\nabla V\bullet p$ appears as $-\nabla^2 V$. We will not explore that further but return to a devẹlopement of C.7. Note that, if one inteprets the quantity in braces as an operator, it is non-hermitian and weakly depends on the eigenvalue. One can, however, identify the spin orbit coupling term and define a zero order operator where it is omitted. And from this point on there are many possible ways of manipulating.

Here, let us follow yet another possible path which is a generalization of what was done by Koelling and Harmon[69] in the case of a spherical potential. Clearly, the spin orbit term would not appear in C.7 if p were always parallel to ∇V. Defining $\hat{v} \equiv \nabla V/|\nabla V|$, we can write C.5b as

$$[\hat{V}\bullet\sigma\ \hat{v}\bullet\sigma\ \sigma \bullet (\hat{v}\ x(\hat{v}xp))]\ U + (V - 2mc^2)\ \Lambda = E\Lambda \qquad (C.8)$$

Clearly, if the double cross product term is dropped, the spin orbit coupling will not appear in C.7. One can thus write a Hamiltonian decomposition in which the spin-orbit coupling does not appear for the large component for the "zero order" Hamiltonian. It still does, however, for the lower component as can be seen by solving C.5a and C.5b for Λ. The relativistic results can be pretty well represented, however, by solving the "spin orbitless" equations and then adding the missing component to the small component using (C.5b). This all works out much nicer when using a spherical potential where $\hat{v} = \hat{r}$ which is what was done by Koelling and Harmon. But the important point to note is that the solution to "H_o" can have a definite spin in the upper component which is of some interest when one wants to consider magnetic exchange interactions. Further it can be used in much heavier systems than the 1/c expansion techniques.

REFERENCES

1. J. P. Desclaux and P. Pyykkö, Chem. Phys. 29, 534 (1974); J. P. Desclaux, 42, 545 (1978); Recherche (Fr.) 11, 592 (1980).
2. J. P. Desclaux, Phys. Scr. 21, 436 (1980).
3. P. Pyykkö, Adv. Quant. Chem. 11, 353 (1978).
4. T. Ziegler, J. G. Snijders and E. J. Baerends, J. Chem. Phys. 74, 1271 (1981).
5. J. C. Slater, Phys. Rev. 51, 846 (1937).
6. T. L. Loacks, Phys. Rev. 139, A1333 (1965).
7. L. F. Mattheiss, Phys. Rev. 151, 450 (1967).
8. J. O. Dimmock, Solid State Physics Advances in Research and Applications 26, 103 (1971).
9. O. K. Andersen, Phys. Rev. B2, 882 (1970).
10. G. O. Arbman and D. D. Koelling, Phys. Scr. 5, 273 (1972).
11. D. D. Koelling, Phys. Rev. 188, 1049 (1969).
12. E. Wigner and F. Seitz, Phys. Rev. 43, 804 (1933); E. Wigner, 46, 509 (1934).
13. O. K. Andersen, Phys. Rev. B12, 3060 (1975).
14. J. C. Slater, Phys. Rev. 92, 603 (1953).
15. J. C. Duthie and D. G. Pettifor, Phys. Rev. Lett. 38, 564 (1977).
16. L. Hodges, H. Ehrenreich and N. D. Lang, Phys. Rev. 152, 505 (1966).
17. F. M. Mueller, Phys. Rev. 153, 659 (1967).
18. H. Hill, Nucl. Met. 17, 2 (1970); E. A. Kmetko and H. Hill, Nucl. Met. 17, 233 (1970).
19. H. Hill and E. A. Kmetko, Heavy Element Properties (W. Miller and H. Blank ed.) North Holland, p 17 (1975).
20. D. D. Koelling, Rep. Prog. Phys. 44, 139 (1981).
21. J. C. Slater and G. F. Koster, Phys. Rev. 94, 1498 (1951).
22. J. Friedel, P. Lenglart and G. Leman, J. Phys. Chem. Solids 25, 781 (1964).
23. T. S. Rahman, J. C. Parlebas and D. L. Mills, J. Phys. F. 8, 2511 (1978).
24. Inversion symmetry will be assumed in most of the following.
25. See for example Y. Yafet, Solid State phys. 13, 1 (1963).
26. N. E. Christensen, J. Phys. F8, L51 (1978).
27. L. F. Mattheiss and R. E. Dietz, Phys. Rev. 22, 1663 (1980).
28. A. H. MacDonald, J. M. Daams, S. H. Vosko and D. D. Koelling, Phys. Rev. B23, 6377 (1981).

29. A. H. MacDonald, J. M. Daams, S. H. Vosko and D. D. Koelling, Phys. Rev. B (in press).
30. A. H. MacDonald unpublished (1981).
31. For a more general theoretical development see P. K. Misra and L. Kleinman, Phys. REv. B5, 4581 (1972) and references therein. The relationship between the present approach and the expressions given by Misra and Kleinman is discussed elsewhere; A. H. MacDonald unpublished (1981).
32. A. H. MacDonald and D. J. W. Geldart, Phys. Rev. B (1981).
33. F. M. Mueller, A. H. Freeman and D. D. Koelling, J. Appl. Phys. 41, 1229 (1970).
34. L. Hodges, D. R. Stone and A. V. Gold, Phys. Rev. Lett. 19, 655 (1967).
35. See for example, Calculated Electronic Properties of Metals, V. L. Moruzzi, J. F. Janak and A. R. Williams (Pergamon, New York, 1978) and references therein.
36. C. Herring in Magnetism, G. T. Rado and H. Suhl, Eds. (Academic Press, New York, 1966).
37. D. E. Ellis A. Rosén and P. F. Walch, Int. J. Quant. Chem. 95, 351 (1975); D. E. Ellis, Int. J. Quant. Chem. 11S, 201 (1977); D. E. Ellis, NATO Adv. Study Ser. B48, 107 (1980).
38. P. Hohenberg and W. Kohn, Phys. Rev. 136, 864 (1964); W. Kohn and L. J. Sham, Phys. Rev. 140A, 1133 (1965).
39. A. K. Rajagopal, J. Phys. C11, L943 (1978).
40. A. H. MacDonald and S. H. Vosko, J. Phys. C12, 2977 (1979).
41. I. A. Akhiezer and Peletminskii, Sov. Phys-JETP 11, 2977 (1979).
42. B. Jancovic, Nuouo Cim. 25, 429 (1962).
43. G. Baym and S. W. Chin, Nucl. Phys. A262, 527 (1976).
44. M. V. Ramana and A. K. Rajagopal, Phys. Rev. A, to be published (1981).
45. G. Breit, Phys. Rev. 34, 553 (1929), Phys. Rev. 36, 383 (1930), Phys. Rev. 39, 616 (1932).
46. J. B. Mann and W. R. Johnson, Phys. Rev. A41, 41 (1971).
47. C. E. Burgess and A. H. MacDonald unpublished.
48. A. H. MacDonald, Ph.D. Thesis, U. of Toronto (1978).
49. K. N. Huang, M. Aoyagi, M. H. Chen and B. Craseman, At. Data and Nuc. Data 18, 243 (1976).
50. M. V. Ramana, A. K. Rajogopal and W. R. Johnson, submitted for publication (1981).
51. A. K. Rajagopal and J. Callaway, Phys. Rev. B7, 1912 (1973).
52. M. V. Ramana and A. K. Rajagopal, J. Phys. C12, L845 (1979).
53. M. V. Ramana and A. K. Rajogopal, J. Phys. C to be published (1981).
54. U. von Barth and L. Hedin, J. Phys. C5, 1629 (1972).
55. In transition metals the main relativistic effect of the $\underline{m}$ dependence of $E_{xc}[n,\underline{m}]$ is likely to be due to the spin-orbit coupling of the single-particle states. This is no analog of spin-orbit coupling in the relativistic electron gas.

56. L. F. Mattheiss, J. H. Wood and A. C. Switendick, Meth. Comp. Phys. 8, 63 (1968).
57. B. N. Harmon and D. D. Koelling, P. Phys. C7, L210 (1974).
58. P. M. Marcus, Int. J. Quant. Chem. 1S, 567 (1967).
59. T. Takeda and J. Kübber, J. Phys. F9, 661 (1979); T. Takeda, J. Phys. F9, 815 (1979).
60. D. D. Koelling and G. O. Arbman, J. Phys. F5, 2041 (1975).
61. J. E. Müller, Ph.D. Thesis, Cornell (1980).
62. Y. Onodera and M. Okazaki, J. Phys. Soc., Japan 21, 1273 (1966).
63. J. Treusch and K. R. Sandroc, Phys. Stat. Solidi 16, 487 (1966).
64. S. Takada, Prog. Theor. Phys. 36, 224 (1966).
65. J. B. Conklin, Jr., L. E. Johnson and G. W. Pratt, Jr., Phys. Rev. 137A, 1282 (1965).
66. M. E. Rose Relativistic Electron Theory, J. Wiley and Sons, p. 123-126.
67. F. Rosicky and F. Mark, J. Phys. B8, 2581 (1975).
68. F. Rosicky, P. Weinberger and F. Mark, J. Phys. B9, 2971 (1976).
69. D. D. Koelling and B. N. Harmon, J. Phys. C10, 3107 (1977).
70. A. H. MacDonald, W. E. Pickett and D. D. Koelling, J. Phys. C13, 2675 (1980).
71. H. Gollisch and L. Fritsche, Phys. Stat. Sol. (b) 86, 145 (1978).
72. T. Takeda, Z. Physik B 32, 43 (1978).
73. J. H. Wood and A. M. Boring, Phys. Rev. B18, 2701 (1978).

RELATIVISTIC HARTREE-FOCK THEORIES FOR MOLECULES AND CRYSTALS IN A LINEAR COMBINATION OF ATOMIC ORBITALS FORM

J. Ladik

Chair for Theoretical Chemistry, Friedrich-Alexander-University Erlangen-Nürnberg, 8520 Erlangen, FRG* and Department of Applied Mathematics, University of Waterloo Waterloo, Ontario, Canada

J. Cizek

Department of Applied Mathematics, University of Waterloo Waterloo, Ontario, Canada* , Guelph-Waterloo Center for Graduate Work in Chemistry, Waterloo Campus, University of Waterloo, Ontario, Canada and Chair for Theoretical Chemistry, Friedrich-Alexander-University Erlangen-Nürnberg, 8520 Erlangen, FRG

P.K. Mukherjee [×]

Chair for Theoretical Chemistry, Friedrich-Alexander-University Erlangen-Nürnberg, 8520 Erlangen, FRG

ABSTRACT

The relativistic Hartree-Fock-Roothaan formalism (ab initio relativistic SCF LCAO method) has been reviewed in the case of molecules with an arbitrary symmetry and for crystals with translational symmetry. In the derivation different coefficients have been introduced for all the four sets of basis functions which constitute the four component Dirac spinor of a molecule or of a crystal (full variation in the framework of the relativistic Hartree-Fock theory) and also for crystals the non local exchange

---

*Permanent address

[×]Permanent address: Optics Department, Indian Association for the Cultivation of Science, Jadavpur, Calcutta-700032, India

has been applied. The problems of admixture of negative energy states and of the choice of the basis sets are discussed.

The DHF equations are shown (again in an LCAO form) also in the case of a time-dependent external perturbation (time-dependent relativistic coupled Hartree-Fock equations).

Finally, in the case of heavy atoms or ions cross terms between relativistic effects and correlation effects, as well the role of corrections to Breit's operator, of three- and more body operators, and of quantum electrodynamical (first and second order) corrections will be discussed.

1. INTRODUCTION

There is an increasing interest in the relativistic treatment of atoms [1], molecules [2] and of solids [3]. A relativistic Hartree-Fock scheme (Hartree-Fock-Dirac(HFD)-method) [4] based on the variation of the total energy expression obtained with a single Slater determinant (in which the one-electron orbitals are four component Dirac spinors) using a Dirac - type Hamiltonian 5 for each electron and the Coulomb interaction among them was developed some time ago [4]. For the remaining interaction terms usually the first order perturbation expression of the Breit interaction operator [6] reduced to large components (Pauli approximation) is taken. This latter procedure is of course rather questionable in the case of the inner shell electrons of heavy atoms, because the Breit operator is correct only in the $(v/c)^2$ approximation and $v \sim Z$. For heavier atoms a modified Breit operator [7] (non-local generalization of the Breit interaction) should be used instead of the original one. Further by heavy atoms also three [8] and more body interactions become non-negligible.

Independently which of the relativistic interaction operators is used one has to generate relativistic one-electron functions (which serve also for the relativistic treatment of correlation). In the case of non-relativistic calculations, the Hartree-Fock approximation provides the best one-electron orbitals from the point of view of minimizing the total energy. In a similar way to find the best relativistic total energy in the one-electron approximation one has to solve the relativistic Hartree-Fock (Hartree-Fock-Dirac(HFD)) equations. These can be used as starting orbitals for the relativistic treatment of a many electron system. It should be mentioned also that <u>only</u> if the Hartree-Fock problem (non-relativistic or relativistic, respectively) is solved, one can define the correlation energy (non-relativistic, or relativistic correlation energy, respectively). This was the motivation to develop a formalism to solve (in any desired accuracy) the relativistic Hartree-Fock problem in the general case (atoms, molecules and crystals).

In the case of atoms and of molecules with central symmetry (when a one-center expansion is feasible) [2, 9] these equations are solved numerically. In the case of molecules with no central symmetry and in the case of crystals this procedure is very likely not possible and one has to expand the molecular orbitals (MO-s) or crystal orbitals (CO-s) as a linear combination of some basis functions (LCAO expansion). For molecules this work has been done [10] and applications can be found for diatomic or for linear molecules [11,12] (the only exception being the H_2CO molecule which was treated by Aoyama et al [13]).

Here we shall rederive the HFD equations in an LCAO form (relativistic Hartree-Fock-Roothaan equations) in the general case (i.e. using different coefficients for all the four components of the atomic spinors (basis functions) taking no advantage of the molecular symmetry and applying the full relativistic molecular or crystal Fock operator, respectively, with the non-local exchange term) which will make the generalization to the case of crystals straightforward.

One should point out that by developing this formalism we did not think first of all on heavy metals (which were treated in the last years successfully with the aid of the relativistic versions of the APW or KKR [3] methods) but on more complicated solids in which we have besides a few heavy atoms several light atoms in the unit cell (like AsF_5, SbF_5, U_3O_8, RbF, KF, etc. crystals). In such cases there are besides large density regions (the regions of the heavy atoms) small density regions (regions between the atoms and regions containing the light atoms) and correspondingly more extended regions with larger density gradients than in the heavy metals. For such systems according to the experience obtained in the non-relativistic case density functional methods (or approximations to them like the KKR method) or methods based on a plane wave (OPW) or modified plane wave (like the APW method) description do not work well. One can expect that a similar situation will arise also in the relativistic case. Therefore, to be able to treat also these systems relativistically, one has to go back to the (compuationally more difficult) but clearly defined Hartree-Fock-Dirac approximation. After having obtained the energy bands of a crystal in the HFD approximation one has the advantage that one can systematically proceed to include the same types of corrections (corrections for the excited states to decrease the gap, short and long range correlation corrections) as it is done in the non-relativistic case [14].

In the final part of this paper we shall give some general remarks about the different - in our opinion - rather serious problems which make difficulties in establishing a well-founded relativistic many electron theory.

2. DERIVATION OF THE RELATIVISTIC HARTREE-FOCK-ROOTHAAN EQUATIONS FOR MOLECULES AND CRYSTALS

Let us assume that we have a crystal of 2N+1 unit cells in each direction and m orbitals in the unit cell. Let us assume further that we have four basis sets $\{\chi^{(t)}\}$ (t=1,2,3,4). With the help of them we can express any four-component relativistic crystal (Bloch) orbital $\tilde{\Psi}_i^{CO}$ as [15]

$$\tilde{\Psi}_i^{CO} = \begin{pmatrix} \tilde{\Psi}_i^{(1)} \\ \tilde{\Psi}_i^{(2)} \\ \tilde{\Psi}_i^{(3)} \\ \tilde{\Psi}_i^{(4)} \end{pmatrix} = \sum_{\vec{q}=-N}^{N} \sum_{g=1}^{m} \begin{pmatrix} c_{i,\vec{q},g}^{(1)} \; \chi_{\vec{q},g}^{(1)} \\ c_{i,\vec{q},g}^{(2)} \; \chi_{\vec{q},g}^{(2)} \\ c_{i,\vec{q},g}^{(3)} \; \chi_{\vec{q},g}^{(3)} \\ c_{i,\vec{q},g}^{(4)} \; \chi_{\vec{q},g}^{(4)} \end{pmatrix} e^{-iW_i t} \tag{2.1}$$

Here the shorthand notation $\vec{q}$ stands for the unit cell characterized by the position vector $\vec{R}_{\vec{q}}=q_1a_1\vec{i}+q_2a_2\vec{j}+q_3a_3\vec{k}$ ($\vec{i},\vec{j},\vec{k}$ are the non-necessarily orthogonal unit vectors) and $\chi_{\vec{q},g}^{(t)} = \chi^{(t)}(\vec{r}-\vec{R}_{\vec{q}}-\vec{g}_A)_g$ stands for the g-th basis functions of the t-th set centered on atom g_A (to which the orbitals g belongs) in the unit cell $\vec{R}_{\vec{q}}$. Finally, W_i is the relativistic one-electron energy belonging to the CO $\tilde{\Psi}_i^{CO}$.

Onc can build up in the usual way a Slater determinant with the aid of the CO-s $\tilde{\Psi}_i^{CO}$. We can now construct the expectation value of the approximate relativistic total Hamiltonian:

$$\hat{H}_{rel} = \sum_{\ell=1}^{M} \hat{H}_D(\ell) + \sum_{k<\ell} 1/r_{k\ell} \tag{2.2}$$

where $\hat{H}^D$ stand for the Dirac operator of the ℓ-th electron and $M=n\ (2N+1)^3$ (n is the number of electrons per unit cell), with this Slater determinant. Performing the variation of this expectation value by taking into account the auxiliary conditions $\langle\tilde{\Psi}_i^{CO} \mid \tilde{\Psi}_j^{CO}\rangle = \delta_{ij}$ and allowing only positive W_i values (in the relativistic sense) to avoid electronic states with negative energies, one arrives in the usual way at the relativistic Hartree-Fock equations of the system (molecule, or, if there is translational symmetry, crystal)

$$\hat{F}^{rel}\underline{\Psi}_i^{CO} = \left[\hat{H}^D + \sum_{\vec{p}=1}^{(2N+1)^3}\sum_{h'=1}^{n^*}(\hat{J}(\vec{p},h') - \hat{K}(\vec{p},h'))\right]\underline{\Psi}_i^{CO} = W_i\underline{\Psi}_i^{CO} \quad (2.3)$$

Here n^* is the number of (in the restricted Hartree-Fock case doubly filled) bands and $\hat{J}$ and $\hat{K}$ are the Coulomb and exchange operators, respectively, formed now with the aid of the four component relativistic HF CO-s instead of the non-relativistic one-component HF CO-s. In deriving (2.3) the time dependent factor e^{-iW_it} of $\underline{\tilde{\Psi}}_i^{CO}$ has been already eliminated by applying the differentiation according to time contained in the original Dirac operator $\hat{H}^D$ [16]

$$-i\left[\frac{1}{i}\frac{\partial}{ic\partial t}\underline{\tilde{\Psi}}^{CO}(\vec{r},t)\right] = -i\left[\frac{1}{i}\frac{\partial}{ic\partial t}(\underline{\Psi}^{CO}(\vec{r})e^{-iW_it})\right]$$
$$= \frac{1}{c}W_i\underline{\Psi}^{CO}(\vec{r})\,e^{-iW_it} \quad (2.4)$$

The scalar potential of the crystal is

$$\Phi = -\sum_{\alpha=1}^{M_A}\sum_{\vec{p}=1}^{(2N+1)^3}\frac{Z_\alpha}{|\vec{r}-\vec{R}_{\alpha,\vec{p}}|} \quad (2.5)$$

where M_A is the number of atomic nuclei in the unit cell, Z_α is the nuclear charge of the α-th nucleus and $\vec{R}_{\alpha,\vec{p}}$ is the position vector of the α-th nucleus in the unit cell characterized by the vector $\vec{p}$. Let us assume further that there is no external magnetic field, so that $\vec{A} = \vec{0}$. Then we can write equ. (2.3) using the detailed form of $\hat{H}^D$ [16] as

$$\left\{\frac{c}{i}\sum_{j=1}^{3}\underline{\alpha}_j\frac{\partial}{\partial x_j} + \Phi + \underline{\alpha}_4 m_0 c + \sum_{\vec{p}'}\sum_{h'}(\hat{J}(\vec{p}';h') - \hat{K}(\vec{p}';h'))\right\}\underline{\Psi}_i^{CO} =$$

$$= \frac{c}{i}\begin{pmatrix}\frac{\partial}{\partial x}\psi^{(4)}(\vec{p},h)\\ \frac{\partial}{\partial x}\psi^{(3)}(\vec{p},h)\\ \frac{\partial}{\partial x}\psi^{(2)}(\vec{p},h)\\ \frac{\partial}{\partial x}\psi^{(1)}(\vec{p},h)\end{pmatrix} + c\begin{pmatrix}-\frac{\partial}{\partial y}\psi^{(4)}(\vec{p},h)\\ \frac{\partial}{\partial y}\psi^{(3)}(\vec{p},h)\\ -\frac{\partial}{\partial y}\psi^{(2)}(\vec{p},h)\\ \frac{\partial}{\partial y}\psi^{(1)}(\vec{p},h)\end{pmatrix} + \frac{c}{i}\begin{pmatrix}\frac{\partial}{\partial z}\psi^{(3)}(\vec{p},h)\\ -\frac{\partial}{\partial z}\psi^{(4)}(\vec{p},h)\\ \frac{\partial}{\partial z}\psi^{(1)}(\vec{p},h)\\ -\frac{\partial}{\partial z}\psi^{(2)}(\vec{p},h)\end{pmatrix} +$$

$$+ m_0 c^2 \begin{pmatrix} \psi^{(1)}(\vec{p},h) \\ \psi^{(2)}(\vec{p},h) \\ -\psi^{(3)}(\vec{p},h) \\ -\psi^{(4)}(\vec{p},h) \end{pmatrix} + \Phi \begin{pmatrix} \psi^{(1)}(\vec{p},h) \\ \psi^{(2)}(\vec{p},h) \\ \psi^{(3)}(\vec{p},h) \\ \psi^{(4)}(\vec{p},h) \end{pmatrix} +$$

$$+ \sum_{\vec{p}'h'} \left\langle \left(\psi^{(1)}(\vec{p}',h';\vec{r}_2)\, \psi^{(2)}(\vec{p}',h';\vec{r}_2)\, \psi^{(3)}(\vec{p}',h';\vec{r}_2)\, \psi^{(4)}(\vec{p}',h';\vec{r}_2) \right) \right|$$

$$\frac{1}{r_{12}}(1-\hat{P}_{12}) \left| \begin{pmatrix} \psi^{(1)}(\vec{p}',h';\vec{r}_2) \\ \psi^{(2)}(\vec{p}',h';\vec{r}_2) \\ \psi^{(3)}(\vec{p}',h';\vec{r}_2) \\ \psi^{(4)}(\vec{p}',h';\vec{r}_2) \end{pmatrix} \right\rangle_2 \begin{pmatrix} \psi^{(1)}(\vec{p},h;\vec{r}_1) \\ \psi^{(2)}(\vec{p},h;\vec{r}_1) \\ \psi^{(3)}(\vec{p},h;\vec{r}_1) \\ \psi^{(4)}(\vec{p},h;\vec{r}_1) \end{pmatrix} = W(\vec{p},h) \begin{pmatrix} \psi^{(1)}(\vec{p},h) \\ \psi^{(2)}(\vec{p},h) \\ \psi^{(3)}(\vec{p},h) \\ \psi^{(4)}(\vec{p},h) \end{pmatrix} \tag{2.6}$$

Here we have introduced instead of the general index i ($i=1,\dots,M$; $M=n\,(2N+1)^3$) of the CO-s double indexing (for future convenience) denoting by h the band indices and by $\vec{p}$ the levels inside the bands. The subscript 2 at the bracket indicates that the integration has to be performed according to the coordinates of the second electron and $\hat{P}_{12}$ stands for the exchange operator. In writing down the detailed form of the relativistic HF equation for a crystal the definitions of the Dirac matrices

$$\underline{\underline{\alpha}}_1 = \begin{pmatrix} 0&0&0&1 \\ 0&0&1&0 \\ 0&1&0&0 \\ 1&0&0&0 \end{pmatrix} \qquad \underline{\underline{\alpha}}_2 = \begin{pmatrix} 0&0&0&-i \\ 0&0&i&0 \\ 0&-i&0&0 \\ i&0&0&0 \end{pmatrix}$$

$$\underline{\underline{\alpha}}_3 = \begin{pmatrix} 0&0&1&0 \\ 0&0&0&-1 \\ 1&0&0&0 \\ 0&-1&0&0 \end{pmatrix} \qquad \underline{\underline{\alpha}}_4 = \begin{pmatrix} 1&0&0&0 \\ 0&1&0&0 \\ 0&0&-1&0 \\ 0&0&0&-1 \end{pmatrix} \tag{2.7}$$

have been applied.

If one introduces the LCAO forms of the four components of

$\underline{\Psi}^{CO}(\vec{p},h)$ (see equ. (2.1)) one arrives in the usual way at the matrix equation

$$\underline{\underline{F}}^{rel} \begin{pmatrix} \underline{C}^{(1)}(\vec{p},h) \\ \underline{C}^{(2)}(\vec{p},h) \\ \underline{C}^{(3)}(\vec{p},h) \\ \underline{C}^{(4)}(\vec{p},h) \end{pmatrix} = W(\vec{p},h)\underline{\underline{S}}^{rel} \begin{pmatrix} \underline{C}^{(1)}(\vec{p},h) \\ \underline{C}^{(2)}(\vec{p},h) \\ \underline{C}^{(3)}(\vec{p},h) \\ \underline{C}^{(4)}(\vec{p},h) \end{pmatrix} \tag{2.8}$$

Here the generalized overlap matrix $\underline{\underline{S}}^{rel}$ is (as it can be shown in complete analogy to the derivation of the non-relativistic Hartree-Fock-Roothaan equations [17])

$$\underline{\underline{S}}^{rel} = \begin{pmatrix} \underline{\underline{S}}^{(1,1)} & & & \underline{\underline{0}} \\ & \underline{\underline{S}}^{(2,2)} & & \\ & & \underline{\underline{S}}^{(3,3)} & \\ \underline{\underline{0}} & & & \underline{\underline{S}}^{(4,4)} \end{pmatrix} \tag{2.9}$$

where for instance the diagonal block $\underline{\underline{S}}^{(3,3)}$ contains only overlap integrals between basis functions belonging to the third basis set:

$$(\underline{\underline{S}}^{(3,3)})_{\vec{q}',g';\vec{q},g} = \left\langle \chi^{(3)}_{\vec{q}',g'} \,\middle|\, \chi^{(3)}_{\vec{q},g} \right\rangle \tag{2.10}$$

($\chi^{(3)}_{\vec{q}',g'}$ stands for the g'-s basis function in the unit cell $\vec{q}'$ belonging to the third set).

To express in detail the left-hand side of equ. (2.8) we can start from equ. (2.6) substituting the (2.1) LCAO form of the CO-s $\Psi^{(t)}$. In this way one arrives to the following matrix equ.-s (again in complete analogy to the non-relativistic derivation [17]):

$$\begin{pmatrix} \underline{\underline{F}}_1^{(1,4)} & & & \underline{\underline{0}} \\ & \underline{\underline{F}}_1^{(2,3)} & & \\ & & \underline{\underline{F}}_1^{(3,2)} & \\ \underline{\underline{0}} & & & \underline{\underline{F}}_1^{(4,1)} \end{pmatrix} \begin{pmatrix} \underline{C}^{(4)}(\vec{p},h) \\ \underline{C}^{(3)}(\vec{p},h) \\ \underline{C}^{(2)}(\vec{p},h) \\ \underline{C}^{(1)}(\vec{p},h) \end{pmatrix} + \begin{pmatrix} \underline{\underline{F}}_2^{(1,4)} & & & \underline{\underline{0}} \\ & \underline{\underline{F}}_2^{(2,3)} & & \\ & & \underline{\underline{F}}_2^{(3,2)} & \\ \underline{\underline{0}} & & & \underline{\underline{F}}_2^{(4,1)} \end{pmatrix} \begin{pmatrix} -\underline{C}^{(4)}(\vec{p},h) \\ \underline{C}^{(3)}(\vec{p},h) \\ -\underline{C}^{(2)}(\vec{p},h) \\ \underline{C}^{(1)}(\vec{p},h) \end{pmatrix} +$$

$$+\begin{pmatrix} \underline{\underline{F}}_3^{(1,3)} & & & \underline{\underline{0}} \\ & \underline{\underline{F}}_3^{(2,4)} & & \\ & & \underline{\underline{F}}_3^{(3,1)} & \\ \underline{\underline{0}} & & & \underline{\underline{F}}_3^{(4,2)} \end{pmatrix} \begin{pmatrix} \underline{c}^{(3)}(\vec{p},h) \\ -\underline{c}^{(4)}(\vec{p},h) \\ \underline{c}^{(1)}(\vec{p},h) \\ -\underline{c}^{(2)}(\vec{p},h) \end{pmatrix} + \begin{pmatrix} \underline{\underline{F}}_4^{(1,1)} & & & \underline{\underline{0}} \\ & \underline{\underline{F}}_4^{(2,2)} & & \\ & & \underline{\underline{F}}_4^{(3,3)} & \\ \underline{\underline{0}} & & & \underline{\underline{F}}_4^{(4,4)} \end{pmatrix} \begin{pmatrix} \underline{c}^{(1)}(\vec{p},h) \\ \underline{c}^{(2)}(\vec{p},h) \\ -\underline{c}^{(3)}(\vec{p},h) \\ -\underline{c}^{(4)}(\vec{p},h) \end{pmatrix} +$$

$$+\begin{pmatrix} \underline{\underline{F}}_5^{(1,1)} & & & \underline{\underline{0}} \\ & \underline{\underline{F}}_5^{(2,2)} & & \\ & & \underline{\underline{F}}_5^{(3,3)} & \\ \underline{\underline{0}} & & & \underline{\underline{F}}_5^{(4,4)} \end{pmatrix} \begin{pmatrix} \underline{c}^{(1)}(\vec{p},h) \\ \underline{c}^{(2)}(\vec{p},h) \\ \underline{c}^{(3)}(\vec{p},h) \\ \underline{c}^{(4)}(\vec{p},h) \end{pmatrix} =$$

(2.11)

$$= W(\vec{p},h) \begin{pmatrix} \underline{\underline{S}}^{(1,1)} & & & \underline{\underline{0}} \\ & \underline{\underline{S}}^{(2,2)} & & \\ & & \underline{\underline{S}}^{(3,3)} & \\ \underline{\underline{0}} & & & \underline{\underline{S}}^{(4,4)} \end{pmatrix} \begin{pmatrix} \underline{c}^{(1)}(\vec{p},h) \\ \underline{c}^{(2)}(\vec{p},h) \\ \underline{c}^{(3)}(\vec{p},h) \\ \underline{c}^{(4)}(\vec{p},h) \end{pmatrix}$$

Here the blocks of the relativistic Fock matrix $\underline{\underline{F}}^{rel}$ are defined through their elements as

$$(\underline{\underline{F}}_1^{(t',t)})_{\vec{q}',g';\vec{q},g} = \frac{c}{i} \langle \chi^{(t')}_{\vec{q}',g'} | \frac{\partial}{\partial x} \chi^{(t)}_{\vec{q},g} \rangle \qquad (2.12a)$$

$(t,t' = 1,2,3,4)$

$$(\underline{\underline{F}}_2^{(t',t)})_{\vec{q}',g';\vec{q},g} = c\left\langle \chi^{(t')}_{\vec{q}',g'} \middle| \frac{\partial}{\partial y}\, \chi^{(t)}_{\vec{q},g} \right\rangle \quad (2.12b)$$

$$(\underline{\underline{F}}_3^{(t',t)})_{\vec{q}',g';\vec{q},g} = \frac{c}{i}\left\langle \chi^{(t')}_{\vec{q}',g'} \middle| \frac{\partial}{\partial z}\, \chi^{(t)}_{\vec{q},g} \right\rangle \quad (2.12c)$$

$$(\underline{\underline{F}}_4^{(t,t)})_{\vec{q}',g';\vec{q},g} = m_o c^2 \left\langle \chi^{(t)}_{\vec{q}',g'} \middle| \chi^{(t)}_{\vec{q},g} \right\rangle = m_o c^2 (\underline{\underline{S}}^{(t,t)})_{\vec{q}',g';\vec{q},g} \quad (2.12d)$$

$$(\underline{\underline{F}}_5^{(t,t)})_{\vec{q}',g';\vec{q},g} = \left\langle \chi^{(t)}_{\vec{q}',g'} \middle| \Phi + \sum_{\vec{p}'} \sum_{h'} (\hat{J}(\vec{p}',h') - \hat{K}(\vec{p}',h')) \middle| \chi^{(t)}_{\vec{q},g} \right\rangle \quad (t',t = 1,2,3,4) \quad (2.12e)$$

The operators $\hat{J}(p',h')$ and $\hat{K}(p',h')$ were written down in more detail in equ. (2.6) and their LCAO form in the crystal case will be given subsequently.

We can write instead of the hypermatrix equation (2.11) the following four matrix equ.-s:

$$(\underline{\underline{F}}_1^{(1,4)} - \underline{\underline{F}}_2^{(1,4)})\underline{C}^{(4)}(\vec{p},h) + \underline{\underline{F}}_3^{(1,3)}\underline{C}^{(3)}(\vec{p},h) + (\underline{\underline{F}}_4^{(1,1)} + \underline{\underline{F}}_5^{(1,1)})\underline{C}^{(1)}(\vec{p},h) = W(\vec{p},h)\underline{\underline{S}}^{(1,1)}\underline{C}^{(1)}(\vec{p},h) \quad (2.13a)$$

$$(\underline{\underline{F}}_1^{(2,3)} + \underline{\underline{F}}_2^{(2,3)})\underline{C}^{(3)}(\vec{p},h) - \underline{\underline{F}}_3^{(2,4)}\underline{C}^{(4)}(\vec{p},h) + (\underline{\underline{F}}_4^{(2,2)} + \underline{\underline{F}}_5^{(2,2)})\underline{C}^{(2)}(\vec{p},h) = W(\vec{p},h)\underline{\underline{S}}^{(2,2)}\underline{C}^{(2)}(\vec{p},h) \quad (2.13b)$$

$$(\underline{\underline{F}}_1^{(3,2)} - \underline{\underline{F}}_2^{(3,2)})\underline{C}^{(2)}(\vec{p},h) + \underline{\underline{F}}_3^{(3,1)}\underline{C}^{(1)}(\vec{p},h) + (\underline{\underline{F}}_4^{(3,3)} - \underline{\underline{F}}_5^{(3,3)})\underline{C}^{(3)}(\vec{p},h) = W(\vec{p},h)\underline{\underline{S}}^{(3,3)}\underline{C}^{(3)}(\vec{p},h) \quad (2.13c)$$

$$(\underline{\underline{F}}_1^{(4,1)} + \underline{\underline{F}}_2^{(4,1)})\underline{C}^{(1)}(\vec{p},h) - \underline{\underline{F}}_3^{(4,2)}\underline{C}^{(2)}(\vec{p},h) + (\underline{\underline{F}}_4^{(4,4)} - \underline{\underline{F}}_5^{(4,4)})\underline{C}^{(4)}(\vec{p},h) = W(p,h)\underline{\underline{S}}^{(4,4)}\underline{C}^{(4)}(\vec{p},h) \quad (2.13d)$$

It is easy to show that all the matrices $\underline{\underline{F}}_\ell^{(t',t)}$ (ℓ=1,2,3,4,5) of order $m(2N+1)^3 \times m(2N+1)^3$ are cyclic hypermatrices if one takes into account the translational symmetry of the crystal and introduces periodic boundary conditions. Therefore, one can apply to all the terms in all the four equations an unitary transformation which blockdiagonalizes all the matrices by multiplying from the left all the equations by $\underline{\underline{U}}^+$ and introduce everywhere the unit matrix $\underline{\underline{I}} = \underline{\underline{U}}\underline{\underline{U}}^+$ between the $\underline{\underline{F}}^{(t',t)}$ matrices and the eigenvectors $\underline{C}^{(t)}$ [18] . So we can write

$$\underline{\underline{U}}^{+}\underline{\underline{F}}_{\ell}^{(t',t)}\underline{\underline{U}}\,\underline{\underline{U}}^{+}\underline{C}^{(t)} = \underline{\underline{F}}_{\ell}^{(t',t)B.D.}\underline{D}^{(t)} \tag{2.14}$$

where $\underline{\underline{F}}_{\ell}^{(t',t)B.D.} = \underline{\underline{U}}^{+}\underline{\underline{F}}_{\ell}^{(t',t)}\underline{\underline{U}}$ and $\underline{D}^{(t)} = \underline{\underline{U}}^{+}\underline{C}^{(t)}$. Further it is easy to show [18] that the unitary matrix $\underline{\underline{U}}$ (defined through its mxm blocks) is of the form

$$\underline{\underline{U}}_{\vec{q}',\vec{q}} = \frac{1}{(2N+1)^{3/2}}\exp\left[i\,\frac{2\pi\vec{q}'\vec{q}}{2N+1}\right]\underline{\underline{1}} \tag{2.15}$$

Since in all four equations (2.13) we have after the unitary transformation only block-diagonal matrices we can easily separate the equations according to their blocks 18 . Assuming further that $N\rightarrow\infty$ and therefore we can introduce instead of the discrete variables $p_i = -N,-N+1,\ldots,0,\ldots,N$ (i=1,2,3) which specify the blocks the continous variables $k_i = 2\pi p_i/a_i(2N+1)$ (the continous variables k_i have thus the range $-\pi/a_i \leqslant k_i \leqslant \pi/a_i$), we arrive at the following relativistic Hartree-Fock-Roothaan equations of a crystal (again in complete analogy to the corresponding non-relativistic derivation [18])

$$\left[\underline{\underline{F}}_1^{(1,4)}(\vec{k})-\underline{\underline{F}}_2^{(1,4)}(\vec{k})\right]\underline{d}^{(4)}(\vec{k})_h + \underline{\underline{F}}_3^{(1,3)}(\vec{k})\underline{d}^{(3)}(\vec{k})_h$$
$$+\left[\underline{\underline{F}}_4^{(1,1)}(\vec{k})+\underline{\underline{F}}_5^{(1,1)}(\vec{k})\right]\underline{d}^{(1)}(\vec{k})_h = W(\vec{k})_h\underline{\underline{S}}^{(1,1)}(\vec{k})\underline{d}^{(1)}(\vec{k})_h \tag{2.16a}$$

$$\left[\underline{\underline{F}}_1^{(2,3)}(\vec{k})+\underline{\underline{F}}_2^{(2,3)}(\vec{k})\right]\underline{d}^{(3)}(\vec{k})_h - \underline{\underline{F}}_3^{(2,4)}(\vec{k})\underline{d}^{(4)}(\vec{k})_h$$
$$+\left[\underline{\underline{F}}_4^{(2,2)}(\vec{k})+\underline{\underline{F}}_5^{(2,2)}(\vec{k})\right]\underline{d}^{(2)}(\vec{k})_h = W(\vec{k})_h\underline{\underline{S}}^{(2,2)}(\vec{k})\underline{d}^{(2)}(\vec{k})_h \tag{2.16b}$$

$$\left[\underline{\underline{F}}_1^{(3,2)}(\vec{k})-\underline{\underline{F}}_2^{(3,2)}(\vec{k})\right]\underline{d}^{(2)}(\vec{k})_h + \underline{\underline{F}}_3^{(3,1)}(\vec{k})\underline{d}^{(1)}(\vec{k})_h$$
$$+\left[\underline{\underline{F}}_4^{(3,3)}(\vec{k})-\underline{\underline{F}}_5^{(3,3)}(\vec{k})\right]\underline{d}^{(3)}(\vec{k})_h = W(\vec{k})_h\underline{\underline{S}}^{(3,3)}(\vec{k})\underline{d}^{(3)}(\vec{k})_h \tag{2.16c}$$

$$\left[\underline{\underline{F}}_1^{(4,1)}(\vec{k})-\underline{\underline{F}}_2^{(4,1)}(\vec{k})\right]\underline{d}^{(1)}(\vec{k})_h - \underline{\underline{F}}_3^{(4,2)}(\vec{k})\underline{d}^{(4)}(\vec{k})_h$$
$$+\left[\underline{\underline{F}}_4^{(4,4)}(\vec{k})-\underline{\underline{F}}_5^{(4,4)}(\vec{k})\right]\underline{d}^{(4)}(\vec{k})_h = W(\vec{k})_h\underline{\underline{S}}^{(4,4)}(\vec{k})\underline{d}^{(4)}(\vec{k})_h \tag{2.16d}$$

In these equations all the mxm matrices $\underline{\underline{F}}^{(t',t)}(\vec{k})$ and the overlap matrices $\underline{\underline{S}}^{(t',t)}(\vec{k})$ have the form of a Fourier transform [18]

$$\underline{\underline{F}}_{\ell}^{(t',t)}(\vec{k}) = \sum_{\vec{q}''}\underline{\underline{F}}_{\ell}^{(t',t)}(\vec{q}'')\,e^{i\vec{k}\vec{R}_{\vec{q}''}} \tag{2.18a}$$
$$\underline{\underline{S}}^{(t',t)}(\vec{k}) = \sum_{\vec{q}''}\underline{\underline{S}}^{(t',t)}(\vec{q}'')\,e^{i\vec{k}\vec{R}_{\vec{q}''}} \tag{2.18b}$$

where $\vec{q}'' = \vec{q}'-\vec{q}$ [19]. The matrices $\underline{\underline{S}}^{(t',t)}(\vec{q}'')$ and $\underline{\underline{F}}_{\ell}^{(t',t)}(\vec{q}'')$ have been defined in detail before through their elements (see equ.-s (2.10) and (2.12)) with the exception of the matrices $\underline{\underline{F}}_5^{(t',t)}(\vec{q}'')$.

To work out the form of the matrices $\underline{\underline{F}}_5^{(t',t)}(\vec{q}'')$ [19] we have to substitute in equ. (2.6) in the operators $\hat{J}$ and $\hat{K}$ the LCAO form of the CO-s $\psi^{(t)}(\vec{p}',h')$. A straightforward derivation after defining the relativistic generalized charge-bond order matrix elements

$$P(\vec{q}_1-\vec{q}_2)^{(j)}_{u,v} = \frac{1}{\omega}\sum_{h'=1}^{n^*}\int_{\omega} d(\vec{k})^{(j)}_{h',u}\; d(\vec{k})^{(j)}_{h',v}\; e^{i\vec{k}(\vec{R}_{\vec{q}_1}-\vec{R}_{\vec{q}_2})}\, d\vec{k} \tag{2.19}$$

(again in complete analogy with the corresponding non-relativistic definitions [18]) leads to the expression

$$(\underline{\underline{F}}_5^{(t,t)}(\vec{q}))_{g',g} = \langle \chi^{(t)}_{\vec{0},g'} | \Phi | \chi^{(t)}_{\vec{q},g} \rangle +$$
$$+ \sum_{\vec{q}_1\vec{q}_2}\sum_{u,v}\sum_{j=1}^{4} P(\vec{q}_1-\vec{q}_2)^{(j)}_{u,v}\, (\langle \chi^{(t)}_{\vec{0},g'}(1)\,\chi^{(j)}_{\vec{q}_1,u}(2) | \frac{1}{r_{12}}(1-\hat{P}_{12}) | \cdot$$
$$\cdot | \chi^{(t)}_{\vec{q},g}(1)\;\chi^{(j)}_{\vec{q}_2,v}(2) \rangle) \tag{2.20}$$

In these expressions ω stands for the volume of the first Brillouin zone, n^* denotes the number of filled bands (in the closed-shell case), Φ has been given for a crystal in equ. (2.5) and all other quantities have been defined before.

Instead of solving the four coupled matrix equations (2.17) we can rewrite them as a hypermatrix equation. To be able to do this we use the relation

$$\underline{\tilde{d}}^{(1)} \equiv \begin{pmatrix} \underline{d}^{(4)}(\vec{k})_h \\ \underline{d}^{(3)}(\vec{k})_h \\ \underline{d}^{(2)}(\vec{k})_h \\ \underline{d}^{(1)}(\vec{k})_h \end{pmatrix} = \underline{\underline{\alpha}}_1 \begin{pmatrix} \underline{d}^{(1)}(\vec{k})_h \\ \underline{d}^{(2)}(\vec{k})_h \\ \underline{d}^{(3)}(\vec{k})_h \\ \underline{d}^{(4)}(\vec{k})_h \end{pmatrix} = \underline{\underline{\alpha}}_1 \underline{\tilde{d}}^{(5)} \tag{2.21}$$

and similar relations for the other three hypervectors $\underline{\tilde{d}}^{(2)}$, $\underline{\tilde{d}}^{(3)}$ and $\underline{\tilde{d}}^{(4)}$: $\underline{\tilde{d}}^{(2)} = \frac{1}{i=2}\alpha\; \underline{\tilde{d}}^{(5)}$, $\underline{\tilde{d}}^{(3)} = \underline{\underline{\alpha}}_3\, \underline{\tilde{d}}^{(5)}$ and $\underline{\tilde{d}}^{(4)} = \underline{\underline{\alpha}}_4 \underline{\tilde{d}}^{(5)}$.
With the aid of these equations we can condense the matrix equ.-s (2.17) (performing the matrix multiplications and adding the corresponding block matrices) into the generalized eigenvalue equation.

$$\underline{\underline{F}}^{rel}(\vec{k})\, \underline{\tilde{d}}^{(5)}(\vec{k}) =$$

$$= \begin{pmatrix} \underline{\underline{F}}_4^{(1,1)}(\vec{k}) + \underline{\underline{F}}_5^{(1,1)}(\vec{k}) & \underline{\underline{0}} & \underline{\underline{F}}_3^{(1,3)}(\vec{k}) & \underline{\underline{F}}_1^{(1,4)}(\vec{k}) - \underline{\underline{F}}_2^{(1,4)}(\vec{k}) \\ \underline{\underline{0}} & \underline{\underline{F}}_4^{(2,2)}(\vec{k}) + \underline{\underline{F}}_5^{(2,2)}(\vec{k}) & \underline{\underline{F}}_1^{(2,3)}(\vec{k}) + \underline{\underline{F}}_2^{(2,3)}(\vec{k}) & -\underline{\underline{F}}_3^{(2,4)}(\vec{k}) \\ \underline{\underline{F}}_3^{(3,1)}(\vec{k}) & \underline{\underline{F}}_1^{(3,2)}(\vec{k}) - \underline{\underline{F}}_2^{(3,2)}(\vec{k}) & \underline{\underline{F}}_4^{(3,3)}(\vec{k}) - \underline{\underline{F}}_5^{(3,3)}(\vec{k}) & \underline{\underline{0}} \\ \underline{\underline{F}}_1^{(4,1)}(\vec{k}) + \underline{\underline{F}}_2^{(4,1)}(\vec{k}) & -\underline{\underline{F}}_3^{(4,2)}(\vec{k}) & \underline{\underline{0}} & \underline{\underline{F}}_4^{(4,4)}(\vec{k}) - \underline{\underline{F}}_5^{(4,4)}(\vec{k}) \end{pmatrix} \underline{\tilde{d}}^{(5)} =$$

$$= W(\vec{k})_h \begin{pmatrix} \underline{\underline{S}}^{(1,1)}(\vec{k}) & & & \underline{\underline{0}} \\ & \underline{\underline{S}}^{(2,2)}(\vec{k}) & & \\ & & \underline{\underline{S}}^{(3,3)}(\vec{k}) & \\ \underline{\underline{0}} & & & \underline{\underline{S}}^{(4,4)}(\vec{k}) \end{pmatrix} \begin{pmatrix} \underline{d}^{(1)}(\vec{k})_h \\ \underline{d}^{(2)}(\vec{k})_h \\ \underline{d}^{(3)}(\vec{k})_h \\ \underline{d}^{(4)}(\vec{k})_h \end{pmatrix} \qquad (2.22)$$

It is easy to show with the aid of the relation $\underline{\underline{F}}_3^{(1,3)} = -(\underline{\underline{F}}_3^{(3,1)})^{tr} = (\underline{\underline{F}}_3^{(3,1)})^{+}$ and of similar ones in the cases of the other corresponding off-diagonal matrix blocks (which all can be proven through integration by parts of their elements; for instance

$$\left\langle \chi^{(t')}_{\vec{0},g'} \middle| \frac{\partial \chi^{(t)}_{\vec{q},g}}{\partial x_i} \right\rangle = - \left\langle \chi^{(t')}_{\vec{q},g} \middle| \frac{\partial \chi^{(t')}_{\vec{0},g'}}{\partial x_i} \right\rangle \qquad (i=1,2,3)\ )$$

that $\underline{\underline{F}}^{rel}$ is a Hermitian matrix. One will obtain therefore after a Löwdin's symmetric orthogonalization procedure [20] applied for a crystal [18]

$$\underline{\underline{S}}^{-1/2}(\vec{k})\, \underline{\underline{F}}^{rel}(\vec{k})\, \underline{\underline{S}}^{-1/2}(\vec{k})\, \underline{\underline{S}}^{+1/2}(\vec{k})\, \underline{\tilde{d}}^{(5)}(\vec{k})_h =$$

$$= \underline{\underline{\tilde{F}}}^{rel}(\vec{k})\, \underline{b}(\vec{k})_h = W(\vec{k})_h \cdot \begin{pmatrix} \underline{b}^{(1)}(\vec{k})_h \\ \underline{b}^{(2)}(\vec{k})_h \\ \underline{b}^{(3)}(\vec{k})_h \\ \underline{b}^{(4)}(\vec{k})_h \end{pmatrix}$$

real (relativistic) one-electron energies $W(\vec{k})_h$. In this way equations (2.22) and (2.23) together with the definitions (2.10) (2.12), (2.19) and (2.20), respectively, fully define the relativistic Hartree-Fock-Roothaan (relativistic ab initio SCF LCAO) equations with different coefficients for all the functions belonging to the four different basis sets (see equ. (2.11)) for a crystal.

The case of a molecule (or an extended system with no translational symmetry) is a special case of this formalism when one does not use a cell index, only an orbital index. The form of all the equations remains in this case the same, one has only to drop the $\vec{k}$ dependence and in the definition of the charge-bond order matrix elements (see equ. (2.19)) one has not to integrate over $\vec{k}$.

3. THE CHOICE OF THE BASIS SETS AND THE PROBLEM OF ADMIXTURE OF NEGATIVE ENERGY STATES

3.A. Basis Sets

To apply the method developed in the previous section into a working computer program one has to have an appropriate choice for the four basis sets [15]. One possibility is to start with relativistic hydrogen-like functions. Since these functions consist essentially of a product of a confluent hypergeometric function with a non-integer power of r, the calculations of the matrix elements (2.12) and especially (2.20) (four-center integrals) may cause very large difficulties in performing a calculation.

Most probably a more realistic way to solve this problem would be to start with the AO-s obtained from an atomic relativistic Hartree-Fock calculation and try to fit the numerical output of this procedure with a linear combination of Gaussians. One can hope that in this way the integral packages of the non-relativistic ab initio molecular programs could be used. To try out this procedure a relativistic HF treatment of the carbon atom has been performed (using the Desclaux program [1,2]) and its output (for the filled $1s_{1/2}$, $2s_{1/2}$, $2p_{1/2}$ states) has been fitted by a non-linear least square procedure using linear combination of Gaussians. It turned out if one wants to fit with an acceptable accuracy the radial part of a relativistic AO, about 30 Gaussians are needed to cover the whole range of r from 0 until 40 atomic units (which is needed in the case of relativistic functions).

To overcome the difficulty the range of r has been divided into four regions (see Table I) which were the same for both the large and small components of the radial part of all the three AO-s ($1s_{1/2}$, $2s_{1/2}$, $2p_{1/2}$). In this way it was possible to fit always with 3 Gaussians a component of the radial part of a relativistic carbon orbital in a given region (see Tables II, III and IV).

Table I. The Regions of the Radial Coordinate Chosen for the Primitive Gaussians in the Approximation of the Numerical Relativistic HF AO-s of the Carbon Atom (These Regions are the Same for all Six Relativistic AO-s).The r Values are Given in a.u.-s

Region	r_{min}	r_{max}
1	3.150×10^{-5}	8.124×10^{-4}
2	8.540×10^{-4}	2.203×10^{-2}
3	2.316×10^{-2}	5.972×10^{-1}
4	6.278×10^{-1}	4.187×10^{1}

Table II. The Linear Coefficients and Orbital Exponents of the Gaussian Expansion of the Numerical Output for the Radial Part of the $1s_{1/2}$ Relativistic Carbon AO

Region	Large component		Small component	
	exponent	coefficient	exponent	coefficient
1	$.709246\times10^{10}$	.0174281	$.115250\times10^{11}$	-.003167
	$.663154\times10^{9}$	.1342063	$.109683\times10^{10}$	-.144682
	$.668763\times10^{8}$	1.0000000	$.119233\times10^{9}$	-1.000000
2	$.547988\times10^{7}$	.009344	$.115413\times10^{8}$	-.002981
	$.454239\times10^{6}$	.096842	$.109590\times10^{7}$	-.053046
	$.346166\times10^{5}$	1.000000	$.113213\times10^{6}$	-1.000000
3	$.128783\times10^{4}$	.214370	$.917609\times10^{4}$	-.007226
	$-.458103\times10^{2}$	.681347	$.922993\times10^{3}$	-.193120
	$-.137436\times10^{2}$	1.000000	$.842388\times10^{2}$	-1.000000
4	$.159355\times10^{1}$	.046071	$.125850\times10^{2}$	-.090276
	$.172433\times10^{1}$	.370211	$.300573\times10^{1}$	-.124683
	$.390551\times10^{0}$	1.000000	$.598104\times10^{0}$	-1.000000

This means that the non-relativistic molecular integral packages using Gaussian basis functions could be used without any change if one takes for each component of each function the linear combination of the corresponding three Gaussian functions. Further one should observe that the orbital exponents of the corresponding large and small components (in contrary to the H atom problem [16]) are not equal (see Tables II-IV). This is obviously caused by the deviation of the simple central Coulomb potential in the H atom from the effective Hartree-Fock potential in the case of atoms with more than one electron.

If it turns out that the relativistic orbitals of other atoms can be expanded in Gaussians in a similar way (thus keeping the number of Gaussians for a given region of r low) it will not be very difficult to develop ab initio relativistic SCF-LCAO programs

Table III. The Linear Coefficients and Orbital Exponents of the Gaussian Expansion of the Numerical Output for the Radial Part of the $2s_{1/2}$ Relativistic Carbon AO

Region	Large component		Small component	
	exponent	coefficient	exponent	coefficient
1	$.854596\times10^{10}$	-0.043226	$.130785\times10^{11}$	.009131
	$.832286\times10^{9}$	-0.163678	$.126895\times10^{10}$	.081642
	$.879549\times10^{8}$	-1.000000	$.140308\times10^{9}$	1.000000
2	$.808547\times10^{7}$	-.017281	$.138757\times10^{8}$	.026423
	$.713038\times10^{6}$	-.097700	$.135426\times10^{7}$	.258674
	$.663103\times10^{5}$	-1.000000	$.144756\times10^{6}$	1.000000
3	$.912482\times10^{4}$	-.187896	$.170530\times10^{5}$	.462386
	$.928049\times10^{3}$	-.552046	$.129597\times10^{4}$	.487012
	$.305059\times10^{2}$	-1.000000	$.118313\times10^{3}$	1.000000
4	$.253196\times10^{1}$	-.08596	$.161403\times10^{2}$	.453691
	$.120361\times10^{0}$	.643077	$.183581\times10^{1}$	.613625
	$.121074\times10^{0}$	1.000000	$.308486\times10^{0}$	1.000000

(at least in the closed-shell case [22]) for molecules with arbitrary symmetry and for crystals. The investigation of these problems especially from the point of view of an optimal choice of the orbital exponents (with the help of a non-linear fitting program) and of the contraction coefficients as well as the development of the corresponding programs is in progress.

Table IV. The Linear Coefficients and Orbital Exponents of the Gaussian Expansion of the Numerical Output for the Radial Part of the $2p_{1/2}$ Relativistic Carbon AO

Region	Large component		Small component	
	exponent	coefficient	exponent	coefficient
1	$.174033 \times 10^{11}$	.001762	131826×10^{11}	.003274
	$.173246 \times 10^{10}$	.018436	$.133796 \times 10^{10}$	.078761
	$.191477 \times 10^{9}$	1.000000	$.149506 \times 10^{9}$	1.000000
2	$.184518 \times 10^{8}$	.016723	$.149050 \times 10^{8}$	.001416
	$.170380 \times 10^{7}$	.158463	$.146798 \times 10^{7}$	.326620
	$.166997 \times 10^{6}$	1.000000	$.158568 \times 10^{6}$	1.000000
3	$.121435 \times 10^{5}$	.115279	$.136523 \times 10^{5}$	.146574
	$.988808 \times 10^{3}$	.464636	$.144676 \times 10^{4}$	.270990
	$.501331 \times 10^{2}$	1.000000	$.128094 \times 10^{3}$	1.000000
4	$.182731 \times 10^{1}$	.199345	$.144363 \times 10^{2}$	.183627
	$.139551 \times 10^{0}$	.877747	$.236021 \times 10^{1}$	.217206
	$.887712 \times 10^{-1}$	1.000000	$.304874 \times 10^{0}$	1.000000

3.B. Admixture of Negative Energy States

Since the eigenvalues of $\hat{H}_{rel.}$ (equ. (2.2)) are not bounded from below, in the course of analytical solutions of the relativistic HF equations one obtains usually an admixture of negative energy states to the states with positive energies $|\underline{\phi}^+\rangle$:

$$\hat{F}\,|\underline{\phi}_i^+\rangle = \varepsilon_i\,|\underline{\phi}_i^+\rangle$$
$$|\underline{\phi}_i^+\rangle = \sum_p c_p\,|\underline{\varphi}_p\rangle + \sum_n c_n\,|\underline{\varphi}_n\rangle\ , \qquad (3.1)$$

where $|\underline{\varphi}_p\rangle$ stand for the positive and $|\underline{\varphi}_n\rangle$ for the negative energy four-component eigenspinors which satisfy the HFD equation. Admixture of these negative-energy states can pull down the lowest positive energy level well below the "true" ground state and this was frequently observed in atomic or molecular calculations [22].

This is not the case in numerical HFD calculations [1,23,24] where the boundary conditions of the eigenspinors effectively eliminate the admixture of the negative-energy states and thus the variational instability does not arise.

To avoid this difficulty one can introduce to the spinors

$\langle \underline{\phi}_i^+ |$ the further constraint

$$\langle \underline{\phi}_i^+ | \underline{\varphi}_n \rangle = 0 \qquad \text{(for all i and n)} \tag{3.2}$$

This leads after some algebra [25] to the expression

$$\hat{P}\,\hat{F}\,\hat{P}\,|\underline{\phi}_i^+\rangle = \varepsilon_i^+ |\underline{\phi}_i^+\rangle \tag{3.3}$$

where the projector operator $\hat{P}$ is defined as

$$\hat{P} = \sum_p |\underline{\varphi}_p^D\rangle\langle\underline{\varphi}_p^D| \tag{3.4}$$

Several authors (like Datta [26]) have suggested that in (3.4) instead of spinors satisfying the Dirac equation in the case of hydrogen atom, spinors which are eigenfunctions of the DHF equations should be used. This procedure, however, would admix to the positive energy solutions again also negative energy states (due to their occurrence in the Coulomb and exchange operators in $\hat{F}_{rel.}$ [27]).

A possible good compromise would be to start with the (3.4) form of $\hat{P}$ and to put at the beginning also Dirac spinors $\underline{\varphi}_p^D$ in the form of an appropriate linear combination of Gaussians in the Coulomb and exchange operators (equ. (2.6)) $\hat{J}(\vec{p},h)$ and $\hat{K}(\vec{p},h)$, respectively. (More precisely in the case of a molecule or a solid one has to start both in the case of $\hat{P}$ and in the cases of $\hat{J}$ and $\hat{K}$ with a linear combination of Dirac spinors). This would secure that at the first iteration one would obtain purely positive energy solutions. As next step not only the operators $\hat{J}$ and $\hat{K}$ but also the projection operator $\hat{P}$ has to be reconstructed with the help of the approximate (but purely positive energy) $|\underline{\phi}_i^{HF+}\rangle$ solutions and the iterations (including the iteration of $\hat{P}$) can be repeated until self consistency.

This problem of course requires further more detailed investigations. First of all with the help of the $\delta^2 = 0$ criterion of Thouless and Cizek and Paldus [28] the stability of the purely positive energy solutions reached in this way should be carefully analysed. Further also the role of the size of the basis sets which influence also the quality of the operator $\hat{P}$ should be investigated.

4. TIME-DEPENDENT RELATIVISTIC COUPLED HARTREE-FOCK SCHEME FOR CLOSED SHELL MOLECULES

The effects of oscillatory fields on non-relativistic atoms and moleulces have been fruitfully studied for quite a long time. In this respect many body techniques have been proved to be very successful in predicting various features of atomic and molecular spectra. Methods like time dependent coupled Hartree-Fock(TDCHF)theory [29] random phase approximation (RPA) [30], Brueckner-Goldstone many body perturbation technique [31] have been successfully applied for studying non-relativistic atomic and molecular spectra, transition properties like oscillator strengths, dynamic polarizabilities, photoionization cross sections and various other properties. Recent astrophysical observations [32] demonstrate the necessity of finding reliable data for highly ionized atoms. Such spectra have also been observed in the laboratory [33]. For such ions with large Z values relativistic effects seem to be important. Recent studies on optical activities of molecules with heavy atoms [34] also indicate the necessity of such a study.

In the present paper we would like to develop a relativistic generalization of TDCHF theory. One can obtain the relevant equations by substituting in the corresponding non-relativistic ones the one-electron part of the Dirac-Fock operator instead of the non-relativistic Fock operator and adopting Dirac spinors instead of non-relativistic orbitals. This was done by Johnson and Lin [35] and applied to He-like systems by Lin et al [36]. They have solved the equations numerically. Here we rederive in a more systematic way the RTDCHF equations and express them in a matrix form applying an analytic basis (LCAO form of the equations). The formulation provides an accurate description of the spectra of highly charged ionic systems.

Let us consider a closed shell molecule in a time-dependent field. The intensity of the field is assumed to be sufficiently low so that the interaction may be treated as a small perturbation. The field free system is described by a relativistic Hamiltonian. The total Hamiltonian is given by

$$\hat{H}(\vec{r},t) = \hat{H}_o(\vec{r}) + \hat{H}'(\vec{r},t) \tag{4.1}$$

where the field free Hamiltonian $\hat{H}_o$ (see equ. 2.2)) satisfies the equation (in a.u.-s)

$$\hat{H}_o(\vec{r})\,\Psi(\vec{r},t) = i\,\frac{\partial \Psi(\vec{r},t)}{\partial t} \tag{4.2}$$

In presence of an oscillatory field described by a vector potential $\vec{A}$ given by

$$\vec{A}(\vec{r},t) = \vec{A}(\vec{r})e^{-i\omega t} + \vec{A}^*(\vec{r})e^{i\omega t} \tag{4.3}$$

the interaction Hamiltonian $\hat{H}'$ may be written as

$$\hat{H}'(\vec{r},t) = G^-(\vec{r})e^{-i\omega t} + G^+(\vec{r})e^{i\omega t} \tag{4.4}$$

with

$$G^{-}(\vec{r}) = \sum_i \underline{\underline{\vec{\alpha}}}_i \cdot \vec{A}(\vec{r}_i) \equiv \sum_i \hat{h}^{-}(\vec{r}_i) \tag{4.5}$$

The operator $c\vec{\alpha}$ behaves as the velocity of the particle. One can, however, work in the length form of the operator. This is essentially choosing a different gauge which does not alter the transition matrix elements [37].

In presence of field the total wavefunction is chosen as

$$\Phi(\vec{r},t) = e^{-iW_0t}\hat{\mathcal{A}} \prod_i \underline{\Phi}_i(\vec{r}_i,t) \tag{4.6}$$

where W_o is the field free energy obtained from the positive energy solution of an appropriate DHF Hamiltonian. The orbitals $\underline{\Phi}_i(\vec{r}_i,t)$ are four component spinors of which a given component can be represented by

$$\Phi_i^{(t)}(\vec{r}_i,t) = \left[\Psi_i^{(t)}(\vec{r}_i) + \delta\Psi_i^{(t)-}(\vec{r}_i)e^{-i\omega t} + \delta\Psi_i^{(t)+}(\vec{r}_i)e^{i\omega t} \right]$$

$$(t = 1,2,3,4) \tag{4.7}$$

Here $\underline{\Psi}_i(\vec{r}_i)$ are the unperturbed spinors and $\delta\underline{\Psi}_i^{\pm}(\vec{r}_i)$ are the changes of $\underline{\Psi}_i$ due to the perturbation. To obtain a solution for Φ we construct a variational functional

$$\tilde{J} = \frac{1}{T}\int_0^T dt \frac{\langle \Phi | \hat{H} - i\frac{\partial}{\partial t} | \Phi \rangle}{\langle \Phi | \Phi \rangle} \tag{4.8}$$

In constructing (4.8) a time average is performed over the period of the external field, which reduces much labour in computation by eliminating several orthogonality constraints. As the operator is not bounded we can obtain only a stationary condition and not necessarily a minimum by varying the functional $\tilde{J}$. This, however, will not cause any problem since $(\hat{H} - i\partial/\partial t)$ is not an energy operator. The choice of such a functional was discussed by Löwdin and Mukherjee [37] and also by Langhoff et al [38]. The orbitals $\underline{\Phi}_i$ can be appropriately normalized with a normalisation constant

$$N_i = \left[1 + \sum_{t=1}^{4} \left\{ \langle \delta\Psi_i^{(t)-} | \delta\Psi_i^{(t)-} \rangle + \langle \delta\Psi_i^{(t)+} | \delta\Psi_i^{(t)+} \rangle \right\} \right]^{-1/2} \tag{4.9}$$

The functional $\tilde{J}$ can be expressed in terms of the one-electron orbitals $\underline{\Phi}_i$. If we retain terms up to quadratic in $\delta\underline{\Psi}_i^{\pm}$, from the requirement

$$\delta\tilde{J} = 0 \tag{4.10}$$

we obtain the RTDCHF equations

$$(\hat{f}_D - \varepsilon_i \mp \omega)|\delta\underline{\Psi}_i^{\mp}\rangle + \hat{h}^{\mp}|\underline{\Psi}_i\rangle$$

$$+ \sum_j \left\{ \langle\underline{\Psi}_j| \frac{1-\hat{P}_{12}}{r_{12}} |\delta\underline{\Psi}_j^{\mp}\rangle + \langle\delta\underline{\Psi}_j^{\pm}| \frac{1-\hat{P}_{12}}{r_{12}} |\underline{\Psi}_j\rangle \right\} |\underline{\Psi}_i\rangle = \underline{0} \qquad (4.11)$$

where

$$\hat{f}_D = \hat{H}^D + \sum_j \langle\underline{\Psi}_j| \frac{1-\hat{P}_{12}}{r_{12}} |\underline{\Psi}_j\rangle \qquad (4.12)$$

Here ε_i is the DHF energy of the ith unperturbed orbital $\underline{\Psi}_i$.

Presently we are interested in studying the transition properties of the system using analytic functions (LCAO expansions). Recently the use of spherical Gaussian functions has been suggested [30]. For general polyatomic molecules it is, however, more advantageous to adopt an LCAO ansatz with a Cartesian Gaussian basis set for the atomic orbitals (see point 3). We propose to take the perturbed part of the orbitals as

$$\delta\underline{\Psi}_i^{(t)}(\omega)^{\pm} = \sum_q C_{iq}^{\pm(t)}(\omega)\,\chi_q^{(t)} \qquad (t = 1,2,3,4) \qquad (4.13)$$

where the C_{iq}'s are real variational parameters, and the $\chi_q^{(t)}$'s are suitable basis functions. Substituting equ. (4.13) into (4.10) and transforming the resulting equations into matrix form we obtain

$$\begin{pmatrix} \underline{\underline{A}}^- & \underline{\underline{B}} \\ \underline{\underline{B}} & \underline{\underline{A}}^+ \end{pmatrix} \begin{pmatrix} \underline{C}^- \\ \underline{C}^+ \end{pmatrix} = \begin{pmatrix} \underline{D}^- \\ \underline{D}^+ \end{pmatrix} \qquad (4.14)$$

where

$$A^{\pm}_{ip,jq} = \langle\underline{\chi}_p| \hat{f}_D - \varepsilon_i \pm \omega |\underline{\chi}_q\rangle + B_{ip,qj}$$

$$B_{ip,jq} = \langle\underline{\chi}_p(1)\,\underline{\chi}_q(2)| \frac{1-\hat{P}_{12}}{r_{12}} |\underline{\Psi}_i(1)\,\underline{\Psi}_j(2)\rangle \qquad (4.15)$$

$$D^{\pm}_{ip} = \langle\underline{\chi}_p| -\hat{h}^{\pm} |\underline{\Psi}_i\rangle$$

The matrices $\underline{A}$ and $\underline{B}$ are hermitian. The system of linear equations can be solved to obtain the coefficients and hence the perturbed orbitals. We should notice that the matrix elements involve four component spinors and all the four components of $\underline{C}^{\pm}$ can be obtained from equ. (4.14) (see the Appendix).

Depending upon the nature of $\hat{h}$ one can study different types of linear response properties of relativistic systems. For example if one is interested in electric multipole transitions, $\hat{h}$ should be chosen so to induce the corresponding multipole excitation. At frequencies corresponding to the natural excitations, the system will go over to its excited states and from a knowledge of the perturbed wavefunctions one can estimate the oscillator strengths. The frequencies at which $\tilde{J}$ becomes singular will furnish excitation energies. Renormalisation of the functions $\delta \underline{\Psi}_i(\omega)$ at resonance frequencies gives an idea of the excited states wavefunctions. Here transitions will occur between states characterized by relativistic quantum numbers and excitations between fine structure levels can be studied. The ground state atomic wavefunctions may be obtained from a non-linear least square fit of the numerically obtained function using either Slater or Gaussian type functions. Once ground state functions are chosen corresponding to the positive energy solutions of the appropriate Dirac Hamiltonian, the perturbation is expected to connect the negative energy states of the initial system very weakly chiefly because of the large energy denominator appearing in the perturbation expansion coefficients. Hence once a good unperturbed wavefunction is chosen most probably one need not to consider the application of suitable projection operators to get rid of the unwanted negative energy states. Numerical analysis on these lines is in progress.

5. BEYOND RELATIVISTIC HARTREE-FOCK THEORY

The total electronic energy of a relativistic many electron system can be written as

$$E_e = E_{rel}^{HF} + E_{rel}^{corr.} + E'_{Breit} + E_{rel}^{(3\text{ or more body})} + E_{QED}^{(1)} + E_{QED}^{(2)} \tag{5.1}$$

Here E_{rel}^{HF} is the total energy of the N electron system obtained from a relativistic Hartree-Fock calculation and E_{rel}^{corr} stands for the correlation energy calculated with the help of the relativistic one-electron functions (four component spinors) obtained from the DHF calculation.

E'_{Breit} denotes the energy obtained with a nonlocal generalization of the original Breit's interaction in the lowest order of perturbation energy (exchange of a single transverse photon) and can be written for 2 electrons as [7]

$$E'_{Breit} = \langle \Psi_{rel} | -\alpha \sum_{i,j=1}^{3} \underline{\underline{\alpha}}_i^{(k)} \underline{\underline{\alpha}}_j^{(\ell)} \Big[\delta_{i,j} \frac{\cos \omega_{k\ell} r_{k\ell}}{r_{k\ell}} + \\ + \frac{\partial^2}{\partial (r_{k\ell})_i \, \partial (r_{k\ell})_j} \frac{\cos(\omega_{k\ell} r_{k\ell}) - 1}{\omega_{k\ell}^2 r_{k\ell}} \Big] | \Psi_{rel} \rangle \tag{5.2}$$

Here $\Psi_{rel} = \Delta_{rel} + \Psi_{rel}^{corr}$ (Δ_{rel} being a Slater determinant of the 2 electron system and a Ψ_{rel}^{corr} is the correlation part of the 2 electron wavefunction), $\alpha = 1/137.03$ is the fine structure constant, $\underline{\underline{\alpha}}_i^{(k)}$ is the i-th (i=1,2,3) Dirac matrix acting on the k-th electron and finally $\omega_{k\ell} = \varepsilon_k - \varepsilon_\ell$ is the difference of DHF orbital energies of the two electrons k and ℓ. It should be pointed out that if one applies the Slater-Condon rules, one finds always a Coulomb and an exchange part of the expectation value (5.2). It is easy to see that in the direct Coulomb term in which $\omega_{k\ell} = 0$, the second term in (5.2) disappears, but in the exchange term ($\omega_{k\ell} \neq 0$) it survives.

In the case of $N > 2$ (5.2) has to be modified. The corresponding expression has been derived with the help of quantum electrodynamics by Grant [41].

It should be pointed out that the expectation value of the operator $\hat{H}'_{Br}$ occurring in (5.2) is in the case of heavy atoms significantly different from that of the original Breit operator.

$$\hat{H}_{Br} = -\frac{\alpha}{2r_{12}}\left[\vec{\underline{\underline{\alpha}}}^{(1)}\,\vec{\underline{\underline{\alpha}}}^{(2)} + \frac{(\vec{\underline{\underline{\alpha}}}^{(1)}\vec{r}_{12})(\vec{\underline{\underline{\alpha}}}^{(2)}\vec{r}_{12})}{r_{12}^2}\right] \qquad (5.3)$$

Mann and Johnson [7] have calculated both expectation values using Dirac spinors as one-electron functions in the case of He-like ions as a function of Z. In the case of Hg (Z=80) the difference $D = \langle \hat{H}'_{Br} \rangle - \langle \hat{H}_{Br} \rangle$ was $\sim$-1 RY = $\sim$-13.6 eV, and in the case of No (Z=102) it is $\sim$-3.3 Ry = $\sim$-45 eV, while for light atoms the difference is negligible (if $Z < 48$, $D \lessdot 1$ eV; see Table I of [7]). It is an open question how large would be these differences if one would use instead of a simple determinant built up from Dirac spinors relativistic HF spinors and construct with the aid of them also the correlation part of the wavefunction.

Turning to the fourth term in equ. (5.1) Mittleman [8] has shown on the basis of quantum electrodynamics that at heavy atoms if someone goes beyond the $(v/c)^2$ approximation also three-body terms arise in an N electron system. (Because of its rather complicated form we do not give here explicitly his three body potential, but rather refer to equations (3.15) and (3.16) with the definition of (2.17) of his paper [8].) The sum of these three body terms he could estimate to be many rydbergs in a heavy atom. So a conservative estimate would be to assume $E_{rel}^{3\ body} \approx D$.

It should be further mentioned that without using quantum electrodynamical methods one can show also with the help of a simple algebra the occurrence of three-body terms in a 3 electron system. Namely one can write down a Breit-type equation for 3 electrons, when Ψ will have $4^3 = 64$ components. So one can write out the matrix equation into 64 scalar equations and one can reduce it to the 8 large components in the $(v/c)^2$ approximation (Pauli approximation) [42]. The resulting scalar equations will contain, as expected the usual one and two electron interaction terms and the latter ones (orbit-orbit, spin-orbit, spin-spin terms) can be summed pairwise. On the other hand if one is inconsistent in the reduction procedure (because the original unreduced Breit's equation is also only in the $(v/c)^2$ approximation correct) and retains terms which are of higher power of v/c than 2, one obtains immediately three-body terms [42].

Since both Mittleman's and our derivations handled explicitly only 3 electrons, one could expect that at least for the highly relativistic inner shell electrons of heavy atoms also 4- and more body terms could play a non-negligible role in accurate calculations. Since in the moment only the calculation of 3-body terms in heavy atoms is in progress [43], we do not have any knowledge about the importance of these higher order terms.

The calculations of the first order quantum electrodynamical (Lamb shift) terms $E^{(1)}_{QED}$ has been performed for many atoms [44] and they have the same order of magnitude as the E_{Breit} contribution. An early calculation on the H_2 molecule [45] has provided a similar result. Therefore, one would expect that the second order quantum electrodynamical correction terms, $E^{(2)}_{QED}$ will be again in the order of D.

$$E^{(2)}_{QED} \approx E'_{Breit} - E_{Breit} \approx E^{3\ body}_{rel.} \qquad (5.4)$$

Until now no attempt has been made to calculate the term $E^{(2)}_{QED}$.

Finally, one should mention that until today instead of trying to calculate E_e defined by equ. (5.1) in most of the heavy atom calculations the expression

$$\tilde{E}_e = E^{HF}_{rel} + E^{corr.}_{n.rel} + E_{Breit} + E^{(1)}_{QED} \qquad (5.5)$$

was used. In this way besides neglecting the difference between E_{Breit} and E'_{Breit}, the terms $E^{3\ body}_{rel}$ and $E^{(2)}_{QED}$ they commit - in the authors' opinion - the most serious error by calculating the correlation energy (and usually also the expectation values of the terms obtained from $\hat{H}_{Breit}$ in the Pauli approximation), with the help of non-relativistic one-electron orbitals. As it is well known $E^{HF}_{n.rel} - E^{HF}_{rel}$ is already at Hg(Z=80) larger than $E^{corr}_{n.rel}$. Therefore, the calculation of the correlation energy with DHF spinors would be absolutely necessary. (We can visualize this also by saying that due to relativistic effects the ls electrons are pulled nearer to the necleus and therefore, being confined to a smaller part of the space the correlation is substantially stronger between them.)

We are fully aware that to perform all the calculations to obtain E_e (instead of $\tilde{E}_e$) requires a formidable effort. On the other hand this as aim should not be lost from sight when we try to develop a better established relativistic many electron theory.

APPENDIX

The matrix $\underline{\underline{A}}^{-}$ may be written explicitly in terms of sixteen components as

$$\underline{\underline{A}}^{-} = \begin{pmatrix} \underline{\underline{A}}_{1+}^{(1,1)} & \underline{\underline{A}}_{2}^{(1,2)} & \underline{\underline{A}}_{3+}^{(1,3)} & \underline{\underline{A}}_{4-}^{(1,4)} \\ \underline{\underline{A}}_{2}^{(2,1)} & \underline{\underline{A}}_{1+}^{(2,2)} & \underline{\underline{A}}_{4+}^{(2,3)} & \underline{\underline{A}}_{3-}^{(2,4)} \\ \underline{\underline{A}}_{3+}^{(3,1)} & \underline{\underline{A}}_{4-}^{(3,2)} & \underline{\underline{A}}_{1-}^{(3,3)} & \underline{\underline{A}}_{2}^{(3,4)} \\ \underline{\underline{A}}_{4+}^{(4,1)} & \underline{\underline{A}}_{3-}^{(4,2)} & \underline{\underline{A}}_{2}^{(4,3)} & \underline{\underline{A}}_{1-}^{(4,4)} \end{pmatrix}$$

where

$$\left(\underline{\underline{A}}_{1\pm}^{(t,t)}\right)_{ip,jq} = \Big\{\langle\chi_p^{(t)}|\pm\hat{F}_4+\hat{F}_5|\chi_q^{(t)}\rangle + \sum_{t'=1}^{4}\sum_k\langle\chi_p^{(t)}\,\psi_k^{(t')}|\chi_q^{(t)}\,\psi_k^{(t')}\rangle$$
$$-\sum_k\langle\chi_p^{(t)}\,\psi_k^{(t)}|\,\psi_k^{(t)}\,\chi_q^{(t)}\rangle\Big\}\delta_{ij} + \langle\chi_p^{(t)}\,\psi_j^{(t)}|\,\psi_i^{(t)}\,\chi_q^{(t)}\rangle$$
$$-\sum_{t'=1}^{4}\langle\chi_p^{(t)}\,\psi_j^{(t')}|\,\chi_q^{(t)}\,\psi_i^{(t')}\rangle$$

$$\left(\underline{\underline{A}}_{2}^{(t,t')}\right)_{ip,jq} = -\sum_k\langle\chi_p^{(t)}\,\psi_k^{(t')}|\,\psi_k^{(t)}\,\chi_q^{(t')}\rangle\delta_{ij} +$$
$$+\ \langle\chi_p^{(t)}\,\psi_j^{(t')}|\,\psi_i^{(t)}\,\chi_q^{(t')}\rangle$$

$$\left(\underline{\underline{A}}_{3\pm}^{(t,t')}\right)_{ip,jq} = \Big\{\langle\chi_p^{(t)}|\pm\hat{F}_3|\chi_q^{(t')}\rangle$$
$$-\sum_k\langle\chi_p^{(t)}\,\psi_k^{(t')}|\,\psi_k^{(t)}\,\chi_q^{(t')}\rangle\Big\}\delta_{ij}$$
$$+\ \langle\chi_p^{(t)}\,\psi_j^{(t')}|\,\psi_i^{(t)}\,\chi_q^{(t')}\rangle$$

$$(\underline{\underline{A}}_{4\pm}^{(t,t')})_{ip,jq} = \Big\{\langle \chi_p^{(t)} | \hat{F}_1 \pm i\hat{F}_2 | \chi_q^{(t')}\rangle$$

$$- \sum_k \langle \chi_p^{(t)} \, \Psi_k^{(t')} | \Psi_k^{(t)} \, \chi_q^{(t')}\rangle\Big\} \delta_{ij}$$

$$+ \langle \chi_p^{(t)} \, \Psi_j^{(t')} | \Psi_i^{(t)} \, \chi_q^{(t')}\rangle$$

The notations used here are as follows:

$$\hat{F}_1 = \frac{c}{i}\frac{\partial}{\partial x} \; ; \; \hat{F}_2 = \frac{c}{i}\frac{\partial}{\partial y} \; ; \; \hat{F}_3 = \frac{c}{i}\frac{\partial}{\partial z} \; ; \; \hat{F}_4 = c^2 ; \; \hat{F}_5 = -\varepsilon_i - \omega + \hat{V}(\vec{r})$$

and $V(\vec{r})$ is the nuclear attraction potential. The matrix $\underline{\underline{A}}^+$ may be constructed in a similar manner only with the redefinition of the operator $\hat{F}_5$ where one has to take $+\omega$ in place of $-\omega$. Further

$$(\underline{\underline{B}}^{(t,t')})_{ip,jq} = \langle \chi_p^{(t)} \, \chi_q^{(t')} | \Psi_i^{(t)} \, \Psi_j^{(t')}\rangle$$

$$- \langle \chi_p^{(t)} \, \chi_q^{(t')} | \Psi_j^{(t')} \, \Psi_i^{(t)} \rangle$$

and finally

$$D_{ip}^{\pm}(t) = \langle \chi_p^{(t)} | - \hat{h}^{\pm} | \Psi_i^{(t)}\rangle$$

Here i,j run over all the orbitals, p,q run over the basis functions and t,t' run over the four spinor components.

ACKNOWLEDGMENTS

This work has been supported by a National Research Council of Canada Grant in Aid of Research, which is hereby gratefully acknowledged. One of us (J.L.) should like to express his gratitude to the Department of Applied Mathematics of the University of Waterloo for inviting him as a Visiting Professor which made for him possible to perform the major part of the work described in this paper and to the "Fond der Chemischen Industrie" for financial support which he gratefully acknowledges.

A part of this work has been completed during the stay of J.C. as Richard-Merton Visiting Professor of the "Deutsche

Forschungsgemeinschaft" at the Lehrstuhl für Theoretische Chemie, Friedrich-Alexander-Universität Erlangen-Nürnberg, Germany which he gratefully acknowledges.

One of us (P.K.M.) should like to express his sincere gratitude to the Alexander von Humboldt Foundation whose "Dozent" Fellowship made his stay in Erlangen possible.

We are further very grateful to Professors M. Mittleman, F. Martino, J. Paldus and G. Malli for very fruitful discussions, to Professor H. Ruder for performing the relativistic HF calculation of the carbon atom and to Dr. S. Suhai for doing the Gaussian expansion of the output of this calculation.

REFERENCES

1. A. Rosen and I. Lindgren, Phys. Rev. 176:114 (1968); I.P. Grant, Adv. Phys. 19:747 (1970); I. Lindgren and A. Rosen, Case Studies in Atomic Physics 4:97,150,199 (1970); J.P. Desclaux, Comp. Phys. Comm. 9:31 (1973); J.P. Desclaux, Act. Nuc. Data Tables 12:311 (1973); L. Armstrong, Jr. and S. Feneuille, Adv. Atomic and Molecular Physics 10:2 (1976); J.P. Desclaux, J. de Physique Colloque C1:40, 109 (1979); J.P. Desclaux: The Status of Relativistic Calculations for Atoms and Molecules, Physica Scripta, Proceedings of the Nobel Symposium on Many Body Theory of Atomic Systems, Göteborg, June 1979 and references therein.
2. J. Ladik, Acta Ac. Sci. Hung. 10:271 (1959); ibid, 13:123 (1961); W.C. Mackrodth, Mol. Phys. 18:697 (1970); J.P. Desclaux and P. Pyykkö, Chem. Phys. Lett. 29:534 (1974); P. Pyykkö and J.P. Desclaux, Chem. Phys. 34:261 (1978) and references therein; see also the review paper by J.P. Desclaux from[1].
3. See for instance: T.L. Loucks, Augmented Plane Wave Method, Benjamin Inc., New York (1967); D.D. Koelling and A.J. Freeman in Plutonium and Other Actinides, J. Blank and R. Lindner, eds., North Holland Publ. Co. Amsterdam (1976) p. 291.
4. See for instance: T.P. Das, Relativistic Quantum Mechanics of Electrons, Harper and Row, New York (1973).
5. H. Bethe and E.E. Salpeter in Encyclopedia of Physics, Vol. 355, S. Flügge, ed., Springer Verlag, Berlin-New York-Heidelberg (1957) p. 140.
6. G. Breit, Phys. Rev. 34:553 (1929).
7. J.B. Mann and W.R. Johnson, Phys. Rev. A4:41 (1971).
8. M.H. Mittleman, Phys. Rev. A4:893 (1971).
9. J.P. Desclaux and P. Pyykkö, Chem. Phys. Lett. 39:300 (1976); P. Pyykkö and J.P. Desclaux, Chem. Phys. Lett. 43:545 (1976); ibid, 50:503 (1977).
10. G. Malli and J. Oreg, J. Chem. Phys. 63:830 (1975).
11. A. Rosen, Int. J. Quant. Chem. 13:509 (1978).

12. J. Oreg and G. Malli, J. Chem. Phys. 61:4349 (1975); ibid 65:1746 (1976); ibid 65:1755 (1976); H.T. Toivonen and P. Pyykkö, Int. J. Quant. Chem. 11:695 (1977).
13. T. Aoyama, H. Yamakowa and O. Matsuoka, J. Chem. Phys. 73:1329 (1980).
14. See for instance: T.C. Collins in "Electronic Structure of Polymers and Molecular Crystals", J.-M. André and J. Ladik, eds., Plenum Press, London-New York (1975) p. 389.
15. A four component relativistic AO has always only two different radial-dependent parts (large and small components). The four different components are then formed with the help of different combinations of these two radial functions with the well-known angular dependent parts of the wave functions (see for instance equ.-s (54), (99), (105) and (107) of reference [16]). Already in the = o j = s = 1/2 case (1s electron) one obtains in this way four different components for the relativistic AO. Therefore, if one wants to perform an LCAO procedure the basis functions have to have four different components which have to be varied independently. This is the reason of the ansatz (2.1.).

Furthermore most probably one could use only 2 different basis sets (one for the large and one for the small component as a function of r, because the angular part of the functions is fixed in case of an H atom). On the other hand to keep open the possibility of basis functions centered not on the atomic nuclei (functions centers at the middle points of chemical bonds, floating Gaussians etc.) we prefer to work with four different basis sets. In a practical case the four basis sets may be equal, $\{\chi^{(1)}\} = \{\chi^{(2)}\} = \{\chi^{(3)}\} = \{\chi^{(4)}\}$, or one can use only one basis $\{\chi^{(1)}\}$ for the large components and another one for the small components $\{\chi^{(2)}\}$.
16. See for instance: N.F. Mott and G. Sneddon, Quantum Mechanics and its Application, Clarendon Press, Oxford (1948), Chapt. XI.
17. C.C.J. Roothaan, Revs. Mod. Phys. 23:69 (1951).
18. G. Del Re, J. Ladik and G. Biczó, Phys. Rev. 155:997 (1967).
19. Due to the translational symmetry of the crystal we can consider instead of the separate $\vec{q}$ and $\vec{q}'$ vectors their difference $\vec{q}'' = \vec{q}-\vec{q}'$. We can put our reference cell at $\vec{q}' = \vec{0}$ and so we have to deal only with $\vec{q}'' = \vec{q}$. In this way in equ.-s (2.10) and (2.12) we can put everywhere instead of $\vec{q}' = \vec{0}$ and we can drop the two primes on $\vec{q}''$ in equ.-s (2.18).
20. P.-O. Löwdin, J. Chem. Phys. 18:365 (1950).
21. It should be mentioned that the whole derivation could be easily repeated for the different orbitals for different spins (DODS) case which would make easy the relativistic HF treatment of open-shell molecules and metals. In this case one has an equ. (2.22) separately for electrons with spin α and β, respectively. Further by defining separate matrix elements $P^{\alpha(j)}_{u,v}$ and $P^{\beta(j)}_{u,v}$ (using the vectors $\underline{d}^{\alpha}_{h}$ and $\underline{d}^{\beta}_{h}$ and omitting the

factor 2) one needs a separate definition for the matrices $\underline{\underline{F}}_5^{(t,t)\alpha}$ and $\underline{\underline{F}}_5^{(t,t)\beta}$, respectively (see equ. (2.20)). In this case the Coulomb term in equ. (2.20)) will remain unchanged and the exchange integral should be always multiplied only by that $P^{\sigma}_{u,v}$ ($\sigma = \alpha, \beta$) matrix element which corresponds to σ in $\underline{\underline{F}}_5^{(t,t)\sigma}$. The definitions of all other quantities occurring in the formalism remain unchanged.

22. Y.K. Kim, Phys. Rev. 154:17 (1967); T. Kagawa, Phys. Rev. A12:2245 (1975); G. Malli, Chem. Phys. Lett. 68:529 (1979); O. Matsuoka, N. Suzuki, T. Aoyama, and G. Malli, J. Chem. Phys. 73:1329 (1980); S. Huzinaga and A. Cantu, J. Chem. Phys. 55:5543 (1971); V. Bonifaci and S. Huzinaga, J. Chem. Phys. 60:2779 (1974).
23. S.J. Rose, I.P. Grant and N.C. Pyper, J. Phys. B11:1711 (1978).
24. P. Pyykkö, Adv. Quant. Chem. P.-O. Löwdin, ed., Academic Press London-New York (1978) p. 353.
25. E.R. Davidson, Chem. Phys. Lett. 21:565 (1973); Y. Ishikawa and G. Malli, Chem. Phys. Lett. 80:111 (1981).
26. S.N. Datta, Chem. Phys. Lett. 74:568 (1978).
27. G.E. Brown and O.G. Ravenhall, Proc. Roy. Soc. A208:552 (1951); M.H. Mittleman, Phys. Rev. A4, 893 (1971)
28. D.J. Thouless, The Quantum Mechanics of Many Body Systems, Academic Press, New York (1961); J. Cizek and J. Paldus, J. Chem. Phys. 47:3976 (1967); J. Paldus and J. Cizek, Phys. Rev. A2:2268 (1970).
29. A. Dalgarno and G. Victor, Proc. Roy. Soc. A291:291 (1966); S. Sengupta and M. Mukherji, J. Chem. Phys. 47:260 (1967); P.K. Mukherjee and R.K. Moitra, J. Phys. B11:2813 (1978).
30. T.H. Dunning and V. McKoy, J. Chem. Phys. 47:1735 (1967); J. Rose, T. Shibuya and V. McKoy, J. Chem. Phys. 58:74 (1973); D.L. Yaeger and V. MacKoy, J. Chem. Phys. 61:755 (1974); M. Ya. Amusia and N.A. Cherepkov, in Case. Stud. At. Phys. 5:47 (1975).
31. H.P. Kelly, Adv. Chem. Phys. I. Prigogine and S.A. Rice, eds., Interscience, John Wiley, New York 14:129 (1969); N.C. Dutty, T. Ishihara, C. Matsubara and T.P. Das, Phys. Rev. Lett. 22:8 (1969); Int. J. Quant. Chem. 3:32 (1969); C. Matsubara, N.C. Dutta, T. Ishihara and T.P. Das, Phys. Rev. A1:561 (1970).
32. A.H. Gabriel and C. Jordan, Case Stud. At. Collision Phys. 2:211 (1972).
33. I. Martinson and A. Gaupp, Phys. Rev. 15C:115 (1974).
34. P.W. Atkins and J.A.M.F. Gomes (private communication).
35. W.R. Johnson and C.D. Lin, Phys. Rev. A14:565 (1976).
36. C.D. Lin, W.R. Johnson and A. Dalgarno, Phys. Rev. A15:154 (1977).
37. P.-O. Löwdin and P.K. Mukherjee, Chem. Phys. Lett. 14:1 (1972).
38. P.W. Langhoff, S.T. Epstein and M. Karplus, Rev. Mod. Phys. 44:602 (1972).
39. G. Malli, Chem. Phys. Lett. 68:529 (1979).

40. J.P. Desclaux, Comp. Phys. Comm. 9:21 (1975).
41. I.P. Grant in this book.
42. D.K. Rai and J. Ladik, Int. J. Quant. Chem. 12:925 (1977).
43. M. Mittleman, personal communication.
44. See for instance: A.M. Desiderio and W.R. Johnson, Phys. Rev. A3:1264 (1971).
45. J. Ladik, J. Chem. Phys. 42:3340 (1965).

RELATIVISTIC SCATTERED-WAVE CALCULATIONS FOR MOLECULES AND CLUSTERS IN SOLIDS

Cary Y. Yang

Surface Analytic Research, Inc.
465A Fairchild Drive, Suite 128
Mountain View, CA 94043

ABSTRACT

A brief review of the origin and development of the self-consistent field-Xα- Dirac-scattered-wave (SCF-Xα-DSW) molecular orbital method is presented. The DSW scheme is the fully relativistic counterpart of the SCF-Xα-SW method, which, in turn, is the molecular analog of the Korringa-Kohn-Rostoker energy band calculation technique. Since the SCF-Xα-SW method follows the same approximations as the nonrelativistic SW method, it suffers from the same severe limitations. However, it incorporates all relativistic effects within the one-electron local exchange framework, yielding four-component wave functions that transform according to the double point group of the given system. Selected DSW results are reviewed to illustrate chronologically the development of the method, and to demonstrate its strength as well as limitations. These examples include molecules I_2 and UF_6, small metal clusters and their use in modeling chemisorption, and cluster complexes of the quasi one-dimensional conductor KCP. They represent the classes of problems where SW calculations will continue to provide useful information. The method should be viewed as complementary to traditional quantum chemical approaches for those calculations where high quality <u>ab initio</u> computations are currently unfeasible.

I. INTRODUCTION

The multiple-scattering or scattered-wave (SW) approach to determining electronic structures of molecules and solids has, in recent years, proven its utilities in many areas of inorganic chemistry and solid-state physics. Combined with the local density approximation for the exchange potential, which effectively reduces the many-electron problem to a much more manageable one-electron one, the SW method becomes a very attractive computational procedure whereby useful information about electronic properties can be obtained. The first derivation of energy band structure using this technique was given by Korringa (1). Subsequently Kohn and Rostoker (2) employed a variational approach with the Green's function technique and arrived at identical results. This so-called KKR method for calculating electronic energy band structure of crystalline solids is similar to the augmented-plane-wave (APW) scheme reported earlier by Slater (3). In fact, their equivalence has been extensively studied by Ziman (4), LLoyd (5), Slater (6), and Johnson (7).

The generalization of the SW technique for finite polyatomic systems with no translational invariance was suggested by Slater (8) and subsequently developed by Johnson (9). Numerous applications to molecules and clusters followed, which can be found in review articles by Johnson (10-12), Connolly (13), Weinberger and Schwarz (14), Rösch (15), Messmer (16), and most recently Case (17). The self-consistent-field-Xα-scattered-wave (SCF-Xα-SW) method, as it is popularly called, has not, and most probably will not replace traditional quantum chemical methods for computing molecular orbitals (MO). However, largely because of its computational tractability, both in bases representation and in numerical procedures, the SCF-Xα-SW method is more suitable in establishing bonding trends for elements across the rows as well as down the columns of the periodic table. Furthermore, molecular properties calculations requiring knowledge of continuum or scattering states (18,19) can be accomplished with reasonable accuracy and with little extra effort. It should be noted that the method suffers from several serious limitations. The principal ones are the local exchange potential and the muffin-tin (MT) approximation, which are discussed at length in Case's review (17). Despite these disadvantages, the inherent flexibility of the method overwhelmingly justifies its extension to the relativistic domain, where systems containing heavy elements can be treated more realistically.

Extension of the SW scheme to include relativistic effects has been explored by many authors (see reviews by Pyykkö (20) and Case (17)). Two approaches have been used. The first employs a simplified Dirac <u>radial</u> equation excluding the spin-orbit term and solves the secular problem in the framework of the non-relativistic SW formalism (21-26). This approach is essentially the one-component

Pauli approximation of the Dirac Hamiltonian. Spin-orbit splittings are subsequently obtained via diagonization of spin-orbit operator matrices for groups of orbitals closely spaced in energy (27,28).

A more direct as well as more accurate approach is to use the full four-component Dirac formalism and double group symmetry throughout. For solids, the Dirac extension of the KKR method was reported by Onodera and Okazaki (29) and by Takada (30). Yang and Rabii (31), Cartling and Whitmore (32), and Rosicky et. al. (33) presented formulations of the SW secular equations for molecules. Yang (34) also developed the complete double point group symmetrization scheme. The SCF-Xα-SW computer codes were adapted by myself to yield a parallel Dirac-Scattered-Wave (DSW) method, which was then applied to a few systems containing heavy elements (35-40). More recently, David Case and I (41,42) coded the self-consistent version of the DSW scheme, and further calculations (42-46) have been performed using the resulting SCF-Xα-DSW method. Some of these results will be highlighted in Section III.

The remaining discussion will be focused on the DSW scheme and its applications in the Xα local exchange framework to molecules and clusters modeling extended systems. Review of the Xα approximation can be found elsewhere (17,47). The overall purpose of this paper is to familiarize the reader with the basic ingredients of a relatively new method, which, despite its limitations, has been shown to be a powerful MO computational technique, and which has definite utilities in interpreting spectra and establishing general bonding trends.

II. DIRAC SCATTERED-WAVE FORMALISM

In this Section, the derivation of the DSW secular equations and the double group symmetrization will be briefly reviewed. Details of the formulation can be found in Yang and Rabii (31) and Yang (34).

(a) The Secular Problem

We start with the one-electron Dirac equation (48)

$$\left(c\vec{\alpha}\cdot\vec{p} + \beta mc^2 + V I_4\right)\psi = W\psi \tag{II. 1}$$

which can be transformed into an integral equation for any given volume ν in space.

$$\int_v \tilde{G}(\vec{r},\vec{r}')V(\vec{r}')\Psi(\vec{r}')d\tau' - ic\int \tilde{G}(\vec{r},\vec{r}')\vec{\alpha}\cdot\hat{n}'\Psi(\vec{r}')dS' = \begin{cases} \Psi(\vec{r}) & \vec{r}\in v \\ 0 & \vec{r}\notin v \end{cases} \tag{II. 2}$$

where $\hat{n}$ is the unit outward normal on the surface enclosing v, and $\tilde{G}(\vec{r},\vec{r}')$ is the Dirac free electron Green's function (48). Within the general SW framework, and in view of (II. 2), the solution is reduced to that of a boundary-value problem. The derivation can proceed once v is specified and boundary conditions are established. With the assumption of space partitioning (MT potential) inherent in the SW approach (9), (II. 2) is reduced to

$$\Psi(\vec{r}) = -ic\int \tilde{G}(\vec{r},\vec{r}')\vec{\alpha}\cdot\hat{n}'\Psi(\vec{r}')dS' \qquad \vec{r}\in v \tag{II. 3}$$

where the intersphere region is the chosen volume. The secular equations will emerge from the matching of trial wave functions across its boundaries.

One assumes a trial wave function in the spherical regions (atomic and outer) with the form

$$\Psi^{\alpha} = \sum_{Q} A_{Q}^{\alpha}\begin{bmatrix} g_{\kappa}^{\alpha}\,\chi_{Q}^{\alpha} \\ i f_{\kappa}^{\alpha}\,\chi_{\bar{Q}}^{\alpha} \end{bmatrix} \tag{II. 4}$$

where α runs over the outer and atomic spheres, and $Q\equiv(\kappa,\mu)$ and $\bar{Q}\equiv(-\kappa,\mu)$. The functions g_{κ}^{α} and f_{κ}^{α} are numerical solutions of the Dirac radial equations. There are two types of boundaries for the intersphere region. The first, S_1, is the outer sphere surface, while S_2 represents the conglomeration of all atomic sphere surfaces. The specific boundary conditions corresponding to (II. 3) are given by

$$\Psi(\vec{r})\Big|_{S_1} = \Psi^{o}(\vec{r}_o)\Big|_{S_1 -} \quad ; \quad \Psi(\vec{r}')\Big|_{S_1} = \Psi^{o}(\vec{r}_o')\Big|_{S_1 +}$$

$$\Psi(\vec{r})\Big|_{S_2} = \Psi^{\alpha}(\vec{r}_\alpha)\Big|_{S_2 +} \quad ; \quad \Psi(\vec{r}')\Big|_{S_2} = \Psi^{\alpha}(\vec{r}_\alpha')\Big|_{S_2 -} \qquad \text{(II. 5)}$$

The final secular equations have been given in detail elsewhere (31). Only the important features are highlighted here. The secular matrix $\mathcal{M}$ consists of a diagonal term, $[T_\kappa^\alpha]^{-1}$, or the inverse t-matrix, which, in terms of ordinary scattering theory, is the cotangent of the phase shift of the κth partial wave scattered by the αth sphere. Hence T_κ^α is potential-dependent. The off-diagonal terms, $\mathcal{G}_{QQ'}^{\alpha\alpha'}$, or structure factors, on the other hand, are purely geometrical entities and are only functions of atomic positions and sphere radii. The structure factors, as in all scattered-wave theories, represent the configuration of the boundary-value problem, while the t-matrices describe the nature of the force fields acting within the given configuration. The complete separation of the two quantities facilitates their computations and in turn the solution of the secular equations.

The t-matrices and structure factors reduce exactly to their usual Schrödinger SW counterparts (31,9) in the limit $c \rightarrow \infty$. The solution of the secular equations $\mathcal{M}\mathcal{C} = 0$ gives rise to a set of one-electron orbitals, each of which has single-center partial-wave expansions in angular momentum in the spherical regions (as in (II. 4)) and a multi-center expansion in the intersphere region. The orbital wave functions can be normalized numerically in the spherical region and analytically in the intersphere region (34). The proper occupation of these orbitals and subsequent generation of a new potential energy represent the onset of the SCF cycle, for which a convergence criterion for the potential is customarily set.

(b) Symmetrization

For most molecules, the secular matrix $\mathcal{M}$ can be "factored out" into smaller matrices with the use of group theory. This symmetrization process is based on the generation of multi-center partial wave bases, each of which transforms as an irreducible representation (irrep) of the double group. The size of each symmetrized matrix is generally at least an order of magnitude less than that of the unsymmetrized one. Thus computational effort is drastically reduced.

We start with the projection operator (49)

$$\mathcal{P}^{\nu}_{\eta\eta} = \frac{l_\nu}{g} \sum_{R} \Gamma^{\nu}(R)^{*}_{\eta\eta}\, O_R \tag{II. 6}$$

and apply it to the wave function in the spherical region given by (II. 4). The double group operator O_R transforms either the upper or the lower component ϕ^{α}_{Q} according to

$$O_R\, \phi^{\alpha}_{\kappa\mu} = \sum_{\alpha'\mu'} \Delta^{\alpha'\alpha}_{\mu'\mu}(R)\ \phi^{\alpha'}_{\kappa\mu'} \tag{II. 7}$$

where the matrix Δ can be expressed as a direct sum

$$\Delta = n_1\Gamma^{1} \oplus n_2\Gamma^{2} \oplus \cdots\cdots \oplus n_\nu\Gamma^{\nu} \oplus \cdots\cdots \tag{II. 8}$$

The Γ^{ν}'s are the double group extra irreps, each of which has a multiplicity of n_ν within Δ. Thus the secular matrix is effectively block-diagonalized. The symmetrized wave function is given by

$$\bar{\Psi} = \sum_{\sigma\kappa n} \bar{A}^{\sigma}_{\kappa n} \begin{bmatrix} {}^{up}\bar{\Phi}^{\sigma}_{\kappa n} \\ {}^{low}\bar{\Phi}^{\sigma}_{\kappa n} \end{bmatrix} \tag{II. 9}$$

where σ represents a set of equivalent atoms. The indices of the symmetrized matrix are now σ, κ, and n instead of α, κ, and μ. The symmetrized $\bar{\Phi}$ is related to its unsymmetrized counterpart by

$$\bar{\Phi}^{\sigma}_{\kappa n} = \sum_{\alpha'\mu'} \mathcal{S}^{\alpha'\mu'}_{\kappa n}\ \phi^{\alpha'(\sigma)}_{\kappa\mu'} \tag{II. 10}$$

where schematically $\mathcal{S} \propto \Gamma^{*}\Delta$. For a given irrep and a set of equivalent atoms, $\mathcal{S}$ contains all of the symmetry coefficients of the spin angular functions corresponding to the atoms of that set. The matrix $\mathcal{S}$ is computed separately in practice. The symmetrized secular matrix $\bar{\mathcal{M}}$ is given by

$$\bar{\mathcal{M}} = \mathcal{S}^{\dagger} \mathcal{M} \mathcal{S} \tag{II. 11}$$

This formulation is completely analogous to the nonrelativistic one given by Diamond (50).

The process of symmetrization is illustrated with the following example. Consider the tetrahedral cluster X_4, where the bases are truncated at $\ell = 4$ for the outer sphere and $\ell = 2$ for the atomic spheres. The size of the unsymmetrized relativistic secular matrix is 122 x 122. However, as one can deduce from Table I, the dimensions of the secular matrices for the two-dimensional irreps E_2 and E_3 and the four-dimensional irrep Q are 11, 10, and 20 respectively.

Table I. Number of Symmetrized Basis Functions for X_4

Group T_d	κ	No. bases in extra irreps E_2	E_3	Q
Outer sphere	-1	1	0	0
	1	0	1	0
	-2	0	0	1
	2	0	0	1
	-3	0	1	1
	3	1	0	1
	-4	1	1	1
	4	1	1	1
	-5	1	0	2
X sphere	-1	1	1	1
	1	1	1	1
	-2	1	1	3
	2	1	1	3
	-3	2	2	4

III. APPLICATIONS OF THE DSW METHOD

This Section is devoted to summaries of DSW results for a few systems. These systems are chosen for two reasons. First, they illustrate chronologically the development of the method. At the same time, they represent the types of problems to which this approach will continue to yield important contributions.

(a) I_2

The first application of the DSW method to system containing heavy elements was the diatomic iodine (Z = 53) molecule (36). The calculation was non-self-consistent and was intended to be a test case of this relativistic technique.

Substantial amount of information on I_2 has appeared in the literature throughout the years (51,52). I_2 has $D_{\infty h}$ symmetry, the ground state valence electronic configuration being

$$(1\sigma_g)^2 \ (1\sigma_u)^2 \ (2\sigma_g)^2 \ (1\pi_u)^4 \ (1\pi_g)^4 \ : {}^1\Sigma_g^+$$

Figure 1 summarizes the results of our calculations on I_2. The first six columns give the corresponding results for I and I_2. The weak sp-hybridization is demonstrated by the large separation in energy ($\sim$ 8 eV) between the s and p complexes. The relativistic effects are very evident by comparing column three with four and column five with six. The σ levels of the 5s complex are shifted down by about 3 eV while the spin-orbit splitting of the $1\pi_u$ level is about 0.9 eV. Since $1\pi_g$ and $2\sigma_g$ are only about 2 eV apart, one would expect a significant spin-orbit mixing of the relativistic $e_{1/2g}$ levels. This accounts for a somewhat smaller splitting of the $1\pi_g$ level ($\sim$ 0.7 eV). In the seventh column, the results of a molecular calculation (53) using the nonrelativistic linear-combination-of-atomic-orbitals (LCAO) method with complete-neglect-of-differential overlap (CNDO) are given for comparison. While the CNDO orbital energies are about 5 eV lower than our results, the separations among the valence levels compare quite well with ours. In addition to the ground state calculations, we have also performed calculations using the transition state concept (54) for the ionization potentials (IP's) of the three upper valence levels. The results are shown in the eighth and ninth columns. The photoemission data of Frost et al. (55) are given in the last column for comparison. We observe that the transition state orbital energies show a downward shift of about 3 eV from their ground state counterparts while the ordering among the three levels remains the same. This shift puts the calculated IP's to within 0.7 eV of the experimental results. The transition state spin-orbit splittings of $1\pi_g$ and $1\pi_u$ are 0.70 eV and 0.88 eV compared to the experimental values of 0.63 eV and 0.81 eV respectively. The IP's of the π levels are slightly overestimated, while that of $2\sigma_g$ slightly underestimated. This might be due to the muffin-tin approximation that we employ. It has been demonstrated with different schemes (56-58), that non-muffin-tin corrections tend to lower the σ levels but raise the π levels.

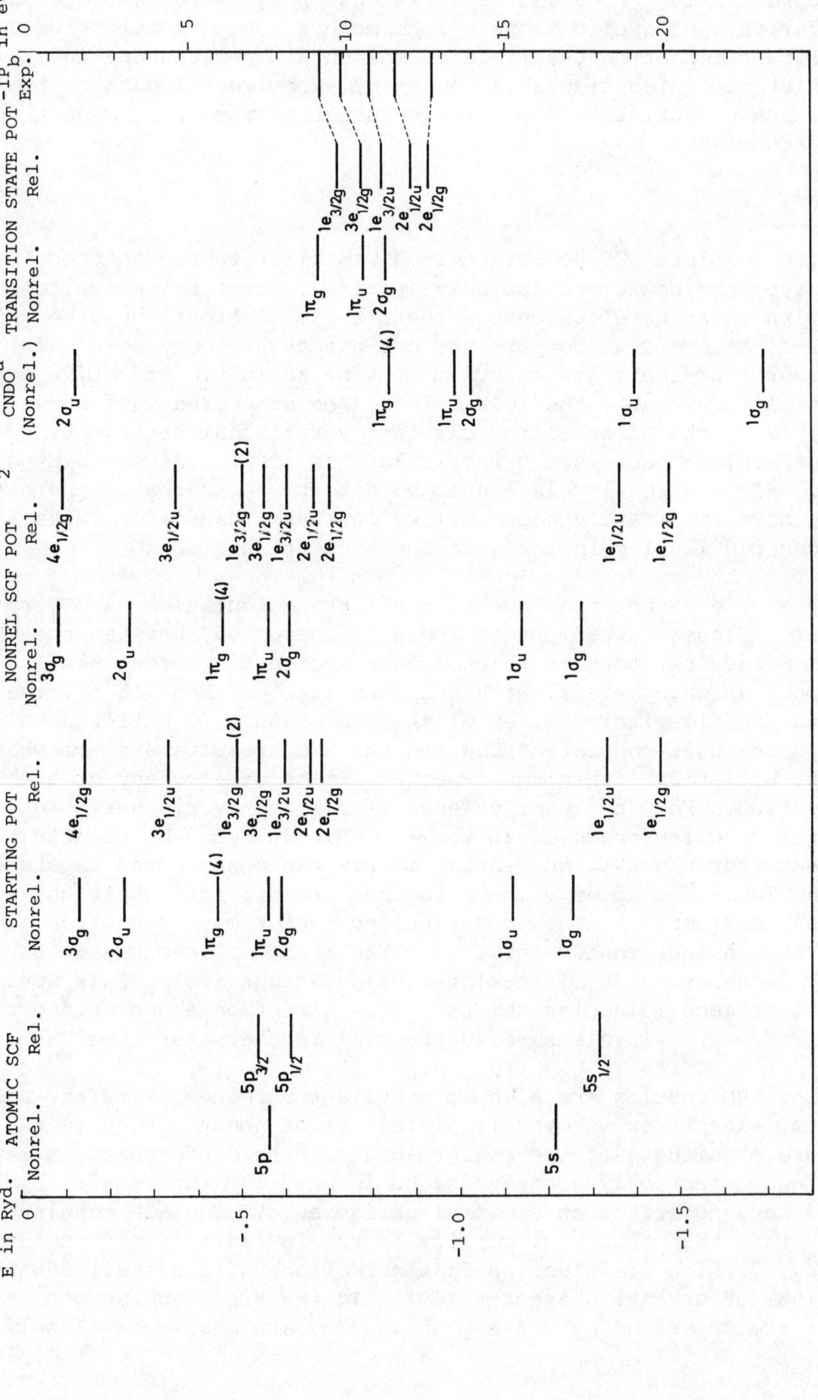

Fig. 1. Valence orbital energies of I and I_2. The highest occupied level in the molecule is designated by its occupation number. (a) Ref. 53; (b) Ref. 54.

The agreement between our results for I_2 and experiment is quite remarkable, considering the physical limitations of the muffin-tin approximation. The DSW calculations on I_2 serve as the ground work for further applications of the method, which should have a significant impact on the field of molecular spectroscopy in view of the existence of considerable amount of experimental data on the ionization potentials and excitation spectra of molecules containing heavy atoms (59).

(b) UF_6

The complete SCF-Xα-DSW method was first tested on the "prototype" molecule uranium hexafluoride so that extensive comparisons with existing electronic structure calculations could be made (42). In Figure 2 we compare the one-electron energy levels for UF_6 from four representative calculations:the ab initio effective core potential (ECP) (60), the relativistic Xα scattered-wave model (RXα) (27), the Dirac-Slater discrete variational method (DVM) (61), and our minimum basis set DSW calculations (DSW). As we pointed out in Ref. 42, the ECP levels cannot be directly compared to the others and we have arbitrarily added 8.1 eV to these in order to make the top occupied level coincide with the present results.

The orderings of the levels in all the calculations are very similar. Closest agreement is found, as expected, between the DSW and RXα results, both of which assume muffin-tin potentials. For the upper valence levels, with energies greater than -15 eV, these results are all within 0.1 eV of the RXα results. Differences between the self-consistent DSW and the RXα results are somewhat larger, but still within the expected limits of accuracy of either calculation. For the lower valence levels, however, there are significant differences of up to ~ 1 eV. This may be related to the fact that the RXα spin-orbit matrix was constructed separately for the lower and upper valence regions and did not couple the two regions together. Another contributing factor may arise from the fact that in the uranium atom, the RXα method places the 6p orbit 0.6 eV higher than do Dirac-slater calculations (26). This would tend to produce molecular shifts in the direction shown, since the $1t_{1u}$ and $2t_{1u}$ levels have substantial 6p character (see Table II).

The DVM results are also in excellent agreement with the DSW results, except for an overall difference of about 0.4 eV in the absolute magnitudes of the energy levels. The differences in level orderings affect only closely spaced levels, and such small changes should have no effect on spectral assignments. The ECP results differ from the others principally in the levels arising from $3t_{1u}$ and $1t_{2g}$, which are fluorine 2p combinations with a small admixture (10%-15%) of uranium character (60). In the ECP calculations, these levels are separated by 0.5 eV, while they are much more closely

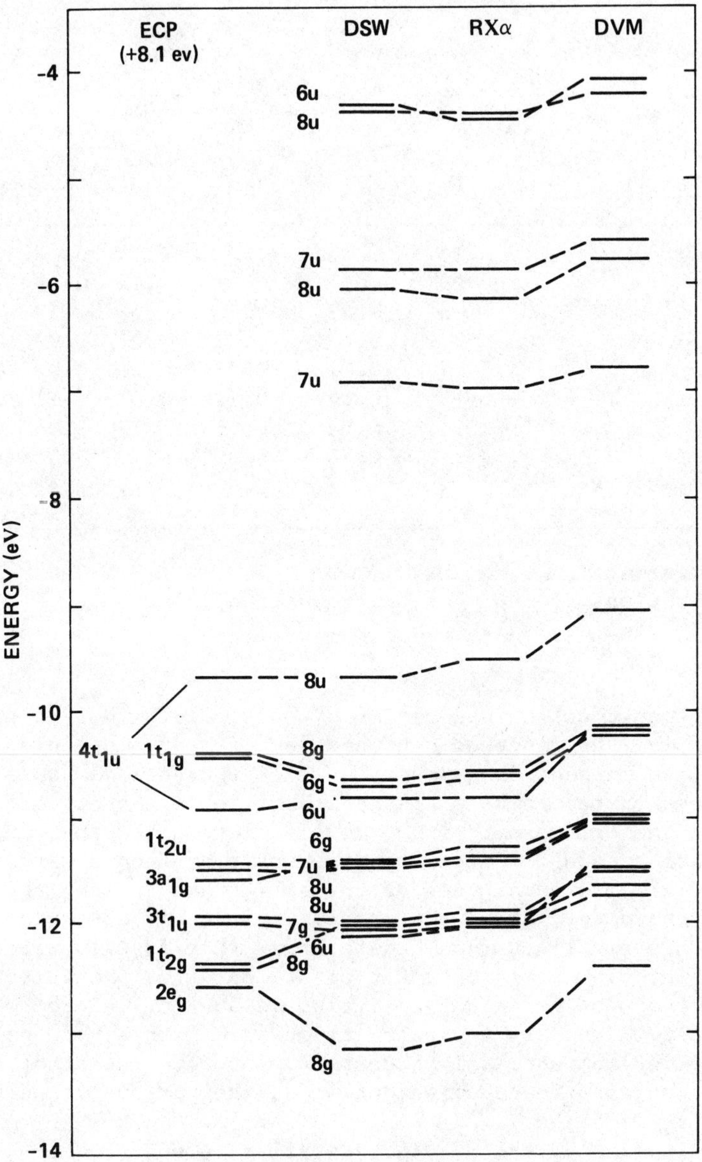

Fig. 2. One-electron energy levels for UF_6.

spaced in the other calculations. As before, these differences are too small to have any effect on conclusions drawn from the calculations.

Table II. Spin-orbit Splittings.[a]

Orbital	%U[b] p	d	f	ECP	RXα	DVM	DSW min	ext
$5t_{1u}$[c]	...	...	84		-0.03	0.14	0.06	0.07
$2t_{2u}$[c]	...	...	95		0.21	0.13	0.19	0.23
$4t_{1u}$	7	...	5	1.23	1.30	1.08	1.13	1.14
$1t_{1g}$	...	0	...	0.02	0.03	0.03	0.03	0.04
$1t_{2u}$	...	...	7	0.02	0.01	0.00	0.02	0.01
$3t_{1u}$	1	...	15	0.07	0.12	0.01	0.11	0.11
$1t_{2g}$	...	11	...	0.04	0.06	0.09	0.05	0.07
$2t_{1u}$	72	...	...	6.75	5.39	5.22	5.57	5.52
$1t_{1u}$	24	...	1	2.52	2.60	3.75	3.48	3.45

[a]Values in eV.
[b]Populations from ECP calculations, Ref. 60.
[c]Unoccupied orbitals.

Spin-orbit splittings can be of particular importance in understanding the electronic properties of heavy molecules. In Table II, we present the amount by which the various nonrelativistic triply degenerate orbitals are split by the spin-orbit effect. (Although the spin-orbit effect couples levels together, to a good approximation each can be treated as deriving from a single parent level (27).) The magnitudes of the splittings reflect the amount of metal character, since the uranium spin-orbit effect (particularly for the 6p orbital) is much greater than that for fluorine. From the point of view of interpretation of the uv or photoelectron spectra, the most important splitting is that of the top occupied level $4t_{1u}$. Since all the calculations place this value at 1.1-1.3 eV, this is most likely a correct picture. The other splittings are also in approximate agreement among the various calculations.

Qualitatively, the results are all in good agreement. The only significant difference between the ECP and the scattered-wave calculations lies in the partitioning of charge between the uranium 5f and fluorine 2p orbitals. Examination of the RXα charge distributions for individual orbitals shows that the occupied $3t_{1u}$, $4t_{1u}$, and $1t_{2u}$ orbitals have more 5f character than the ECP results, while the unoccupied $2t_{2u}$ and $5t_{1u}$ orbitals have less 5f character.

This leads to a less positive metal in the scattered-wave calculations ($U^{+1.5}$) than in the ECP ($U^{+2.4}$). This is most likely due in large part to the different ways of partitioning the overlap charge in the two calculations; it is a general rule that scattered-wave calculations yield a less positive metal than do Hartree-Fock calculations (62,63). In spite of this difference, though, the scattered-wave calculations are remarkably similar to those obtained from the ECP calculations, and give the same picture for the molecular orbital structure of UF_6.

From the point of view of computational efficiency, the scattered-wave methods (RXα and DSW) are clear superior to the linear combination of atomic orbitals (LCAO) methods, and this discrepancy will be even larger for bigger molecules. The DVM results were limited to a near-minimal basis set, and even at this level have significant residual numerical errors in the one-electron energies. Improvements in computer codes may be able to alleviate this situation somewhat. By making optimal use of symmetry, Hay et al. (60) were able to calculate the two-electron integrals needed for the ECP calculation in 4 min on a CDC 7600 computer. Some additional time is required for the self-consistent-field (SCF) iterations, and the generation of the core potential itself is a major undertaking, albeit one that needs to be done only once. By contrast, the RXα method requires about 3 sec and the DSW about 8 sec per iteration, again on a CDC 7600. About 20-30 iterations are required to achieve self-consistency, so that total computational times are on the order of 1-4 min. The extended basis set DSW calculations are about twice as time consuming as the minimal basis set calculations. Both the RXα and DSW timings could be decreased by ~40% by making better use of symmetry (based on unpublished improvements in nonrelativistic octahedral complexes). For larger molecules, or for calculations including polarization functions, the advantage of the scattered-wave method becomes much more pronounced.

(c) Small Metal Clusters

The demand for a computationally tractable MO method in electronic structure studies of systems containing heavy transition and noble metal atoms has been growing along with increased sophistication in spectral measurements for particles and surfaces. Improved understanding of the latter systems will stimulate further advances in the important area of catalysis. State-of-the-art electronic structure calculations on models of these systems (12, 16,18,38,43-45,64-72) have been focusing on qualitative description of bonding trends and interpretation of photoelectron spectra. Despite the qualitative usage of such calculations, relativistic effects remain crucial to correct descriptions of valence as well as core electronic structures of heavy-atom-containing systems. This point is illustrated in the next example (45).

In order to establish the importance of relativistic effects in heavy d-electron systems, a comparison of the Pt_2 valence orbitals calculated with the nonrelativistic SCF-Xα-SW and the relativistic SCF-Xα-DSW schemes is presented in Figure 3. The nonrelativistic results show a d^{10} configuration, with virtually no sd-hybridization. The relativistic orbitals, on the other hand, reveal a d^9s^1 configuration and consist of sd-hybrids. Aside from the large spin-orbit splitting of the d-complex into $d_{3/2}$ and $d_{5/2}$ sub-complexes and the admixture of the s with the d orbitals, there are no substantial reordering of d orbital energies. However, because of this sd-hybridization, the nature of the chemical bond described relativistically (which is more accurate) differs considerably from that of its nonrelativistic counterpart. Hence, in studies of systems involving heavy transition metals, one must take into consideration relativistic effects to obtain realistic bonding information.

Further relativistic studies of the electronic structure of metal clusters are ongoing. Some examples are given in Figure 4 (45). The valence orbitals for each of the tetrahedral clusters, Ag_4, Au_4, Pt_4, and Pd_4 exhibit band-like complexes. As one would expect, the noble metals have filled d-bands and an unfilled s-band, while the $d_{5/2}$- and s-bands are both unfilled in the transition metals. The sd-hybridization increases considerably going from Ag_4 to Au_4 and from Pd_4 to Pt_4. This is primarily a relativistic effect, manifested by stabilization of the s orbitals and destabilization of the d orbitals as the element goes down the column of the periodic table. Increased overlap of the s and d orbitals results.

Because of the unfilled $d_{5/2}$-band and the larger d bandwidths (implying more d delocalization), the transition metals Pd and Pt are more likely than the noble metals Ag and Au to have participation of d-electrons in chemisorptive bonds. For the transition metal clusters, the more diffused and lower s-band in Pt_4 allows most of the charge exchange with an adsorbate through Pt s-electrons. In comparison, for Pd_4, significantly more charge transfers are effected through Pd d orbitals. Depending on the adsorption geometry, this fundamental difference in metal-adsorbate bonding between Pd and Pt could give rise to different preferred sites and different dissociation probabilities for an adsorbed species on their respective surfaces.

(d) Cluster Modeling of Chemisorptive Systems

The tetrahedral clusters studied in the previous sub-section can serve as models of two different systems. It can be viewed as a small fragment of the (111) surface, on which adsorption on different sites can be modeled with the four metal atoms and one adsorbate atom or molecule. This model for the surface is probably too small to provide useful quantitative information. On the other hand, the same cluster is the smallest tetrahedral particle and as

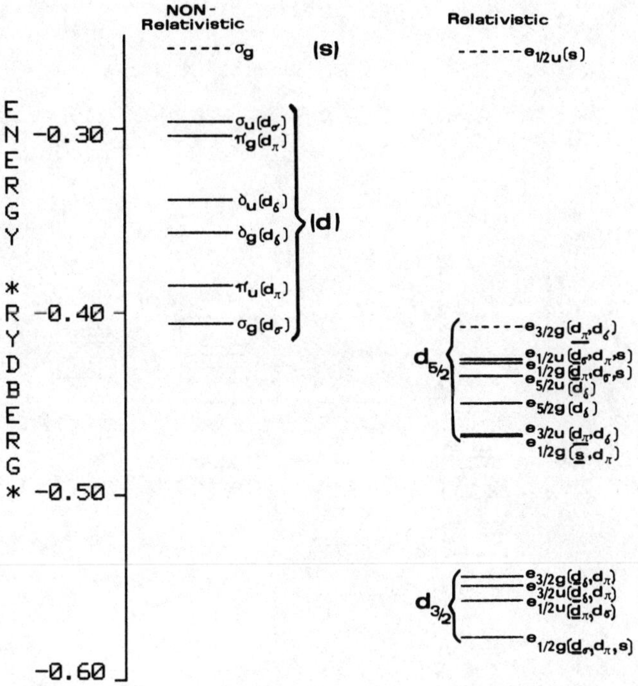

Fig. 3. Scattered-wave calculations of Pt_2 valence electronic structure. Broken lines indicate unoccupied orbitals. The dominant nonrelativistic character for each relativistic orbital is underlined.

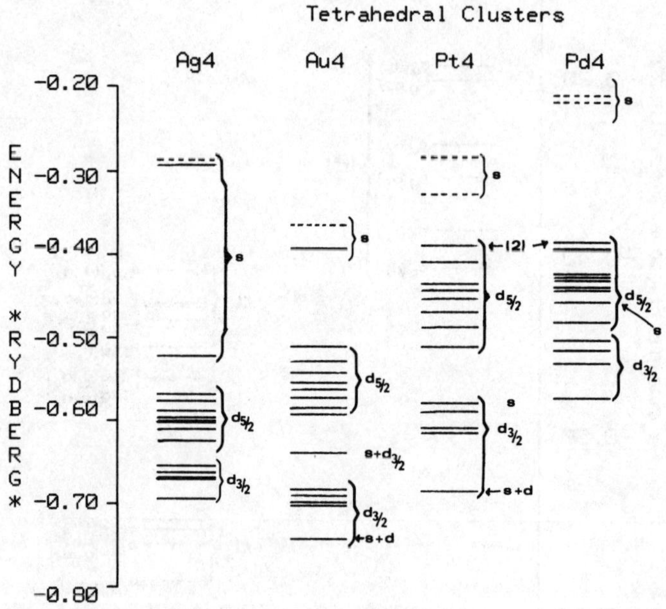

Fig. 4. Comparison of heavy transition and noble metal clusters via SCF-Xα-DSW calculations of their tetrahedral valence electronic structure.

such, can provide better descriptions of the inherent particle electronic structure. DSW calculations with the tetrahedral cluster modeling adsorption on supported particles are in progress. Preliminary results were reported recently (73). Here some previous chemisorption calculations are reviewed (43,44).

We model adsorption of carbon monoxide by calculating the electronic structure of Pt and CO/Pt clusters as shown in Figure 5a, chosen for different surfaces and adsorption sites. Each cluster consists of at least one shell of nearest metal atoms and hence provides the local bonding picture between the adsorbate and the metal atoms. The parameters chosen for CO are identical to those used for cluster calculations of CO on copper and nickel surfaces (66).

The results are summarized in Figure 5b. The $d_{3/2}$ - $d_{5/2}$ spin-orbit splitting is evident in all Pt-containing clusters. Strong $d_{3/2}$ - $d_{5/2}$ hybridization is observed only in sd or spd orbitals, which are largely responsible for bonding among the Pt atoms. Upon inclusion of the CO molecule, the metal "bands" undergo different changes in each model cluster. In all cases, the s-"band", which "overlaps" the d-"band", is partially depleted in the d-"band" region. In Pt_2CO, the density of states at the Fermi level (E_F) is smaller than that in Pt_2, in agreement with most UPS results (74-77). However, this trend is reversed in the Pt_5CO and Pt_5 clusters. The extent to which the basic limitations of our calculational method affect this finding is difficult to assess. Nevertheless, the results, combined with those below, do suggest that fourfold hole sites are not significantly occupied in these experiments.

In general, the orbitals responsible for bonding between CO and the Pt surface atoms are 5σ and $2\pi^*$. The 5σ orbital donates electrons to some spd components of the surface Pt atoms. Its energy position with respect to the centroid of the d-band remains fixed for all three clusters. The CO $2\pi^*$ orbital receives charge from the Pt s- and d-bands. Table III summarizes the overall bonding picture for all three model clusters. We expect that with larger clusters and different C-surface spacings h, the actual magnitude of these charges would change somewhat, but that trends among the three sites are reliable. Our Pt_5 calculation yields an electron-deficient surface by .06 electron per atom, while each "surface" atom in Pt_5CO is electron-rich by .02. Therefore, the net charge transfer from Pt to CO in the fourfold hole site seems to originate from the bulk instead of the surface.

The C-Pt bond gives rise to at least one antibonding orbital, as shown in Figure 5b. In the Pt_5CO cluster, but not the other two, one of the orbitals with the antibonding character is occupied. Calculations on Ni/CO (66) suggest that this result should not be

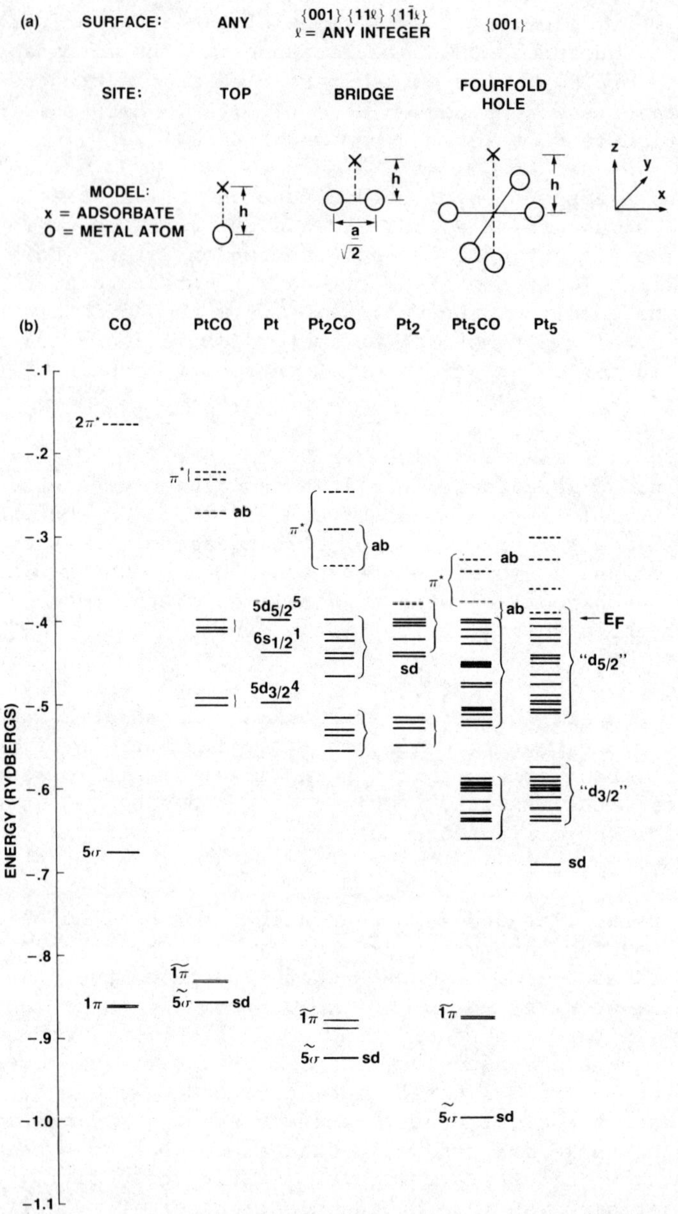

Fig. 5. (a) Cluster models for adsorption on fcc surfaces.
(b) SCF-Xα-DSW calculations using these model clusters for the CO/Pt system. The highest occupied orbital in each Pt-containing cluster is aligned in energy with that in PtCO. Orbitals with substantial Pt sd-hybridization are so indicated. "ab" denotes C-Pt antibonding character. Those unoccupied orbitals with significant amounts of CO π^* character are also indicated.

Table III. Summary of charge transfer in the CO/Pt chemisorption bond for the three high symmetry sites. Positive charge transfer (in electrons) occurs from CO to Pt. The Pt orbitals of primary importance to bonding are shown in parentheses. The coordinate systems are shown in Figure 5a.

	5σ (Bonding)	$2\pi^*$ (Back-bonding)	Others	Total
Top	0.51 (sp_σ d_σ)	-0.30 (sd_σ d_π)	-0.02	0.19
Bridge	0.67 (sp_x d_{xz})	-0.72 (sd_{yz} $d_{x^2-y^2}$)	0.05	0.00
Fourfold hole	0.63 (sp_π d_π $d_{x^2-y^2}$)	-0.76 (sd_π d_{xy})	0.07	-0.06

affected significantly by changes in the value of h. Thus qualitatively, we expect the fourfold hole site to be the least binding of the three high-symmetry sites for CO/Pt(100).

Despite the obvious shortcomings of our cluster sizes in achieving bulk valence band properties, the study of the Pt-CO chemisorbed bond using these clusters has appeared to be quite successful. The ionization energies (IE) of all CO orbitals have been calculated using the transition state procedure (78). The results are shown in Figure 6. Comparisons with UPS and XPS data are made by aligning the 4σ levels from the experiments with that in the gas-phase spectrum. In each cluster, $5\sigma/1\pi$ reversal from the gas phase is obtained as predicted by calculations for CO on other metals (79-81). The calculated core IE's differ from the XPS results by about 5 eV. However, the energy difference between the calculated C 1s core IE's for PtCO and Pt_2CO follows the same trend as the experimentally resolved chemical shift between the linearly and bridge bonded CO (75). Similar agreement is obtained for the O 1s core IE's. Despite generally good quantitative agreement between our calculations with such small clusters and photoemission data, it is premature to interpret the differences among IE's in the three clusters as characteristic of the adsorption site. Variations in the cluster size and the C-surface spacing h may result in significant relative changes for the IE's among the three sites. However, even with our minimum size clusters modeling CO adsorption on platinum surfaces, substantial progress has been made with the SCF-Xα-DSW method in the direct interpretation of photoemission spectra.

(e) Quasi One-Dimensional Conductors

Compounds containing the square-planar tetracyanoplatinate ion, $Pt(CN)_4^{2-}$, have been the subject of considerable experimental and

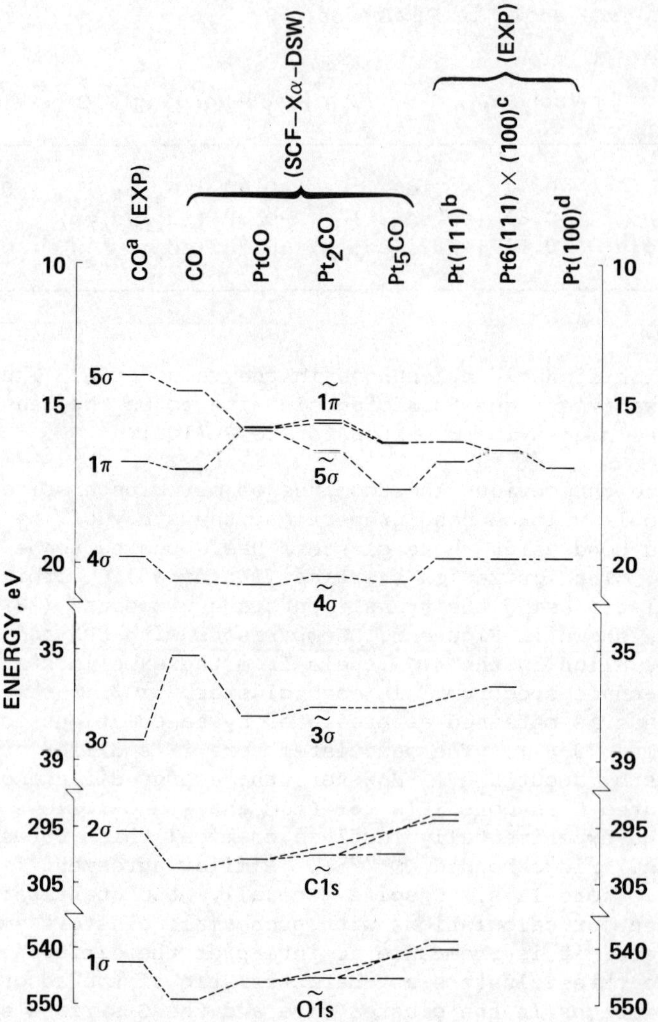

Fig. 6. Ionization energies of CO. Comparison of cluster calculations and photoemission results. All experimental 4σ levels are aligned with their gas-phase counterpart. (a) Ref. 90; (b) Ref. 75; (c) Ref. 76; (d) Ref. 77

theoretical study in recent years (82,83). In these compounds the $Pt(CN)_4^{2-}$ units are generally parallel stacked in a staggered conformation forming linear chains with a Pt backbone in each chain. Although the compounds are insulating in the pristine form, they acquire anisotropic conductivity (along the Pt back-bone) when they are doped with oxidizing agents such as halogens. Experiments have shown that two phenomena occur simultaneously upon doping: a partial oxidation of each stack and a decrease of the distance between the stacks (82,83).

The current theoretical interpretation of these results is that the important interaction between stacks must involve the dz^2-like molecular orbitals of the individual $Pt(CN)_4$ units, with the z-axis being the Pt back-bone. In the infinite chain the dz^2-like orbitals broaden into a dz^2-like band, the top of which is at the Fermi level. When electrons are depleted from this band by oxidizing dopants, a directional conductivity is created. Simplified band calculations that incorporate these ideas have been presented by Whangbo and Hoffman (84) and by Messmer and Salahub (85).

Interrante and Messmer (86) have presented nonrelativistic SCF-Xα-SW calculations for the monomer $Pt(CN)_4^{2-}$ and the staggered dimer $[Pt(CN)_4^{2-}]_2$. For the dimer they performed calculations for a Pt-Pt distance of 2.89Å, which corresponds to the one found in the bromine-doped compound KCP(Br), $K_2Pt(CN)_4Br_{0.3}\cdot 3H_2O$ (82). They found for the monomer that the dz^2-like orbital is 0.8 eV below the highest occupied molecular orbital (HOMO). For the dimer, however, one of the dz^2-like orbitals becomes the HOMO. They have not reported calculations for larger inter-stack separations, but they suggest that for larger separations, the HOMO may not be dz^2-like. Viswanath and coworkers (87) have performed Extended Huckel MO (88) calculations for the clusters $[Pt(CN)_4^{2-}]_n$ with n = 1, 2 and 3, for the purpose of interpreting optical data. Both eclipsed and staggered conformations have been used for Pt-Pt separations of 3.27 and 3.48Å. They found that for all cases the HOMO is dz^2-like.

Hoping to obtain a better understanding of the electronic structure of tetracyanoplatinate complexes, we have initiated relativistic SCF-X -DSW calculations for clusters up to the trimer (89). A representative MO diagram is shown in Figure 7.

Result for the monomer unit (or infinite Pt-Pt separation) calculated with a basis set equivalent to that used for the trimer is also shown for comparative purposes. We focus our attention on only two aspects of the results, namely, the position of the dz^2-like orbital and the nature of the HOMO. The nature of the HOMO changes with decreasing separation between stacks. While for the monomer and for the trimer at 3.5Å it is of $(d_\delta + d_\pi)$ character, for the trimer at 2.89Å it is clearly dz^2-like. Particularly intersesting is the examination of the $(d_\delta + d_\pi)$-like orbital at 3.5Å. Orbital

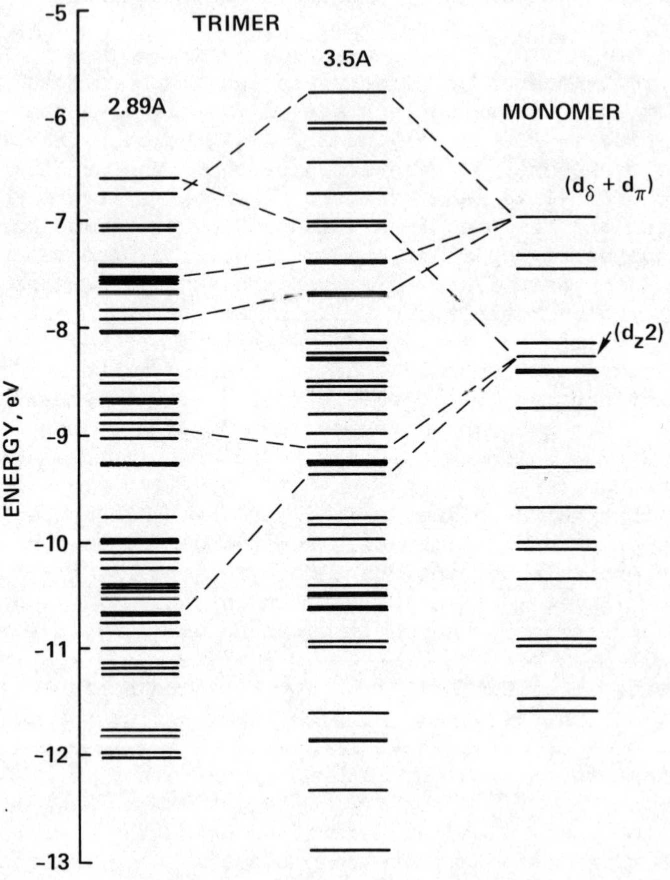

Fig. 7. Molecular orbital energy diagram for the cluster $[Pt(CN)_4{}^{2-}]_3$ at two different inter-stack separations, and for the monomer $Pt(CN)_4{}^{2-}$.

density contour maps show that all the electronic charge is localized at the central unit of the cluster. However, strong metal-ligand mixing is observed within the central unit. At 2.89Å, the HOMO is dz^2-like. This dz^2-like orbital is delocalized throughout the entire cluster, and is of antibonding character between Pt atoms. The dz^2 character at the central atom is 30.1%, and the total dz^2 character of both peripheral Pt atoms is 27.0%. In general this orbital is much more metal-like compared with the $(d_\delta + d_\pi)$ orbital. It is seen from Figure 7 that the energy level position of the dz^2-like orbitals depends critically on the separation between the stacks, and this is reasonable since they are concentrated along the Pt backbone.

These results lead to the following scenario for the insulator-conductor transition. The oxidizing dopant removes electrons from a $(d_\delta + d_\pi)$-like orbital of the insulator, generating an unstable, partially oxidized structure. This structures is stabilized by a process which results in a decrease of the inter-stack distance, accompanied by the reordering of the dz^2 and $(d_\delta + d_\pi)$ bands at the Fermi level, resulting then in the partial occupation of the dz^2-like band. The stabilization mechanism is suggested below.

Partial oxidation or removal of electrons from the highly localized $(d_\delta + d_\pi)$ band will create a heterogeneous distribution of charges along the $Pt(CN)_4$ units. This decrease in negative charges must in turn produce a decrease in the inter-stack electrostatic repulsion and consequently a shortening of the inter-stack distance. Hence, the antibonding dz^2-like orbital will be pushed up in energy, becoming the new valence band edge. The resulting valence band is partially filled as before. Two competing factors resulting from the partial oxidation of the dz^2-band are principally responsible for maintaining the equilibrium prescribed above. Since the dz^2 valence band is antibonding between neighboring Pt atoms, its partial oxidation results in a strengthening of the Pt-Pt bond. However, further bond-length contraction is prevented by the net positive charge created around each Pt atom.

To summarize, our results show that the oxidation process is responsible for drastic changes in the electronic structure of the insulator, particularly in the nature of the bands near the Fermi level. Based on the cluster model, we infer that at the valence band edge of the insulator, there exists a set of nonbonding states highly localized about each stack of the linear chain. Each of these states represents substantial metal-ligand mixings. Partial depopulation of these states leads to a structure where the partially filled valence band has antibonding states delocalized along the Pt backbone. These states have much less metal-ligand mixings, since they consist essentially of dz^2 character along the Pt backbone. Hence, electron transport occurs principally through the metal atoms.

IV. CONCLUSIONS

In this narrative, we hope to have demonstrated the utilities of the SCF-Xα-DSW method. Further development and applications of this method will most likely lead to even better understanding of the bonding trends and more concrete interpretation of experimental spectra for systems containing heavy elements. While it is unreasonable to expect precise quantitative information from scattered-wave calculations, or to argue that they can describe correctly all types of bonding, it does appear that the reliability of this method is sufficient to recommend its use for molecules and clusters too large or too heavy for high quality *ab initio* computations.

ACKNOWLEDGEMENTS

I would like to thank NASA-Ames Research Center for support of this work under Contracts NAS2-10187, 10735, and 10789. I am grateful to all whom I have collaborated with on relativistic calculations. Special thanks are extended to David Case, whose contributions to the SCF-Xα-DSW methodology have been indispensible.

REFERENCES

1. J. Korringa, *Physica* 13:392 (1947).
2. W. Kohn and N. Rostoker, *Phys. Rev.* 94:1111 (1954).
3. J.C. Slater, *Phys. Rev.* 51:846 (1937).
4. J.M. Ziman, *Proc. Phys. Soc.* (London) 86:337 (1965).
5. P. Lloyd, *Proc. Phys. Soc.* (London) 86:825 (1965).
6. J.C. Slater, *Phys. Rev.* 145:599 (1966).
7. K.H. Johnson, *Phys. Rev.* 150:429 (1966).
8. J.C. Slater, *J. Chem. Phys.* 43:S228 (1965).
9. K.H. Johnson, *J. Chem. Phys.* 45:3085 (1966).
10. K.H. Johnson, *Adv. Quantum Chem.* 7:143 (1973).
11. K.H. Johnson, *Ann. Rev. Phys. Chem.* 26:39 (1975).
12. K.H. Johnson, *C.R.C. Critical Reviews in Solid State and Materials Science* 7:101 (1978).
13. J.W.D. Connolly, p. 105, *in*: "Semiempirical Methods of Electronic Structure Calculation, Part A:Techniques," G.A. Segal, ed., Plenum, New York (1977).
14. P. Weinberger, K. Schwarz, p. 257, *in*: "International Review of Science, Physical Chemistry, Series Two, Vol. 1, Theoretical Chemistry ," A.D. Buckingham, C.A. Coulson, eds.,Butterworth, London (1975).
15. N. Rösch, p. 1, *in*: "Elctrons in Finite and Infinite Structures," P. Phariseau, L. Scheire, eds., Plenum, New York (1977).
16. R.P. Messmer, *Surf. Sci.* 106:225 (1981).
17. D.A. Case, *Ann. Rev. Phys. Chem.* (1982), in press.

18. J.L. Dehmer, D. Dill, p. 225, in: "Electron-Molecule and Photon-Molecule Collisions," T. Rescigno, V. McKoy, B. Schneider, eds., Plenum, New York (1979).
19. N.F. Lane, Rev. Mod. Phys. 52:29 (1980).
20. P. Pyykkö, Adv. Quantum Chem. 11:353 (1978).
21. Y. Onodera and M. Okazaki, J. Phys. Soc. Jpn. 21:1273 (1966).
22. D.S. Choo, Ph.D. dissertation, M.I.T. (1975).
23. R.D. Cowan, D.C. Griffin, J. Opt. Soc. Am. 66:1010 (1976).
24. D.D. Koelling, B.N. Harmon, J. Phys. C: Solid St. Phys. 10:3107 (1977).
25. W.V.M. Machado, L.G. Ferreira, Chem. Phys. Lett. 37:51 (1976).
26. J.H. Wood, A.M. Boring, Phys. Rev. B18:2701 (1978).
27. A.M. Boring, J.H. Wood, J. Chem. Phys. 71:32 (1979).
28. A.M. Boring, J.H. Wood, J. Chem. Phys. 71:392 (1979).
29. Y. Onodera and M. Okazaki, J. Phys. Soc. Jpn. 21:1273 (1966).
30. S. Takada, Prog. Theor. Phys. 36:224 (1966).
31. C.Y. Yang and S. Rabii, Phys. Rev. A12:362 (1975).
32. B.G. Cartling and D.M. Whitmore, Int. J. Quantum Chem. 10:393 (1976).
33. F. Rosicky, P. Weinberger, and F. Mark, J. Phys. B 9:2971 (1976).
34. C.Y. Yang, J. Chem. Phys. 68:2626 (1978).
35. C.Y. Yang, Ph.D. dissertation, University of Pennsylvania (1975).
36. C.Y. Yang, Chem. Phys. Lett. 41:588 (1976).
37. C.Y. Yang and S. Rabii, Int. J. Quantum Chem. S10:313 (1976).
38. R.P. Messmer, D.R. Salahub, K.H. Johnson, and C.Y. Yang, Chem. Phys. Lett. 51:84 (1977).
39. C.Y. Yang, K.H. Johnson, and J.A. Horsley, J. Chem. Phys. 68: 1001 (1978).
40. C.Y. Yang and S. Rabii, J. Chem. Phys. 69:2497 (1978).
41. D.A. Case and C.Y. Yang, Int. J. Quantum Chem. 18:1091 (1980).
42. D.A. Case and C.Y. Yang, J. Chem. Phys. 72:3443 (1980).
43. C.Y. Yang, H.L. Yu, and D.A. Case, Chem. Phys. Lett, 81:170 (1981).
44. C.Y. Yang and D.A. Case, Surf. Sci. 106:523 (1981).
45. C.Y. Yang, in: "Proceedings of Electrocatalysis Symposium, Minneapolis, May 1981," in press.
46. S. Rabii and C.Y. Yang, submitted to Chem. Phys. Lett.
47. J.C. Slater, Adv. Quantum Chem. 6:1 (1972).
48. For a review of the background formalisms, see, for example, M.E. Rose, in: "Relativistic Electron Theory," Wiley, New York (1961).
49. For a discussion of double groups, see, for example, E.P. Wigner, in: "Group Theory and its Applications to Quantum Mechanics of Atomic Spectra," Academic, New York (1959).
50. J.B. Diamond, Chem. Phys. Lett.20:63 (1973).
51. G. Herzberg, in: "Molecular Spectra and Molecular Structure, Vol. 1. Spectra of Diatomic Molecules," Van Nostrand, Princeton (1960).
52. R.S. Mulliken, J. Chem. Phys. 55:288 (1971).
53. M.S. Nakhmanson, V.I. Baranovskii and Tu.M. Zaitsev, Teor. i

Eksperim. Khim. 7:240 (1971).
54. J.C. Slater, J.B. Mann, T.M. Wilson and J.H. Wood, Phys. Rev. 184:672 (1969).
55. D.C. Frost, C.A. McDowell and D.A. Vroom, J. Chem. Phys. 46: 4255 (1967).
56. N. Rösch, W.G. Klemperer and K.H. Johnson, Chem. Phys. Lett. 23:149 (1973).
57. J.B. Danese, private communication.
58. E.J. Baerends and P. Ros, Chem. Phys. 8:412 (1975).
59. D.W. Turner, C. Baker, A.D. Baker and C.R. Brundle, in: "Molecular Photoelectron Spectroscopy," Wiley-Interscience, New York (1970).
60. P.J. Hay, W.R. Wadt, L.R. Kahn, R.C. Raffenetti, and D.C. Phillips, J. Chem. Phys. 71:1767 (1979).
61. D.D. Koelling, D.E. Ellis, and R.J. Bartlett, J. Chem. Phys. 65:3331 (1976).
62. D.A.Case and M. Karplus, J. Am. Chem. Soc. 99:6182 (1977).
63. D.A. Case, B.H. Huynh, and M. Karplus, J. Am. Chem. Soc. 101: 4433 (1979), and references therein.
64. R.P. Messmer, Vol. 8, Chapter 6, in: "Semiempirical Methods of Electronic Structure Calculations," G.A. Segal, ed., Plenum, New York (1977).
65. G.A. Ozin, Catal. Rev. - Sci. Eng. 16:191 (1977).
66. H.L. Yu, J. Chem. Phys. 69:1755 (1978).
67. A. Rosén, P. Grundevik, and T. Morovic, Surf. Sci. 95:477 (1980).
68. A.Rosén, E.J. Baerends, and D.F. Ellis, Surf. Sci. 82:139 (1979).
69. J.W. Davenport, Phys. Rev. Lett. 36:945 (1976); J. Vac. Sci. Tech. 15:433 (1978).
70. F.W. Kutzler, C.R. Natoli, D.K. Misemer, S. Doniach, and K.O. Hodgson, J. Chem. Phys. 73:3274 (1980).
71. F.W. Kutzler, R.A. Scott, J.M. Berg, K.O. Hodgson, S. Doniach, S.P. Cramer, and C.H. Chang, J.Am. Chem. Soc. 103:6083 (1981).
72. A.C. Balazs and K.H. Johnson, Surf. Sci. 114:197 (1982).
73. C.Y. Yang and D.A Case, APS March Meeting, Dallas (1982).
74. G. Apai, P.S. Wehner, R.S. Williams, J. Stohr, and D.A. Shirley, Phys. Rev. Lett. 37:1497 (1976).
75. P.R. Norton, J.W. Goodale, and E.B. Selkirk, Surf. Sci. 83:189 (1979).
76. J.N. Miller, D.T. Ling, I. Lindau, P.M. Stefan, and W.E. Spicer, Phys. Rev. Lett. 38:1419 (1977).
77. H.P. Bonzel and T.E. Fischer, Surf. Sci. 51:213 (1975).
78. J.C. Slater, J.B. Mann, T.M. Wilson and J.T. Wood, Phys. Rev. 84:672 (1969).
79. N. Rösch and D. Menzel, Chem. Phys. 13:243 (1976).
80. I.P. Batra and P.S. Bagus, Solid State Commun. 16:1097 (1975).
81. K. Hermann and P.S. Bagus, Phys. Rev. B16:4195 (1977).
82. J.S. Miller and A.J. Epstein, Progress in Inorg. Chem. 20:1 (1976).
83. J.M. Williams and A.J. Schultz, p. 337, in: "Molecular Metals," W.E. Hatfield, ed., Plenum, New York (1979).

84. M. Whangbo and R. Hoffmann, J. Amer. Chem. Soc. 100:6093 (1978).
85. R.P. Messmer and D.R. Salahub, Phys. Rev. Lett. 35:533 (1975).
86. L.V. Interrante and R.P. Messmer, Chem. Phys. Lett. 26:225 (1974); p.382, in: "Extended Interactions Between Metal Ions," L.V. Interrante, ed., American Chemical Society Symposium Series No. 5, American Chemical Society, Washington, D.C. (1974).
87. A.K. Viswanath, M.B. Krogh-Jespersen, J. Vetuskey, C. Baber, W.D. Ellenson, and H.H. Patterson, Molecular Phys. 42:1431 (1981).
88. R. Hoffmann, J. Chem. Phys. 39:1397 (1963).
89. J.P. Lopez, C.Y. Yang, D.A. Case, APS March Meeting, Dallas (1982); manuscript in preparation.
90. U. Gelius, E. Basilier, S. Svensson, T. Bergmark, and K. Siegbahn, J. Electron Spec. 2:405 (1973).

FULLY RELATIVISTIC EFFECTIVE CORE POTENTIALS (FRECP)

Yasuyuki Ishikawa[+] and G.L. Malli*

Department of Chemistry
*Department of Chemistry and Theoretical Sciences Institute
Simon Fraser University, Burnaby, B.C.
Canada V5A 1S6

ABSTRACT

A method is developed for obtaining fully relativistic effective core potentials from numerical Dirac-Fock self-consistent-field calculations. Analytical forms for the effective core potentials are derived for Ca and Tl, and the results of valence-only Dirac-Fock calculations are presented.

INTRODUCTION

During the last decade, the Dirac-Fock-Roothaan(DFR) SCF theories have been developed[1-4], and calculations on atoms[1,2] and molecules[4,5] containing light atoms have been reported. A major difficulty with the DFR SCF theories is that the schemes do not guarantee upperbounds to the true energy. During the NATO Advanced Study Institute, Davidson et al.[6] proposed a simple basis set expansion scheme which guarantees upperbound to the true energy, and the scheme was verified by test calculations on the Be atom.[6]

Though the relativistic SCF schemes are especially useful for heavy atoms and for molecules containing heavy atoms, their direct applications to such systems are very expensive in computer time because of the very large basis set necessary to obtain reasonable results. Indeed, even in the numerical Dirac-Fock (DF) one-center expansion calculations[7], the numerical calculations are very costly in computer time because of the large number of occupied core spinors in heavy elements. A fully relativistic Effective Core Potential (ECP) method would be especially useful, therefore, if it is incorporated into the DFR and DF one-center calculations.

Recently much work has been done on quasi-relativistic ECP formulations which incorporate relativity as a perturbation to the non-relativistic Schroedinger Hamiltonian.[8-14] In practical applications, these quasi-relativistic ECP schemes seem to be able to reproduce major relativistic effects in heavy atomic systems.[8,10] The first steps towards developing rigorous quasi-relativistic ECP theories which incorporate relativistic effects as lowest order perturbations to a Schroedinger-like Hamiltonian, were taken recently by Pyper.[15] He showed that the validity of straightforward incorporation of the ECP which represents the mass-velocity and Darwin terms, etc., into the non-relativistic Schroedinger equation is theoretically questionable. Our recent study on a series of rare-gas atoms also showed that the lowest order perturbation treatment is unable to account for relativistic energies for Kr and heavier elements.[16]

In the present paper, we report a formulation and the applications of our fully relativistic ECP method for use in the DF or DFR SCF calculations on atoms and molecules. This method is based directly on a frozen core approximation of fully relativistic DF SCF scheme: we constructed the ECP so that it reproduces both the energies and the shapes of the outer region of valence spinors obtained from all-electron atomic DF SCF calculations.

In the next section, we discuss our fully relativistic ECP scheme. Applications of the ECP scheme to Calcium and Thallium will be given in the third section, and the results for various excited states calculations, using the valence-only DF scheme, will be compared with those obtained from the full DF calculations.

FULLY RELATIVISTIC EFFECTIVE CORE POTENTIALS

Fock, Wesselov, and Petraschen[17] first formulated the approach for core-valence separation of electronic groups in atoms and molecules. They treated the special case of two electrons outside a core electronic group. Their approach can be generalized to relativistic many electron systems in the following way.

For an atomic system with n_c core electrons and n_v valence electrons, an approximate total wave function can be written as:

$$\Psi = MA_p\{\ \Phi_c\Phi_v\ \}$$

where the core and the valence wave functions Φ_c, Φ_v consist of Slater determinants of four component Dirac-Fock atomic spinors:

$$\phi_{n\kappa m} = \frac{1}{r}\begin{bmatrix} P_{n\kappa}(r)\chi_{\kappa m}(\theta,\varphi) \\ Q_{n\kappa}(r)\chi_{-\kappa m}(\theta,\varphi) \end{bmatrix}$$

where

$$\chi_{\kappa m}(\theta,\varphi) = \sum_{\sigma=\pm\frac{1}{2}} C(\ell\tfrac{1}{2}j;m-\sigma,\sigma)Y_{\ell}^{m-\sigma}(\theta,\varphi)\phi_{\frac{1}{2}}^{\sigma} .$$

$Y_{\ell}^{m-\sigma}$ is a spherical harmonic. $\phi_{\frac{1}{2}}^{\sigma}$ are Pauli spinors and are given as: $\phi_{\frac{1}{2}}^{\frac{1}{2}} = \binom{1}{0}$, and $\phi_{\frac{1}{2}}^{-\frac{1}{2}} = \binom{0}{1}$. $C(\ell\tfrac{1}{2}j;m-\sigma,\sigma)$ are the Clebsch-Gordan coefficients.

If we assume that all the core and valence spinors are mutually orthogonal, the total energy of the system can be written in the form:

$$E = \langle\Phi_v|H_R^v|\Phi_v\rangle + E_{core}$$

where E_{core} takes the standard form for a closed shell single determinant wave function:

$$E_{core} = 2\sum_c \langle\phi_c|\hat{h}_D|\phi_c\rangle + \sum_{c,c'}(2J_{cc'} - K_{cc'}) .$$

The relativistic valence Hamiltonian H_R^v in atomic units (a.u.) is given as:

$$H_R^v = \sum_{v=1}^{n_v}\{\hat{h}_D(v) + \sum_c (J_c(v) - K_c(v))\} + \tfrac{1}{2}\sum_{v\neq v'} 1/r_{vv'} . \qquad (1)$$

In (1), the Dirac Hamiltonian h_D has the form:

$$h_D = c\alpha\cdot p + \beta' c^2 - Z/r$$

where Z is the nuclear charge, and α is given in terms of Pauli matrix σ^p as:

$$\alpha = \begin{pmatrix} 0 & \sigma^p \\ \sigma^p & 0 \end{pmatrix} \quad \text{and} \quad \beta' = \begin{pmatrix} 0 & 0 \\ 0 & -2I \end{pmatrix} .$$

In our ECP scheme, the relativistic valence Hamiltonian H_R^v in (1) is replaced by the model relativistic Hamiltonian H_R^m, in which all the effects of core spinors on the valence spinors, and core-valence orthogonality constraints are folded into the ECP:

$$H_R^m = \sum_v \{c\alpha_v p_v + \beta_v' c^2 - Z_{eff}/r_v + U^{EP}_{(v)}\} + \tfrac{1}{2}\sum_{v\neq v'} 1/r_{vv'} , \qquad (2)$$

where U^{EP} is the effective potential (EP) given in a 4 x 4 matrix form, and $Z_{eff} = Z - n_c$.

We assume that radial parts of the EP are the same for all the spinors having higher angular momentum quantum numbers than are present in the core. Using the closure property of the projection operator, we can then write the U^{EP} in the matrix form:

$$U^{EP} = \begin{bmatrix} U_K^P(r) & & & 0 \\ & U_K^P(r) & & \\ & & U_{-K}^Q(r) & \\ 0 & & & U_{-K}^Q(r) \end{bmatrix} + \begin{bmatrix} \Sigma U_{\kappa-K}^P(r)\,|\kappa m\rangle\langle\kappa m| & 0 \\ 0 & \Sigma U_{-\kappa+K}^Q(r)\,|-\kappa m\rangle\langle-\kappa m| \end{bmatrix} \tag{3}$$

The validity of this approximation will be justified by the success of the EP calculations in reproducing the full DF results in its applications to molecules.

In (3), $|\kappa m\rangle$ is $\chi_{\kappa m}(\theta,\varphi)$, the two-component angular function given previously, and

$$\begin{aligned} U_{\kappa-K}^P(r) &= U_\kappa^P(r) - U_K^P(r) \\ U_{-\kappa+K}^Q(r) &= U_{-\kappa}^Q(r) - U_{-K}^Q(r) \end{aligned} \tag{4}$$

The summations in (3) run over all the quantum numbers κ and m of the atomic spinors that are present in the core. If L is the smallest quantum number of the spinor which is not present in the core, then the K in (3) and (4) is defined by:

$$K = J + 1/2 \qquad \text{if } J = L - 1/2$$

$$K = -(J + 1/2) \quad \text{if } J = L + 1/2 \;.$$

The Hamiltonian in (2), together with the EP in (3), leads to the differential equations for the radial valence pseudospinors:

$$\frac{d}{dr}\begin{bmatrix} P_A^v(r) \\ Q_A^v(r) \end{bmatrix} = \begin{bmatrix} -\kappa/r & (2/\alpha) + \alpha\left[\varepsilon_A^v - V_A^Q(r)\right] \\ -\alpha\left[\varepsilon_A^v - V_A^P(r)\right] & \kappa/r \end{bmatrix} \begin{bmatrix} P_A^v(r) \\ Q_A^v(r) \end{bmatrix}$$

$$+\begin{bmatrix} X_A^Q(r) \\ \\ X_A^P(r) \end{bmatrix} , \qquad (5)$$

where α is the fine-structure constant. $P_A^V(r)$ and $Q_A^V(r)$ are the upper and lower components of the valence pseudo-spinor having quantum number κ. The $V_{A(r)}^{(P\ or\ Q)}$ which represents the sum of the EP due to the core electrons and the coulomb interactions among the valence electrons has the form:

$$V_A^P(r) = -Z_{eff}/r + U_K^P(r) + U_{\kappa-K}^P(r) + \sum_{B,k} a^k(A;B)Y^k(A;B)$$

$$V_A^Q(r) = -Z_{eff}/r + U_{-K}^Q(r) + U_{-\kappa+K}^Q(r) + \sum_{B,k} a^k(A;B)Y^k(A;B)$$

The exchange term, $X_{(r)}^{(P\ or\ Q)}$ is defined by[18,19]:

$$(r/\alpha)X_A^{(P\ or\ Q)}(r) = \sum_{B\neq A} \{ \varepsilon_{AB}^v + \sum_k b^k(A;B)Y^k(A;B) \}P_B^v \ (\text{or } Q_B^v)$$

$$+ \sum c^k(A,B;C,D)Y^k(C,D)P_B^v(\text{or } Q_B^v) ,$$

where $\{\varepsilon_{AB}^v\}$ are the off-diagonal Lagrange multipliers that arise due to the orthogonality constraints of occupied valence pseudo-spinors. The a^k, b^k and c^k are angular coefficients, and $Y^k(A;B)$ is defined by[18,19]:

$$Y^k(A;B) = 1/r^k \int_0^r F(S)S^k ds + r^{k+1} \int_r^\infty F(S)\frac{ds}{S^{k+1}} , \quad \text{and}$$

$$F(s) = P_A^v(S)P_B^v(S) + Q_A^v(S)Q_B^v(S)$$

Equation (5) has the desirable form for the valence-only numerical DF calculations which can easily be incorporated into the existing programs.

EFFECTIVE POTENTIAL PARAMETRIZATION

We solve the valence-only numerical DF SCF equation given in (5) where the components of U^{EP} in (3) are expressed analytically in a few terms of the form[8-10]:

$$\sum_i C_{ki} r^{n_i} \exp(-\zeta_{ki} r^2)$$

We then determine the parameters $\{C_{\kappa i}\}$ and $\{\zeta_{\kappa i}\}$ so that the valence one-electron energy ε_A^v and the outer region of the radial valence pseudo-spinors $P_A^V(r)$ and $Q_A^V(r)$ reproduce the corresponding valence energy and radial parts of the spinors of the all-electron single-configuration DF calculations. This technique has been successfully applied by Huzinaga et al. in their non-relativistic EP scheme[20].

In the non-relativistic and quasi-relativistic ECP schemes, many workers have used the Phillips-Kleinman(PK) pseudo-orbital transformation method, which was originally proposed for pseudo-potential study of solids, in order to generate the EP's.

This procedure introduces negative contributions (a long range tail) to the EP which arises due to the difference between the normalized PK and the Hartree-Fock orbitals in the valence region[10]. The procedure which Christiansen, Lee, and Pitzer have recently proposed[10] is also based on the frozen core approximation and avoids the PK pseudo-orbital transformation to eliminate the long range tail in their EP's. Their scheme is capable of precisely reproducing the outer region of the all-electron valence orbitals. However, their method has so far been applied to generate the EP's only from the nodeless pseudo-orbitals.

In the present study, we have chosen the same analytical EP for both the upper and lower components of U_K and $U_{\kappa-K}$'s which are analytically expanded as:

$$U^P_{\kappa-K}(r) = B_{\kappa 1}\exp(-\beta_{\kappa 1} r^2) + B_{\kappa 2}\exp(-\beta_{\kappa 2} r^2) + B_{\kappa 3}\exp(-\beta_{\kappa 3} r^2)/r^2$$

$$U^P_K(r) = U^Q_{-K}(r) = A\exp(-\alpha r^2)/r \qquad (6)$$

and

$$U^P_{\kappa-K}(r) = U^Q_{-\kappa+K}(r) .$$

In equation (4), this leads to:

$$V^P_A(r) = V^Q_A(r) \qquad (7)$$

Further, we have chosen the upper components of the lowest valence pseudo-spinors $P_A^V(r)$ to be nodeless. We did not impose such a condition on the lower component $Q_A^V(r)$, since we have assumed the equality in (7), which cannot, in general, eliminate all the nodes simultaneously from upper and lower components.

Whether or not the pseudo-orbitals should possess nodes to reproduce the reliable results, especially for the multiplet states, will be examined in the future.

In order to solve the equation (5) using the form of $U_K^P(r)$, etc. in (6) and (7), we have modified the numerical DF program of Desclaux[18]. The proposed scheme has been applied to a series of atomic systems[21]. In this paper, we present the results on Ca and Tl as test cases for the applications of our fully relativistic EP scheme. Full DF reference calculations were performed with point nucleus approximation; and for openshell states, we have optimized average energy of configuration[18] throughout this work.

(a) Calcium

Calcium is among the lighter atoms, and its relativistic effects are known to be small. We have chosen this system to compare our results with the non-relativistic and quasi-relativistic ECP studies available in the literature. In the present study, Ca has been treated as an atom comprised of two valence electrons with the atomic core of Ar. The parameters A, α, $\{B_{\kappa i}\}$ and $\{\beta_{\kappa i}\}$ for each symmetry of the atomic core are determined so that they reproduce the valence-electron energy ε_κ and the outer region of the corresponding DF spinor of its alkali-like ion, Ca^+. Table 1 shows the EP parameters determined in such a manner.

In Table II, we give, for each symmetry, the r^n expectation values ($n>0$) and the one-electron energy for the valence electron of Ca^+ for our ECP and full DF calculations. The agreement with the full DF calculations is quite clearly satisfactory for all the symmetry species. This indicates that the outer region of the valence pseudo-spinors correctly reproduces the corresponding region of valence spinors obtained from the full DF calculations.

Simons[22] constructed Fues-type EP's in the non-relativistic scheme, and Hafner et al. have reported quasi-relativistic ECP's for this system[14]. They obtained their EP parameters so that the valence-only calculations reproduce the experimental valence electron spectra.

We have examined these non-relativistic and quasi-relativistic EP's to see if these give reasonable results when used in the fully relativistic ECP scheme. In order to do this, we have simply replaced U^{EP} in (6) with their ECP's and solved the valence-only DF

Table I. Effective Potential Parameters for Ca as Defined in (6)

i	$B_{s1/2,i}$	$\beta_{s1/2,i}$	$B_{p1/2,i}$	$\beta_{p1/2,i}$	$B_{p3/2,i}$	$\beta_{p3/2,i}$	A	α
1	9.347	0.752	10.632	0.719	10.634	0.719		
2	0.040	0.250	0.098	0.170	0.095	0.170	-4.30	0.593
3	3.075	0.520	2.298	0.490	2.950	0.490		

Table II. One-Electron Energy and r^n Expectation Values (a.u.) for the Valence Electron of Ca^+ from our E.P., Full DF, Simons' EP (SEP) and Hanfer's EP (HEP) Schemes.

	$4s_{1/2}$				$4p_{1/2}$				$4p_{3/2}$			
	EP	DF	SEP	HEP[a]	EP	DF	SEP	HEP	EP	DF	SEP	HEP
$-\varepsilon_\kappa$	0.4175	0.4179	0.4362	-	0.3102	0.3102	0.3209	0.3215	0.3089	0.3093	0.3209	0.3205
$<r>$	3.730	3.714	2.828	-	4.555	4.536	3.741	4.407	4.582	4.551	3.741	4.423
$<r^2>$	15.52	15.55	9.512	-	23.13	23.16	16.33	21.67	23.38	23.31	16.33	21.82
$<r^4>$	363.6	366.9	164.4	-	805.3	811.5	457.2	708.6	821.1	821.7	457.2	717.8

[a] $4S_{1/2}$ state calculation with this potential diverged.

Table III. One-Electron Energy and r^n Expectation Values for Neutral Ca from our EP and Full DF Calculations[a] (a.u.).

	$(4s_{1/2})^2$		$(4s_{1/2})^1(4p_{1/2})^1$				$(4s_{1/2})^1(4p_{3/2})^1$			
	$4s_{1/2}$		$4s_{1/2}$		$4p_{1/2}$		$4s_{1/2}$		$4p_{3/2}$	
	EP	DF	EP	DF	EP	DF	EP	DF	EP	DF
$-\varepsilon_\kappa$	0.1953	0.1963	0.2456	0.2458	0.1317	0.1316	0.2462	0.2462	0.1309	0.1309
$\langle r \rangle$	4.223	4.202	4.017	4.006	5.426	5.424	4.014	4.005	5.458	5.446
$\langle r^2 \rangle$	20.36	20.30	18.27	18.33	33.70	33.84	18.25	18.31	34.08	34.12
$\langle r^4 \rangle$	685.4	682.0	536.3	540.5	1882.	1899.	534.7	539.5	1921.	1928.

[a] Average energy of configuration is optimized as in Ref. 19.

equation given in (5). The energies and the r^n expectation values obtained from these ECP calculations are also listed in Table II.

Both Simons' and Hafner's EP's give reasonable orbital energies but the r^n expectation values obtained with Simons' EP are considerably smaller than those obtained from other EP and all-electron DF calculations. However, Hafner's EP's give consistently smaller, but still reasonable r^n expectation values. The r^n expectation values obtained from his ECP's are also consistently smaller for other systems studied, indicating that both the non-relativistic and the quasi-relativistic ECP's are too attractive in the valence region. As reported elsewhere,[23] the preliminary quasi-relativistic molecular calculations using Hafner's ECP's give bond lengths approximately 0.5 a.u. too short, compared with the experimental values for many heavy atom molecules. This also suggests that Hafner's EP's are too attractive in the valence region.

The quasi-relativistic ECP scheme of Lee et al.[9] is similar to ours in the sense that they constructed the EP to reproduce the full DF results. Since the analytic expressions of the EP's for Xe and Au are available,[9] we have examined them to see if their EP can reproduce reasonable results when used in the full DF calculations. The r^n expectation values obtained show some deviations, but this may result from the use of the Phillips-Kleinman scheme to generate their ECP's.

Using the ECP parameters in Table I determined for the Ca^+ system, we have also performed valence-only DF calculations on the various valence states of neutral Ca.

In Table III, the one-electron energies and r^n expectation values for the valence pseudo-spinors are given, as well as those obtained from the full DF calculations. The r^n expectation values calculated from the pseudo-spinors agree fairly well with the values from all-electron DF calculations. For the two-valence electron system, the present ECP scheme clearly generates valence pseudo-spinors whose outer region reproduces the corresponding region of the valence spinors obtained from all-electron DF calculations.

Fig. 1 and 2 show the radial parts of the large and small components obtained from the ECP calculation and from the full DF calculation on the $(4s1/2)^1(4p1/2)^1$ state of Ca.

The radial part of the large component of the pseudo-spinor is nodeless as expected, but that of the small component possesses one node. Later, this feature will be discussed in more detail.

(b) Thallium

The relativistic effects in Tl atom have been thoroughly studied

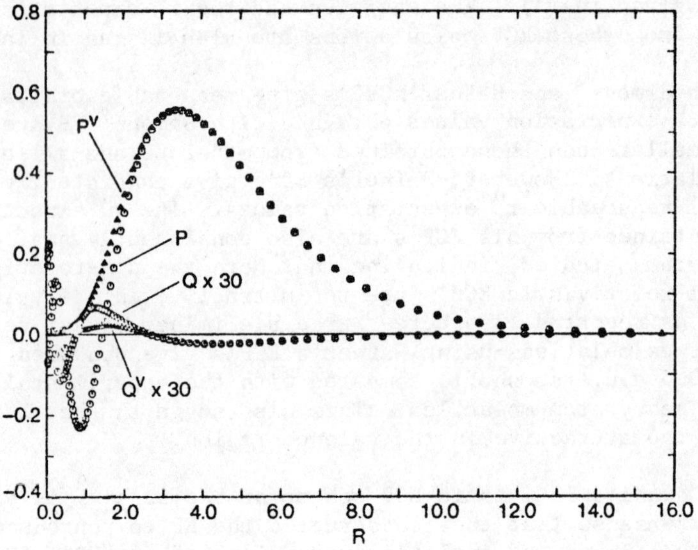

FIG. 1. Radial components of 4s1/2 pseudospinor (Δ) and DF spinor (O) of Ca. P^V and Q^V are the large and the small component of pseudospinor; P and Q are the large and the small component of DF spinor.

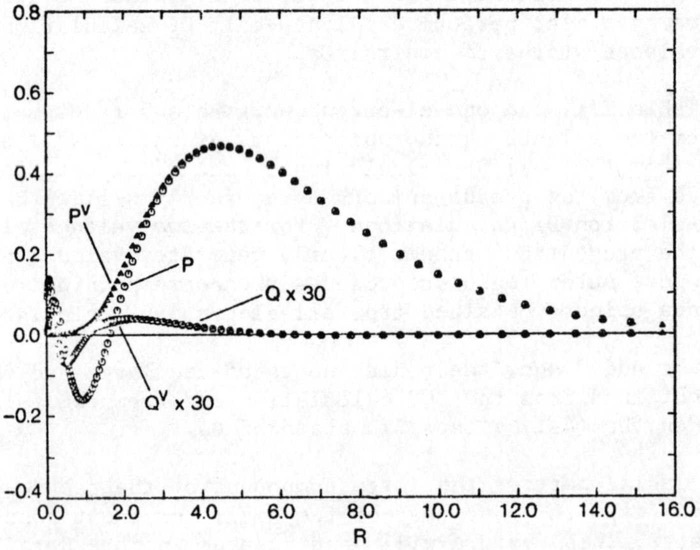

FIG. 2. Radial component of 4p 1/2 pseudospinor (Δ) and DF spinor (⊙) of Ca. P^V and Q^V are the large and the small component of pseudospinor; P and Q are the large and the small component of DF spinor.

by Grant et al.[24] using numerical DF scheme. The TlH molecule has been studied by Pyykko et al. using one-center DF scheme,[25] and by Pitzer et al.[10] using quasi-relativistic ECP scheme. In the present paper, Tl is treated as an atom comprised of three valence electrons having the core of Xe $4f^{14}5d^{10}$. The "non-frozen core" effects increase for such a heavy atom system, and the EP's constructed for the one-valence electron Tl^{2+} system gave rather poor results for the Calculations on the Tl neutral system[26]. In their non-relativistic EP scheme, Kahn et al. have also noted that the EP generated from highly ionized ions may not give satisfactory results when used in the calculations on its neutral atom and molecules. In the present study, the EP were determined directly from the neutral Tl, which consists of three valence electrons. First, we have determined the parameters A, α, $B_{s1/2,i}$ and $\beta_{s1/2,i}$ from the calculations on configuration of Tl(Core) $(6s\,1/2)^2(5g\,7/2)^1$. With these parameters fixed, the $\{B_{\kappa i}\}$ and $\{\beta_{\kappa i}\}$ for p 1/2, 3/2 potentials, etc., were determined from the calculations on electronic configurations Tl(Core)$(6s\,1/2)^2(6p\,1/2, 3/2)^1$, etc.

Table IV gives the EP parameters determined in such a manner. In Table V, we list reference configurations used to determine the EP parameters; and for each configurations, we also present $\{\varepsilon_\kappa\}$ and $\{\langle r^n\rangle\}$ for the valence spinors obtained from the EP's and the full DF calculations. It can be seen that the r^n and energy expectation values from the EP calculations reproduce those of the all-electron DF calculations satisfactorily, except for the 6s 1/2 pseudo-spinors, where slight deviations occur. The present scheme gives good results on Ca, but the results for Tl are less accurate. This indicates possible difficulties when the non-frozen core effects increase for heavy atoms. For such heavier systems as Tl, it may be necessary to include the 5d 3/2, 5/2 spinors in the valence shell to obtain better agreement.

In the full DF calculations, we have observed that the energy and the r^n expectation values for the 5f 5/2 and 5f 7/2 spinors are identical for Tl. Thus, the EP's for the f 5/2 and f 7/2 are also identical.

In order to test our EP's, calculations have also been performed for a few other states of Tl and Tl^+. Table VI shows the results of the EP's, as well as the full DF calculations; and one can see that the EP scheme satisfactorily reproduces the valence electron energies and the r^n expectation values also for these states.

In fig. 3 and 4, the radial components of 6s1/2 and 6p1/2 pseudo-spinors obtained from the EP calculations on the (Core)$(6s1/2)^2(6p1/2)^1$ state of Tl are plotted together with the radial parts of the spinors from full DF calculations. One can see that the essential profiles of the outer regions of large and small components have been satisfactorily reproduced. The large components of the pseudo-spinors

Table IV. Effective Core Potential Parameters for Tl.

i	$B_{s1/2,i}$	$\beta_{s1/2,i}$	$B_{p1/2,i}$	$\beta_{p1/2,i}$	$B_{p3/2,i}$	$\beta_{p3/2,i}$	$B_{d3/2,i}$	$\beta_{d3/2,i}$	$B_{d5/2,i}$	$\beta_{d5/2,i}$
1	141.2	2.514	61.30	1.671	62.60	1.490	55.40	0.890	56.20	0.842
2	0.027	0.132	0.023	0.140	0.015	0.142	0.010	0.141	0.008	0.140
3	2.96	0.468	1.402	0.401	1.580	0.375	1.120	0.371	1.120	0.371

i	$B_{f5/2,i}$	$\beta_{f5/2,i}$	$B_{f7/2,i}$	$\beta_{f7/2,i}$	A	α
1	4.00	0.850	4.00	0.850		
2	0.009	0.140	0.009	0.140	-6.450	0.520
3	0.600	0.550	0.600	0.550		

Table V. Valence Electron Configurations of Tl used to Determine the EP, and the Expectation Values for the Spinors Obtained from our EP and Full DF Schemes (a.u.).

	$(6s1/2)^2(6p1/2)^1$				$(6s1/2)^2(6p3/2)^1$				$(6s1/2)^2(6d3/2)^1$			
	6s1/2		6p1/2		6s1/2		6p3/2		6s1/2		6d3/2	
	EP	DF	EP	DF	EP	DF	EP	DF	EP	DF	EP	DF
$-\varepsilon_\kappa$	0.4335	0.4380	0.2123	0.2134	0.4529	0.4554	0.1752	0.1759	0.5796	0.5796	0.0557	0.0557
$\langle r\rangle$	2.627	2.590	3.511	3.502	2.607	2.572	4.032	4.021	2.512	2.475	10.48	10.47
$\langle r^2\rangle$	7.763	7.696	14.29	14.35	7.631	7.585	18.88	18.91	6.998	6.952	125.5	125.4
$\langle r^4\rangle$	97.31	97.72	361.8	365.5	93.57	94.64	630.4	633.2	74.54	75.77	2.528×10^4	2.526×10^4

Table V (Cont'd)

	$(6s1/2)^2(6d5/2)^1$				$(6s1/2)^2(5f5/2)^1$				$(6s1/2)^2(5g7/2)^1$			
	6s1/2		6d5/2		6s1/2		5f5/2		6s1/2		5g7/2	
	EP	DF	EP	DF	EP	DF	EP	DF	EP	DF	EP	DF
$-\varepsilon_\kappa$	0.5806	0.5805	0.0555	0.0555	0.6272	0.6273	0.0313	0.313	0.6498	0.6499	0.0200	0.0200
$\langle r \rangle$	2.511	2.474	10.55	10.55	2.508	2.471	17.98	17.97	2.508	2.471	27.50	27.50
$\langle r^2 \rangle$	6.997	6.951	127.0	127.1	6.976	6.927	359.2	359.0	6.975	6.927	825.0	824.9
$\langle r^4 \rangle$	74.49	75.72	2.580×10^4	2.586×10^4	73.78	74.94	1.894×10^5	1.892×10^5	73.76	74.92	9.384×10^5	9.383×10^5

Table VI. The Expectation Values Determined from our EP and Full DF Calculations of Tl and Tl^+ (a.u.).

	Tl $(6s1/2)^1(6p1/2)^1(6p3/2)^1$						Tl^+ $(6s1/2)^2$	
	6s1/2		6p1/2		6p3/2		6s1/2	
	EP	DF	EP	DF	EP	DF	EP	DF
$-\varepsilon_\kappa$	0.5313	0.5280	0.2408	0.2395	0.1817	0.1826	0.6898	0.6899
$\langle r \rangle$	2.546	2.518	3.344	3.351	3.957	3.947	2.508	2.471
$\langle r^2 \rangle$	7.234	7.237	12.85	13.05	18.14	18.17	6.975	6.927
$\langle r^4 \rangle$	82.11	84.53	285.0	295.8	578.0	581.1	73.76	74.92

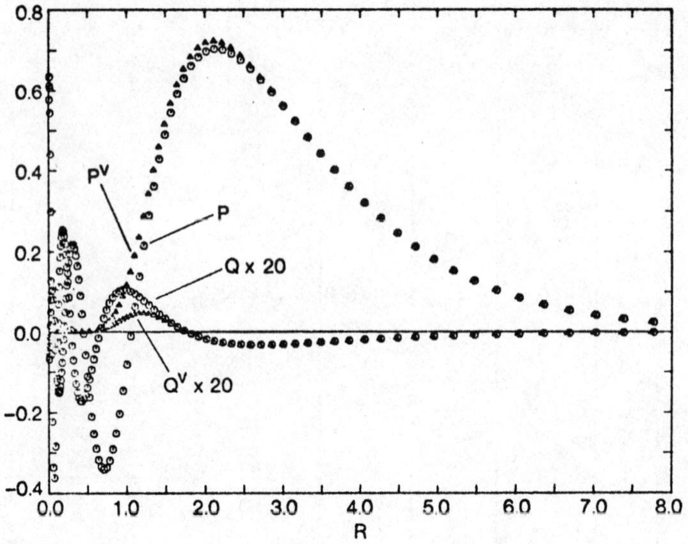

FIG. 3. Radial components of 6s1/2pseudospinor (Δ) and DF spinor (Θ) of Tl. P^V and Q^V are the large and the small component of pseudospinor; P and Q are the large and the small component of DF spinor.

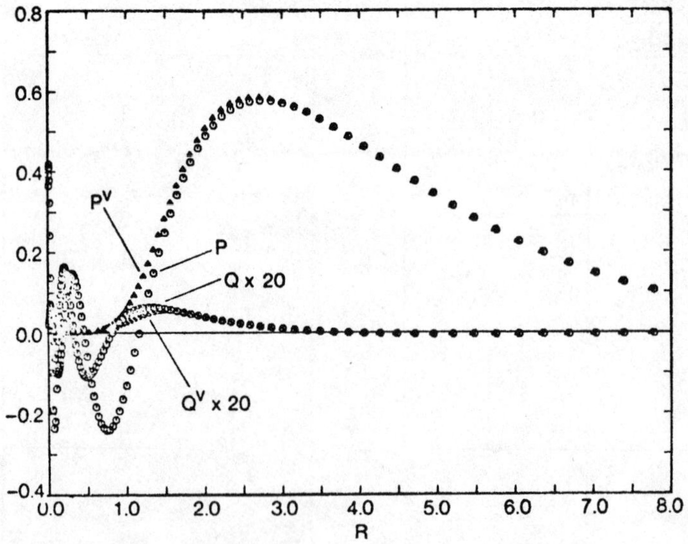

FIG. 4. Radial components of 6p1/2 pseudospinor (Δ) and DF spinor (Θ) of Tl. P^V and Q^V are the large and the small component of pseudospinor; P and Q are the large and the small component of DF spinor.

are nodeless, while the small components possess one node. We have also found that there occurs one node each in the small components of Ca and other systems. This results from the assumption of a common EP for the large and small components for each spinor, and constraint that only the large component of the lowest valence pseudo-spinor be nodeless. The presence of a node in the small component would require more basic functions in order to represent it accurately in the DFR SCF calculations on atoms and molecules.

In general, however, it may not be possible to eliminate all the nodes from the large and small components simultaneously.

ACKNOWLEDGMENTS

One of the authors (Y.I.) thanks Professors J. Podrasky and H. Jacoby for comments on the manuscript.

---

*This work was supported in part by the Natural Sciences and Engineering Research Council (NSERC) of Canada through grant no. A3598.

+Present address: Department of Chemistry
University of Wisconsin
Marathon County Center
Wausau, Wisconsin 54401
U.S.A.

References

1. Y.K. Kim, Phys. Rev. 154, 17 (1967).
2. T. Kagawa, Phys. Rev. A12, 2245 (1975).
3. G.L. Malli and J. Oreg, J. Chem. Phys. 63, 830 (1975).
4. O. Matsuoka, N. Suzuki, T. Aoyama and G. Malli, J. Chem. Phys. 73, 1320 (1980).
5. F. Mark and F. Rosicky, Chem. Phys. Lett. 74, 562 (1980).
6. E. R. Davidson, Y. Ishikawa and G. L. Malli, presented during the panel discussion of this NATO ASI.
7. P. Pyykko and J. P. Desclaux, Acc. Chem. Res. 12, 276 (1979).
8. L. R. Kahn, P. J. Hay and R. D. Cowan, J. Chem. Phys. 68, 2386 (1978).
9. Y.S. Lee, W.C. Ermler and K.S. Pitzer, J. Chem. Phys. 67, 5861 (1977).
10. P.A. Christiansen and K.S. Pitzer, J. Chem. Phys. 73, 5160 (1980); P.A. Christiansen, Y.S. Lee and K.S. Pitzer, J. Chem. Phys. 71, 4445 (1979).
11. G. Das and A.C. Wahl, J. Chem. Phys. 69, 53 (1978).
12. P.J. Hay, W.R. Wadt, L.R. Kahn and F.W. Bobrowicz, J. Chem. Phys. 69, 984 (1978).
13. S.N. Datta, C.S. Ewig and J.R. Van Wazer, Chem. Phys. Lett. 57, 83 (1978).

14. P. Hafner and W.H.E. Schwarz, J. Phys. B11, 217 (1978).
15. N.C. Pyper, Mol. Phys. 39, 1327 (1980).
16. E.R. Davidson, Y. Ishikawa, and G.L. Malli, Chem. Phys. Lett. in print.
17. V. Fock, W. Wesselov and M. Petraschen, Zh. Eksp. Teor. Fiz. 10 723 (1940).
18. J.P. Desclaux, Comput. Phys. Commun. 9, 31 (1975).
19. I.P. Grant, Adv. Phys. 19, 747 (1970).
20. V. Bonifacic and S. Huzinaga, J. Chem. Phys. 60, 2779 (1974); O. Gropen, S. Huzinaga and A.D. Mclean, J. Chem. Phys. 73, 402 (1980).
21. Y. Ishikawa (unpublished).
22. G. Simons, J. Chem. Phys. 55, 756 (1971).
23. P. Hafner, Y. Ishikawa and W.H.E. Schwarz, "Quasi-Relativistic Model Potential Method: Theory and Applications", Abstracts of Third International Congress of Quantum Chemistry, Kyoto, Japan (1979).
24. S.J. Rose, I.P. Grant, and N.C. Pyper, J. Phys. B11, 1171(1978)
25. P. Pyykko and J.P. Desclaux, Chem. Phys. Lett 42, 545 (1976)
26. L.R. Kahn, P. Baybutt and D.G. Truhlar, J. Chem. Phys. 65, 3826 (1976).

ELECTRONIC STRUCTURE OF MOLECULES USING RELATIVISTIC EFFECTIVE CORE POTENTIALS

P. Jeffrey Hay

Los Alamos National Laboratory
Los Alamos, New Mexico USA

INTRODUCTION

The application of ab initio techniques to molecular electronic structure has enabled the quantum chemist to make reliable predictions of molecular geometries and spectroscopic properties, to calculate excitation and ionization energies, and to characterize transition states and energy barriers in chemical reactions. Investigations of the electronic properties of molecules containing heavier atoms (such as transition-metal or actinide compounds) has been hampered by (1) the increase in computational effort (by roughly N^4) with the number of electrons (N) in heavier atoms and (2) the importance of relativistic effects, even on valence electrons, with increasing Z.

In this review we outline our approach in studying molecules containing heavy atoms with the use of relativistic effective core potentials (RECP's). These potentials play the dual roles of (1) replacing the chemically inert core electrons and (2) incorporating the mass velocity and Darwin terms into a one-electron effective potential. This reduces the problem to a valence-electron problem and avoids computation of additional matrix elements involving relativistic operators. The spin-orbit effects are subsequently included using the molecular orbitals derived from the RECP calculation as a basis.

The present approach retains the one-component formalism familiar to chemists for electronic structure calculations while, at the same time, implicitly includes the dominant molecular relativistic effects. The resulting simplicity of this approach enables one to use without further modification the well-developed techniques of quantum chemistry for molecular problems--including normal evaluation of 1- and 2-electron integrals, SCF, and MC-SCF techniques for obtaining orbitals, and configuration interaction (CI) methods for treating electron correlation. The computational expense is considerably reduced by the drastically fewer number of basis functions needed to describe the valence orbitals of the system and the concomittant reduction in number of two-electron integrals. The only additional requirements in carrying out a RECP calculation are (1) evaluation of the multi-center one-electron integrals between atomic basis functions (χ_a) of the potentials, $\langle\chi_a|V_c|\chi_b\rangle$; and (2) evaluation of matrix elements between molecular orbitals $\langle\phi_i s_i|V_{so}|\phi_j s_j\rangle$ where s_i denotes the associated electron spin (α or β).

In the following sections we discuss one-component relativistic Hartree-Fock orbitals, the derivation of RECP's from these orbitals, and various levels of treatment for the spin-orbit coupling operator. Applications of RECP's to calculation of molecular potential energy curves, excitation energies and geometries are then discussed.

The work discussed here was carried out by the author and his collaborators Willard R. Wadt, Robert D. Cowan, and Richard L. Martin at Los Alamos and Luis R. Kahn at Battelle Columbus Laboratories.

METHODS

Relativistic Hartree-Fock Wavefunctions

The traditional approach to relativistic effects in atoms, the Dirac-Hartree-Fock (DHF) method, describes each atomic orbital in terms of 4-component spinors--a "large" and "small" component for $j = \ell - 1/2$ and a similar pair for $j = \ell - 1/2$. Thus a 6p shell in an atom would be described by four radial functions, large and small components associated with $6p_{1/2}$ and $6p_{3/2}$. Cowan and Griffin[1] have developed an alternative 1-component approach--relativistic Hartree-Fock (RHF)--which still retains the important relativistic mass-velocity and Darwin terms. Spin-orbit coupling parameters are then computed from the RHF wavefunctions using the Blume-Watson method. The use of a single-component radial function for each shell (i.e., one radial function for the 6p shell) results in a significant simplification in the atomic case and also makes the RHF wavefunctions more amenable for use in molecular calculations.

The RHF equations are derived as follows. From the local potential approximation to the DHF equations by Desclaux one obtains

$$P_k' = -\frac{k}{r} P_k + \frac{\alpha}{2} (\varepsilon - V(r) + \frac{4}{\alpha^2}) Q_k \quad (1)$$

$$Q'_k = \frac{\alpha}{2} (V(r) - \varepsilon) P_k + \frac{k}{r} Q_k \quad (2)$$

where P and Q are the "large" and "small" components, respectively, and

$$k = \begin{cases} \ell & \text{when } j = \ell - \frac{1}{2} \\ -\ell-1 & \text{when } j = \ell + \frac{1}{2} \end{cases}$$

From solving (1) for Q_k, evaluating Q'_k, and substituting in (2), one obtains an equation in terms of P and P'. Replacing the k-dependent term by the (2j+1)-weighted average, - 1/r, yields the RHF equations

$$[H_{NR} + H_{MV} + H_D] P_i(i) = \varepsilon_i P_i(r) \quad (3)$$

$$H_{NR} = - \frac{d}{dr^2} + \frac{\ell_i(\ell_i+1)}{r^2} + V_i(r) \quad (4)$$

$$H_{MV} = - \frac{\alpha^2}{4} [\varepsilon_i - V_i(r)]^2 \quad (5)$$

$$H_D = - \delta_{\ell,0} \frac{\alpha^2}{4} [1 + \frac{\alpha^2}{4} (\varepsilon_i - V_i(i))(\frac{dV_i}{dr}) (\frac{dr^{-1}P_i}{dr}) \quad (6)$$

The potential $V_i(r)$ in H_{NR} in Eq. (4) is the coulomb and non-local exchange potential normally used in nonrelativistic Hartree-Fock calculations; in the mass-velocity and Darwin terms [Eqs. (4) and (5)] V_i is replaced by the "local exchange" form using the $\rho^{1/3}$ density expression.

The radial characteristics and energies of the U atom valence orbitals are compared for nonrelativistic Hartree-Fock (HF), RHF and DHF wavefunction in Table 1. The relativistic shifts apparent in the j-weighted average DHF wavefunctions are well reproduced by the RHF wavefunctions. One sees the typical orbital contraction and stabilization for low ℓ (s orbitals) and the orbital expansion and destabilization for high ℓ (d and f orbitals). Comparisons of excitation energies in the Au atom from HF, RHF, and averaged DHF results (Table 2) reveal virtually identical excitation energies for RHF and DHF calculations. The large relativistic shifts are evident for these valence electron transitions, since the nonrelativistic results give the incorrect ordering (compared to RHF and experiment) for the 5d→6s and 6s→6p transitions.

Relativistic Effective Core Potentials

The numerical relativistic orbitals of Cowan and Griffin form the starting point for obtaining the relativistic effective core potentials (RECP) according to the procedure of Kahn, Hay and Cowan.[2]

Table 1. Comparison of Orbital Sizes and Energies for the Uranium Atom. (Ref. 1)

	HF[a]	RHF[b]	DHF[c]	DHF (average)
		$\langle r^2\rangle$ (bohr^2)		
$5f_-$ $5f_+$	1.94	2.57	2.53 2.67	2.61
$6d_-$ $6d_+$	9.63	12.4	11.1 13.7	12.7
7s	28.8	21.5	21.8	21.8
$6p_-$ $6p_+$	4.11	3.73	3.13 4.08	3.76
		Orbital Energy (a.u.)		
$5f_-$ $5f_+$	-0.634	-0.331	-0.352 -0.297	-0.320
$6d_-$ $6d_+$	-0.267	-0.188	-0.208 -0.172	-0.186
7s	-0.167	-0.201	-0.199	-0.199
$6p_-$ $6p_+$	-1.04	-1.086	-1.363 -0.959	-1.094

[a] Nonrelativistic Hartree-Fock.
[b] Relativistic Hartree-Fock (Cowan-Griffin).
[c] Dirac-Hartree-Fock.

One seeks to obtain valence "pseudo-orbitals," $\phi_{n\ell}$, and RECP's V_ℓ, which satisfy the valence Hartree-Fock equation.

$$[-\frac{1}{2}\nabla^2 - \frac{Z}{r} + U_\ell(r) + \tilde{W}_{val}]\,\tilde{\phi}_{n\ell} = \varepsilon_{n\ell}\,\tilde{\phi}_{n\ell}$$

which may be compared to the original equation satisfied by the RHF orbitals, ϕ_{nl}

$$[-\frac{1}{2}\nabla^2 - \frac{z}{r} + W_{core} + W_{rel} + W_{val}]\phi_{n\ell} = \varepsilon_{n\ell}\,\phi_{n\ell}$$

In these equation W_{val} ($\tilde{W}_{val}$) represents the coulomb and exchange contributions of the other valence orbitals (pseudo-orbitals); W_{rel}

Table 2. Excitation Energies for the Au Atom. (Ref. 13)

	Excitation Energy (eV)			
	HF	RHF	DHF (avg)	Expt
$5d^{10}\ 6s^{1}(^{2}S)$	0.00	0.00	0.00	0.00
$5d^{9}\ 6s^{2}\ (^{2}D)$	5.13	1.86	1.86	1.74
$5d^{10}\ 6p^{1}\ (^{2}P)$	2.71	4.24	4.24	4.95

represents the relativistic mass-velocity and Darwin terms. Once the pseudo-orbitals are defined, the RECP can be obtained by inverting Eq. (7). The ϕ_{nl} have been defined in two different manners: two different manners:

(1) Linear combinations of RHF orbitals. In this approach[3,4] a smooth, nodeless pseudo-orbital $\tilde{\phi}_{nl}$ is defined in terms of the RHF orbitals

$$\tilde{\phi}_{n\ell}(r) = \sum_k b_{k\ell}\, \phi_{k\ell}(r) \tag{9}$$

where the coefficients b_{kl} are defined by minimizing a "smoothness" functional related to the kinetic energy.

(2) Conservation of valence density. In this approach,[5] the pseudo-orbital is required to be as similar as possible to the RHF orbital in the valence region ($r > R_c$)

$$\tilde{\phi}_{n\ell}(r) = \begin{cases} \phi_{n\ell}(r) & r > R_c \\ \chi_{n\ell}(r) & r < R_c \end{cases} \tag{10}$$

where $X_{n\ell}$ is a polynomial which is smoothly matched to $\phi_{n\ell}$ and its derivatives across R_c. The latter approach has been more successful in reproducing all-electron molecular results with nonrelativistic ECP's, apparently by virtue of providing a more faithful representation of the valence charge density, and therefore should also provide a more reliable basis for molecular results using relativistic ECP's.

After obtaining RECP's for each angular momentum value ℓ = 0,1,... L, where L is the lowest value not present in the core (e.g., 5g for Au), the total potential is given as

$$U_{core}(r) = U_L(r) + \sum_{\ell=0}^{L-1} |\ell\rangle[U_\ell(r) - U_L(r)]\langle\ell| \tag{11}$$

where the closure property of projection operators has been invoked and an implicit sum over $-\ell < m < \ell$ is also implied in the above equation.

The potentials are fit to analytic Gaussian forms

$$\sum_k d_k r^{n_k} e^{-\alpha_k r^2} \qquad n_k = 0,1,2 \tag{12}$$

which are convenient for evaluation of multi-center matrix elements[3] $\langle\chi_a|V_c|\chi_b\rangle$ needed in molecular calculations.

Spin-Orbit Coupling

The effects of the spin-orbit coupling operator are included using the total energies E_I and wavefunctions Φ_I of the electronic states calculated using the RECP's,

$$H_{I,J} = \delta_{I,J}E_I + \langle\Phi_I|V_{s-o}|\Phi_J\rangle \tag{13}$$

where Φ_I can represent either a single-configuration Hartree-Fock wavefunction or a multi-configuration MC-SCF or configuration interaction (CI) wavefunction comprised of one-electron orbitals ϕ_i. The Hamiltonian in Eq. (13) is then diagonalized to obtain the energies and wavefunctions in the presence of the spin-orbit operator.

The spin-orbit operator can be treated in three increasingly more sophisticated (and more computationally demanding) levels.

(1) Atoms-in-molecules method. When the molecular states retain, to a good approximation, their atomic parentage, the semi-empirical atoms-in-molecules approach can be easily applied. The spin-orbit Hamiltonian matrix can be written as

$$\underset{\sim}{H}_{s-o}(\underset{\sim}{R}) = \underset{\sim}{\delta}\, E_I(R) + \underset{\sim}{V}_{s-o} \tag{14}$$

where $E_I(R)$ is the potential energy curve of the Ith molecular state, R denotes the nuclear coordinates and V_{s-o} depends only on the spin-orbit splittings of the constituent atomic fragments. For example, for the states of XeF dissociating to $Xe^+ + F^-$ (see Fig. 1), H_{s-o} has the form[8]

$$H_{s-o} = \begin{bmatrix} E_1(R) & -\sqrt{2}\lambda & 0 \\ -\sqrt{2}\lambda & E_2(R)+\lambda & 0 \\ 0 & 0 & E_1(R) - \lambda \end{bmatrix} \tag{15}$$

where E_1 and E_2 are the total energies of the $2\ ^2\Sigma^+$ and $2\ ^2\Pi$ states, respectively, and λ is one-half the empirical $Xe^+\ ^2P_{1/2} - {}^2P_{3/2}$ splitting. Good agreement between all-electron[6] and ECP calculations[7] and also between theory and experiment is evident from the spectroscopic properties given in Table 3.

(2) Effective spin-orbit operator. For the majority of molecules, the delocalization, hybridization and charge transfer effects occurring in chemical bonding will distort the orbitals from their atomic character and require a more sophisticated spin-orbit treatment. These effects can be successfully incorporated by the effective spin-orbit operator approach, in which matrix elements over the one-electron orbitals, $\langle\phi_i|V^{eff}_{s-o}|\phi_j\rangle$ are used to construct the spin-orbit matrix in Eq. (13) over state functions $\langle\Phi_I|V_{s-o}|\Phi_J\rangle$. The effective operator V^{eff}_{s-o} has the form[8]

$$V^{eff}_{s-o} = \frac{\alpha^2}{2} \sum_{i_k,k} \frac{Z^{eff}}{r^3_{ki_k}}\ \underset{\sim}{\ell}_{ki_k} \cdot \underset{\sim}{s}_{i_k} \tag{16}$$

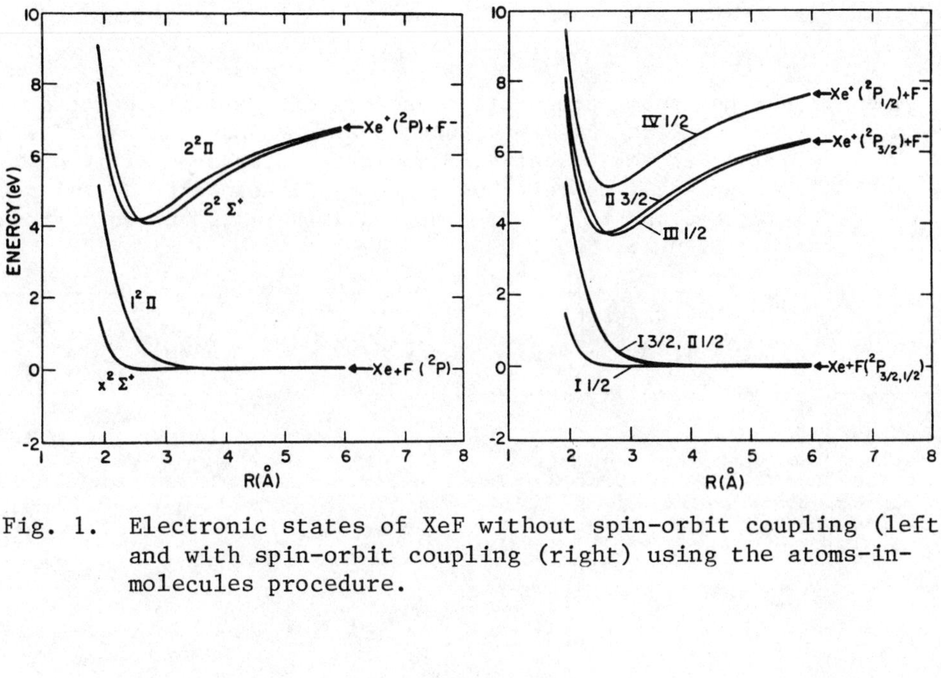

Fig. 1. Electronic states of XeF without spin-orbit coupling (left) and with spin-orbit coupling (right) using the atoms-in-molecules procedure.

Table 3. Spectroscopic properties of the $III_{1/2}$ state of XeF. (Ref. 7)

	D_e(eV)	R_e(Å)	ω_e(cm^{-1})	Emission energy (eV)
AE	5.07	2.68	303	3.65
VE(NR)	5.17	2.63	311	3.56
VE(R)	5.15	2.63	311	3.52
Expt	5.31	(2.49)	309	3.52

where α is the fine-structure constant, k indexes the nuclei and i_k indexes the electrons on center k. The parameter Z_k^{eff} is adjusted to reproduce the relevant atomic spin-orbit constants ζ_k. Also, only 1-center terms are used to evaluate the above 1-electron operator. In all-electron applications Z_k^{eff} can be viewed as an effective nuclear charge. In ECP calculations, where the valence orbitals are nodeless with zero amplitude at the nucleus, the resulting values of Z_k^{eff} must be taken to be very large (compared to the actual Z) to match the experimental atomic spin-orbit parameters and hence lose any physical significance. Nevertheless the effective operator approach has proved extremely useful in accounting for the effects of spin-orbit coupling for valence electrons in molecules.

Table 4. Silicon atom spin-orbit constants (in cm^{-1}) calculated using Hartree-Fock wavefunctions over Gaussian basis sets and an effective one-electron one-center spin-orbit operator adjusted to reproduce the Si(^3P) constant, VE=valence electron, AE(FC) = all-electron with $1s^2 2s^2 2p^6$ core frozen from Si(^3P). (Ref. 9)

	AE(FC) (11s8p)/[4s4p]	VE (3s4p)	VE (5s7p)	Expt
Si [$3s^2 3p^2$] ^3P	148.9	148.9	148.9	148.9
Si^+[$3s^2 3p^1$] ^2P	192.6	179.81	179.8	191.3
Si^+[$3s^2 4p^1$] ^2P	27.2	27.3	32.7	40.0
Si^{3+}[$3p^1$] ^2P	295.0	240.2	250.2	306.9

In Table 4 recent calculations by Wadt[9] compare the use of the effective operator method in all-electron and valence-electron calculations on the Si atom and its ions. The values of Z^{eff} for the HF orbitals and pseudo-orbitals, respectively, were chosen to match the quantity $Z^{eff} \langle r^{-3} \rangle_{3p}$ to the Si spin-orbit constant, $\zeta = 2/3\ [{}^3P_2 - {}^3P_0]$. Since the calculations covered large changes in the radial extent of the Si 3p orbital they provide an excellent test of the method. The good agreement (within 20 percent) between all-electron, valence-electron and experimental results supports the validity of the approach; in the next section the method is tested in a molecular example.

(3) Breit-Pauli spin-orbit operator. The spin-orbit term in the Breit-Pauli Hamiltonian is given by

$$V_{so} = \frac{\alpha^2}{2} \sum_{i,k} \frac{Z_k}{\sigma_{ki}^3}\, \underset{\sim}{\ell}_{ki} \cdot \underset{\sim}{s}_i + \sum_{i,j}{}' \left(\frac{\underset{\sim}{r}_{ij}}{r_{ij}^3} \times \underset{\sim}{p}_i \right) \cdot (s_i + 2s_j) \tag{17}$$

A rigorous application of this operator to molecular problems requires multi-center integral evaluation of 1-electron and 2-electron operators and is applicable only to all-electron calculation to a few molecular systems. Langhoff[10] has used the full Breit-Pauli operator to calculate rare gas-oxygen matrix elements and to compare to the all-electron results using the 1-center, 1-electron effective spin-orbit operator discussed above. The results are shown in Fig. 2 for the coupling matrix elements $\langle {}^3\Sigma^- | V_{s-o} | {}^1\Sigma^+ \rangle$ and $\langle {}^3\Pi | V_{s-o} | {}^1\Sigma^+ \rangle$ for ArO, KrO and XeO. Excellent agreement is obtained between the results using the effective operator and the Breit-Pauli Hamiltonian and enhance the validity of the simpler approach

APPLICATIONS

Relativistic Effects in Transition Metal Atoms

A recent comprehensive study of relativistic effects in transition metal atoms and ions by Martin and Hay[11] revealed large effects in excitation and ionization energies. The relativistic shifts for excitation energies involving the s^2d^n, s^1d^{n+1}, and d^{n+2} states of the neutral species computed using Cowan-Griffin RHF wave functions are shown in Fig. 3. The simultaneous stabilization of s orbitals and destabilization of d orbitals arising from the relativistic terms in the Hamiltonian becomes increasingly apparent in the larger relativistic shifts observed as one proceeds from the first through third transition series. The effects of relativity were found to be several eV in the third transition series and surprisingly significant (as large as 0.6 eV) for the first transition series.

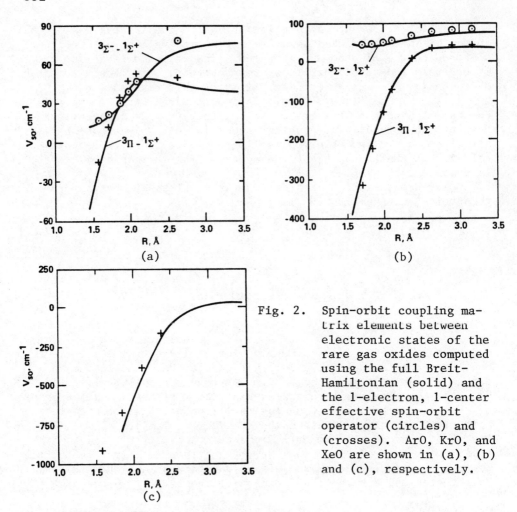

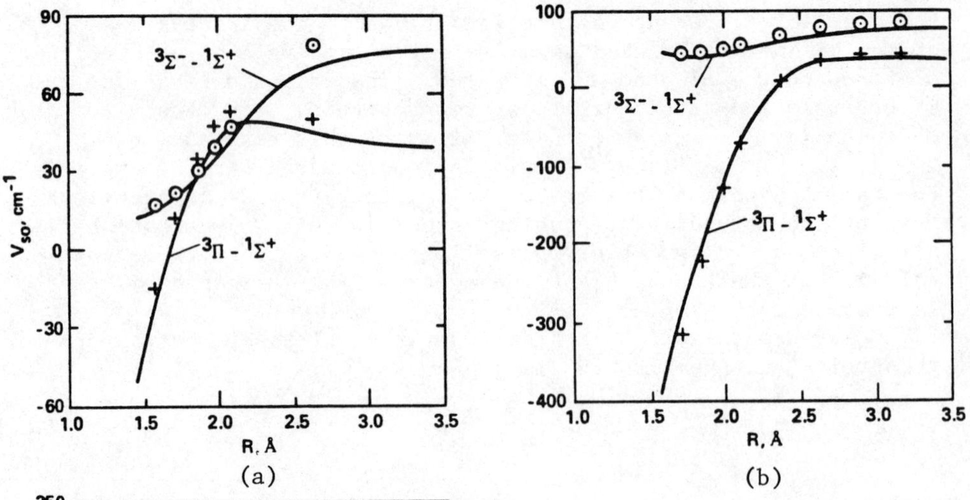

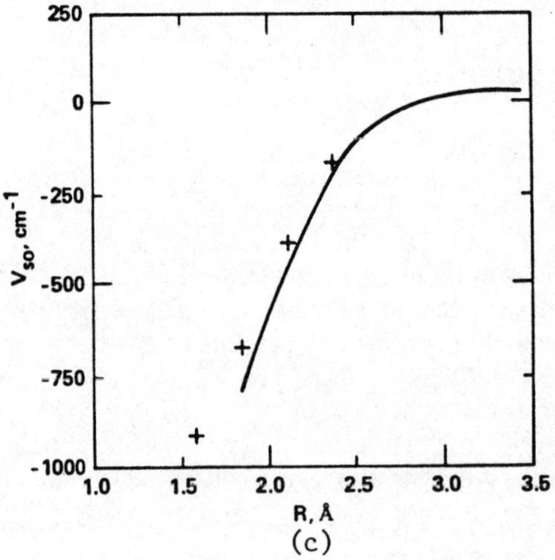

Fig. 2. Spin-orbit coupling matrix elements between electronic states of the rare gas oxides computed using the full Breit-Hamiltonian (solid) and the 1-electron, 1-center effective spin-orbit operator (circles) and (crosses). ArO, KrO, and XeO are shown in (a), (b) and (c), respectively.

Relativistic Effects in Chemical Bonds

The importance of relativistic effects in chemical bonding of molecules has been studied using a variety of theoretical approaches including 1-center Dirac-Fock,[12] relativistic ECP,[13] and full Dirac-Fock[14] methods. Hay et al.[13] used RECP's to investigate the ground states of the AuH and AuCl molecules and to compare with the results using nonrelativistic ECP's (Fig. 4). With a properly dissociating 2-configuration (GVB-1) wavefunction the predicted bond length from the RECP calculations differed by only 0.01Å from the experimental value (1.52Å) as shown in Table 5. By contrast the nonrelativistic ECP calculation gave a bond length over 0.3Å too long! A relativistic contraction of 0.16Å was also observed in the ionic AuCl species (Table 6). One interpretation ascribes the shortening of the bond in AuH to the relativistic contraction of the

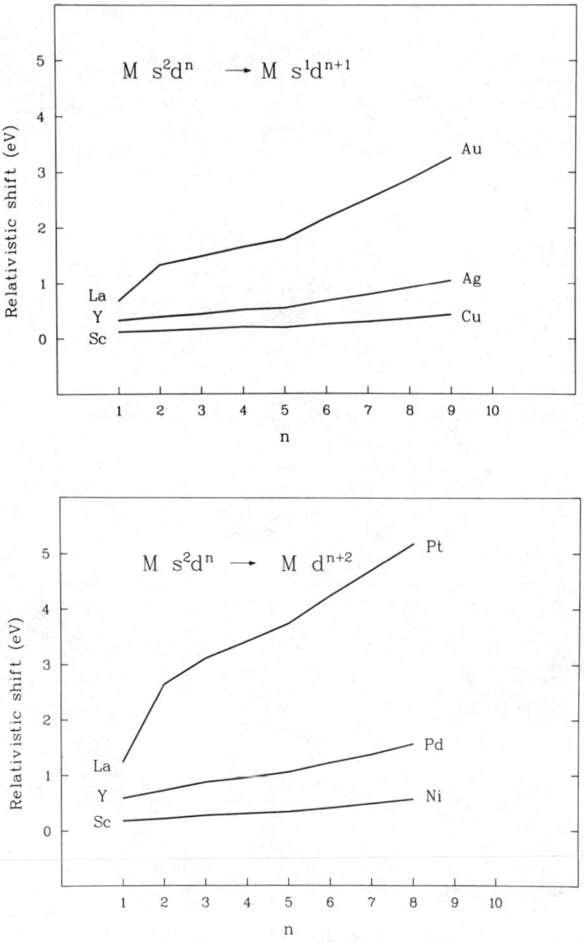

Fig. 3. Relativistic shifts in excitation energies for transition metal atoms obtained from comparing HF and relativistic HF calculations.

6s orbital in Au, as $\langle r \rangle_{6s}$ is reduced from 3.7 to 3.0 a_0 and the $5d^{10}$ core of Au does not participate in the bonding.

A similar relativistic contraction (0.16Å) is observed in AuCl (Table 6). The relativistic calculations give a larger dissociation energy for AuH compared to the nonrelativistic results, while the reverse is true in AuCl. The decrease in the bond energy of AuCl can be correlated with the increase in energy of the ionic Au^+ + Cl^- asymptote relative to the covalent limit upon inclusion of relativistic effects in the atom.

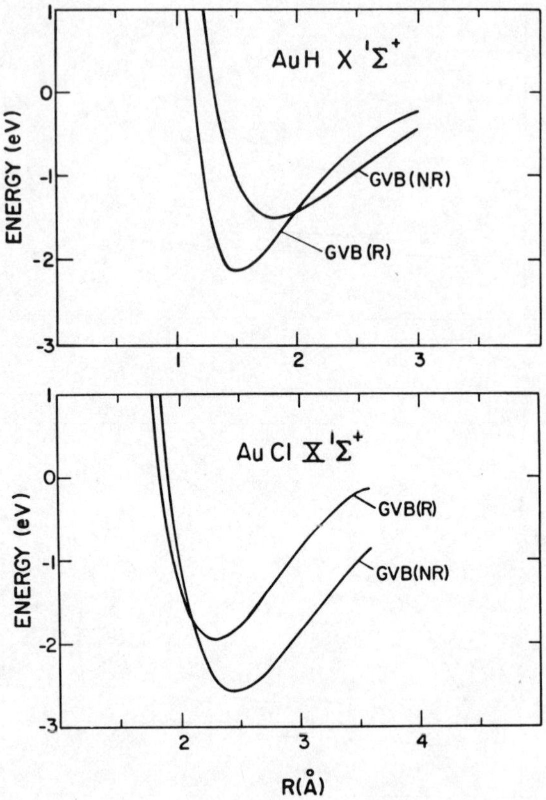

Fig. 4. Potential energy curves for the ground states of AuH and AuCl computed using nonrelativistic (NR) and relativistic (R) ECP calculations.

Electronic Structure of Transition-Metal Complexes

The advent of relativistic ECP's makes calculations feasible for the first time on third-row transition metal complexes. Theoretical studies of platinum complexes using RECP's have explored (1) the structures and energetics of Pt(II) hydride and chloride complexes,[15] (2) the oxidative addition of H_2 to a Pt(0) complex,[16] and (3) the binding of ethylene in Zeise's salt,[17] $PtCl_3(C_2H_4)^-$. The geometric parameters of Zeise's salt computed from Hartree-Fock

Table 5. Spectroscopic Properties of AuH. (Ref. 13)

AuH $(X^1\Sigma^+)$	R_e(Å)	D_e(eV)	ω_e(cm^{-1})
Nonrel. ECP			
HF	1.763	0.99	1387
GVB-1	1.820	1.52	1203
POL-CI	1.807	1.57	12.17
Rel ECP			
HF	1.508	1.55	2014
GVB-1	1.514	2.14	1891
POL-CI	1.522	2.23	1871
Exptl	1.5237	3.37	2305
1-Center Dirac-Fock			
Nonrel.	1.745	--	2296
Rel.	1.659	--	2178

wavefunctions using a RECP on Pt are compared with the values obtained from neutron diffraction in Fig. 5. The perpendicular oreintation of the ethylene ligand relative to the $PtCl_3^-$ plane is correctly predicted to be the stable form, with the planar conformation calculated to lie 15 kcal/mol higher in energy. The other geometrical parameters, including the bending of the CH_2 groups away from the metal, are also well reproduced by the calculations.

Both Zeise's salt and the related $PtCl_4^{2-}$ species have $(5d)^8$ formal configurations, where the $5d_{x^2-y^2}$ orbital oriented along

Table 6. Spectroscopic Properties of AuCl. (Ref. 13)

AuCl $(X^1\Sigma^+)$	R_e(Å)	D_e(eV)	ω_e(cm^{-1})
Nonrel. ECP			
GVB-1	2.447	2.58	277
Rel. ECP			
GVB-1	2.283	1.96	298
POL-CI	2.291	2.39	306
Exptl	--	3.5 ± 0.1	382

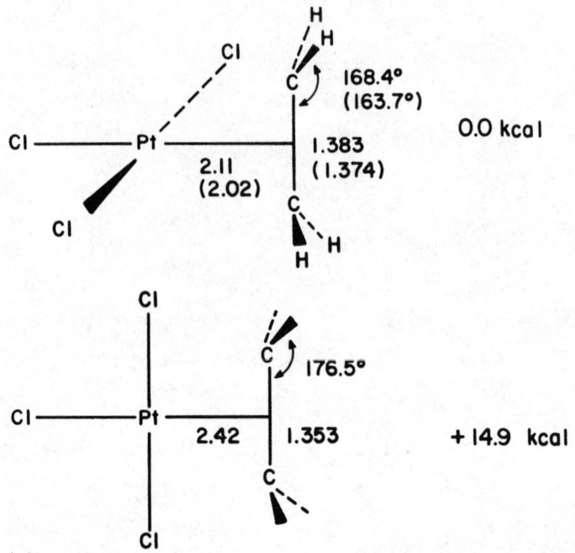

Fig. 5. Calculated geometrical parameters (with experimentally observed values in parentheses) for Zeise's salt.

the metal-ligand axes is nominally vacant. The lowest excited states in both systems arise from d-d excitations from the (xz, yz), xy or z^2 orbitals into the x^2-y^2 orbitals leading to states of E_g, A_{2g}, and B_{1g} symmetry, respectively, in $PtCl_4{}^{2-}$. The results of CI calculations[18] are shown in Fig. 6 where triplet and singlet states of these symmetries are identified. Spin-orbit coupling has been included using the effective operator method discussed earlier, and in this case the procedure involves setting up and diagonlizing the 17 x 17 spin-orbit Hamiltonian matrix over the above set of singlet and triplet states. The CI + spin-orbit results are compared with experimentally observed levels, which are in reasonably good agreement with the predicted levels, although there is still considerable controversy over some of the assignments.

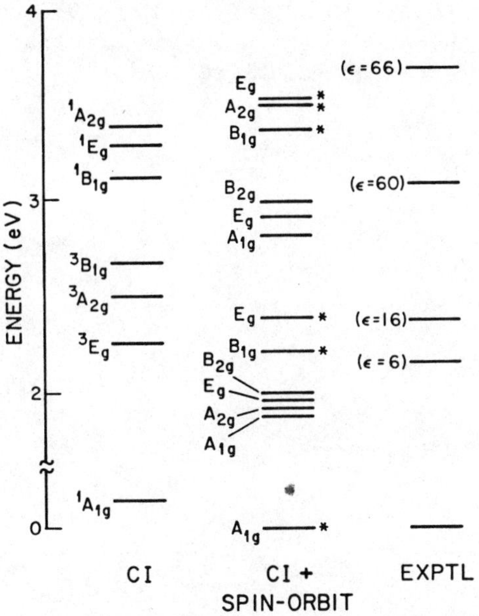

Fig. 6. Electronic states of $PtCl^{2-}$ obtained from CI calculations with and without spin-orbit coupling.

Actinide Compounds

The electronic structure of actinide compounds including UF_6,[19] UF_5,[20], UO_2^{2+},[21] and ThO_2[21] has also been explored with the aid of the RECP method. In Figs. 7-9 some aspects of the electronic structure of UF_6 are depicted. To a zeroth-order approximation, one may regard UF_6 as having formal charges of U^{+6} and F^{-1} and thus having a $(5f)^0$ configuration in the ground state. (Mulliken population analyses of the ground state wavefunction of UF_6 actually show considerable "back-bonding" into the 5f orbitals.) The highest occupied levels arise from symmetry combinations of the F 2p orbitals (Fig. 7) while the lowest unoccupied levels correspond to the seven U 5f levels. Spin-orbit coupling has only a modest effect on the level ordering with the exception of the $2t_{1u}$ orbital which contains significant U 6p character and is split into two components--8u and 6u in the octaheral double group representation--separated by over 2 eV. Excited states may then be viewed in terms of charge-transfer excitations from the 36 F 2p levels (including spin-degeneracies) into the 14 U 5f levels--leading to 504 states in all. These transitions account for the entire uv absorption spectrum of UF_6 between 3 and 10 eV (Figs. 8 and 9), where the results ofCI + spin-orbit calculations[22] are shown for these 504 states classed according to the dominant excitations.

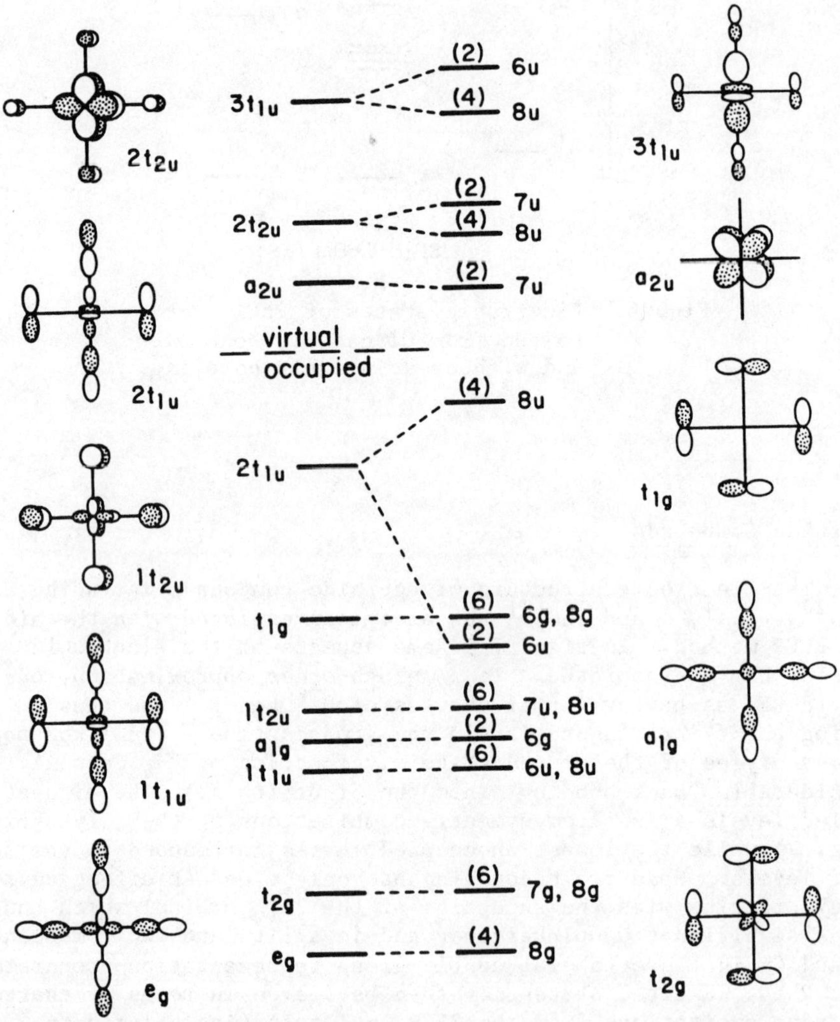

Fig. 7 Schematic diagram of the highest occupied and lowest unoccupied orbitals of UF_6.

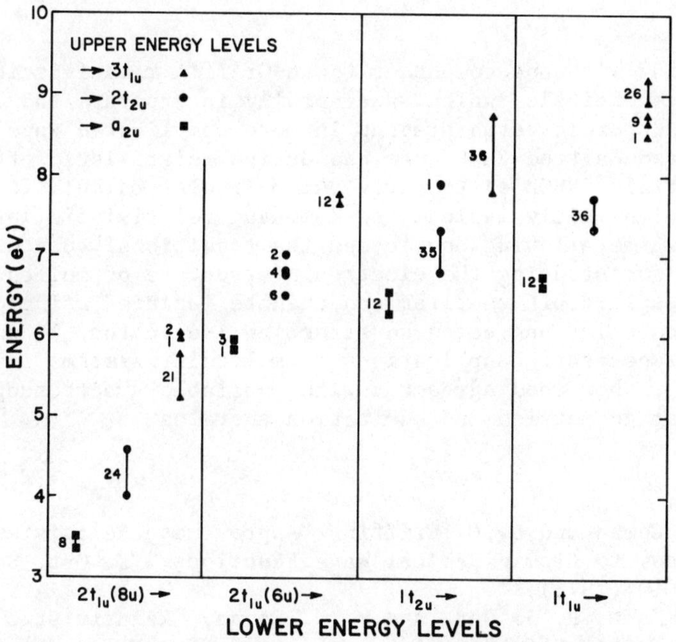

Fig. 8. Excited states of UF_6 with spin-orbit coupling (g states) grouped according to orbital parentage.

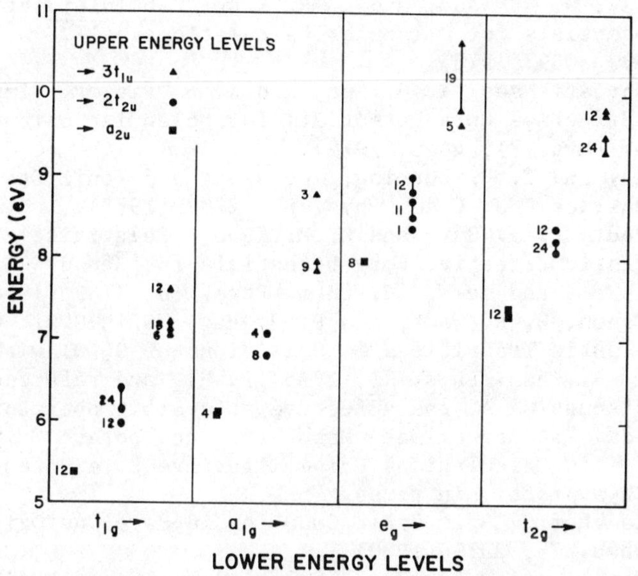

Fig. 9. Excited states of UF_6 with spin-orbit coupling (u states) grouped according to orbital parentage.

SUMMARY

Starting with one-component Cowan-Griffin relativistic Hartree-Fock orbitals, which successfully incorporate the mass-velocity and Darwin terms present in more complicated wavefunctions such as Dirac-Hartree-Fock, one can derive relativistic effective core potentials (RECP's) to carry out molecular calculations. These potentials implicitly include the dominant relativistic terms for molecules while allowing one to use the traditional quantum chemical techniques for studying the electronic structure of molecules. The effects of spin-orbit coupling can then be included using orbitals from such calculations using an effective 1-electron, 1-center spin-orbit operator. Applications to molecular systems involving heavy atoms, show good agreement with available spectroscopic data on molecular geometries and excitation energies.

REFERENCES

1. R. D. Cowan and D. C. Griffin, "Approximate Relativistic Corrections to Atomic Radical Wave Functions," J. Opt. Soc. Am. 66, 1010 (1976).
2. L. R. Kahn, P. J. Hay, and R. D. Cowan, "Relativistic Effects in Ab Initio Effective Core Potentials for Molecular Calculations. Application to the Uranium Atom," L. R. Kahn, P. J. Hay, and R. D. Cowan, J. Chem. Phys. 68, 2368 (1978).
3. L. R. Kahn, P. Baybutt, and D. Truhlar, "Ab Initio Effective Core Potentials," J. Chem. Phys. 65, 3826 (1976).
4. P. J. Hay, W. R. Wadt, and L. R. Kahn, "Ab Initio Effective Core Potentials for Molecular Calculations. II," J. Chem. Phys. 68, 3059 (1978).
5. P. A. Christiansen, Y. S. Lee, and K. S. Pitzer, "Improved Ab Initio Effective Core Potentials for Molecular Calculations," J. Chem. Phys. 71, 4445 (1979).
6. P. J. Hay and T. H. Dunning, "Covalent and Ionic States of the Xenon Halides," J. Chem. Phys. 69, 2209 (1978).
7. W. R. Wadt, P. J. Hay, and L. R. Kahn, "Relativistic and Non-Relativistic Effective Core Potentials for Xenon. Applications to XeF, Xe_2, and Xe_2^+," J. Chem. Phys. 68, 1752 (1978).
8. J. S. Cohen, W. R. Wadt, and P. J. Hay, "Spin-Orbit Coupling and Inelastic Transitions in Collisions of $O(^1D)$ with Ar, Kr, and Xe," J. Chem. Phys. 71, 2955 (1979), and references therein for a discussion of the effective spin-orbit operator.
9. W. R.Wadt, "An Approximate Method to Incorporate Spin-Orbit Effects into Calculations Using Effective Core Potentials," Chem. Phys. Lett., in press.
10. S. R. Langhoff, "Spin-Orbit Coupling in Rare-Gas Oxides," J. Chem. Phys. 73, 2379 (1980).
11. R. L. Martin and P. J. Hay, "Relativistic Contributions to the Low-Lying Excitation Energies and Ionization Potentials of the Transition Metals," J. Chem. Phys. 75, 4539 (1981).

12. J. P. Desclaux and P. Pyykko, "Dirac-Fock One-Center Calculations," Chem. Phys. Lett. 39, 300 (1976).
13. P. J. Hay, W. R. Wadt, L. R. Kahn, and F. W. Bobrowicz, "Ab Initio Studies of AuH, AuCl, HgH, and $HgCl_2$ Using Relativistic Effective Core Potentials," J. Chem. Phys. 69, 984 (1978).
14. Y. S. Lee and A. D. McLean, "Relativistic Effects on R_e and D_e in AgH and AuH From All-Electron Dirac-Hartree-Fock Calculations," J. Chem. Phys. 76, 735 (1982).
15. J. O. Noell and P. J. Hay, "Ab Initio Studies of the Structures of Square Planar $Pt(PH_3)_2XY$ Species (X,Y=HCl) and Bonding of Hydrido- and Using Relativistic Effective Core Potentials," Inorg. Chem. 21, 14 (1982).
16. J. O. Noell and P. J. Hay, "An Ab Initio Study of the Oxidative Addition of H_2 to a Pt(0) Complex," J. Amer. Chem. Soc., in press.
17. P. J. Hay, "The Binding of Ethylene to Platinum and Palladium. An Ab Initio Study of the $MC\ell_3(C_2H_4)^-$ Species," J. Amer. Chem. Soc. 103, 1390 (1981).
18. P. J. Hay, "The Electronic States of $PtC\ell_4{}^{2-}$," in preparation.
19. P. J. Hay, W. R. Wadt, L. R. Kahn, R. C. Raffenetti, and D. H. Phillips, "Ab Initio Studies of the Electronic Structure of UF_6, $UF_6{}^+$, and $UF_6{}^-$ Using Relativistic Effective Core Potentials," J. Chem. Phys. 70, 1767 (1979).
20. W. R. Wadt and P. J. Hay, "Ab Initio Studies of the Electronic Structure and Geometry of UF_5 Using Relativistic Effective Core Potentials," J. Amer. Chem. Soc. 101, 5198 (1979).
21. W. R. Wadt, "Why $UO_2{}^{++}$ is Linear and Isoelectronic ThO_2 is Bent," J. Am. Chem. Soc. 103, 6053 (1981).
22. P. J. Hay, "Electronic States of UF_6," in preparation.

ELECTRON STRUCTURE OF MOLECULES WITH VERY HEAVY ATOMS USING EFFECTIVE CORE POTENTIALS

Kenneth S. Pitzer

Department of Chemistry and Lawrence Berkeley Laboratory
University of California
Berkeley, California 94720

INTRODUCTION

One of the primary objectives of quantum chemical theory is the calculation of the energy of arrays of atoms (molecules, activated complexes, etc.) as a function of their geometrical arrangement. One seeks to make these calculations from first principles (i.e., without empirical adjustments) but approximations are necessary. The first approximation is implicit in the statement of the problem - that of Born and Oppenheimer in treating electronic motion separately with fixed nuclei. For elements of low atomic number, the nonrelativistic Schrödinger Hamiltonian is commonly assumed. One of our principal interests concerns the difference between results calculated on the basis of the Dirac, relativistic Hamiltonian from those calculated nonrelativistically. Before proceeding to that topic, however, it is desirable to review further the other approximations in the usual nonrelativistic treatments. Most serious are the approximations in the expression of the wavefunction: (1) the basis functions and (2) the terms for electron correlation (usually via configuration interaction (CI)). For atoms of small atomic number, all electrons are treated explicitly, but even for atoms of intermediate atomic number (e.g., chlorine) the calculational burden associated with inner-shell electrons has become substantial. Since these inner-shell orbitals are practically unaffected in molecule formation, it is useful to attempt to simplify the calculation by some process averaging the net effect of core electrons on valence electrons. Under given computer limitations this allows the use of more extended basis functions or more complete CI for the valence electrons.

As one turns ones attention to atoms of very high atomic number, the problem of inner-shell electrons becomes more severe and their removal from the detailed calculations is essential for most work with current computers. Also, relativistic effects are now very significant and the calculations must be based on the Dirac equation. These two aspects will be the primary topics of this paper.

Additional points to be noted include (1) the size and shape of the nucleus and (2) many particle relativistic effects. While these are important for some purposes, their effect on valence electrons is very small and will be ignored.

The general method for the removal of inner-shell electrons from the detailed calculations is a frozen core, effective potential approximation. Pertinent theory related to such an approximation will be considered on both nonrelativistic and relativistic bases. But the eventual verification is comparison with accurate all-electron calculations. Such comparisons will be made for nonrelativistic examples. All-electron, relativistic calculations on appropriate molecules with very heavy atoms are needed as standards of comparison but are not yet available.

EFFECTIVE POTENTIALS

There are several ways of formulating a frozen-core, effective potential (EP) approximation.[1] The basic criterion of merit is agreement with all-electron calculations for the properties of primary interest. The widely used Phillips-Kleinman[2] method was designed to yield accurate orbital energies. The initial emphasis was band energies for crystals. In effect the Phillips-Kleinman method transfers the orbital energy for core orbitals to the valence-electron orbital energy; hence, molecular or crystal orbitals in E.P. calculations yield relatively accurate orbital energies at the fixed, experimental geometry. But it has been found that the Phillips-Kleinman procedure is not satisfactory for calculation of dissociation energies or for determination of the potential minima which determine bond distances if there is more than one valence electron in the atom. Christiansen, et al.,[3] explained the cause of this difficulty and proposed a greatly improved alternative for the purposes of bond-distance and dissociation-energy calculations.

As shown in detail by Christiansen, et al., the basic requirement is that the valence pseudo-orbital from which the EP is derived, must be exactly the true atomic valence orbital in the outer or valence portion of the atom. In the core region the oscillations of the true atomic orbital are eliminated by a smoothing process, the details of which may be varied somewhat, but the total electron population in the core region must be the same for the pseudo-orbital as for the true atomic orbital. The Phillips-Kleinman method transfers some electron population from the valence to the core region; this is the cause of the difficulties with that procedure.

The particular improved procedure for definition of a valence pseudo-orbital used in this laboratory involves adoption of the exact atomic orbital outside a radius r_{match}. Inside r_{match} the pseudo-orbital is chosen to be a five-term polynomial in r with a leading power $\ell + 2$. At r_{match}, the amplitude and first three derivatives must agree. Also the total pseudo-orbital must be normalized, have no more than two inflexions nor more than three inflexions in the first derivative. One chooses the smallest r_{match} at which all of these conditions can be fulfilled. For a particular angular symmetry, the EP is derived from the radial factor χ_v of a valence pseudo-orbital by the expression

$$U_v^{EP}(r) = [(\varepsilon_v + 1/2\,\nabla^2 + Z/r - W_{val}^{PS})\chi_v]/\chi_v \tag{1}$$

where ε_v is the atomic orbital energy and W_{val}^{PS} is the potential (comprising the usual coulomb and exchange terms) arising from the interaction of an electron in χ_v with all other valence electrons in their pseudo-orbitals.

The general form of effective potential expression of Phillips and Kleinman[2] and of Kahn and Goddard[4] is retained. On the non-relativistic basis it is

$$U^{EP} = U_L^{EP}(r) + \sum_{\ell=0}^{L}\sum_{m=-\ell}^{\ell}[U_\ell^{EP}(r) - U_L^{EP}(r)]|\ell m\rangle\langle\ell m| \tag{2}$$

where L is an angular quantum number larger than the ℓ values represented in the core, $U_\ell^{EP}(r)$ is the effective potential for angular symmetry ℓ in the atom of interest and the final factor is the projection operator for angular symmetry ℓ, m. Lee, et al.,[5] showed that the substitution of the Dirac Hamiltonian for the Schrödinger Hamiltonian led in a straightforward manner to the relativistic EP

$$U^{EP} = U_{LJ}^{EP}(r) + \sum_{\ell=0}^{L}\sum_{j=|\ell-1/2|}^{\ell+1/2}\sum_{m=-j}^{j}[U_{\ell j}^{EP} - U_{LJ}^{EP}]|\ell jm\rangle\langle\ell jm| \tag{3}$$

Now the $U_{\ell j}^{EP}(r)$ are obtained from relativistic pseudo-orbitals and are different for $j = \ell - 1/2$ and $j = \ell + 1/2$. This difference with j for a given ℓ is just the spin-orbit effect. Again L, J are angular quantum numbers exceeding those represented in the core; it is found that there is no significant change of $U_{\ell j}^{EP}(r)$ for values of ℓ and j higher than this. The projection operators are now two-component angular bases that are eigenfunctions of the Pauli approximation to the Dirac Hamiltonian.

In addition to the spin-orbit effect, the relativistic EP will differ from the nonrelativistic EP numerically. In the Pauli approximation these differences are ascribed to the mass-velocity and the Darwin terms. But in our calculations we use the full Dirac operator rather than the Pauli approximation.

The EP can be expressed numerically or by expansions in appropriate mathematical functions. Since they are ordinarily derived from orbitals expressed numerically, we have found it convenient also to express the EP numerically.

Of course, one does not ordinarily have exact atomic orbitals as an input to the generation of effective potentials. Usually the orbitals from numerical Hartree-Fock (HF) or Dirac-Fock (DF) calculations are used. In addition to ground state atomic calculations, one must have results for appropriate excited states in which other orbitals of interest are occupied. If the energies calculated for these excited states agree reasonably well with the experimental values, one presumes that the various orbitals, pseudo-orbitals, and effective potentials will be quite accurate. This has been the case for the atoms of greatest interest in our recent work, e.g., gold, thallium, and lead.

But there are cases where the HF or DF calculations are in serious error with respect to the energy differences between various low-energy atomic states. This is well-known for the elements of the first transition series. For example, for nickel the $3d^84s^2(^3F)$ and $3d^94s(^3D)$ states actually differ in energy by only 0.03 eV whereas HF calculations place the ^3D state higher by 1.28 eV. The error for the $3d^{10}(^1S)$ state is even larger. Also these errors are increased somewhat for relativistic DF calculations. Martin[6] discusses this problem and attempts its resolution by consideration of electron correlation. For the molecule Ni_2, fairly accurate calculations[7] can be made by considering only the $3d^94s$ state and suppressing the $3d^84s^2$ state of the atom. Such molecular calculations are hazardous, however, and it is desirable to obtain atomic calculations which accurately reproduce all of the important atomic states as a basis for the generation of the effective potentials.

The Christiansen effective potentials have been employed in comparison with all-electron (AE) calculations for F_2, $C\ell_2$, and $LiC\ell$ in the original paper[3] proposing that method and more recently[8] for the ground states of Ar_2, Kr_2, and Xe_2 and for the $^2\Sigma_u^+$ states of Ar_2^+, Kr_2^+, and Xe_2^+. In the first series the AE calculations are by Hay, et al;[9] in the latter by Wadt.[10] The basis sets and the extent of CI were chosen in each case to be effectively identical for the AE and the EP treatments. In all cases the AE and EP potential curves are essentially identical; this is true, not only at radii from the potential minima outward, but also at distances well up the repulsive curves where the frozen-core, EP approximation would be expected to be poorest. Figure 1 shows this comparison for the ground state of Ar_2 and the $^2\Sigma_u^+$ state of Ar_2^+. Thus for a variety of molecules the comparison of nonrelativistic EP and AE calculations confirms the accuracy of the EP results with the procedures described above. It is highly desirable that accurate, relativistic, AE calculations be completed for a few molecules where relativistic effects are substantial. Such standards can then be used to check relativistic EP calculations.

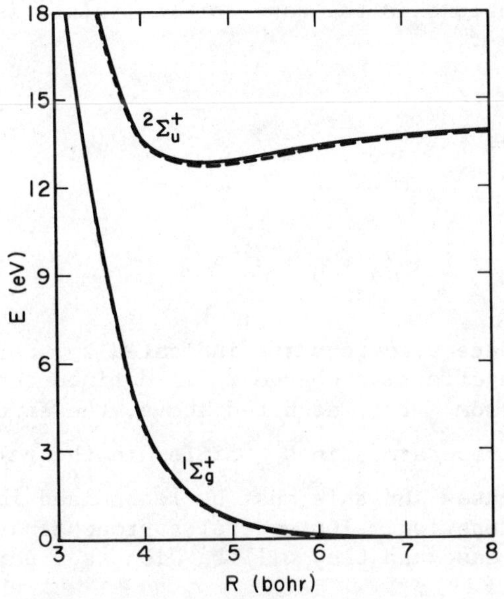

Figure 1. Comparison of AE (dashed) and EP (solid) dissociation curves for Ar_2 and Ar_2^+ from POL-CI calculations.

HAMILTONIAN FOR VALENCE-ELECTRON MOTION

The Schrödinger Hamiltonian is adequate for valence-electron motion in the outer or valence region of atoms or molecules. This is established[5] most easily by the smallness of the small component relative to the large component in the valence-level Dirac spinors for every heavy atoms. Thus it is an adequate approximation to simply ignore the small component and adopt the large component as the pseudo-orbital in the outer region of the atom. An alternate procedure, theoretically more exact, is to make the Foldy-Wouthuysen transformation of the DF orbital, but it has no significant effect on the result. Relativistic effects are important, in heavy atoms, on the motion of electrons near the nucleus - even of valence electrons of low angular momentum which do approach the nucleus. But all of these effects are incorporated in the effective potentials: both the indirect effects of core electrons and the direct effects on valence electron motion near the nucleus. Thus the use of the nonrelativistic Hamiltonian is adequate for molecular calculations but the relativistic properties of the EP, i.e., the difference for $j = \ell + 1/2$ and $j = \ell - 1/2$, impose relativistic symmetry on the molecular wavefunctions.

MOLECULAR CALCULATIONS: THEORY

Given the EP and the adequacy of the nonrelativistic Hamiltonian for valence electron motion, the form of the Hamiltonian for molecular problems is the same for the relativistic or non-relativistic basis.

$$H = \sum_{\mu=1}^{n_v} h_\mu + \sum_{\mu>\nu} (r_{\mu\nu})^{-1} \tag{4}$$

$$h_\mu = -1/2\, \nabla_\mu^2 + \sum_{\alpha}^{N} (-Z_\alpha/r_{\alpha\mu} + U_\alpha^{EP}) \tag{5}$$

where the n_v valence electrons are indicated by μ or ν and the N nuclei by α. The effective charge Z_α is defined consistently with the EP for that atom. But, as noted above, the angular symmetries of the projection operators in U_α^{EP} differ in the relativistic and nonrelativistic cases and this must be recognized in the formulation of the wavefunction. Two general approaches are possible in the relativistic case and they will be discussed serially.

ω-ω Coupling

The most straightforward procedure for a relativistic problem is to formulate the molecular wavefunction as linear combinations of relativistic atomic spinors. For valence electrons the small components of the four-component Dirac spinors may be neglected, leaving two-component spinors. The matrix elements of the EP on the same atom are very simple since the projection operators involve the angular factors of these same two-component spinors. The radial factors can be expressed in either Slater or Gaussian basis functions. This procedure is given in detail (for Slater basis functions and linear molecules) by Lee, et al.,[12] for single configuration, self-consistent-field (SCF) calculations. It was extended to multiconfiguration SCF (MCSCF) calculations by Christiansen and Pitzer.[13]

For linear molecules this method is reasonably satisfactory since a relatively small basis of Slater functions is adequate and the various matrix elements are calculated without particular difficulty. For most cases, however, a single configuration is inadequate - even more inadequate than for the nonrelativistic examples with light atoms. The reason is that the ground atomic states of even the heaviest atoms of interest are in intermediate coupling rather than very close to j-j coupling. In other words the valence-level, electron-repulsion integrals are of the same magnitude as the spin-orbit (SO) terms. In ω-ω coupling the SO terms are included in the single configuration treatment. But it is not a good approximation to regard the electron-repulsion terms as a small perturbation; hence an appropriate MCSCF calculation is required. To properly account for electron correlation a large configuration interaction (CI) calculation is required, and this has not yet been accomplished in ω-ω coupling.

Molecules containing the thallium atom were chosen as examples for early treatment since that atom has only one 6p electron, but its SO interaction is large. Calculations were made for TℓH[13,14] and for several low-energy states of $T\ell_2^+$ and $T\ell_2$.[15]

There are serious limitations to the method starting in ω-ω coupling. Programs for CI calculations have not been prepared. Extensions from linear to nonlinear molecules will require new programs of considerable complexity. Also, one has been trained to think about molecules in Λ-S rather than ω-ω coupling, and it is easier, conceptually, to add SO terms to a calculation initiated in Λ-S coupling than to add electron repulsion terms to the ω-ω treatment. Thus we turn now to the alternate approach.

Λ-S Coupling

If one eliminates, for the moment, the spin-orbit term, the relativistic EP have the same symmetry as the nonrelativistic EP (but the numerical values of the EP still differ). This can be accomplished by taking the appropriate weighted average of $U^{EP}_{\ell j}$ for $j = \ell + 1/2$ and $j = \ell - 1/2$ and using that averaged relativistic EP (AREP) with the nonrelativistic projection operators in Equation (2). Specifically the AREP are

$$U^{AREP}_{\ell} = (2\ell+1)^{-1} \, [(\ell+1) \, U^{EP}_{\ell,\ell+1/2} + \ell \, U^{EP}_{\ell,\ell-1/2}] \qquad (6)$$

In the particular case of s elections there is no SO effect and no averaging is involved. Thus molecules such as Au_2, where the bonding involves primarily s orbitals, can be treated[16,17] easily in Λ-S coupling.

Alternatively, approximate AREP have been obtained[18] by the use of atomic calculations in which the mass-velocity and Darwin terms in the Pauli approximation are added to the nonrelativistic Hamiltonian. Since the SO terms is not included, the orbitals remain the same for $j = \ell + 1/2$ and $j = \ell - 1/2$. While this method is less accurate than the averaging of results from DF atomic calculations, the difference does not appear to be significant in work published to this time.

Given the AREP, the molecular calculation is set up with a wavefunction expressed in spin-orbitals and can be completed at the SCF, MCSCF, or CI level by the same methods used in nonrelativistic calculations. Either Slater or Gaussian basis functions can be used and programs are available for nonlinear as well as linear structures.

But for accurate results in most cases the SO term must be included at the MCSCF or CI level (or as a perturbation if it is small). In most work presently available this SO term is introduced empirically[18,19] with an operator related to the experimental SO splitting in the spectra. In many cases this appears to be a good approximation. Nevertheless, one prefers a nonempirical method with a sound theoretical basis and this was recently developed by Ermler, et al.[20] The spin orbit operator for use with molecular pseudo-orbitals is simply the difference in the EP for $j = \ell + 1/2$ and $j = \ell - 1/2$ multiplied by the appropriate projection operator.

$$H^{SO} = \sum_{\ell=1}^{L-1} \Delta U^{EP}_{\ell} \, \{ \frac{\ell}{2\ell+1} \sum_{-\ell-1/2}^{\ell+1/2} |\ell,\ell+1/2,m\rangle\langle\ell,\ell+1/2,m| - \frac{\ell+1}{2\ell+1} \sum_{-\ell+1/2}^{\ell-1/2} |\ell,\ell-1/2,m\rangle\langle\ell,\ell-1/2,m| \} \qquad (7)$$

with

$$\Delta U_{\ell} = U^{EP}_{\ell,\ell+1/2}(r) - U^{EP}_{\ell,\ell-1/2}(r). \tag{8}$$

The matrix elements of H^{SO} with respect to the atomic spin-orbital basis set will have the form

$$H^{SO}_{pq}(\rho_r \rho_s) = \langle \chi_p \rho_r | H^{SO} | \chi_q \rho_s \rangle \tag{9}$$

where χ_p and χ_q are spacial basis functions and the Pauli spinors ρ_i define the α and β spins of the electrons such that $\rho_i = \alpha = \binom{1}{0}$ or $\rho_i = \beta = \binom{0}{1}$. The matrix elements of H^{SO} between various Λ-S states for a given molecule can then be obtained as a sum of these terms with appropriate expansion coefficients.

If real spin-orbitals are chosen for the basis, some of the SO matrix elements may be imaginary. This introduces some complications in CI procedures, but the roots are still real. Calculations following this procedure are in progress in this laboratory for $T\ell H$, $T\ell_2$ and Pb_2; enough results have been obtained to establish the effectiveness of the method with relatively large CI.

It appears to be relatively straightforward to extend these methods to nonlinear molecules using existing programs for all but the SO terms and this is also in progress in this laboratory.

MOLECULAR CALCULATIONS: RESULTS

In view of the fact that the bonding electron in the gold atom is an s electron without SO effect, Au_2 was chosen for early study.[16,17] The bond in this molecule is anomalously strong, stronger than in either Cu_2 or Ag_2. It is found that this bond is stronger by about one electron volt on the real, relativistic basis than on a nonrelativistic basis. It is also of interest to note that Hg_2^{++} is isoelectronic with Au_2 and the stability of this anomalous dimeric, doubly charged ion can be ascribed to this 1 eV relativistic strengthening of the bond. Numerous excited states were calculated for Au_2. Good agreement was obtained for all experimentally known quantities, for both ground and excited states, although the calculated bond distances are somewhat too short. These Au_2 calculations were made with Phillips-Kleinman EP which are now known to yield bond distances that are too short. It would be desirable to repeat the Au_2 calculations with more reliable EP.

$T\ell_2$ and $T\ell_2^+$ represent particularly interesting examples. The $T\ell$ atom has a single electron in a $6p_{1/2}$ spinor. The $p_{1/2}$ spinor is 2/3 p_π and 1/3 p_σ and if one combines these to form a diatomic molecular spinor it is either σ bonding and π antibonding (if of g symmetry) or σ antibonding and π bonding (if of u symmetry). The lighter analogs of Tl_2 (B_2, Al_2, etc.) show σ bonding which is expected from the nonrelativistic p_σ orbitals. This state can be obtained for Tl_2 but it requires promotion of the $6p_{1/2}$ electrons to p_σ orbitals which are 2/3 $p_{3/2}$ + 1/3 $p_{1/2}$ and this requires almost 2/3 eV per electron. Of course, one expects partial rather than full promotion.

The results[15] for Tl_2^+ are shown in Figure 2. These are single configuration calculations in ω-ω coupling. Since there is a single bonding electron, the correlation correction for the change in energy on dissociation should be small. We see that the $(1/2)_g$ state of Tl_2^+ is significantly bound with D_e = 0.58 eV and R_e = 3.84 Å. There is substantial promotion from $p_{1/2}$ toward p_σ spinors in this state. For the $(1/2)_u$ state of Tl_2^+ there is only a very shallow potential minimum, but it does lie at a short bond distance (3.50 Å) as would be expected for a π bond. At longer distances the σ antibonding effect yields a broad maximum in the energy curve for the $(1/2)_u$ state. Experimentally Tl_2^+ is a known species but its exact parameters (R_e, D_e) have not been determined.

Discussion of the low-lying states of Tl_2 can best begin with consideration of the situation without the SO effect as shown in the upper curves[15] of Figure 3. There are three Λ-S terms $^3\Sigma_g^-$, $^3\Pi_u$, and $^1\Sigma_g^+$ which correspond to π^2, πσ, and σ^2 bonding, respectively. The potential minima lie at about the same level. When the spin-orbit effect is included, the energy of the dissociated atoms drops far below the minima of the curves without SO. The $^3\Sigma_g^-$ term splits into 0_g^+ and 1_g states while the $^3\Pi_u$ splits into 0_u^-, 0_u^+, 1_u, and 2_u states. The $^1\Sigma_g^+$ is a single state now called 0_g^+. Among these states in ω-ω coupling only 0_g^+, 0_u^-, and 1_u dissociate to yield two ground state atoms ($^2P_{1/2}$). The potential curves[15] for these states are shown on Figure 3; none is strongly bound but, with more adequate CI, all would doubtless show significant potential minima. The 0_u^- state is lowest in our calculations but the differences are small. These calculations (made in ω-ω coupling) include only the required number of configuration for dissociation to neutral atoms. This requires two configurations for 0_g^+ but a single configuration sufficed for 0_u^- and 1_u. Thus electron corre-

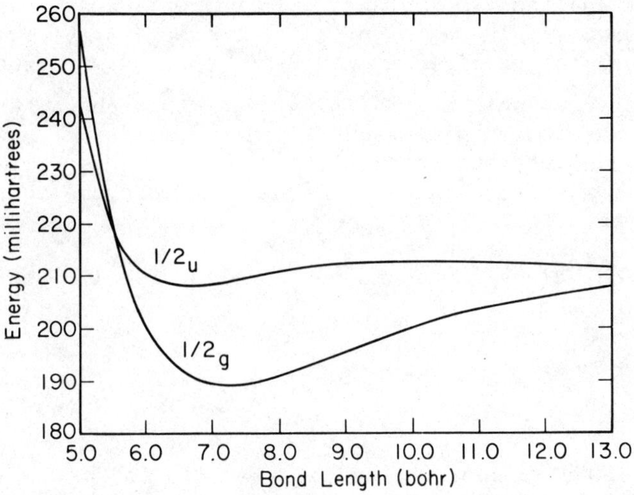

Figure 2. Potential curves for the lowest $(1/2)_g$ and $(1/2)_u$ states of $T\ell_2^+$.

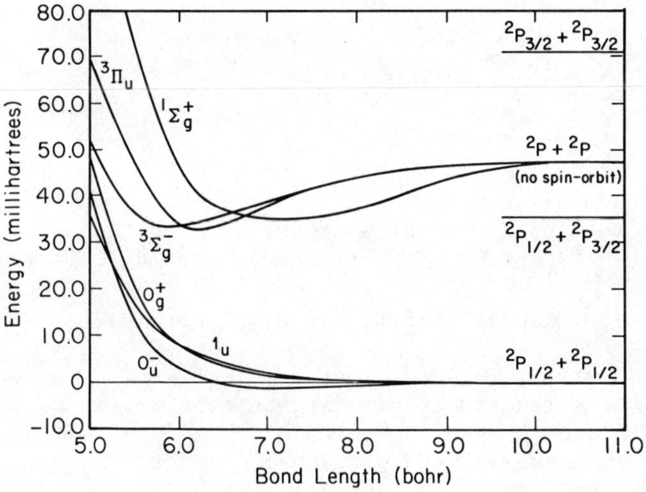

Figure 3. Potential curves for the 0_g^+, 0_u^-, and 1_u states of $T\ell_2$ with $^3\Sigma_g^-$, $^3\Pi_u$, and $^1\Sigma_g^+$ curves (computed without spin-orbit coupling) for comparison.

lation for the two bonding electrons is not well-described in these calculations and the true potential curves will be somewhat lower at bond distances. There is very little experimental evidence for $T\ell_2$ molecules; one preliminary report[21] appeared very recently. The bond distance, spin multiplicity, and other characteristics assumed in that report will need revision. It is not now clear whether revised interpretation of these experiments will yield agreement with accurate calculations giving more consideration to electron correlation.

The calculations[13] made in ω-ω coupling for TℓH yielded the results in Table 1.

Table 1. Spectroscopic Parameters for TℓH in the Ground State Calculated by the ω-ω Coupling Method

	Calculated		
	SCF	MCSCF	Experimental
R_e (A)	1.93	1.96	1.87
D_e (eV)	0.93	1.66	1.97
ω_e (cm^{-1})	1450	1330	1391

In view both of basis set inadequacies and the small number of configurations (5) in the MCSCF treatment, the agreement is good.

RESULTS OF CI-SO CALCULATIONS IN Λ-S COUPLING

This supplement, written after completion of the preceding sections, reports results of CI calculations for several states of TℓH which were made by the Λ-S coupling procedure described above. The REP were the same as used previously[13,15] for Tℓ with 10d 2s 1p valence electrons. After the SCF calculation with 14 valence electrons in TℓH, the 10 primarily d orbitals were frozen and the CI calculation included explicitly only the outer three electrons of thallium and the 1s electron of hydrogen.

Our CI wavefunction for the 0^+ state was generated from seven reference configurations with occupations (ignoring $1\sigma^2$, σ^2,

$\sigma\pi_x\alpha\alpha$, $\sigma\pi_x\beta\beta$, $\sigma\pi_y\alpha\alpha$, $\sigma\pi_y\beta\beta$, $\pi_x\pi_y\alpha\beta$, and $\pi_x\pi_y\beta\alpha$. All normal single and double promotions were allowed from the first five references. The sixth and seventh were allowed only limited single and double promotions. This results in a total of approximately 1700 determinants. These seven references are required to allow the wavefunction the flexibility of intermediate coupling. The wavefunction formed in this manner will not give a fully balanced description of the separated atoms relative to the molecule; hence, the bond energy was not computed from this wavefunction alone. Instead, the energy for the separated atoms was computed for comparison. For thallium a CI wavefunction was generated using all single and double promotions from the three references $6s^2 6p_\sigma\alpha$, $6s^2 6p_x\beta$ and $6s^2 6p_y\beta$. For the $^2P_{1/2}$ state the total energy was -50.6827 a.u. The $^2P_{3/2}$ state was higher in energy by .0339 a.u. or .92 eV, which is in reasonable agreement with the experimental splitting of .97 eV.[22]

For the 0^- state the first reference listed above (for the 0^+ state) is eliminated and certain sign relationships between the other terms are reversed. Also a $\sigma\sigma'\alpha\beta$ reference was added. Similar methods yield the appropriate references for the 1 and 2 states. From a Λ-S coupling basis the $^1\Sigma^+$ state relates to the lowest 0^+ state and the $^3\Pi$ term is split to yield the second 0^+ state and the lowest 0^-, 1, and 2 states. The $^1\Pi$ term yields the second 1 state while the highly repulsive $^3\Sigma^+$ term yields the second 0^- and third 1 states.

The calculated energies, relative to ground-state atoms, are shown in figure 4. Included are results for the first excited states of 0^+ and 1 symmetry. These states are related to the $^3\Pi$ and $^1\Pi$ terms and should be reasonably well described by the basis of these calculations.

The experimental evidence for TℓH was discussed by Ginter and Battino[23] whose potential curves for the two 0^+ states are compared in figure 5 with our calculations. Other data and references are summarized by Huber and Herzberg.[24] Calculated and experimental spectroscopic constants are given in Table II.

The calculated potential curve for the ground state is somewhat too high at short interatomic distances. The cause is probably the absence of intershell correlation involving thallium d-shell electrons together with valence-shell electrons. Expansion of the CI to include all d-shell excitations of this type would exceed the capacity of the present program. Also, to properly include these effects, one would have to expand the basis by the addition of f orbitals. The very recent nonrelativistic calculations of McLean[25] for AgH with very extensive CI lend support to this view. He finds about 0.2 bohr shortening of R_e from that for an MCSCF calculation to the values for any of a number of calculations with high order

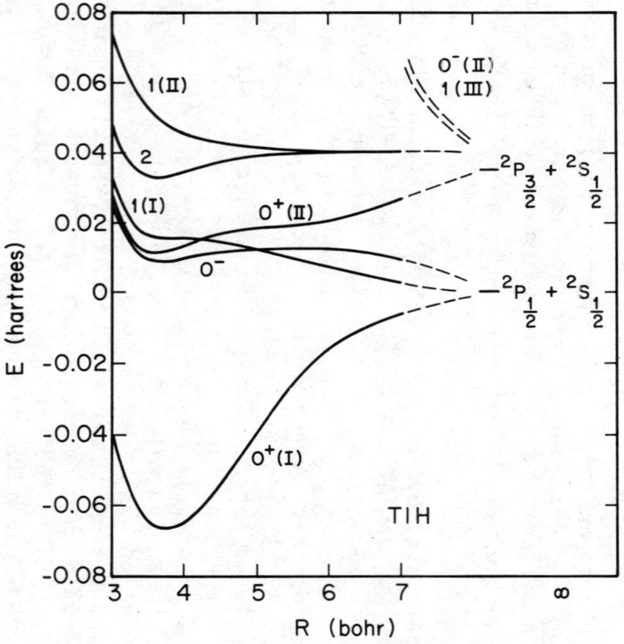

Figure 4. Computed bonding curves for the two lowest 0^+ and 1 states and the lowest 0^- and 2 states of TℓH. The dashed curves at large R are estimates.

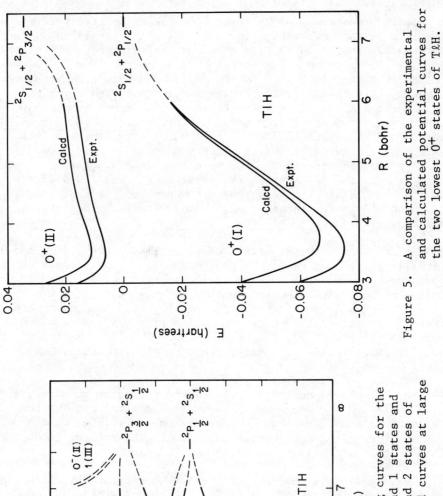

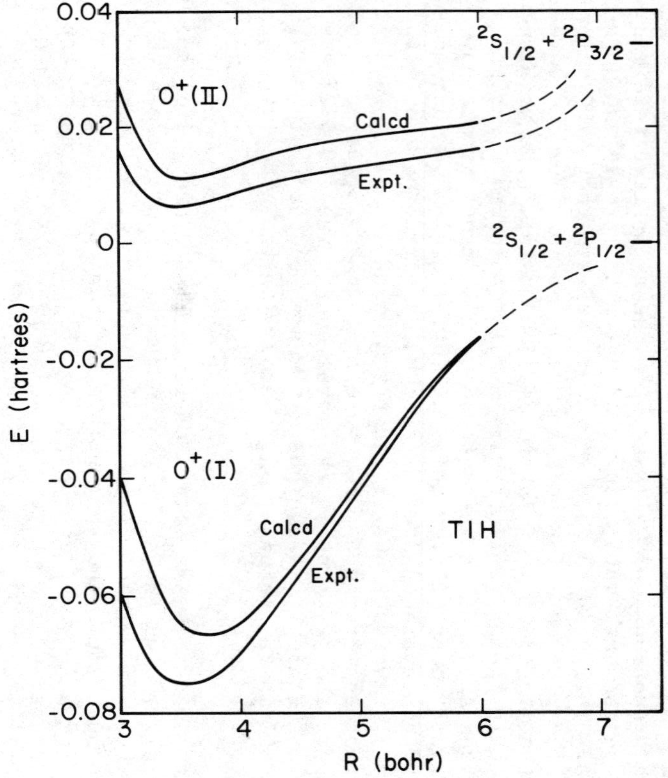

Figure 5. A comparison of the experimental and calculated potential curves for the two lowest 0^+ states of TℓH.

Table II. Spectroscopic Constants for Some Low Lying Bound States of TℓH Calculated with CI by the Λ-S Coupling Method

State	R_e (A)	D_e (eV)	ω_e (cm^{-1})	T_e (cm^{-1})
0^+(I) Theory	1.99	1.81	1300	0
0^+(I) Experiment	1.87	1.97	1391	0
0^- Theory	1.95	-	795	16600
0^+(II) Theory	1.91	0.61	1000	17100
0^+(II) Experiment	1.91	0.74	760	17723
2 Theory	1.90	-	740	21800
1(II) Theory	3.1	-	200	23400
1(II) Experiment	2.9	-	140	24180

CI including these intershell correlation terms. McLean also reports similar but less extensive results for AuH. It is clear that our wavefunction for TℓH is somewhat deficient at these short interatomic distances but further work will be required to remedy this situation. For distances greater than about 4.5 bohr, where d-electron effects on the potential curve should be negligible, the agreement is excellent.

The wavefunctions for the two 0^+ states, as expected, are dominated by singlet sigma and triplet pi character. In the bonding region the molecular ground state is essentially singlet sigma. However, at very large distances the triplet pi slightly dominates since the Tℓ atom is 2/3 p_π. The reverse is true for the excited state; at shorter distances the wavefunction is heavily dominated by triplet pi character, with the singlet sigma slightly dominating at very large distances. This interchange of sigma and pi character is apparently responsible for the peculiar behavior of the excited state around 5 to 7 bohr. Figure 5 shows the striking agreement of the shapes of the calculated and experimental curves for this 0^+(II) state.

With this substantial confirmation of these calculations for the two 0^+ states where the experimental evidence is unambiguous, it is interesting to consider the predictions for the 0^-, 1, and 2 states in relationship to the minimal experimental data for these states for TℓH and in comparison with the data for InH where the spin-orbit splitting is much smaller but still significant. First, one notes that the inner well at about 3.5 bohr in the 0^+(II), 0^-, 1(I), and 2 states appears to be at least partially the result of an avoided crossing which has been previously observed for the lowest $^3\Pi$ state of BH.[26] In the region outside the inner well the wavefunction is dominated by configurations which correspond roughly to the s^2p isolated thallium atom. However, in the region of the inner well, there is considerable sp^2 character, thereby allowing substantial sigma bonding of H with the s orbital on thallium. As noted above, this unusual shape of the excited 0^+ state agrees very well with the experimentally known potential.[23]

The inner portion of the potential curves for the 0^-, 0^+(II), 1(I), and 2 states are all very similar, hence their relationship to the $^3\Pi$ state in Λ-S coupling is pertinent. This is confirmed by an examination of the wavefunctions which are dominantly $^3\Pi$ in the range 3.0 to 3.5 bohr. The spin-orbit energies simply shift the absolute energies in this region, and the pattern is similar to that found for InH where the order is the same and the spacings also increase in the same sequence $(0^+-0^-)<(1-0^+)<(2-1)$. But the very large spin-orbit separation of the atomic energies for Tℓ has a profound effect at larger R. The curve for the 1(I) state has no significant minimum; this agrees with the failure to observe discrete spectra for this state in TℓH (in contrast to InH where it is observed).

Selection rules make direct observation of the 0^- state difficult, and it has not been measured for any of the molecules GaH, InH, or TℓH. The relative shapes of the 0^- and 1(I) curves in the vicinity of 6 bohr can be understood from the details of the wavefunctions. At long distance the Tℓ atom must approach a $s^2p_{1/2}$ configuration where the $p_{1/2}$ spinor is 1/3 p_σ and 2/3 p_π. For the 1 state the s orbital on H can immediately have a bonding interaction with the p_σ on Tℓ whereas this is not possible for the 0^- state. Thus the initial interaction of the atoms is more repulsive in the 0^- state than in the 1 state.

There are several spectral lines observed by Larsson and Neuhaus[27] for TℓH and TℓD which have been interpreted as arising from transitions from the ground state to the 2 and 1(II) states. They conclude that their "explanations are largely conjectural" and that further experiments are needed. There is little doubt that these lines lie close to the dissociation limit to $^2P_{3/2}$ and $^2S_{1/2}$ atoms. Larsson and Neuhaus find for the 1(II) state the remarkably low and anharmonic sequence of vibrational spacings of 98 and 56 cm^{-1} with ω_e = 140 cm^{-1} and an R_e value about 2.9 Å or 5.5 bohr. The calculated curve for the 1(II) state shows a nearly flat region from 5 to 7 bohr; a cubic equation through the four points in this region yields the results given in Table II, $\omega_e \cong 200$ cm^{-1} and $R_e \cong 3.1$ Å. Thus the agreement is remarkably good for such a sensitive feature in an excited state.

Larsson and Neuhaus also assign a few lines to transitions $2 \leftarrow 0^+(I)$ appearing in violation of selection rules because of a perturbation with the 1(II) state. Indeed our calculations yield a crossing of the 2 and 1(II) curves at 7 bohr. However, the curves are so flat that a very small shift in their relative energy would cause a large change in the R-value of the crossing.

Thus these calculated results are fully consistent with the experimental measurements, if allowance is made for uncertainties in accuracy, and indicate the potential of this method to calculate rather complex features in molecular potential surfaces.

ACKNOWLEDGMENTS

Great credit is due to my research associates in relativistic quantum chemical studies: Yoon S. Lee, Walter C. Ermler, and Phillip A. Christiansen. Much of our research was supported by the Division of Chemical Sciences, Office of Basic Energy Science, U. S. Department of Energy under Contract No. W-7405-ENG-48.

REFERENCES

1. R. N. Dixon and I. L. Robertson, "Specialist Periodical reports on Theoretical Chemistry, The Chemical Society, London, 3:100 (1978).
2. J. C. Phillips and L. Kleinman, Phys. Rev. 116:287 (1959).
3. P. A. Christiansen, Y. S. Lee, and K. S. Pitzer, J. Chem. Phys. 71:4445 (1979).
4. L. R. Kahn and W. A. Goddard III, Chem. Phys. Lett. 2:667 (1968).
5. Y. S. Lee, W. C. Ermler, and K. S. Pitzer, J. Chem. Phys. 67:6851 (1977).
6. R. L. Martin, Chem. Phys. Lett. 75:290 (1980); see also R. L. Martin and P. J. Hay, J. Chem. Phys. in press.

7. J. O. Noell, M. D. Newton, P. J. Hay, R. L. Martin, and F. W. Bobrowicz, J. Chem. Phys. 73:2360 (1980).
8. P. A. Christiansen, K. S. Pitzer, Y. S. Lee, J. H. Yates, W. C. Ermler, and N. W. Winter, Lawrence Berkeley Laboratory Report LBL-13035, J. Chem. Phys. in press.
9. P. J. Hay, W. R. Wadt, and L. R. Kahn, J. Chem. Phys. 68:3059 (1978).
10. W. R. Wadt, J. Chem. Phys. 68:402 (1978).
11. L. L. Foldy and W. A. Wouthuysen, Phys. Rev. 78:29 (1950).
12. Y. S. Lee, W. C. Ermler, and K. S. Pitzer, J. Chem. Phys. 73:360 (1980).
13. P. A. Christiansen, and K. S. Pitzer, J. Chem. Phys. 73:5160 (1980).
14. K. S. Pitzer and P. A. Christiansen, Chem. Phys. Lett. 77:589 (1981).
15. P. A. Christiansen and K. S. Pitzer, J. Chem. Phys. 74:1162 (1981).
16. Y. S. Lee, W. C. Ermler, K. S. Pitzer, and A. D. McLean, J. Chem. Phys. 70:288 (1979).
17. W. C. Ermler, Y. S. Lee, and K. S. Pitzer, J. Chem. Phys. 70:293 (1979).
18. L. R. Kahn, P. J. Hay, and R. D. Cowan, J. Chem. Phys. 68:2386 (1978).
19. P. J. Hay, W. R. Wadt, L. R. Kahn, R. C. Raffenetti, and D. H. Phillips, J. Chem. Phys. 71:1767 (1979).
20. W. C. Ermler, Y. S. Lee, P. A. Christiansen, and K. S. Pitzer, Chem. Phys. Lett. 81:71 (1981).
21. G. Balducci and V. Piacente, J.C.S. Chem. Comm. 1980:1287.
22. C. E. Moore, "Atomic Energy Levels", N.B.S. Circular 467, Vols. I-III.
23. M. L. Ginter and R. Battino, J. Chem. Phys. 42, 3222 (1965).
24. K. P. Huber and G. Herzberg, "Molecular Spectra and Molecular Structure, IV, Constants of Diatomic Molecules" (Van Nostrand Reinhold Co., New York, 1979).
25. A. D. McLean, private communication.
26. R. J. Blint and W. A. Goddard III, Chem. Phys. 3:297 (1974).
27. T. Larsson and H. Neuhaus, Arkiv f. Physik, 23:461 (1963); 31:299 (1966).

CALCULATION OF BONDING ENERGIES BY THE HARTREE-FOCK SLATER TRANSITION STATE METHOD, INCLUDING RELATIVISTIC EFFECTS

Tom Ziegler

Department of Chemistry
University of Calgary
Calgary, Alberta, Canada T2N 1N4

1. INTRODUCTION

Baerends *et al.* (1) proposed in 1973 a computational scheme based on the Hartree-Fock-Slater method, originally suggested by Slater (2). This scheme has recently been extended by Snijders *et al.* (3) to include relativiscic effects.

One of the features of the HFS-DVM method by Baerends *et al.* as well as the R-HFS-DVM method by Snijders *et al.* was the accurate representation of electronic potentials and densities. Thus the HFS-DVM scheme is within the HFS framework an *ab initio* method.

We shall in section 2 discuss the general methodology (4) behind HFS calculations on bond energies and bond distances, and compare the results with HF *ab initio* calculations as well as experiment. The inclusion of relativistic effects is outlined in section 3 and section 4, and the importance of such effects on the chemical bond is discussed in section 5.

2.0 THE NON-RELATIVISTIC TOTAL ENERGY EXPRESSION

In this section a brief discussion is given of the Hartree-Fock-Slater method, as well as the related statistical total energy expression. The method is introduced by comparison to the well known Hartree-Fock scheme.

The calculation of energy differences within the Hartree-Fock-Slater method is often carried out by a transition state procedure. This procedure is discussed in section 2.3 in connection with calculations of bonding energies.

The statistical energy expression can be viewed as a modification of the energy in Equation (2.3), and the statistical energy has the form:

$$E_x^{o} = \sum_i \int \phi_i^{o}(1)\,|-\tfrac{1}{2}\nabla_1^{2} + V_N(\vec{r}_1)|\,\phi_i^{o}(1).d\vec{\tau}_1$$

$$+ \frac{1}{2}\sum_i\sum_j \int \phi_i^{o}(1)\phi_i^{o}(1)\,|1/r_{12}|\,\phi_j^{o}(2)\phi_j^{o}(2).d\vec{\tau}_1.d\vec{\tau}_2 \qquad (2.5)$$

$$+ \frac{3}{4}\sum_i \int \phi_i^{o}(1)\phi_i^{o}(1)\; V_x(\sum_i \phi_i^{o}(1)\phi_i^{o}(1)).d\vec{\tau}_1$$

where

$$V_x(1) = -3\alpha_{ex}\left[\frac{3}{8\pi}\sum_i \phi_i^{o}(1)\phi_i^{o}(1)\right]^{1/3} \qquad (2.6)$$

Here α_{ex} is the exchange scale factor (2). The statistical energy expression depends only on the one-electron density matrix

$$\rho^{o}(1,1') = \sum_i \phi_i^{o}(1).\phi_i^{o}(1') \qquad (2.7)$$

and it might be written in a compact form as:

$$E_x^{o}[\rho^{o}] = \int_{1\to 1'} \rho^{o}(1,1')\left[-\frac{1}{2}\nabla_1^{2} + V_N(1) + \frac{1}{2}V_c(\rho^{o}(1)) + \frac{3}{4}V_x(\rho^{o}(1))\right].d\vec{\tau}_1 \qquad (2.8)$$

The diagonal elements of $\rho^{o}(1,1')$ have been written as $\rho^{o}(1)$, and $\rho^{o}(1)$ integrated over spin as $\rho^{o}(\vec{r}_1)$. The requirement that $E[\rho^{o}]$ shall be minimized with respect to the single-determinantal wave function ϕ^{o} leads to n one-electron equations from which the set $\{\phi_i^{o}\}$ can be determined (1).

$$f^{o}(1)\phi_i^{o}(1) = \varepsilon_i^{o}\phi_i^{o} \qquad (2.9)$$

Here $f^{o}(1)$ is the Hartree-Fock-Slater operator given by:

$$f^{o}(1) = \left\{-\frac{1}{2}\nabla_1^{2} + V_N(\vec{r}_1)\right\} + \left\{V_c(\rho^{o}) + V_x(\rho^{o})\right\}$$
$$= h^{o}(1) + V_{el}^{o}(1) \qquad (2.10)$$

2.1 THE NON-RELATIVISTIC HARTREE-FOCK METHOD

The non-relativistic n-electron Hamiltonian is given in atomic units by:

$$H^o = -\frac{1}{2}\sum_i \nabla_i^2 + \sum_i V_N(\vec{r}_i) + \frac{1}{2}\sum_{ij} 1/|\vec{r}_i - \vec{r}_j| \tag{2.1}$$

Here the three terms in Equation (2.1) from left to right represent the electronic kinetic energy, the electron-nucleus attraction, and the electron-electron repulsion. The operator $V_N(\vec{r}_i)$ is given by $\sum_A Z_A/|\vec{r}_i - \vec{R}_A|$.

The Hartree-Fock method affords the single determinantal wave function

$$\phi^o = |\phi_1{}^o(1)\phi_2{}^o(2) \cdots \phi_m^o(m)| \tag{2.2}$$

with respect to which, the total energy

$$\begin{aligned} E^o = {} & \sum_i \int \phi_i{}^o(1)| -\frac{1}{2}\nabla_1{}^2 + V_N(\vec{r}_1)|\phi_i{}^o(1).d\vec{\tau}_1 \\ & + \frac{1}{2}\sum_i\sum_j \int \phi_i{}^o(1)\phi_i{}^o(1)|1/r_{12}|\phi_j{}^o(2)\phi_j{}^o(2).d\vec{\tau}_1.d\vec{\tau}_2 \\ & - \frac{1}{2}\sum_i\sum_j \int \phi_i{}^o(1)\phi_j{}^o(1)|1/r_{12}|\phi_i{}^o(2)\phi_j{}^o(2).d\vec{\tau}_1.d\vec{\tau}_2 \end{aligned} \tag{2.3}$$

has a minimum.

The requirement, that E^o shall be minimized with respect to ϕ^o, gives rise to n one-electron equations, from which the optimal set of one-electron orbitals $\{\phi_i{}^o\}$ can be determined:

$$\left\{-\frac{1}{2}\nabla_1{}^2 + V_N(\vec{r}_1) + \sum_j \int \frac{\phi_j{}^o(2)(1-P_{12})\phi_j{}^o(2)d\vec{\tau}_2}{r_{12}}\right\}\phi_i(1) = \varepsilon_i{}^o\phi_i{}^o(1) \tag{2.4}$$

2.2 THE NON-RELATIVISTIC HARTREE-FOCK-SLATER METHOD

The Hartree-Fock-Slater method first suggested by J.C. Slater (2) is based on the statistical total energy expression. It affords a single determinantal wave function for which the statistical energy is minimized.

2.3 CALCULATIONS OF BOND ENERGIES BY THE TRANSITION STATE METHOD

Let us consider the chemical process:

$$A + B \rightarrow AB \tag{2.11}$$

The corresponding bonding energy for the process in Equation (2.1) is given by

$$\Delta E = E^{o}[\rho_{AB}] - E^{o}[\rho_{A}] - E^{o}[\rho_{B}] \tag{2.12}$$

Here ρ_A, ρ_B, and ρ_{AB} are the non-relativistic electronic densities for the three electronic systems A, B, AB.

In their implementation of the HFS-method, Baerends *et al.* (1) make extensive use of numerical integration. Thus bonding energies can not be evaluated with confidence as a relative small difference between large numbers (total energies). This problem can be circumvented by the derivation of a direct expression for the bonding energy (4) ΔE in the following way. Neglecting for the moment terms containing V_x, and writing ρ_{AB} as $\rho_A + \rho_B + \Delta\rho_{AB}$, we have:

$$\Delta E = \Delta E_{elstat} + \int_{1\rightarrow 1'} \Delta\rho_{AB}(1,1') \left| -\frac{1}{2}\nabla_1^2 + V_N(\vec{r}_1) + V_c(\vec{r}_1) \right| . d\vec{\tau}_1 \tag{2.13}$$

Here

$$\Delta E_{elstat} = \sum_{\alpha}^{A}\sum_{\beta}^{B} \frac{Z_\alpha Z_\beta}{R_{\alpha\beta}} + \int \frac{\rho_A(\vec{r}_1)\rho_B(\vec{r}_2)}{r_{12}} d\vec{r}_1 d\vec{r}_2$$
$$- \sum_{\alpha}^{A} \int \frac{Z_\alpha Z_\beta(\vec{r}_1)}{|\vec{r}_1 - R_\alpha|} dr_1 - \sum_{\beta}^{B} \int \frac{Z_\beta \rho_A(\vec{r}_1)}{|\vec{r}_1 - R_\beta|} d\vec{r}_1 \tag{2.14}$$

is the electrostatic interaction between A and B.

Including now the exchange terms we have in addition:

$$I = \frac{3}{4}\int \rho_{AB}(\vec{r}_1)V_x(\rho_{AB})d\vec{\tau}_1 - \frac{3}{4}\int \rho_A(\vec{r}_1)V_x(\rho_A)d\vec{\tau}_1$$
$$- \frac{3}{4}\int \rho_B(\vec{r}_1)V_x(\rho_B)d\vec{\tau}_1 \tag{2.15}$$

By rearrangement:

$$I = \Delta E_{exch} + \frac{3}{4}\int \rho_{AB}(\vec{r}_1)V_x(\rho_{AB})\cdot d\vec{\tau}_1 - \frac{3}{4}\int\left[\rho_A(\vec{r}_1) + \rho_B(\vec{r}_1)\right] V_x\left(\left[\rho_A(\vec{r}_1) + \rho_B(\vec{r}_1)\right]\right)\cdot d\vec{r}_1 \quad (2.16)$$

Here

$$\Delta E_{exch} = \frac{3}{4}\int (\rho_A + \rho_B)V_x(\rho_A + \rho_B)d\vec{r}_1 - \frac{3}{4}\int \rho_A V_x(\rho_A)d\vec{r}_1 - \frac{3}{4}\int \rho_B V_x(\rho_B)d\vec{r}_1 \quad (2.17)$$

is the exchange interaction between A and B.

The difference between the last two terms in Equation (2.16) can be evaluated by a Taylor expansion (4) to any order required in $\Delta\rho_{AB}$. To second order we have:

$$\frac{3}{4}\int \rho_{AB}(1)V_x(\rho_{AB})d\vec{\tau}_1 - \frac{3}{4}\int\left[\rho_A(1) + \rho_B(1)\right]V_x\left(\left[\rho_A(1) + \rho_B(1)\right]\right)d\vec{\tau}_1 \quad (2.18)$$
$$\cong \int \Delta\rho_{AB}\cdot V_x(\rho_A + \rho_B + \frac{1}{2}\Delta\rho_{AB})\cdot d\vec{\tau}_1$$

The final expression for the bonding energy takes the form:

$$\Delta E = \Delta E_{elstat} + \Delta E_{exch} + \int \Delta\rho_{AB}(1)f^{o}(\rho_A + \rho_B + \frac{1}{2}\Delta\rho_{AB})\cdot d\vec{\tau}_1 \quad (2.19)$$

Here $f^{o}(\rho)$ is just the Hartree-Fock-Slater operator defined in Equation (2.10). The procedure outlined above is often referred to as the transition state method since the operator f is defined with respect to the intermediate density $\rho = \frac{1}{2}\{\rho_{AB} + \rho_A + \rho_B\}$. The transition state method can also be used to analyse the chemical bond as discussed in References(6,7). It is worth to point out that the method in itself only is a technique by which energy differences can be evaluated in a numerically stable way.

Comparison between bonding energies calculated by the HF and the HFS method respectively.

We compare in table 1 bond distances (R_e), bond energies (D_e) and stretching frequencies (ω_e) for some diatomic molecules, calculated by either the HF-method or the HFS-method. The same extensive basis-set was used in both cases. The HFS results compare at

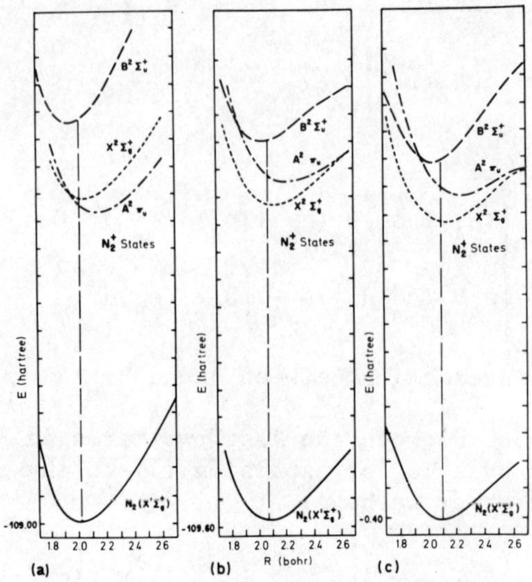

Figure 1. The energy of N_2 and the first three states of N_2^+ as a function of the inter-atomic distance. Fig. 1a is from HF-calculations, Fig. 1b from experiment and Fig. 1c is from HFS-calculations. The unit of the ordinate is .05 au,.

Table I. Calculation of bond distance (R_e), bond energies (D_e) and vibrational frequencies (ω_e) from HFS calculations and HF-calculations compared with experiment.

(R_e) (au.)			(D_e) (ev.)			(ω_e) (cm^{-1})		
HFS	HF [a]	EXP. [b]	HFS	HF	EXP.	HFS	HF	EXP.
2.61	2.51	2.68	3.54	-1.63	1.67	1207	1257	892
2.33	2.18	2.28	4.14	1.43	5.22	1563	2000	1580
2.11	2.01	2.07	10.86	5.27	9.91	2456	2730	2358
2.16	2.08	2.13	9.41	7.89	11.22	2248	2431	2170
2.19	----	2.17	6.42	----	6.62	1903	----	1904

a Ref. 8

b Ref. 9

least as well with experiment as the results from *ab initio* HF calculations.

In Figure 1 we compare the experimental energy surface of N_2 and the first three states of the N_2^+ ion with respectively $E_{HF}(r)$ and $E_{HFS}(r)$. Experiment, Fig. 1b, reveals convergence of the two first states, X and A, of N_2^+ at large inter-atomic distances, and a crossover of A and B at small distances. Both trends are reproduced by HFS but not by HF where, in addition the A state is placed below X. A more extensive comparison of the two methods is given in Refs. (4,5).

3. THE RELATIVISTIC TOTAL ENERGY EXPRESSION

Non-relativistic calculations are usually carried out in the 2x2 Pauli representation where each electron has two spin components.

In the relativistic case one can use either the 4x4 Dirac representation or the 2x2 Pauli representation. We shall in this section discuss the relation between the two relativistic formulations.

3.1 THE RELATIVISTIC TOTAL ENERGY EXPRESSION IN THE 4x4 DIRAC REPRESENTATION

It is not possible to write down a particle conserving many-electron Hamiltonian. However, neglecting terms containing retardation as well as some magnetic effect, one can write an approximate Hamiltonian as

$$H^D = \sum_i \left[c\, \vec{\alpha}_i \vec{p}_i + c^2\beta + V_N(\vec{r}_i) \right] + \frac{1}{2} \sum_{ij} {}^1/_{r_{ij}} \qquad (3.1)$$

Here $\vec{\alpha}_j$, j = 1,2,3, and β are the 4x4 Dirac matrices, $\vec{p}$ is the momentum operator and c the velocity of light. Minimization of the total energy

$$E^D = \langle \Psi^D | H | \Psi^D \rangle \qquad (3.2)$$

where Ψ^D is a single determinantal wave function of four-component one electron spinor $\{\Psi_i\}$

$$\Psi^D = |\Psi_1(1) \ldots\ldots \Psi_n(n)| \qquad (3.3)$$

leads to n one-electron equations from which the optimal set $\{\Psi_i\}$ can be determined

$$\left[c\vec{\alpha}\cdot\vec{p} + c^2\beta + V_N(r_1) + \sum_j \int \frac{\Psi_j(2)(1-P_{12})\Psi_j(2)}{r_{12}} \right] \Psi_i(1) = \varepsilon_i \Psi_i(1) \qquad (3.4)$$

3.2 THE RELATIVISTIC TOTAL ENERGY EXPRESSION IN THE 2x2 PAULI REPRESENTATION

Foldy and Wouthuysen (10) have shown that it is possible in the 2x2 Pauli representation to find a Hamiltonian equivalent to H^D by a unitary transformation.

$$H^D \rightarrow H^P = \sum_i H\ (i) + \frac{1}{2} \sum_{ij} G(i_1 j) \tag{3.5}$$

The optimal total energy

$$E = \langle \phi^P | H^P | \phi^P \rangle \tag{3.6}$$

with respect to a single determination wave function

$$\phi^P = |\phi_1(1)\phi_2\ (2).....\phi_m(m)| \tag{3.7}$$

is now determined by the set of solutions $\{\phi_i\}$ to the one-electron equation

$$\{H(1) + \int \sum_j \left[\phi_j(2)G(1,2)(1-P_{12})\phi_j{}^*(2)\cdot d\vec{\tau}_2\right]\}\phi_i(1) = \varepsilon_i\phi_i(1) \tag{3.8}$$

The procedure outlined in Eq. (3.6) to Eq. (3.8), although formally correct, has little practical value, since both of the operators H(1) and G(1,2) in fact consist of an infinite sum of terms (10,11), ordered in powers of α^2

$$H(1) = \sum_{n=o}^{\infty} \alpha^{2n}\ h^{(n)}(1);\ G(1,2) = \sum_{n=o}^{\infty} \alpha^{2n} g^{(n)}(1.2) \tag{3.9}$$

where α is the fine structure constant. Thus relativistic calculations in the Pauli representation are only practical, in as much as H(1) and G(1,2) can be approximated by a finite sum of terms.

With an eye on such an truncation at a later stage, one can reformulate the procedure in Eq. (3.6) to Eq. (3.8) in a complitely equivalent way using perturbation theory. The virture of such a formulations is, that when the sum in Eq. (3.9) is truncated, then the energy as well as the wave function is evaluated at the same level of approximation. In the perturbation theory α^2 is considered as a perturbational parameter. Each of the one-electron orbitals $\phi_i(1)$ are formally expressed as an infinite sume of terms ordered in powers of α^2:

$$\phi_i(1) = \sum_{n=0}^{\infty} \alpha^{2n}\ \phi_i{}^{(n)}(1) \tag{3.10}$$

A substitution of Eq. (3.10) into Eq. (3.8) followed by a collection of terms to the same power in α^2, affords instead of Eq. (3.8) a set of n equations

$$\sum_{r=0}^{n} \left\{ h_i^{(n-r)}(1) + \left[\sum_{j}^{n-r}\sum_{s} \sum_{t}^{n-r-s} \int \phi_j^{(s)}(2)\, g(1,2)^{(n-r-s-t)} (1-P_{12}) (\phi_j^{(t)}(2)^{*} \cdot d\vec{\tau}_2 + cc.\right] \phi_i^{(r)}(1) = \varepsilon_i \phi_i^{(r)}(1); \right. \tag{3.11}$$

$n = 0,1,2....\infty.$

from which $\phi_i(1)$ can be determined by normal perturbational procedures.

The zero order equation (n=o) thus afford $\{\phi_i^o(1)\}$, the non-realitivistic orbital set. The subsequent solution of the first order equation (n=1), the second order equation (n=2) as well as higher order equations provide $\{\phi_i^1(1)\}$ and $\{\phi_i^2(1)\}$ as well as higher order contributions. One can finally from the orbital set $\{\phi_i\}$ write down the total energy, now ordered in powers of α^2:

$$E = \sum_{n=0}^{\infty} \alpha^{2n} E^{(n)} \tag{3.12}$$

Carried to the limit ($n\to\infty$), the energy expression of Eq. (3.12) is identical to that of Eq. (3.6).

4. THE REALTIVISTIC HARTREE-FOCK-SLATER METHOD.

Snijders *et al* (3) have extended the implementation of the HFS-method to the relativistic case. In their method terms are retained to first order in Eq. (3.10). Thus,

$$\phi_i(1) \simeq \phi_i^o(1) + \alpha^2 \phi_i^{(1)}(1) \tag{3.13}$$

The zero-order orbitals $\{\phi_i^o\}$ are solutions to the non-relativistic HFS-equation already given in section 2.2 as:

$$f^o(1)\phi_i^o(1) = \{h^o(1) + V_{el}^o(1)\}\phi_i^o(1) = \varepsilon_i^o\phi_i^o(1) \tag{2.9}$$

The first order corrections $\{\phi_i^1\}$ are on the other hand solutions to the first order equation (n=1):

$$(f^o-\varepsilon_i^o)\phi_i^1 = (\varepsilon_i^1 - f^1)\phi_i^o \tag{4.14}$$

where (3)

$$f^1(1) = h^1(1) + V^1_{el}(1) \qquad (4.15)$$

In Eq. (4.15)

$$V^1_{el}(1) = \int \rho^1(\vec{r}_2)/r_{12} \cdot d\vec{r}_2 - \alpha_{ex}\left(\frac{3}{8\Pi}\right)^{1/3}\left[\rho^o(\vec{r}_1)\right]^{-2/3} \cdot \rho^1_1(\vec{r}_1) \qquad (4.16)$$

is the first order correction to the electron-electron Coulomb interaction, due to the change in density

$$\rho^1_1(1) = \sum_i \phi^1_i(1) \cdot \phi^{o\,*}_i(1) + \phi^o_i(1) \cdot \phi^{1\,*}_i(1) \qquad (4.17)$$

induced by relativity. On the other hand

$$h^1(1) = -\alpha^2/8\ \nabla^4 + \alpha^2/8\ \nabla^2 V_N(1) + \alpha^2/8\ \vec{\sigma} \cdot \left[\vec{\nabla}(V_N + V^{core}_{el}) x\vec{p}\right] \qquad (4.18)$$

The first term in Eq. (4.18) is the mass-velocity correction to the non-relativistic kinetic energy (11). The second term is the relativistic Darwin correction to the electron-nucleus attraction due to the "Zitter-bewegung" of the electron (11). The last term represents the spin-orbit interaction. In the spin-orbit term $\vec{\delta}$ are the Pauli-matrices, and

$$V^{core}_{el} = \int \rho^o_{core}(2)/r_{12} \cdot dr_2 - 3\alpha_{ex}\left[\left(\frac{3}{8\Pi}\right)\rho^o_{core}\right]^{1/3} \qquad (4.19)$$

Here ρ^o_{core} (1) is the non-relativistic density due to the core electrons. The relativistic total energy expression, consistent with the R-HFS method due to Snijders *et al.* (3) is given by:

$$E \simeq E^o_x[\rho^o] + \alpha^2 E^1[\rho^o] + \alpha^4 E^2[\rho^1] \qquad (4.20)$$

Here $E^o_x[\rho^o]$ is just the non-relativistic statistical energy expression of Eq. (2.6). Further,

$$\alpha^2 E^{(1)}[\rho^o] = \int_{1\to 1'} h^1(1)\rho^o(1,1')d\vec{\tau}_1 \qquad (4.21)$$

and

$$\alpha^4 E^{(2)}[\rho^{(1)}] = \frac{1}{2}\int_{1\to 1'} h^1(1)\rho^1(1,1')d\vec{\tau}_1 \qquad (4.22)$$

It is worth noting that the first order relativistic correction to the total energy $E^{(11)}[\rho^o]$ only depends on the non-relativistic density. This is a general result, and would apply in the HF-case as well. The first order change in the density $\rho^{(1)}$, induced by relativity, enters the total energy to second order. The reader is refered to Ref. (12) for a more detailed derivation of Eq. (4.21) and Eq. (4.22).

4.1 RELATIVISTIC CORRECTIONS TO THE BONDING ENERGY

The relativistic corrections to the bonding energy for the process A + B → AB can now be written as

$$\Delta E \simeq \Delta E^{(1)} + \Delta E^{(2)} = \alpha^2\{E^{(1)}[\rho^o_{AB}]-E^{(1)}[\rho^o_A]-E^{(1)}[\rho^o_B]\} + \alpha^4\{E^{(2)}[\rho^{(1)}_{AB}] - E^{(2)}[\rho^{(1)}_A] - E^{(2)}[\rho^{(1)}_B] \tag{4.23}$$

Here

$$\Delta E^{(1)} = \frac{\alpha^2}{8}\int (\nabla^2 V^B_N)\rho^o_A \cdot d\vec{x} + \frac{\alpha^2}{8}\int (\nabla^2 V^A_N)\rho^o_B \cdot d\vec{x} + \frac{\alpha^2}{4}\int \vec{\sigma}\cdot[\nabla(V^B_N + V^B_{core}) \times \vec{p}]\rho^o_A \cdot d\vec{x} \tag{4.24}$$

$$\frac{\alpha^2}{4}\int \vec{\sigma}\cdot[\nabla(V^A_N + V^A_{core}) x\vec{p}]\rho^o_B \cdot d\vec{x} + \int_{1\to 1'} h^1(1)\Delta\rho^o_{AB}(1,1')\cdot d\vec{x}_1$$

and

$$\Delta E^{(2)} = \frac{\alpha^2}{16}\int (\nabla^2 V^B_N)\rho^1_A \cdot d\vec{x} + \frac{\alpha^2}{16}\int (\nabla^2 V^A_N)\rho^1_B \cdot d\vec{x} + \frac{\alpha^2}{8}\int \vec{\sigma}[\nabla(V^B_N + V^B_{core}) x\vec{p}]\rho^{(1)}_A \cdot d\vec{x}$$

$$\frac{\alpha^2}{8}\int \vec{\sigma}\cdot[\nabla(V^A_N + V^A_{core}) x\vec{p}]\rho^{(1)}_B \cdot d\vec{x} + \frac{1}{2}\int_{1\to 1'} h^1(1)\Delta\rho^{(1)}_{AB}(1,1')\cdot d\vec{x} \tag{4.25}$$

All but the last term in Eq. (4.24) and Eq. (4.25) are in practice neglectable, although all terms have been retained in the numerical calculations represented in Ref. (12) and Table 2.

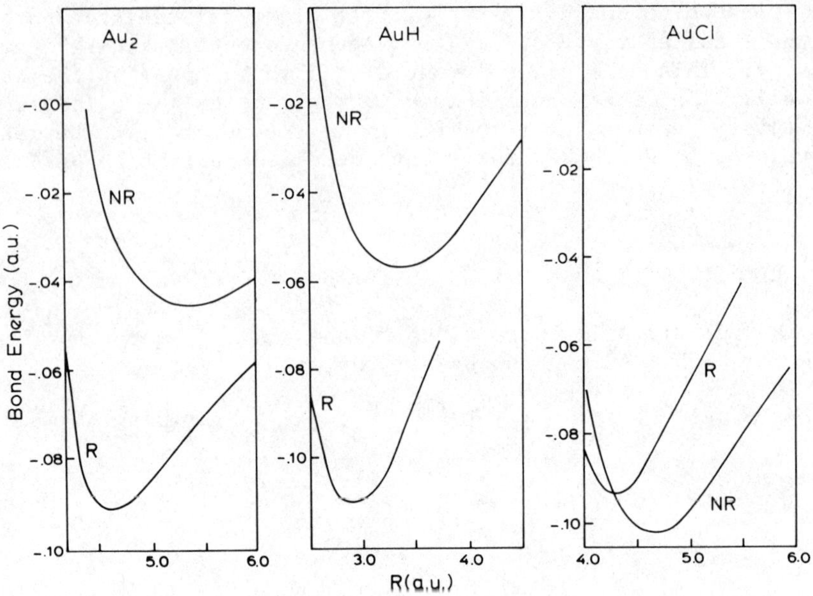

Figure 2. The bonding energy of Au_2, AuH and AuCl as a function of the inter-atomic distance R from non-relativistic HFS calculations (NR) as well as relativistic HFS calculations (R).

Table II. Calculations on bond distances (R_e), dissociation energies ($D_e = -\Delta E$), and vibrational frequencies (ω_e) based on the relativistic HFS method and the REP-HF-CI procedure, compared with the experimental data. Results from nonrelativistic calculations are in parentheses.

	R_e (Å)			D_e (kcal/mol)			ω_e (cm^{-1})		
Compound	HFS	HF-CI	EXP	HFS	HF-CI	EXP	HFS	HF-CI	EXP
AuH	1.55(1.78)	1.52(1.81)[a]	1.52	68(37)	51(36)	74 3	2241(1704)	1871(1217)	2305
AuCl	2.31(2.44)	2.29(2.45)[a]	....	58(64)	45(60)	69 15	386(363)	298(277)	383
Au_2	2.44(2.90)	2.47(2.83)[b]	2.47	58(27)	52(...)	52 2	201(93)	165(105)	919
$HgCl_2$	2.36(2.45)	2.30(2.41)[a]	2.25	107(128)	60(100)	106	329(249)	...	360

a Ref. 13

b Ref. 14

5. COMPUTATIONAL RESULTS FROM RELATIVISTIC RHFS CALCULATIONS ON BOND ENERGIES AND BOND DISTANCES

5.1 COMPOSITION WITH RELATIVISTIC HARTREE-FOCK CALCULATIONS (REP -HF)

REP-HF-CI calculations have recently been published on Au_2 by Lee *et al.* (13) and on AuH, AuCl, and $HgCl_2$ by Kahn *et al.* (14). Their results are compared in Table 2 to results from relativistic HFS calculations (RHFS) on the same four molecules.

The relativistic corrections calculated by the two methods follow closely the same trend in Au_2, AuH, AuCl, and $HgCl_2$.

The bond length and stretching frequencies evaluated nonrelativistically are, respectively, decreased and increased by relativistic corrections in such a way that the final results (including relativisitic effects) are in better agreement with experiment.

The relativistic corrections to the bonding energies increase the strength of the bond in AuH and particularly Au_2, and decrease the strength in AuCl and $HgCl_2$. For the results based on the RHFS method this adjustment brings about an improved agreement with experiment (compared to nonrelativistic HFS results) for all four molecules. The same is true in the case of the REP-HF-Cl method for AuH as well as Au_2 but not for AuCl and $HgCl_2$.

Probably, the poor agreement here is due to a too low bonding energy in the nonrelativistic case rather than an inadequate relativistic treatment. The relatively good agreement of the HFS results with experiment finds its origin partly in the difference in the nonrelativistic treatments, as documented extensively in Refs. 4 and 5, and section 2.3.

5.2 RELATIVISTIC BOND CONTRACTION

The bonding energies of Au_2, AuH and AuCl are shown in Fig. 2 as functions of the interatomic distances R from non-relativisitic HFS calculations as well as relativistic HFS calculations.

The bond distance is contracted in all three molecules irrespectively of whether the bond strength is increased as in Au_2, AuH, or decreased as in AuCl. Similar contractions have been obtained from calculations on various other molecules using a number of different methods (12, 13, 14, 15). The non-relativistic bonding energy can be written from Eq. (2.9) as

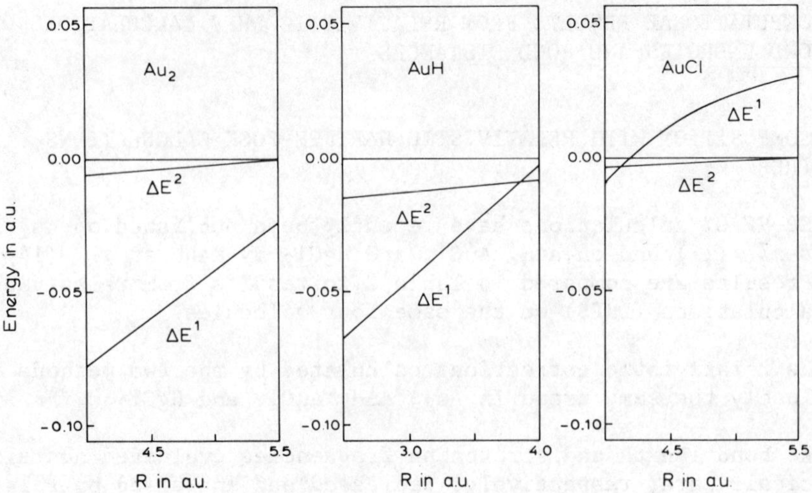

Figure 3. First- and second-order relativistic corrections to the binding energies of Au_2, AuH and AuCl as a function of the interatomic distance R.

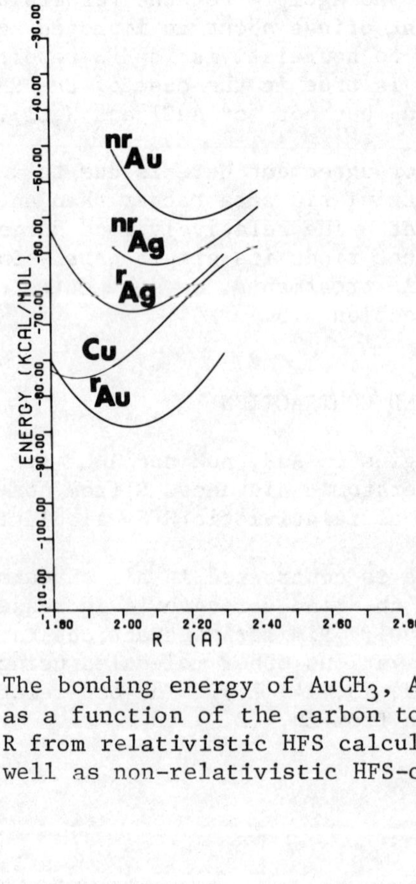

Figure 4. The bonding energy of $AuCH_3$, $AgCH_3$ and $CuCH_3$ as a function of the carbon to metal distance R from relativistic HFS calculations (R) as well as non-relativistic HFS-calculations.

$$\Delta E^{o} = -\frac{1}{2}\int_{1\to 1'} \nabla^2(1)\Delta\rho^{o}_{AB}(1,1')\cdot d\vec{\tau}_1 + \int \Delta\rho^{o}_{AB}(1)[V_x(\rho^{o}_A + \rho^{o}_B + \frac{1}{2}\Delta\rho^{o}_{AB}) + V_c(\rho^{o}_A + \rho^{o}_B + \tfrac{1}{2}\Delta\rho^{o}_{AB})]d\vec{\tau}_1 + \Delta E^{o}_{elstat} + \Delta E^{o}_{exch} \quad (5.1)$$

We have in addition for the first order relativistic correction

$$\Delta E^1 = -\frac{\alpha^2}{8}\int_{1\to 1'} \nabla^4(1)\Delta\rho^{o}_{AB}(1,1')\, d\vec{\tau}_1 + \frac{\alpha^2}{8}\int(\nabla^2 V_N(1))\Delta\rho^{o}_{AB}\cdot d\vec{\tau}_1 \quad (5.2)$$

and for the second order relativistic correction, see Eq. (2.24) or Eq. (4.25)

$$\Delta E^2 = -\frac{\alpha^2}{16}\int_{1\to 1'} \nabla^4(1)\Delta\rho^{1}_{AB}(1,1')d\vec{\tau}_1 + \frac{\alpha^2}{16}\int(\nabla^2 V_N(1))\Delta\rho^{1}_{AB}\cdot d\vec{\tau}_1 \quad (5.3)$$

Note that spin orbit-corrections are absent in Eq. (5.2) and Eq. (5.3) since all three molecules considered are closed shells. The first order corrections as well as the second order corrections are shown for Au_2, AuH and AuCl as a function of R in Fig. 3.

It is clear from Fig. 3 that it is the first order correction that primarily is responsible for the relativistic bond contraction Au_2, AuH and AuCl. The same conclusion holds for other molecules on which relativistic calculations have been carried out. Pyykko (15) was one of the first to call attention to the relativistic bond contraction. He explained it as a second order effect caused by changes in the density (s-orbital contraction) induced by relativity (16). However, it is clear from Fig. 3 that second order effects have a minor influence on bond lengths and bond distances. For a recent discussion of this point see Refs. (17,18). It is not difficult to understand the physical reason for the contraction caused by ΔE^1 of Eq. (5.2).

Non-relativistically (Eq. (5.1)) the kinetic energy rises as the atoms are pushed together, in accordance with the uncertainty principle, since the electrons are forced to occupy a smaller volume (of course the potential energy falls at first leading to a minimum in the energy curve). The most important contribution to ΔE^1 of Eq. (5.2) is the mass-velocity correction to the kinetic energy $-\alpha^2/8\cdot P^4$, a negative definite quantity, which will tend to become more and more negative as the positive non-relativistic kinetic energy $P^2/2$ goes up. It thus partly relaxes the kinetic repulsion and will cause the minimum in the bonding energy curve to shift to smaller internuclear distances, thus explaining the relativistic bond-length contraction. A more extensive discussion of the contraction is given in Refs. (12, 17, 18).

5.3 INFLUENCE OF RELATIVITY ON PERIODIC TRENDS

The influence of relativity on periodic trends has been the subject of two recent review papers (16, 19). We shall in this final section illustrate the influence of relativity by considering the energy curves of $CuCH_3$, $AgCH_3$ and $AuCH_3$ as shown in Fig. 4.

In the non-relativistic case the order of stability is $CuCH_3 > AgCH_3 > AuCH_3$. The inclusion of relativistic effects results in the new order of stability $AuCH_3 > CuCH_3 > AgCH_3$ We note again in Fig. 4 the by now well known relativistic bond contraction.

REFERENCES

1. E.J. Baerends, D.E. Ellis, and P. Ros, Chem. Phys., 2, 41 (1973)
2. J.C. Slater, Phys. Rev., 81, 385 (1951)
3. J.G. Snijders and E.J. Baerends, Mol. Phys., 36, 1789 (1978)
   J.G. Snijders, E.J. Baerends and P. Ros, Mol. Phys., 38, 1909 (1979)
4. T. Ziegler and A. Rauk, Theor. Chim. Acta., 46, 1 (1977)
5. E.J. Baerends and P. Ros, Int. J. Quantum Chem. Symp. 12, 169 (1978)
6. T. Ziegler and A. Rauk, Inorg. Chem., 18, 1558 (1979)
7. T. Ziegler and A. Rauk, Inorg. Chem., 18, 1755 (1979)
8. P.D. Cade, K.D. Sales and W.C. Wahl, J. Chem. Phys., 44, 1973 (1966).
9. Tables of Interatomic Distances and Configurations in Molecules and Atoms, edited by L.E. Sutton, Chem. Soc. Spec. Publ., 11 (1958)
10. L.L. Foldy and S.A. Wouthuysen, Phys. Rev., 78, 29 (1950)
11. R.E. Moss, Advanced Molecular Quantum Mechanics (Chapman and Hall, London (1973))
12. T. Ziegler, J.G. Snijders and E.J. Baerends, J. Chem. Phys., 1271, 74 (1981)
13. Y.S. Lee, W.C. Ermler, K.S. Pitzer and A.D. McLean, J. Chem. Phys., 70, 293 (1979)
14. P.J. Hay, W.R. Wadt, L.R. Kahn, and F.W. Bobrowicz, J. Chem. Phys., 69, 984 (1978)
15. P. Pyykko, Adv. Quantum Chem., 11, 353 (1978)
16. P. Pyykko and J.P. Desclaux, Acc. Chem. Res., 12, 276 (1979)
17. T. Ziegler, J.G. Snijders and E.J. Baerends, Chem. Phys. Lett., 75, 1 (1980)
18. J.G. Snijders and P. Pyykko, Chem. Phys. Lett., 75, 5 (1980)
19. K.S. Pitzer, Acc. Chem. Res., 12, 271 (1979)

RELATIVISTIC CALCULATIONS FOR ATOMS, MOLECULES AND IONIC SOLIDS: FULLY AB-INITIO CALCULATIONS AND THE FOUNDATIONS OF PSEUDO-POTENTIAL AND PERTURBATION THEORY METHODS

N.C. Pyper

University Chemical Laboratory
Lensfield Road
Cambridge CB2 1EW
ENGLAND

INTRODUCTION

It need hardly be pointed out at this conference that there is currently much interest in elucidating the role played by relativity in determining the electronic structure of atoms, molecules and solids. The subject of my two talks is the behaviour of the valence electrons in both atoms and molecules because the chemical properties of an element as well as the low frequency region of its electronic spectrum are essentially determined by these electrons. Thus the core electrons are only of interest in as far as they influence the behaviour of the valence electrons. There is evidence (eg. the appendix of |1|) that the purely quantum electrodynamic (QED) effects of Lamb shift and vacuum polarization, although important for core electrons |2|, are negligible for valence electrons. Consequently, these talks will be concerned solely with relativistic effects, defined as those contained with the Dirac equation and its many electron generalization (the Brown Hamiltonian including the Breit interaction), and will not consider the 'one-body' QED Lamb shift and vacuum polarization effects. The relativistic Dirac-Fock and non-relativistic Hartree-Fock Koopmans therem predictions of atomic ionization potentials shown in Table 1 provide ample evidence for the importance of relativistic effects in the valence shells of heavy atoms. Indeed the doubling of the predicted first ionization potential of E112 caused by relativity shows that one should not hope using non-relativistic calculations to make any sensible predictions about the chemistry of the 7p series of super heavy elements whose nuclear changes lies within the predicted island of nuclear stability |3|.

It seems that there are two distinct methods for investigating relativistic effects. The first is the direct use of the Dirac hamiltonian and four component orbitals thereby incorporating the major one-electron like relativistic effects to all orders in $(z/c)^2$ (z = the nuclear charge and c is the velocity of light†). The second method is the use of non-relativistic wavefunctions, the effects of relativity being introduced by perturbing operators. It might be hoped that the complexities of full relativistic calculations based on the Dirac equation could be avoided by adopting the second approach and it might further be hoped that even such quasi-relativistic calculations could be further simplified by using pseudo-potential methods to simulate the effect of the cores which would not therefore need to be considered explicitly.

These two lectures are divided into two major sections. The first is concerned with theoretical basis and limitations of the quasi-relativistic schemes and pseudo-potential methods whilst the second outlines a method for performing fully relativistic, fully ab-initio calculations for diatomic molecules. Applications to ionic compounds which might be formed by superheavy elements and to one diatomic molecule containing a superheavy element are described.

PERTURBATION THEORY METHODS USING THE SCHRODINGER KINETIC ENERGY OPERATOR

A. Background and Overview

The objective of this part of the two talks is to probe the theoretical foundations and thus establish the limitations of the various quasi-relativistic schemes which have been used to perform molecular calculations including those introducing pseudo-potentials. A number of different justifications for these schemes have been advanced, those given some papers being contradicted by those appearing in others. Thus it has been claimed either that valence electron dynamics is non-relativistic |4,5| or that, although relativistic, this can be adequately described |6| by a first order perturbation treatment of relativity. The most common justification given for the pseudo-potential methods using the Schrodinger kinetic energy operator |5,7-11| is that all relativistic effects can be incorporated in the pseudo-potential |7, 9, 11| although no rigorous theoretical basis is provided to support this claim. Thus these theories appear to be some-empirical is as much as the pseudo-potential is chosen to reproduce the Dirac-Fock orbital eigenvalues |5, 7, 9| or even experimental data 11 rather than being derived ab-initio from the Dirac-Fock equations. The approaches of |4| and |9| are not entirely ab-initio because spin-orbit coupling has to be inserted in a somewhat ad-hoc fashion using atomic spin-orbit coupling constants.

---

† atomic units are used throughout this article

Furthermore in the approach |9|, the Phillips-Kleinman pseudo-potential |12| employed, which is derived |9| assuming that the core orbitals are all eigenfunctions of the same hamiltonian and hence orthogonal, is constructed from a set of non-orthogonal atomic orbitals calculated by a method |13| which is known to be inconsistent and theoretically dubious |6,14|.

In this section of the talks these problems and inconsistencies are clarified by first deriving (in section 2) the lowest order ($1/c^2$) relativistic corrections to both the exact (section B.1.a) and the Hartree-Fock (section B.1.b) wavefunctions. These latter corrections are then used to show (section B.3.a) that although lowest order ($1/c^2$) perturbation theory is adequate for elements of small and medium nuclear charges, it fails for s and p valence orbitals in elements heavier than the third transition series. It further shows (section B.3.b) that divergence difficulties prevent this failure from being readily repaired by going to higher order in perturbation theory. The theory presented in section B.1. also uncovers an important difference between the relativistic corrections to the energies of the exact and Hartree-Fock wavefunctions. This is used to show in section B.2.a, that previous discrepancies |15| between observed and calculated fine structure in light atoms were not caused by electron correlation but resulted from use of an incomplete perturbating hamiltonian. The theory derived in section B.1.b is also used (section B.2.b) to show that the fine structure inversions occurring in the excited states of alkali atoms can be explained simply in physical terms.

The theoretical basis of the pseudopotential method is investigated in section C by first deriving, from the fully relativistic generalization of Phillips-Kleinman pseudo-potential theory based on the Dirac equation, the lowest order ($1/c^2$) relativistic corrections to non-relativistic Phillips-Kleinman pseudo-potential theory. It is then shown in section C.2 that the divergence difficulties besetting the non-pseudo-potential theories do not prevent the derivations of higher order corrections to the pseudo-potential. This result taken in conjunction with the numerical results reported in section C.1 showing that the direct relativistic corrections are largely transferred to the pseudo-potential provides the theoretical basis needed to establish relativistic pseudo-potential theories on a firm foundation.

B. Lowest Order Theories Excluding Pseudo-Potentials

1. Derivation of relativistic connections

(a) The exact wavefunction.

The object of this subsection is to define as precisely as possible the relationship between the exact non-relativistic wavefunction for an N-electron system and its relativistic analogue. The exact non relativistic wavefunction for level α, denoted $|\psi_\alpha^{NR}(\underline{r}_1, \underline{r}_2 \ldots \underline{r}_N)\rangle$ is simply defined as the solution of the Schrodinger equation

$$\hat{\mathcal{H}}_T^{NR}|\psi_\alpha^{NR}(\underline{r}_1\ldots\underline{r}_N)\rangle = E_\alpha^{NR}|\psi_\alpha^{NR}(\underline{r}_1\ldots\underline{r}_N)\rangle \tag{1}$$

with

$$\hat{\mathcal{H}}_T^{NR} = \sum_{i=1}^{N}\left(\frac{\hat{p}_i^2}{2} + \hat{V}_{nuc}(\underline{r}_i)\right) + \sum_{i=1}^{N-1}\sum_{j=i+1}^{N} r_{ij}^{-1} \tag{2}$$

where $\hat{V}_{nuc}(\underline{r}_i)$ is the operator for the purely electrostatic interaction between electron i and all the nuclei in the system. Although the definition of the exact non relativistic wavefunction is quite unambiguous, it is only possible to construct a relativistic analogue of (1) by abandoning the requirement that the corresponding eigenvalues yield the exact experimental energies. The closest relativistic generalisation of (1) is the equation determining the eigenkets $|\psi_\alpha(\underline{r}_1..\underline{r}_2..\underline{r}_N)\rangle$ and energies E_α of the Brown Hamiltonian |16,17,18|

$$\hat{\mathcal{H}}_{Br}|\psi_\alpha(\underline{r}_1\ldots\underline{r}_N)\rangle = \hat{\rho}^{(+)}\hat{\mathcal{H}}_T\hat{\rho}^{(+)}|\psi_\alpha(\underline{r}_1\ldots\underline{r}_N)\rangle = E_\alpha|\psi_\alpha(\underline{r}_1\ldots r_N)\rangle \tag{3}$$

with

$$\hat{\mathcal{H}}_T = \sum_{i=1}^{N}(\hat{\mathcal{H}}_{KE}(i) + \hat{V}_{nuc}(\underline{r}_i)) + \sum_{i=1}^{N-1}\sum_{j=i+1}^{N} r_{ij}^{-1} \tag{4}$$

and

$$\hat{\mathcal{H}}_{KE}(i) = c\underline{\alpha}(i).\hat{\underline{p}}(i) + c^2(\beta(i) - 1) \tag{5}$$

$$\hat{\rho}^{(+)} = \sum_{r^{(+)}}|\Phi_r^{(+)}(\underline{r}_1,\underline{r}_2\ldots\underline{r}_N)\rangle\langle\Phi_r^{(+)}(\underline{r}_1,\underline{r}_2\ldots\underline{r}_N)| \tag{6}$$

In the relativistic kinetic energy operator $\hat{\mathcal{H}}_{KE}(i)$, which is expressed in atomic units with c the velocity of light (="137"), the zero of energy is defined to correspond to that of a stationary free electron and $\alpha(i)$ and $\beta(i)$ are the 4 x 4 Dirac matrices |19|. The quantities $\hat{\rho}^{(+)}$ are projection operators onto the subspace spanned by the infinite number of antisymmetrized Hartree products

$$|\Phi_r^{(+)}(\underline{r}_1\ldots\underline{r}_N)\rangle = \mathcal{A}\left(\prod_{i=1}^{N}|\Phi_{a_i}(\underline{r}_i)\rangle\right) \tag{7}$$

which can be constructed from the solutions $|\phi_a(\underline{r})\rangle$ of the Dirac-Fock equation

$$\hat{F}|\Phi_a(\underline{r})\rangle = (\hat{\mathcal{H}}_{KE} + \hat{V}_{nuc} + \hat{V}_{e\ell})|\Phi_a(\underline{r})\rangle = \varepsilon_a|\Phi_a(\underline{r})\rangle \tag{8}$$

which have energies $\varepsilon_a > -2c^2$ and therefore describe electrons. The Brown Hamiltonian is not unique because the electronic potential $\hat{V}_{e\ell}$ used to define the $|\Phi_a\rangle$ and hence the $\hat{\rho}^{(+)}$ is in principle arbitrary. This non-uniqueness would not carry over into the prediction of experimentally observable quantities if full account was taken, in a quantum electro-dynamic calculation of the solutions of (8) which, having $\varepsilon_a < -2c^2$ describe positrons. Furthermore results of atomic calculations |2| show that the physically sensible choice for $V_{e\ell}$, namely the Dirac-Fock orbitals occupied in the single manifold |20| description of the N electron system, ensures that even the leading quantum electrodynamic corrections are a small faction of ε_a. Hence the total quantum electrodynamic correction can be calculated to a good approximation by simply adding the energies arising from the leading corrections, namely the self-energy and vacuum polarization terms |21| to the energy E_α predicted by equation (3). It should further be pointed out that these QED corrections are of order $(1/c^3)$ so that they need not be considered when deriving the lowest order $(1/c^2)$ relativistic corrections to the energy of the exact non-relativistic wavefunction.

The description provided by equation (3) is incomplete not only because it neglects quantum electrodynamic effects but also because the r_{ij}^{-1} terms are not even approximately Lorentz covariant. However it is well known that to lowest order in $1/c^2$ the latter deficiency can be rectified by introducing the Breit Hamiltonian as a first order perturbation. This procedure predicts the total energy $E_{\alpha T}$ of the state $|\psi_\alpha>$ to be

$$E_{\alpha T} = E_\alpha + <\Psi_\alpha(\underline{r}_1 \ldots \underline{r}_N)|\hat{\rho}^{(+)} \hat{\mathcal{H}}_{Brei} \hat{\rho}^{(+)}|\psi_\alpha(\underline{r}_1 \ldots \underline{r}_N)> \tag{9}$$

$$\hat{\mathcal{H}}_{Brei} = \sum_{i=1}^{N-1} \sum_{j=i+1}^{N} - \left[\underline{\alpha}(i).\underline{\alpha}(j) r_{ij}^{-1} + (\underline{\alpha}(i).\underline{r}_{ij})(\underline{\alpha}(j).\underline{r}_{ij}) r_{ij}^{-3}\right]/2 \tag{10}$$

Results of atomic Dirac-Fock calculations |22| have shown that use of the low frequency form (10) of the Breit Hamiltonian rather than Mittleman's more general expression (2.17) of |18| is an excellent approximation except for core electrons in atoms of very large nuclear charge. These considerations show that the solutions of (3) with the potential in (8) taken to be in the Dirac-Fock potential generated by the orbitals occupied in the single manifold description of the N electron system, are the relativistic analogues of the non relativistic wavefunctions $|\psi_\alpha^{NR}>$ whilst the energies E_α are the corresponding analogues of the non relativistic ones E_α^{NR}. Since the Breit hamiltonian is correct only to order $1/c^2$ because it is derived by treating the interaction with the quantized electromagnetic field as a second order perturbation on the Fock space equivalents of the solutions of (3), it is consistent to calculate the total energy $(E_{\alpha T})$ from (9). Higher order quantum electrodynamic corrections to the Breit hamiltonian will produce energy changes some of which could be described by inserting $\hat{\mathcal{H}}_{Brei}$ into the equation (3) determining the wavefunction $|\psi_\alpha>$. However it is consistent to regard such higher order energy corrections in the same way as the self energy and vacuum polarization, which are small except for core electrons in heavy atoms, as corrections to the energy $(E_{\alpha T})$. Hence the solutions of (3), like those of (1), can be regarded as taking full account of electron correlation.

The method of Foldy and Wouthuysen |23| is used throught this work to derive relativistic corrections to theories based on the Schrodinger hamiltonian from equivalent fully relativistic theories based on the Dirac equation. This method is chosen because it is very systematic enabling the corrections both to the Schrodinger hamiltonian and to the wavefunctions, which latter are especially important in the Hartree-Fock and Phillips-Kleinman pseudo-potential cases, to be discussed on an equal footing. Furthermore it has the added advantage that all the corrections to the hamiltonian are guaranteed to be hermitian because they are derived from hermitian operators by performing a series of unitary transformations. As discussed in section B.3.b, the divergence of many of the corrections of higher order than $1/c^2$ should not be regarded as a valid objection to using the Foldy-Wouthuysen method to derive the leading terms.

The relativistic corrections to (1) are derived by applying the Foldy-Wouthuysen method to its relativistic analogue (3). Application of the transformation defined by

$$\hat{T} = \exp(i\hat{S}_T) \qquad \hat{S}_T = -\frac{1}{2c}\sum_{i=1}^{N} \beta(i)\underline{\alpha}(i).\hat{p}(i) \tag{11}$$

to (3), shows by using standard techniques that this equation becomes

$$\hat{\rho}^{(+)\prime}\Big[\sum_{i=1}^{N} \Big(c^2(\beta(i) - 1) + \beta(i)\frac{\hat{p}_i^{\,2}}{2} + \hat{V}_{nuc}(\underline{r}_i)\Big) + \sum_{i=j} r_{ij}^{-1} + \hat{\mathcal{H}}_1\Big]\hat{\rho}^{(+)\prime}\,|\psi'_\alpha\rangle = E_\alpha|\psi'_\alpha\rangle \tag{12}$$

where

$$\hat{\mathcal{H}}_1 = \sum_{i=1}^{N}\Big[\beta(i)\hat{\mathcal{H}}^M(i) + \hat{\mathcal{H}}^{DN}(i) + \frac{1}{4c^2}\Sigma(i).(\nabla_i\hat{V}_{nuc}(\underline{r}_i) \times \underline{\hat{p}}_i)\Big] - \frac{1}{4c^2}\sum_{i\neq j}\Big|\underline{\Sigma}(i).\frac{\underline{r}_{ij} \times \underline{\hat{p}}_i}{r_{ij}^{\,3}} + 2\pi\delta(\underline{r}_{ij})\Big] \tag{13}$$

$$\hat{\rho}^{(+)\prime} = \hat{T}\hat{\rho}^{(+)}\hat{T}^{-1} = \sum_{r^{(+)}}|\Phi_r^{(+)\prime}(\underline{r}_1\ldots\underline{r}_N)\rangle\langle\Phi_r^{(+)\prime}(\underline{r}_1\ldots\underline{r}_N)| \tag{14}$$

$$|\Phi_r^{(+)\prime}(\underline{r}_1\ldots\underline{r}_N)\rangle = \hat{A}\Big(\prod_{i=1}^{N}\exp(i\hat{S}(i))|\Phi_{a_i}(\underline{r}_i)\rangle\Big) = \hat{A}\Big(\prod_{i=1}^{N}|\Phi_{a_i}(\underline{r}_i)\rangle\Big) \tag{15}$$

$$|\psi'_\alpha(\underline{r}_1\ldots\underline{r}_N)\rangle = \hat{T}|\psi_\alpha(\underline{r}_1\ldots\underline{r}_N)\rangle \tag{16}$$

In (13), $\hat{\mathcal{H}}^M(i)$ and $\hat{\mathcal{H}}^{DN}(i)$ are the mass-velocity and nuclear Darwin corrections for the i th electron written in the notation of |24| equation (55). The third one electron term is the nuclear contribution to the spin-orbit coupling whilst the two electron terms are the interelectronic spin-orbit coupling and interelectronic Darwin operators. The matrices $\underline{\Sigma}$ are the 4 x 4 equivalents of the Pauli matrices defined by

$$\Sigma_q = \begin{pmatrix} \sigma_q^{\,p} & 0 \\ 0 & \sigma_q^{\,p} \end{pmatrix} \qquad q = x,y,z \tag{17}$$

where $\sigma_q^{\,p}$ is a 2 x 2 Pauli matrix. It is shown in the next section where the relativistic corrections to Hartree-Fock theory are derived that to order $1/c^2$ the transformed orbitals $|\Phi_a'(\underline{r})\rangle$ having $\varepsilon_a > -2c^2$ are given by

$$|\Phi_a'(\underline{r})\rangle = \begin{pmatrix} (1 + \frac{\hat{p}^2}{8c^2})\Phi_a^{\,L}(\underline{r}) \\ 0 \end{pmatrix} \tag{18}$$

whilst application of the charge conjugation operator to (18) shows that the large components of the positron like solutions ($\varepsilon_a < -2c^2$) of (8) vanish. These results taken in conjunction with the absence of terms in (12) which couple large and small components shows that

$$[\hat{\rho}^{(+)'}, \{\sum_{i=1}^{N} c^2(\beta(i) - 1) + \beta(i)\frac{\hat{p}_i^{\,2}}{2} + \hat{V}_{nuc}(\underline{r}_i) + \sum_{i=j} r_{ij}^{-1} + \hat{\mathcal{H}}_1\}] = 0 \tag{19}$$

This result shows that the large components of the solutions of (12) having non zero E_α satisfy (20) whilst the small components vanish.

$$(\hat{\mathcal{H}}_T^{NR} + \hat{\mathcal{H}}_T^{(1)})|\psi_\alpha^{L'}(\underline{r}_1 \ldots \underline{r}_N)\rangle = E_\alpha|\psi_\alpha^{L'}(\underline{r}_1 \ldots \underline{r}_N)\rangle \tag{20}$$

$$\hat{\mathcal{H}}_T^{(1)} = \sum_{i=1}^{N} (\hat{\mathcal{H}}^M(i) + \hat{\mathcal{H}}^{SON}(i) + \hat{\mathcal{H}}^{DN}(i)) - \frac{1}{4c^2}\sum_{i\neq j}(\underline{\sigma}^p(i).(\frac{\underline{r}_{ij} \times \hat{\underline{p}}_i}{r_{ij}^3}) + 2\pi\delta(\underline{r}_{ij})) \tag{21}$$

$$\hat{\rho}^{(+)'}|\psi_\alpha'(\underline{r}_1 \ldots \underline{r}_N)\rangle = |\psi_\alpha'(\underline{r}_1 \ldots \underline{r}_N)\rangle \quad E_\alpha \neq 0 \tag{22}$$

It should be stressed that, although the non-relativistic wavefunction and $|\psi_\alpha^{L'}(\underline{r}_1 \ldots \underline{r}_N)\rangle$ both have unit norm to order $1/c^2$, there is no reason to believe that $|\psi_\alpha^{L'}(\underline{r}_1 \ldots \underline{r}_N)\rangle$ is the non-relativistic wavefunction. Indeed the function $|\psi_\alpha^{L'}(\underline{r}_1 \ldots \underline{r}_N)\rangle$ is exactly defined from the relativistic wavefunction by (16). It is thus necessary to define the difference ($|\psi_\alpha^{(1)}(\underline{r}_1 \ldots \underline{r}_N)\rangle$) between these two wavefunctions through

$$|\psi_\alpha^{L'}(\underline{r}_1 \ldots \underline{r}_N)\rangle = |\psi_\alpha^{NR}(\underline{r}_1 \ldots \underline{r}_N)\rangle - |\psi_\alpha^{(1)}(\underline{r}_1 \ldots \underline{r}_N)\rangle \tag{23}$$

where the equality of the norms of $|\psi_\alpha\rangle$ and $|\psi_\alpha^{NR}\rangle$ to order $1/c^2$ ensures that

$$\langle\psi_\alpha^{L'}(\underline{r}_1 \ldots \underline{r}_N)|\psi_\alpha^{(1)}(\underline{r}_1 \ldots \underline{r}_N)\rangle = 0 \tag{24}$$

It then follows by right multiplication of (20) by $<\psi_\alpha^{NR}(\underline{r}_1 \ldots \underline{r}_N)|$ and invoking (1) that

$$E_\alpha = E_\alpha^{NR} + <\psi_\alpha^{NR}|\hat{\mathcal{H}}_T^{(1)}|\psi_\alpha^{NR}> \quad (25)$$

By means of the identity

$$<\psi_\alpha|\hat{\rho}^{(+)}{}_{Brei}\hat{\rho}^{(+)}|\psi_\alpha> = <\psi_\alpha|\hat{\rho}^{(+)}\hat{T}{}_{Brei}\hat{T}^{-1}\hat{\rho}^{(+)}|\psi_\alpha> \quad (26)$$

it follows by invoking (22) and then using standard techniques that to order $1/c^2$

$$E_{\alpha T} = E_\alpha^{NR} + <\psi_\alpha^{NR}|\hat{\mathcal{H}}_T^{(1)} + \hat{\mathcal{H}}_T^{SOO} + \hat{\mathcal{H}}_T^{OO} + \hat{\mathcal{H}}_T^{SS}|\psi_\alpha^{NR}> \quad (27)$$

The result (27) shows that the relativistic correction to the energy of an exact non relativistic wavefunction is given by simply taking the expectation value of the standard Darwin mass-velocity, spin-orbit coupling, spin-other-orbit ($\hat{\mathcal{H}}_T^{SOO}$), orbit-orbit ($\hat{\mathcal{H}}_T^{OO}$) and spin-spin ($\hat{\mathcal{H}}_T^{SS}$) interaction perturbations |21|. Although this result is not surprising, it needs to be explicitly derived because the relativistic correction to the energy of an approximate non-relativistic wavefunction is not necessarily given by replacing the exact non-relativistic wavefunction in (27) by the approximate one. For example it is shown in the next section that (27) is incorrect for the most commonly used Hartree-Fock description of an open shell system having valence electrons of non-zero orbital angular momentum. The derivation presented here, by explicitly introducing the projection operators $\hat{\rho}^{(+)}$ removes the objection which can be raised against that given in |21| that their starting point, equation (38.1), has no physically acceptable solutions |16|. The present approach, unlike that of |25| which appears to be based on a plane wave expansion of the electron-positron field, is immediately seen to be correct for the exact non relativistic wavefunction. It is also less immediately obvious why the particular quantum electrodynamic approach used by Itoh who obtained the result (27) is correct for some non-relativistic wavefunctions but fails for others. The purely position space methods used here are not only simple and conceptually straightforward but also yield results in the most suitable form for computational applications.

(b) The Hartree-Fock wavefunction.

(i) The eigenvalue connection for closed shell systems. The first step in deriving the relativistic connections to Hartree-Fock theory is to derive the connections to the orbital eigenvalues. The derivation of these corrections for a closed shell system will first be sketched stressing both the physical ideas and the differences from the case of exact wavefunctions discussed in the last section. The full mathematical details can be found in |26| and |27|.

In the closed shell and unrestricted Dirac-Fock theories each orbital $|a\rangle$ is a four component spinor satisfying

$$\hat{F}|a\rangle = \varepsilon_a |a\rangle \tag{28}$$

It is useful to introduce the notation

$$|a(\underline{r})\rangle = \begin{pmatrix} a^L(\underline{r}) \\ a^S(\underline{r}) \end{pmatrix} \tag{29}$$

where the large and small components $a^L(\underline{r})$ and $a^S(\underline{r})$ are both two component column vectors having adjoints $a^L(\underline{r})^\dagger$ and $a^S(\underline{r})^\dagger$ which are two component row vectors. All kets and operators without superscripts are four component spinors or operators involving 4 x 4 matrices or the unit matrix whilst superscripted kets and operators involve two components. In (28) $\hat{F}$ is the Dirac-Fock hamiltonian which becomes in atomic units defining the zero of energy to correspond to that of a free electron

$$\hat{F} = c\underline{\alpha}.\hat{p} + c^2(\beta - 1) + \hat{V}_{LOC} - \hat{K}_T \tag{30}$$

$\hat{V}_{LOC}$ is the total local potential composed of the nuclear potential $\hat{V}_{NUC}$ plus an electronic contribution $\hat{J}_T$ whilst $\hat{K}_T$ is the total exchange contribution

$$\hat{V}_{LOC} = \hat{V}_{NUC} + \hat{J}_T \tag{31}$$

$$\hat{J}_T = \sum_a \hat{J}_a \tag{32}$$

$$\hat{K}_T = \sum_a \hat{K}_a \tag{33}$$

where the sums over a are over all the occupied orbitals. The coulomb and exchange operators $\hat{J}_a$ and $\hat{K}_a$ derived from orbital a are defined by their action on the arbitrary ket $|\phi(\underline{r})\rangle$.

$$\hat{J}_a(\underline{r}_1)|\phi(\underline{r}_1)\rangle = \hat{J}_a(\underline{r}_1)\begin{pmatrix} \phi^L(\underline{r}_1) \\ \phi^S(\underline{r}_1) \end{pmatrix}$$

$$= \int (a^L(\underline{r}_2)^+ r_{12}^{-1} a^L(\underline{r}_2) + a^S(\underline{r}_2)^+ r_{12}^{-1} a^5(\underline{r}_2))d\underline{r}_2 \begin{pmatrix} \phi^L(\underline{r}_1) \\ \phi^S(\underline{r}_1) \end{pmatrix} \tag{34}$$

$$\hat{K}_a(\underline{r}_1)|\phi(\underline{r}_1)\rangle = \hat{K}_a(\underline{r}_1)\begin{pmatrix} \phi^L(\underline{r}_1) \\ \phi^S(\underline{r}_1) \end{pmatrix}$$

$$= \int (a^L(\underline{r}_2)^+ r_{12}^{-1} \phi^L(\underline{r}_2) + a^S(\underline{r}_2)^+ r_{12}^{-1} \phi^S(\underline{r}_2))d\underline{r}_2 \begin{pmatrix} a^L(\underline{r}_1) \\ a^S(\underline{r}_1) \end{pmatrix} \tag{35}$$

It is useful to divide the orbitals $|a>$ into core orbitals denoted by $|c>$ and the most loosely bound or valence orbitals denoted by $|v>$.

The first step in deriving the relativistic corrections is to subject (28) to the Foldy-Wouthuysen transformation defined by

$$\exp(i\hat{S}) = \exp(i(-i\beta\underline{\alpha}\cdot\hat{\underline{p}})) \tag{36}$$

to remove the term $c\underline{\alpha}.\underline{p}$ in (28) which couples the large and small components. A straightforward calculation using standard techniques then converts (28) to

$$\Big[c^2(\beta - 1) + \frac{\beta\hat{p}^2}{2} + \hat{V}_{\ell oc} + \hat{\mathcal{H}}^{DN} + \tfrac{1}{2}\hat{\mathcal{H}}^{DEd} + \hat{\mathcal{H}}^{SON} + \hat{\mathcal{H}}^{SOEd1} - \exp(i\hat{S})\hat{K}_T\exp(-i\hat{S})\Big]|v'> = \varepsilon_v|v'> \tag{37}$$

where

$$|a'> = \exp(iS)|a> \tag{38}$$

for all orbitals $|a>$. Here $\hat{\mathcal{H}}^{DN}$ and $\hat{\mathcal{H}}^{SON}$ are the nuclear-Darwin and nuclear spin-orbit coupling operators whilst $\hat{\mathcal{H}}^{DEd}$ and $\hat{\mathcal{H}}^{SOEd1}$ defined mathematically by (55c) and (55f) of |26| are Darwin and spin-orbit coupling operators which arise from the presence of the electrons and are one-electron (in the same sense as the coulomb operator $\hat{J}_a$ is one electron operator) operators acting on any one-electron function. The label d denotes that these are direct (as opposed to exchange) inter-electronic terms, so that $\hat{\mathcal{H}}^{DEd}$ and $\hat{\mathcal{H}}^{SOEd1}$ are Darwin and spin-orbit coupling corrections experienced by an electron by virtue of its motion in the direct potential generated by the electrons.

The relation (37) differs from the corresponding relation (12) for the exact N-electron wavefunction in that the transformed hamiltonian in (37) contains the orbitals whereas in (12) the transformed hamiltonian does not contain (disregarding the projection operators) any wavefunctions. Furthermore both the direct and exchange potentials $\hat{J}_T$ and $\hat{K}_T$ are constructed from relativistic orbitals and therefore differ from their non-relativistic counterparts $\hat{J}_T{}^{NR}$ and $\hat{K}_T{}^{NR}$ built from non-relativistic Hartree-Fock orbitals. This shows that the relativistic correction $\varepsilon_V{}^{(1)}$ (= $\varepsilon_V - \varepsilon_V{}^{NR}$) to the valence orbital eigenvalue will contain one term arising from the difference between the relativistic and non-relativistic direct electronic potentials plus a corresponding term arising from the difference between the relativistic and non-relativistic exchange potentials. Before these corrections can be derived it is necessary to show that the small components of the transformed orbitals $|a'>$ vanish to order $1/c^2$. This is achieved by first calculating the exchange terms in (37) and then expanding the resultant equation as a power series in 1/c. The details of this procedure, fully described on P1331-P1335 of |26|, need not be presented here. To order $1/c^2$ the result is

$$a^S(\underline{r}) = \underline{\sigma}^P \cdot \hat{\underline{p}} a^L(\underline{r})/2c \tag{39}$$

whence it follows that

$$\begin{aligned} |a^L(\underline{r})\rangle &= (1+\frac{\hat{p}^2}{8c^2})|a^L(\underline{r})\rangle \\ |a^S(\underline{r})\rangle &= 0 \end{aligned} \tag{40}$$

The relativistic corrections to the potentials $\hat{J}_T$ and $\hat{K}_T$ are calculated by expressing both these operators in terms of the $|a'\rangle$ by using the inverse of (38)

$$|a\rangle = \exp(-i\hat{S})|a'\rangle \tag{41}$$

and then introducing the non-relativistic wavefunction and its first order correction through some definition such as

$$|a^{L'}\rangle = |a^{NR}\rangle - |a^{(1)}\rangle \tag{42}$$

or

$$\langle a^{L'}|a^{L'}\rangle^{-1/2}|a^{L'}\rangle = |a^{NR}\rangle - |a^{(1)}\rangle \tag{43}$$

The non-relativistic orbitals $|a^{NR}\rangle$ are taken to be the limit of $|a^{L'}\rangle$ as $c \to \infty$ as discussed in the next section. Although (42) is adequate for the core orbitals of very light atoms and for valence orbitals in most atoms, it fails for core orbitals of heavy atoms |27,28| because the norm of $|a^{L'}\rangle$ is only unity to order $1/c^2$, this norm deviating from unity at order $1/c^4$. This causes the coulomb potential predicted by constructing the density $\int a^{L'}(\underline{r}_2)^+ r_{12}^{-1} a^{L'}(\underline{r}_2) d\underline{r}_2$ to be incorrect at large distances thereby causing the theory to fail catastrophically |28|. The approach (43) is one of number |27| yielding very similar results |28| in all of which this deficiency is rectified. The use of the relations (40)-(43) to calculate the corrections to the potentials appearing in (37), which is fully described elsewhere |27|, will be illustrated by considering the local term $\hat{J}_T$. Use of (41) shows that

$$\begin{aligned} \hat{J}_T(\underline{r}_1) &= \sum_a \int a^+(\underline{r}_2) r_{12}^{-1} a(\underline{r}_2) d\underline{r}_2 \\ &= \sum_a \langle a'|a'\rangle^{-1} \{ \int a'(\underline{r}_2)^+ r_{12}^{-1} a'(\underline{r}_2) d\underline{r}_2 \\ &\quad + \int a'(\underline{r}_2)^+ [\exp(i\hat{S}_2), r_{12}^{-1}] \exp(-i\hat{S}_2) a'(\underline{r}_2) d\underline{r}_2 \} \end{aligned} \tag{44}$$

After calculating the commutator by expanding $\exp(i\hat{S}_2)$, (36) where the subscript 2 denotes that this operator acts on the co-ordinates of electron 2, and introducing (43), (44) becomes

$$\hat{J}_T(\underline{r}_1) = \hat{J}_T^{NR}(\underline{r}_1) + \sum_a\left[-\hat{J}_a^{(1)}(\underline{r}_1) + \hat{J}_a^{(2)}(\underline{r}_1)\right] + \hat{\mathcal{H}}^{SOEd2} + \tfrac{1}{2}\hat{\mathcal{H}}^{DEd} \quad (45)$$

Here $\hat{\mathcal{H}}^{SOEd2}$ defined by (55g) of |26| is a further spin-orbit coupling term arising from electron-electron repulsion. After carrying out a similar calculation for the exchange terms in (37) as described on p. 1063 of |27|, envoking the result $(\hat{J}_a - \hat{K}_a)|a\rangle = 0$ to remove self interaction, and substituting the result and (45) without the term a = v into (37), this equation becomes

$$(\hat{F}^{NR} + \hat{F}^{(1,Dir)} + \hat{F}^{(I,Ind)})(|v^{NR}\rangle - |v^{(1)}\rangle) = \varepsilon_v(|v^{NR}\rangle - |v^{(1)}\rangle) \quad (46)$$

where

$$\hat{F}^{(1,Dir)} = \hat{\mathcal{H}}^{M} + \hat{\mathcal{H}}^{DN} + \hat{\mathcal{H}}^{DEd} + \hat{\mathcal{H}}^{DEx} + \hat{\mathcal{H}}^{SON} + \hat{\mathcal{H}}^{SOEd1} + \hat{\mathcal{H}}^{SOEd2} + \hat{\mathcal{H}}^{SOEx1} + \hat{\mathcal{H}}^{SOEx2} \quad (47)$$

$$\hat{F}^{(1,Ind)} = \sum_{a\neq v}(-\hat{J}_a^{(1)} + \hat{J}_a^{(2)} + \hat{K}_a^{(1)} - \hat{K}_a^{(2)}) \quad (48)$$

where $|v^{L'}\rangle$ has been decomposed according to (42). Left multiplication of (46) by $\langle v^{NR}|$ shows, after noting that the non-relativistic Hartree-Fock orbital $|v^{NR}\rangle$ is an eigenfunction of Fock operator $\hat{F}^{NR}$, that the relativistic correction $\varepsilon_v^{(1)}$ to the valence orbital eigenvalue is given by

$$\varepsilon_v^{(1)} = \langle v^{NR}|\hat{F}^{(1,Dir)} + \hat{F}^{(1,Ind)}|v^{NR}\rangle \quad (49)$$

The perturbing operators whose expectation values over the non-relativistic valence wavefunction yield the eigen-value correction $\varepsilon_v^{(1)}$ fall into these classes:

(i) The standard relativistic corrections arising even if the wavefunction is not anti-symmetric which therefore do not originate from the exchange term $\hat{K}_T$ in (30). These operators are the mass-velocity connection $\hat{\mathcal{H}}^{M}$, the nuclear spin orbit coupling ($\hat{\mathcal{H}}^{SON}$) and nuclear Darwin ($\hat{\mathcal{H}}^{DN}$) corrections plus two spin-orbit coupling corrections ($\hat{\mathcal{H}}^{SOEd1}$ and $\hat{\mathcal{H}}^{SOEd2}$) and one Darwin correction ($\hat{\mathcal{H}}^{DEd}$) which arise from the motion of the each election in the direct electrostatic field of the electrons.

(ii) Terms analogous the last three in (i) which originate from the exchange term $\hat{K}_T$ in (30) thus arising from the anti-symmetrization of the wavefunction. These consist of two spin-orbit coupling corrections $\hat{\mathcal{H}}^{SOEx1}$ and $\hat{\mathcal{H}}^{SOEx2}$) plus one Darwin correction ($\hat{\mathcal{H}}^{DEx}$) arising from the electronic motion in the exchange field of the electrons the label x denotes that these terms arise from the motion of each electron in the exchange field of the other electrons

(iii) Terms arising because the charge distribution of the relativistic orbitals differs from that of the corresponding non-relativistic orbitals. This difference causes both the direct and the exchange potentials arising from the purely coulombic repulsion (r_{ij}^{-1}) experienced by an electron to differ from the non-relativistic potentials. The sum $\sum_{a\neq v}(-\hat{J}_a^{(1)} + \hat{J}_a^{(2)})$ constitutes the correction to the direct potential whilst $\sum_{a\neq v}(\hat{K}_a^{(1)} - \hat{K}_a^{(2)})$ is the exchange potential correction.

Since the corrections (i) and (ii) arise because the electron dynamics is intrinsically relativistic the expectation value $\langle v^{NR}|\hat{F}^{(1,Dir)}|v^{NR}\rangle$ constitutes the direct relativistic correction, the sum of the operators (i) and (ii) being denoted $\hat{F}^{(1,Dir)}$. This correction vanishes if the valence electrons do not move relativistically. Since these corrections are already of order $1/c^2$ the operators $\mathcal{R}^{SOEd1}$, $\mathcal{R}^{SOEd2}$, $\mathcal{R}^{SOEx1}$, $\mathcal{R}^{SOEx2}$, $\mathcal{R}^{DEd}$ and $\mathcal{R}^{DEx}$ are defined in terms of non-relativistic orbitals although they arise earlier in the derivation expressed in terms of relativistic wavefunctions. The four operators in (48) constituting the term (iii) are defined by

$$\hat{J}_a^{(1)}(\underline{r}_1) = \int (a^{NR}(\underline{r}_2)^+ r_{12}^{-1} a^{(1)}(\underline{r}_2) + a^{(1)}(\underline{r}_2) r_{12}^{-1} a^{NR}(\underline{r}_2)) d\underline{r}_2 \tag{50a}$$

$$\hat{J}_a^{(2)}(\underline{r}_1) = \int a^{(1)}(\underline{r}_2)^+ r_{12}^{-1} a^{(1)}(\underline{r}_2) d\underline{r}_2 \tag{50b}$$

$$\hat{K}_a^{(1)} = \int (a^{NR}(\underline{r}_2)^+ r_{12}^{-1} \hat{P}_{12} a^{(1)}(\underline{r}_2) + a^{(1)}(\underline{r}_2)^+ r_{12}^{-1} \hat{P}_{12} a^{NR}(\underline{r}_2)) d\underline{r}_2 \tag{50c}$$

$$\hat{K}_a^{(2)} = \int a^{(1)}(\underline{r}_2)^+ r_{12}^{-1} \hat{P}_{12} a^{(1)}(\underline{r}_2) d\underline{r}_2 \tag{50d}$$

where $\hat{P}_{12}$ is the operator interchanging labels 1 and 2. With the definition (43) for the orbital corrections $|a^{(1)}\rangle$ the operators $\hat{J}_T^{(2)}$ and $\hat{K}_T^{(2)}$ although of order $1/c^4$ have to be included to ensure that the coulomb potential is correct assymptotically |27,28|. The corrections (iii), unlike those (i) and (ii) would be non-zero even if the valence electron moved non-relativistically and originate as shown by the forms (50) from the change in potential generated by the other electrons. The sum (48) of the four operators (50) is therefore denoted $\hat{F}^{(1,Ind)}$, the expectation value $\langle v^{NR}|\hat{F}^{(1,Ind)}|v^{NR}\rangle$ constituting the indirect relativistic effect.

Although the decomposition of the eigenvalue correction $\varepsilon_v^{(1)}$ into the direct and indirect effects presented here has been based on perturbation theory, it should be stressed that these concepts transcend perturbation theory. Indeed the indirect effect was first noted |29| in a Dirac-Hartree calculation on the Hg atom whilst a quantitative estimate of these two effects, described in section B.3.a. below has been given |30| within a fully relativistic framework.

(ii) The total energy correction in closed shell Hartree-Fock theory. The relativistic correction to the total N-electron Hartree-Fock energy for atomic level α is just the difference between the total Dirac-Fock energy ($E_{DF\alpha}$) and the total non-relativistic Hartree-Fock energy ($E_{HF\alpha}$). For the closed shell systems of interest here, this difference can be expressed as

$$E_{DF\alpha} - E_{HF\alpha} = \sum_a (\varepsilon_a - \varepsilon_a^{NR}) - \tfrac{1}{2}\sum_{a\neq b} \left(\langle ab|r_{12}^{-1}(1-\hat{P}_{12})|ab\rangle - \langle a^{NR}b^{NR}|r_{12}^{-1}(1-\hat{P}_{12})|a^{NR}b^{NR}\rangle\right) \tag{51}$$

The interelectronic repulsions appearing explicitly in this relation correct the double counting of these repulsions in the sum over the orbital eigenvaules. The form (51) is convenient because $(\varepsilon_a - \varepsilon_a^{NR})$ is the relativistic correction to the eigenvalue of orbital a which has already been calculated (equation (49)). These relativistic repulsions in (51) are first expressed in terms of the orbitals $|a'\rangle$ by envoking (41) and the resultant expression is then related to the non-relativistic repulsions by means of (43). Thus

$$\begin{aligned}
\tfrac{1}{2}\sum_{a\neq b} \langle ab|r_{12}^{-1}(1-\hat{P}_{12})|ab\rangle
&= \tfrac{1}{2}\sum_{a\neq b} \langle a'b'|\exp(i\hat{S}_1)\exp(i\hat{S}_2)|r_{12}^{-1}(1-\hat{P}_{12})|\exp(-i\hat{S}_1)\exp(-i\hat{S}_2)|ab\rangle \\
&= \tfrac{1}{2}\sum_{a\neq b} \Big(\langle a^{NR}b^{NR}|r_{12}^{-1}(1-\hat{P}_{12})|a^{NR}b^{NR}\rangle \\
&\quad + \langle a^{NR}|-\hat{J}_b^{(1)}+\hat{J}_b^{(2)}+\hat{K}_b^{(1)}-\hat{K}_b^{(2)}|a^{NR}\rangle + \langle b^{NR}|-\hat{J}_a^{(1)}+\hat{J}_a^{(2)}+\hat{K}_a^{(1)} \\
&\quad -\hat{K}_a^{(2)}|b^{NR}\rangle\Big) + \tfrac{1}{2}\sum_a \langle a^{NR}|\hat{\mathcal{H}}^{SOEd1}+\hat{\mathcal{H}}^{SOEd2}+\hat{\mathcal{H}}^{DEd}+\hat{\mathcal{H}}^{SOEx1} \\
&\quad +\hat{\mathcal{H}}^{SOEx2}+\hat{\mathcal{H}}^{DEx}|a^{NR}\rangle
\end{aligned} \tag{52}$$

where only terms up to order $1/c^2$ are retained in the second step. Substitution of this result and that (49) for the eigenvalue correction, noting that the correction $\varepsilon_a^{(1)}$ has $\sum_{b\neq a}$ in (48), into (51) shows that

$$\begin{aligned}
E_{DF\alpha} - E_{HF\alpha} &= \sum_a \langle a^{NR}|\hat{\mathcal{H}}^{M} + \hat{\mathcal{H}}^{SON} + \hat{\mathcal{H}}^{DN} \\
&\quad + \tfrac{1}{2}(\hat{\mathcal{H}}^{SOEd1} + \hat{\mathcal{H}}^{SOEd2} + \hat{\mathcal{H}}^{OEd} + \hat{\mathcal{H}}^{SOEx1} + \hat{\mathcal{H}}^{SOEx2} + \hat{\mathcal{H}}^{DEx})|a^{NR}\rangle \\
&= \sum_a \langle a^{NR}|\hat{\mathcal{H}}^{M} + \hat{\mathcal{H}}^{SON} + \hat{\mathcal{H}}^{DN} + \hat{\mathcal{H}}^{SOEd1} + \hat{\mathcal{H}}^{SOEx1} + \tfrac{1}{2}(\hat{\mathcal{H}}^{DEd} + \hat{\mathcal{H}}^{DEx})|a^{NR}\rangle
\end{aligned} \tag{53}$$

This result contains only the standard mass-velocity, spin-orbit coupling and Darwin type terms because the potential modifications $\sum_{b=a} \langle a^{NR}| - \hat{J}_b(1) + \hat{J}_b(2) + \hat{K}_b(1) - \hat{K}_b(2)|a^{NR}\rangle$ entering the eigenvalue corrections (49) cancell with the corresponding terms in (52).

The result (53) which has exactly the same form as (25) shows that the relativistic correction to the total energy of the N-electron system is given to order $1/c^2$ by simply taking the expectation value of the non-relativistic Hartree-Fock wavefunction over the perturbing operator $\mathfrak{K}_T^{(1)}$ (21). It should not be supposed that this result is so obvious as not to require proof and that the potential corrections have no real physical significance because it is shown in the next section that (25) does not hold for an open shell Hartree-Fock wavefunction of non-zero total orbital angular momentum if the orbitals used to construct it satisfy the usual symmetry and equivalence restrictions |31|.

(iii) Relativistic eigenvalue corrections in atoms containing one valence electron. The relativistic correction to the energy of the exact N-electron wavefunction for a state corresponding to a relativistic configuration consisting of closed sub-shells plus a single valence electron is still given by (25) since no assumptions about the details of the electronic structure had to be made when deriving (25). Since this is not the case for the usual Dirac-Fock or Hartree-Fock description of such a state, within this framework the relativistic corrections need to be explicitly derived.

In the Dirac-Fock description of a system having one valence electron, this occupies the valence orbital, denoted $|V\rangle$, whilst the remaining electrons occupy core orbitals denoted $|c\rangle$. It is usual to demand that all these orbitals are eigenfunctions of the operators $\hat{j}^2$ and $\hat{j}_z$ as well as of the Dirac parity operator $\beta\hat{i}$ ($\hat{i}$ the spatial inversion) |32|. They therefore take the standard form |32| (equation 2.12)

$$|a_{nkjm}\rangle = \frac{1}{r}\begin{pmatrix} P_{nk}(r)\chi_{kjm}(\theta,\phi) \\ iQ_{nk}(r)\chi_{-kjm}(\theta,\phi) \end{pmatrix} \tag{54}$$

where n and k are the principal and kappa quantum numbers |32|, $P_{nk}(r)$ and $Q_{nk}(r)$ are purely radial functions and $\chi_{kjm}(\theta,\phi)$ is a vector coupled space-spin functions which is an eigenfunction of the operators $\hat{j}^2$ and $\hat{j}_z$ of eigenvalue $j(j+1)$ and m respectively. Furthermore the radial functions of all the $(2j+1)$ orbitals, said to constitute a sub-shell |20|, which differ only in the m_j quantum number m are taken to be identical. Thus the core is composed of closed subshells in each of which each of the $2j+1$ degenerate orbitals is occupied by one electron. The imposition of these symmetry and equivalence conditions is equivalent to seeking orbitals which render stationary subject to the usual orthonormality conditions the average energy of all the determinants which differ only in the m_j quantum number of the valence electron |33|. However if the valence electron has non-zero orbital angular momentum (ℓ) in the non-relativistic limit there are two Dirac-Fock subshells denoted $\bar{j}$ and j^+ having total angular momentum of $\bar{j} = \ell - \frac{1}{2}$ and $j^+ = \ell + \frac{1}{2}$ respectively composed of $(2\bar{j}+1)$

and $(2j^+ + 1)$ orbitals $|\bar{v}\,\bar{m}>$ and $|v^+m^+>$. The labels $\bar{m}$ and m^+ denote the m_j quantum number whilst expressions involving kets of the type $|v>$ in which m_j does not appear are valid for any of the $2(2 + 1)$ orbitals $|\bar{v}\,\bar{m}>$ or $|v^+m^+>$. Calculations where the atomic states $|\text{core}\ \bar{v}>$ and $|\text{core}\ v^+>$ are considered separately are merely special cases of the extended average level (EAL) method |20,34| in which the average energy E_{aV} is considered.

$$E_{aV} = W_{\bar{j}}[\bar{j}]^{-1}\sum_{\bar{m}}<(\text{core}\ \bar{v}\bar{m})\hat{A}|\hat{\mathcal{H}}_T|\hat{A}(\text{core}\ \bar{v}\bar{m})>$$
$$+ W_{j^+}[j^+]^{-1}\sum_{m^+}<(\text{core}\ v^+m^+)\hat{A}|\hat{\mathcal{H}}_T|\hat{A}(\text{core}\ v^+m^+)> = W_{\bar{j}}E(\bar{j}) + W_{j^+}E(j^+) \tag{55}$$

with $W_{\bar{j}} + W_{j^+} = 1$

In (79) $W_{\bar{j}}$ and W_{j^+} are arbitrary real positive weights usually chosen to be both $\frac{1}{2}$ or $[\bar{j}]/(2(2\ell + 1))$ and $[j^+]/(2(2\ell + 1))$ and $[j]$ is $(2_j + 1)$. The equations satisfied by the Dirac-Fock orbitals are derived by the well known procedure of first demanding that the functional

$$\mathcal{L} = W_{\bar{j}}(E(\bar{j}) - [\bar{j}]^{-1}\varepsilon_{\bar{v}}\sum_{\bar{m}}<\bar{v}\bar{m}|\bar{v}\bar{m}>) + W_{j^+}(E(j^+) - [j^+]^{1}\varepsilon_{v^+}\sum_{m^+}<v^+m^+|v^+m^+>)$$
$$- \sum_{cvm}(\varepsilon_{cv}<vm|c> + \varepsilon_{vc}<c|vm>) - \sum_{cd}\varepsilon_{cd}<d|c> \tag{56}$$

presented here solely to fix the notation for the off-diagonal Lagrange multipliers, be stationary with respect to all variations of the orbitals and then eliminating the terms ε_{cd} having $c \neq d$. The valence orbitals which are found to satisfy

$$\hat{F}_{TC}|v> = \varepsilon_v|v> + [j(v)](W_{j(v)})^{-1}\sum_c \varepsilon_{cv}|c> \tag{57}$$

where

$$\hat{F}_{TC} = c\underline{\alpha}.\hat{\underline{p}} + c^2(\beta - 1) + \hat{V}_{nuc} + \sum_c(\hat{J}_c - \hat{K}_c) \tag{58}$$

are symmetry adapted because $\hat{F}_{TC}$ involves sums over all the orbitals in any subshell. Although the equations (57) are valid for any of the $2(2\ell + 1)$ valence orbitals $|v>$, ε_{cv} is zero unless $|c>$ has the same angular quantum numbers as $|v\ m>$ and $j(v)$ is the j value of orbital $|v>$ namely $\bar{j}$ or j^+. The eigenvalues are physically significant becauce the energy of $\hat{A}|\text{core}\ vm>$ is $E_{core} + \varepsilon_v$.

The non-relativistic orbitals satisfying the non-relativistic analogue of (57) are two component functions which can be written

$$|a_{nkjm}^{NR}> = \frac{1}{r}R_{nk}(r)\chi_{kjm}(\theta,\phi) \tag{59}$$

Orbitals (59) differing only in the j quantum number are degenerate. The choice (59) in which the $|a^{NR}\rangle$ are eigenfunctions of $\hat{j}^2$ and $\hat{j}_z$ as well as of $\hat{\ell}^2$ ensures that these orbitals are identical to those obtained by considering the $|a'\rangle$ to be functionals of the velocity of light and taking the large components in the limit $c \to \infty$. Hence $|a^{L'}\rangle$ passes smoothly into $|a^{NR}\rangle$ and $|a^{(1)}\rangle$ tends to zero in this limit thereby enabling relativity to be treated as a small perturbation on the non-relativistic orbitals.

The relativistic corrections to the eigenvalues ε_v^{NR} are calculated by applying the theory described in section (i) above to (57) which thereby becomes transformed to

$$(\hat{F}_{TC}^{NR} + \hat{F}_{TC}^{(1,Dir)} + \hat{F}_{TC}^{(1,Ind)})|v^L\rangle = (\varepsilon_v^{NR} + \varepsilon_v^{(1)})|v^L\rangle + [j(v)](W_{j(v)})^{-1}\sum_c(\varepsilon_{cV}^{NR} + \varepsilon_{cV}^{(1)})|c^L\rangle \tag{60}$$

with

$$\hat{F}_{TC}^{(1,Dir)} = \hat{\mathcal{H}}^{M} + \hat{\mathcal{H}}^{DN} + \hat{\mathcal{H}}_{TC}^{DEd} + \hat{\mathcal{H}}_{TC}^{DEx} + \hat{\mathcal{H}}^{SON} + \hat{\mathcal{H}}_{TC}^{SOEd_1} + \hat{\mathcal{H}}_{TC}^{SOEd_2} + \hat{\mathcal{H}}_{TC}^{SOEx_1} + \hat{\mathcal{H}}_{TC}^{SOEx_2} \tag{61}$$

$$\hat{F}_{TC}^{(1,Ind)} = \sum_c(-\hat{J}_c^{(1)} + \hat{J}_c^{(2)} + \hat{K}_c^{(1)} - \hat{K}_c^{(2)}) \tag{62}$$

Here the inter-electronic Darwin and spin-orbit coupling operators bearing the label TC differ from the previous ones defined by (55) of |26| only in that the sums ($\sum_a$) are replaced by sums ($\sum_c$) over just the core orbitals c. Before extracting $\varepsilon_v^{(1)}$ from (60) by left multiplication by $\langle v^{NR}|$ it is necessary to note the result that to order $1/c^2$

$$\langle c^{NR}|v^{(1)}\rangle = -\langle c^{(1)}|v^{NR}\rangle \tag{63}$$

which is proved by substituting (42) into $\langle c'|v'\rangle = 0$. Left multiplication of (60) by $\langle v^{NR}|$ followed by use of the non-relativistic equivalent of (57), (42) and (63) shows that

$$\varepsilon^{(1)} = \langle v^{NR}|\hat{F}_{Tc}^{(1,Dir)} + \hat{F}_{Tc}^{(1,Ind)}|v^{NR}\rangle + [j(v)](W_{j(v)})^{-1}\sum_c(\varepsilon_{cv}^{NR}\langle v^{NR}|c^{(1)}\rangle + \varepsilon_{vc}^{NR}\langle c^{(1)}|v^{NR}\rangle) \tag{64}$$

It is possible to perform further manipulations to express this correction as the expectation value of an operator (|26| equation (88)) although this need not be presented here.

The result (64) shows that the eigenvalue correction contains not only the direct and indirect corrections arising in the closed shell case but also further terms involving the off-diagonal Lagrange multipliers. The data reported in Table 2 confirm that the

Table 1. Dirac-Fock and Non-relativistic Hartree-Fock Koopman's theorem first ionization potentials and valence orbital radii in heavy atoms.

	Hg($6s^2$)		E112($7s^2$)		E113($7\bar{p}$)	
	rel	NR	rel	NR	rel	NR
Ionization Potential (e.v)	8.93	7.10	12.52	6.48	7.22	5.01
<r> (a.u.)	2.84	3.33	2.47	3.64	3.04	4.21

Table 2. Perturbation theory predictions and components of relativistic corrections to valence orbital eigenvalues in light atoms (cm^{-1}).

	Li	Na	Al $3\bar{p}$	Al 3p
Direct[(a)]	-4.14	-61.44	-131.59	-31.92
Indirect[(b)]	0.56	9.43	132.40	152.15
ODLM[(c)]	0.04	0.43	1.48	-0.44
Total $\varepsilon_v^{(1)}$	-3.53	-51.58	2.29	119.80
Exact $\varepsilon_v - \varepsilon_v^{NR}$	-3.52	-51.58	2.07	119.70

(a) $= \langle v^{NR} | \hat{F}_{TC}^{(1,Dir)} | v^{NR} \rangle$

(b) $= \langle v^{NR} | \hat{F}_{TC}^{(1,Ind)} | v^{NR} \rangle$

(c) Off-diagonal Lagrange multiplier contribution = the remaining terms of (64).

theory is correct because the perturbation theory corrections $\varepsilon_v^{(1)}$ agree almost perfectly with the exact eigenvalue corrections $(\varepsilon_v - \varepsilon_v^{NR})$. Although the contribution from the term involving the off-diagonal Lagrange multipliers is small, it must be included if the perturbation theory is to agree with the exact results. The data also show the importance of the indirect corrections, the perturbation theory results for the $\bar{p}$ and p orbitals being quite wrong unless the indirect terms are included.

Further data and the details of how these calculations were carried out are reported in |28|.

(iv) The total energy correction for single valence electron atoms. The relativistic correction to the total N-electron energy of an atom containing a single valence electron can be derived by methods analogous to those used to derive the total energy correction in closed shell systems. This derivation can be found in |1|, the result being equation (38) of that paper. However, it need not be presented here because the interesting result that, unlike the closed shell case, the potential corrections, do not always disappear can be obtained by a very simple argument.

Consider the two configurations $|\text{core } \bar{v}\rangle$ and $|\text{core } v^{+}\rangle$ constructed from a common set of core orbitals as described in the last section. The difference ΔE $(=E(j^{+}) - E(\bar{j}))$ between the total N-electron energies of these two levels is simply equal to the difference $(\varepsilon_{v+} \quad \varepsilon_{\bar{v}})$ in the valence orbital eigenvalues. Since the two levels $|\text{core } \bar{v}\rangle$ and $|\text{core } v^{+}\rangle$ are degenerate in the non-relativistic limit it follows that this difference equals that between the relativistic corrections to the eigenvalues. Thus

$$\Delta E = E(j^{+}) - E(\bar{j}) = \varepsilon_{v+} - \varepsilon_{\bar{v}} = \varepsilon_{v+}^{(1)} - \varepsilon_{\bar{v}}^{(1)} \tag{65}$$

Since the eigenvalue corrections $\varepsilon_v^{(1)}$ contain the indirect terms $\langle v^{NR}| F_{TC}(1,\text{Ind}) |v^{NR}\rangle$ which are not identical for the two orbitals $|\bar{v}^{NR}\rangle$ and $|v^{+NR}\rangle$, it follows that the excitation energy correction ΔE has a contribution from the indirect relativistic effect. This immediately shows that the relativistic correction to the total N-electron energy cannot be calculated by taking the expectation value over just the standard operators $\hat{\mathcal{H}}_T^{(1)}$ (21) because (21) contains no potential corrections. Thus it has been shown that equation (25) with $|\psi_\alpha\rangle$ replaced by the open-shell non-relativistic Hartree-Fock function is not correct for such functions in which the valence electron has a non-zero orbital angular momentum. This argument clearly does not apply if the valence electron occupies an s orbital. For these systems it can be shown |1| using the methods described in section B.1.b (ii) that (25) with $|\psi_\alpha\rangle$ replaced by the Hartree-Fock function is correct. This seems to show that the failure of (25) for systems where the valence electron has orbital angular momentum arises from the fact that there are two different relativistic configurations which map into the same non-relativistic configuration.

2. Application to atomic spin-orbit excitation energies

(a) Relation to previous theory.

For an atom described in non-relativistic theory as having one valence electron of orbital angular momentum ℓ outside closed shells, the fine structure excitation energy ΔE_T is the difference between the total energies $E_T(j^+)$ and $E_T(\bar{j})$ of the two levels having respective total angular momenta $j^+ = \ell + \frac{1}{2}$ and $\bar{j} = \ell - \frac{1}{2}$. In the fully relativistic theory, each of the total energies $E_T(j)$ is (c.f. (9)) the sum of the appropriate Dirac-Fock energy $E(j)$ plus the Briet energy calculated as the expectation value over the Briet hamiltonian (10).

In the EAL variant of the MCDF method both the CSFs $|\text{core } \bar{v}\rangle$ and $|\text{core } v^+\rangle$ are constructed from a common set of orbitals chosen to optimize the quantity $(W_{\bar{j}}E(\bar{j}) + W_{j^+}E(j^+))$ (55). The core contribution to the excitation energy therefore vanishes because both CSFs contain the identical core orbitals. Consequently the fine structure excitation energy ΔE equals the sum of the difference $(\varepsilon_{v^+} - \varepsilon_{\bar{v}})$ between the valence orbital eigenvalues plus the difference ΔE^B_{CV} in the core-valence Breit energy. $(\Delta E^B_{CV} = -\sum_c \langle cv|\mathcal{H}_B(1,2)|vc\rangle)$. The direct core-valence Breit energy $(= \sum_c \langle cv|\mathcal{H}_B(1,2)|cv\rangle)$ vanishes |35,36|. Thus

$$\Delta E = \varepsilon_{V^+} - \varepsilon_{\bar{V}} + \Delta E^B_{CV} \tag{66}$$

The weight factors $W_{\bar{j}}$ and W_{j^+} entering the functional (see above) used to optimize the orbitals were chosen to be $[j]/(2[\ell])$ and $|j^+|/(2[\ell])$ respectively. This choice ensures that the large components if pairs of orbitals having the same principal quantum number and orbital angular momentum associated with the large components become identical as $c \to \infty$. This shows that the non-relativistic limit of such a calculation is the standard non-relativistic Hartree-Fock function in which the radial parts of orbitals having the same n and ℓ but different j $(= \ell \pm \frac{1}{2})$ quantum numbers are constrained to be identical. Consequently the results of these EAL calculations, unlike those of OL calculations |20|, are directly comparable with those obtained by adding to the non-relativistic Hartree-Fock energy the open shell analogue, containing potential connections, of the N-electron perturbation energy (53). The expression is given as equation (38) of |1| but is not needed here because the energy difference determining the fine structure can alternatively be expressed in terms of the valence orbital eigenvalues. The results obtained by this perturbation treatment will agree with those of the Dirac-Fock EAL calculations for atoms in which the relativistic effects are sufficiently small that a first order $(1/c^2)$ perturbation treatment is adequate.

The perturbation theory predictions of the fine structure splittings are given as the sum of quantity $(\varepsilon_{v^+}{}^{(1)} - \varepsilon_{\bar{v}}{}^{(1)})$ plus the contribution (ΔE^B_{CVPT}) from the Breit interaction. Thus the fine structure in perturbation theory ΔE_{PT} is given by

$$\Delta E_{PT} = \varepsilon_{v^+}{}^{(1)} - \varepsilon_{\bar{v}}{}^{(1)} + \Delta E^{B}_{CVPT} \tag{67}$$

The only contributions to $\varepsilon_v{}^{(1)}$ (64) which need to be considered are those from the spin-orbit coupling operators $\hat{\mathcal{H}}^{SON}$, $\hat{\mathcal{H}}_{TC}{}^{SOEd1}$ and $\hat{\mathcal{H}}_{TC}{}^{SOEx1}$, the relativistic correction $(\hat{K}_{TC}{}^{(1)} - \hat{K}_{TC}{}^{(2)})$ to the core exchange potential and the off-diagonal Lagrange multiplier correction because the expectation values $\langle v^{NR}|\hat{\mathcal{H}}_{TC}{}^{SOEd2}|v^{NR}\rangle$ and $\langle v^{NR}|\hat{\mathcal{H}}_{TC}^{SOEx2}|v^{NR}\rangle$ vanish |37| whilst the mass velocity, Darwin and direct potential $(-\hat{J}_{TC}{}^{(1)} + \hat{J}_{TC}{}^{(2)})$ corrections are the same for both the orbitals $|\bar{v}^{NR}\rangle$ and $|v{+}^{NR}\rangle$. The first order corrections to the core orbitals needed to construct $\hat{K}_{TC}{}^{(1)}$ and $\hat{K}_{TC}{}^{(2)}$ are computed from the definition (43). The contributions from the Breit interaction are evaluated by replacing $\psi_\alpha{}^{NR}$ by the Hartree-Fock function in (26). The contributions of the orbit-orbit $\hat{\mathcal{H}}_T{}^{OO}$ and spin-spin $\hat{\mathcal{H}}_T{}^{SS}$ terms do not need to be considered because they are the same for both the levels, whilst the direct $(\sum_{a=b}\langle a^{NR}b^{NR}|\hat{\mathcal{H}}^{SOO}|a^{NR}b^{NR}\rangle)$ contribution to the spin-other orbit interaction is zero |37|. The only non-zero exchange $\sum_{a=b}\langle a^{NR}b^{NR}|\hat{\mathcal{H}}^{SOO}|b^{NR}a^{NR}\rangle$ contribution to the spin-other orbit expectation value comes from terms involving the valence orbitals. For a system containing one electron outside closed shells, these are simply related to the exchange contribution to the spin-orbit coupling

$$\langle\psi_{HF\alpha}|\hat{\mathcal{H}}_T^{SOO}|\psi_{HF\alpha}\rangle = 2\langle v^{NR}|\hat{\mathcal{H}}_{TC}^{SOEx1}|v^{NR}\rangle \tag{68}$$

so that

$$\Delta E^{B}_{CVPT} = 2(\langle v^{+NR}|\hat{\mathcal{H}}_{TC}^{SOEx1}|v^{+NR}\rangle - \langle\bar{v}^{NR}|\hat{\mathcal{H}}_{TC}^{SOEx1}|\bar{v}^{NR}\rangle) \tag{69}$$

(b) Ground state spin-orbit excitation energies.

The results shown in Table 3 compare with experiment the fine structure predictions for atomic ground states made by both the Dirac-Fock EAL method and perturbation theory (equation (67)). For Boron and Aluminium both the Dirac-Fock EAL and the perturbation theory predictions show excellent agreement with experiment. The data of Table 4 show that this good agreement depends on including the Breit interaction. Estimates |1| of the quantum electrodynamic contribution to the fine structure show this to be negligible compared with the discripancies between the Dirac-Fock results and experiment for all the atoms reported in Table 3. This coupled with the result for the heavier atoms that the Dirac-Fock method underestimates the fine structure shows that core-valence correlation energy of the level $|\text{core}\ \bar{v}\rangle$ is greater in magnitude than that of the level $|\text{core}\ v^+\rangle$. Since these two levels are degenerate in the non-relativistic limit,

Table 3. Dirac-Fock and perturbation predictions of fine structure in atoms containing one valence electron (included the Breit term) (cm^{-1}).

	B	Al	Ga	Tl	Sc	Lu
$\Delta E_{PT}^{(a)}$	15.18	111.4	696		144	
$\Delta E^{(b)}$	15.14	111.4	774	7509	138	1312
$expt^{(c)}$	16.0	112.0	826	7793	168	1994

(a) Perturbation theory (order $1/c^2$) equation (67).
(b) Dirac-Fock EAL plus Breit equation (66) - see |1| for further results.
(c) See |1| for sources of experimental data.

Table 4. Contributions of the Breit interaction to fine structure in atoms containing one valence electron (cm^{-1}).

	B	Al	Ga	Tl	Sc	Lu
ΔE_{CVPT}^{B}	-5.22	-6.1	-15	-38	-32	-60
ΔE_{CV}^{B}	-5.22	-6.2	-17	-130	-31	-72

(a) Perturbation theory (order $1/c^2$) equation (69).
(b) Fully relativistic difference between expectation values of Dirac-Fock wavefunctions over the Breit hamiltonian (10).

Table 5. Components of perturbation theory (order $1/c^2$) predictions of fine structure in light atoms (cm^{-1}).

	B	Al	Gu	Sc
$\Delta E_{SB}^{(a)}$	15.18	93.6	696	193
$\Delta E_{K}^{(b)}$	0.0	19.8	71	-49
$\Delta E_{ODLM}^{(c)}$	0.0	-1.9	-7	0
ΔE_{PT}	15.18	111.4	759	144

(a) Contribution from spin-orbit coupling operators including Breit term (69). $\Delta E_{SB} = \langle v^{+NR}|F_{TC}^{(1,Dir)}|v^{+NR}\rangle - \langle \bar{v}^{NR}|F_{TC}^{(1,Dir)}|\bar{v}^{NR}\rangle + \Delta E_{CVPT}^{B}$
(b) Contribution from relativistic connection to core exchange potential. $\Delta E_K = \langle v^{+NR}|K_{TC}^{(1)} - K_{TC}^{(2)}|v^{+NR}\rangle - \langle \bar{v}^{NR}|K_{TC}^{(1)} - K_{TC}^{(2)}|\bar{v}^{NR}\rangle$.
(c) Difference between off-diagonal lagrange multiplier terms.

this result shows that the relativistic contribution to the correlation energy in the level $|core\ \bar{v}>$ is greater than that in the level $|core\ v^{+}>$.

The perturbation theory predictions of the fine structure reported in Table 3 are decomposed into their component terms in Table 5. Only for Boron and Aluminium does the fine structure predicted by perturbation theory agree with that predicted by the equivalent fully relativistic theory based on the Dirac equation, namely the Dirac-Fock EAL method with $[j(V)]/(2[\ell])$ weighting factors. For the heavier atoms for which no results are reported in Tables 3 and 5 the agreement between the perturbation theory predictions and the Dirac-Fock EAL ones is even less satisfactory than it is for Scandium and Gallium. Since the perturbation theory is derived from Dirac-Fock theory by neglecting terms of higher order than $1/c^2$, the predictions of the former theory cannot be credited with much validity if they fail to reproduce the Dirac-Fock results. The data of Table 3 similarly show that for the heavier atoms the spin-other-orbit term does not reproduce the Breit contribution to the fine structure splittings. Previous perturbation theoretic calculations of fine structure in both atoms |15,37-39| and molecules |40,41| considered only the spin-orbit coupling operators but included neither the relativistic correction to the core exchange potential nor the off-diagonal Lagrange multiplier corrections. The core exchange potential corrections to the energy of the CSF are never zero although the expression (16) of |28| shows that they do not contribute to the difference between the energies of the CSFs $|core\ \bar{v}>$ and $|core\ v^{+}>$ if the core is composed entirely of s orbitals. The results for Aluminium (Table 5) confirm the importance of including both this potential correction and the off diagonal Lagrange multiplier term because perturbation theory only reproduces the Dirac-Fock result, which agrees with experiment, if these terms are included. The fine structure excitation energy of 93.6 cm^{-1} predicted from just the spin-orbit coupling operator which is very similar to the value of 90.8 cm^{-1} |37| clearly does not agree satisfactorily with experiment. This result shows that it would be quite wrong to ascribe the difference between experiment and the result calculated solely from the spin-orbit coupling operators to electron correlation. The Dirac-Fock wavefunction whose predictions agree with experiment does not contain electron correlation. Further the use of just the standard perturbing operators $\mathcal{H}_T^{(1)}$ (21) with a open shell Hartree-Fock wavefunction of non-zero orbital angular momentum is not fully correct as discussed in section B.1.b.iv. These results for the Aluminium atom strongly suggest that the omission of the $\hat{K}_{TC}^{(1)} - \hat{K}_{TC}^{(2)}$ operator and the Lagrange multiplier correction is responsible for the discrepancies between experiment and the predictions of non-relativistic Hartree-Fock theory for the fine structure in second row hydrides |41|. In particular the experimental value of 108 cm^{-1} and the Hartree-Fock prediction of 91.1 cm^{-1} for the fine structure splitting in the $^2\pi$ state of AlH^+ are remarkably similar to the corresponding results (Table 5) for the Aluminium atom which suggests that the discrepancy for AlH^+ would be largely removed if the exchange potential and off diagonal Lagrange multiplier corrections were included. Similarly the accuracy of the atomic spin-orbit coupling constants predicted from non-relativistic Hartree-Fock wavefunctions |38,39| is questionable because neither of these two terms was considered. The theory presented here shows that one should not hope to predict the fine structure accurately by taking the expectation value of the standard perturbing operators $\mathcal{H}_T^{(1)}$ (21) over a non-relativistic Hartree-Fock wavefunction. There are two ways of attempting to rectify the deficiency of this approach. One has the choice of either adding the potential and off-diagonal Lagrange multiplier terms to the perturbing Hamiltonian or of transcending the Hartree-Fock description by using a much closer approximation to the exact non-relativistic wavefunction. Since this wavefunction neither contains closed core shells nor need make any explicit reference to orbitals, it is hardly surprising that the relativistic corrections to the energy of $|\psi_\alpha^{NR}\rangle$ can be calculated from (21).

It is interesting to note that perturbation theory fails to reproduce the Dirac-Fock EAL results for just those atoms of intermediate or high atomic number for which single manifold Dirac-Fock theory fails to reproduce the experimental fine structure. These failures are probably related because for light atoms the core-valence correlation energies of the two levels $|\text{core } \bar{v}\rangle$ and $|\text{core } v^+\rangle$ will be almost identical because the charge distributions of the orbitals $|\bar{v}\rangle$ and $|v^+\rangle$ are not significantly different. For a heavier atom, however the orbital $|\bar{v}\rangle$ being contracted relative to the orbital $|v^+\rangle$ has a greater density in the core spatial regions and consequently the core-valence correlation energy of the level $|\text{core } \bar{v}\rangle$ is of greater magnitude than that of the level $|\text{core } v^+\rangle$. Hence single manifold Dirac-Fock theory underestimates the fine structure splitting. However a significant difference between the charge distributions of the $|\bar{v}\rangle$ and $|v^+\rangle$ orbitals implies a significant relativistic correction, for which a first order perturbation description may well be inadequate. The mean radii reported in |1| of both the relativistic and non-relativistic valence orbitals enable the onset of these failures for Scandium and Gallium to be rationalised as discussed elsewhere |1|.

(c) Fine structure inversions in excited states.

It is now well established that the fine structure in certain excited ^2D and ^2F states of the alkali atoms is inverted. References to experimental work and a discussion of previous calculations are given in |42| and do not need to be repeated here. This section has two main objectives which are

(i) To show that the physical origins of these inversions can be explained by using the perturbation theory for Hartree-Fock wavefunctions described above.

(ii) To demonstrate as a consequence of achieving the first objective that the fine structure inversions can be explained within the framework of central field models of the Hartree-Fock or Dirac-Fock types, so that these inversions should not be regarded as a consequence of electron correlation.

The fine structure excitation energies ΔE_{PT} predicted by equation (67) using the perturbation theory described above are compared in Table 6 with experiment for excited d and f states of the alkali atoms. Many further results yielding the same conclusions are reported in |42|. Although the calculations do not always quantitatively reproduce the observed fine structure, the agreement between theory and experiment is sufficiently good that one can conclude that the theory presented above in section B.2.a contains the essential physics responsible for the fine structure inversions. It should be pointed out that the sign of the fine structure is always predicted correctly. The decomposition of the predicted fine structure ΔE into the spin-orbit ΔE_{SO}, Breit ΔE^{B}_{CVPT} and core exchange potential correction ΔE_K contributions shows that the latter term is responsible for any inversions which may occur. The sum ΔE_{SB} of the spin-orbit and the Breit corrections always contributes positively to the fine structure. It is explained in |42| why the fine structure of the Na f levels is not inverted whilst that of both the Na d levels and the Rb and Cs f levels is inverted.

These calculations reported in Table 6 differ from those for the atomic ground states (Tables 2-5) because purely numerical difficulties prevented the computation of highly excited self consistent field valence orbitals. The non-relativistic valence wavefunctions needed were therefore constructed by assuming them to be hydrogenic, with an effective nuclear charge chosen to reproduce the experimental ionization potentials of the excited state of interest. This technique can only be used for highly excited d or f orbitals, for which the effective nuclear charges deduced from experiment do not deviate from unity by more than 0.005. All the states studies here satisfy this criterion. Further it was verified that there were no significant differences between calculations using this prescription for the effective nuclear charge and those in which this effective charge was taken to be unity. The approximation of assuming the valence orbital to by hydrogenic would be expected to be more accurate for the f levels than for the corresponding d ones. This could explain why the calculations for the former levels agree better with experiment than those for the latter. It should also be pointed out that the calculation for the 4f level in Cesium in which the valence orbital was computed by solving the Schrodinger equation in the local field generated by the non-relativistic Hartree-Fock core of the first positive ion of the atom under consideration predicts ΔE to be -188.4 mK. Both the relativistic and non-relativistic core orbitals required to compute the fine structure were taken from single configuration Dirac-Fock or Hartree-Fock calculations on the first positive ions of the

Table 6. Perturbation theory (order $1/c^2$) predictions of fine structure in excited states of alkali atoms (mk).[a]

	Na 3d	Na 4f	K 20d	Cs 4f
ΔE_{SB}	32.60	7.60	0.05	7.44
ΔE_K	-68.47	-0.13	-5.48	-182.19
ΔE_{PT}	-35.87	7.47	-5.43	-174.75
expt[b]	-49.43	7.64	-5.08	-181.3

(a) See notes to table 5.
(b) See |42| for sources of experimental data.

atoms under consideration.

Previous non-relativistic calculations of the fine structure inversions, which are discussed in |42|, considered only the standard spin-orbit coupling operators but used more accurate non-relativistic wavefunctions taking account of electron correlation. Although such calculations are in principle capable of predicting the exact fine structure if a sufficiently elaborate non-relativistic wavefunction is used, it can be argued that such a procedure is far from ideally suited to answering the question why do such inversions occur. The answer to this as to other questions seeking reasons rather than purely numerical predictions must be given by the simplest calculation which explains the phenomenon in question. The simplest calculations predicting fine structure inversion, which do also yield reasonable agreement with experiment, would appear to be those presented in this section. Thus it can be concluded that these inversions arise from the lowest order ($1/c^2$) relativistic connection to the core exchange potential

For relativistic wavefunctions the correlation energy can be defined |1| as the difference between the "exact" relativistic energy (9) including the Breit term and the corresponding energy predicted by a single manifold |20| Dirac-Fock wavefunction. This definition shows that the predictions of the fine structure calculated as the difference between the total energies of the two Dirac-Fock wavefunctions $|\text{core } n\ell\bar{j}\rangle$ and $|\text{core } n\ell j^+\rangle$ augmented by the Breit contribution does not include any effects of electron correlation because electron correlation is not included in a single determinant Dirac-Fock wavefunction. Consequently the result (67) for the fine structure obtained by expressing the Dirac-Fock prediction as the lowest order ($1/c^2$) relativistic correction to the non-relativistic Hartree-Fock energy does not include electron correlation. This shows that the fine structure inversions can not be ascribed to electron correlation.

The orbitals used to construct the Dirac-Fock wavefunctions, from which the perturbing operators (50c) and (50d) are derived, have the standard central field form |32| being eigenfunctions of $\hat{j}^2$ and $\hat{j}_z$, with the radial parts of the 2j + 1 orbitals differing only in the m_j quantum number taken to be identical. This observation taken in conjunction with the results presented in Table 6 shows that the fine structure inversions can be explained entirely within the central field model.

It should also be pointed out that the prediction of the fine structure inversions by the perturbation expression (67) shows that such inversions do not arise because the valence electron polarizes the core. Thus the core wavefunctions and hence the relativistic correction $\hat{K}_{TC}^{(1)} - \hat{K}_{TC}^{(2)}$ to the non-relativistic core exchange potential is computed from Dirac-Fock and Hartree-Fock calculations on the first positive ion which is a closed shell system. Since the valence electron is not present in the calculation of $\hat{K}_{TC}^{(1)} - \hat{K}_{TC}^{(2)}$ the prediction of fine structure inversions when a hydrogenic valence orbital is placed in the field of this fixed core shows that the inversions do not arise because the valence electron polarizes the core. This mechanism causing fine structure inversions is therefore quite different from that generating the negative spin densities at the nuclei in excited states such as Li $1s^2 2p$. In the latter case no spin-density is predicted if a 2p orbital is placed in the field of the $Li^+ 1s^2$ core, negative spin density being predicted only when the core wavefunctions are calculated in the presence of the exchange field of the 2p valence electron.

It is shown in |42| how the present approach can be very conveniently used to understand the dependence of the fine structure inversion on the position of the atom in the periodic table.

3. Breakdown of perturbation theory

(a) Failure of lowest order ($1/c^2$) theory.

The perturbation theory expressions (49) and (64) for the eigenvalue connections are only correct to the lowest order ($1/c^2$) in the relativistic perturbation. It should therefore be suspected that these expressions might break down for systems in relativistic effects are large.

The data reported in Table 7 compare the lowest order ($1/c^2$) perturbation theory predictions ($\varepsilon_v^{(1)}$) of the valence orbital eigenvalue connection with their exact counterparts ($\varepsilon_v - \varepsilon_v^{NR}$). For very light atoms this comparison has already been presented (Table 2). The results (Table 7) show that the perturbation results become steadily worse as the nuclear charge increases, the errors being about 1% for K and Sc, increasing to 10% for p, $\bar{d}$ and d orbitals in the heavier systems. For the valence s and $\bar{p}$ orbitals, however, the quality of the perturbation results deteriorates much more rapidly with increasing nuclear charge. For these types of orbital in elements heavier

Table 7. Comparison between lowest order ($1/c^2$) perturbation theory predictions ($\varepsilon_v^{(1)}$) of relativistic connections to valence orbital eigenvalues and the exact results $\varepsilon_v - \varepsilon_v^{NR}$ (cm^{-1}) (a)

	K(4s)	In(5$\bar{p}$)	In(5p)	La(5$\bar{d}$)	La(5d)	Lu(5$\bar{d}$)	Lu(5d)
$\varepsilon_v^{(1)}$	-120.4	-702.6	1265.0	6791	8000	11947	13708
$\varepsilon_v - \varepsilon_v^{NR}$	-121.6	-879.4	1248.8	6714	7747	11418	12804
	Au(6s)	Tl(6$\bar{p}$)	Tl(6p)	Fr(7s)	E111(7s)	E113(7$\bar{p}$)	E113(7p)
$\varepsilon_v^{(1)}$	-10587	-1884	3786	-2183	-16286	+512	10568
$\varepsilon_v - \varepsilon_v^{NR}$	-15560	-4158	3481	-3250	-48069	-17845	7322

than the third transition series, the errors have become so large (30% for Au and Fr) that it can only be concluded that the perturbation theory has broken down. Indeed for the 6$\bar{p}$ orbital in Tl, perturbation theory recovers less than half of the energy connection caused by relativity. The conclusion that perturbation theory breaks down for s and $\bar{p}$ valence orbitals in heavy elements is underlined by the result for the 7$\bar{p}$ orbital in E113 (eka-Tl) for which the perturbation method is even unable to predict the sign of the large relativistic connection.

The cause of the failure of perturbation theory can be located more precisely by comparing the perturbation predictions of the direct and indirect effects with estimates obtained from a fully relativistic method |30|. The difference between the full Dirac-Fock eigenvalue (ε_v) and that (ε_v^{NRRC}) obtained by solving the non-relativistic Hartree-Fock equation ((21) of |28|) for the valence electron moving in the field of the relativistic core provides one estimate of the direct effect. This approach disentangles the effect of the relativistic motion of the valence electron itself (i.e. of replacing the Schrodinger kinetic energy operator by the Dirac one) from that of the core potential which is held constant. The estimates thus obtained for Lu, Au and Tl are reported in the row labelled rel core in Table 8. The difference between the non-relativistic eigenvalue (ε_v^{NR}) and that (ε_{VNRc}) resulting from solving the Dirac-Fock equation for the valence orbital in the potential generated by non-relativistic core provides another estimate of the direct effect reported as the row labelled NR core in Table 8. These four different valence orbital eigenvalues also provide two estimates ($\varepsilon_V - \varepsilon_{VNRc}$) and ($\varepsilon_V^{NRRc} - \varepsilon_V^{NR}$) of the indirect effect. These two differences, labelled rel motion and NR motion in Table 8, show the effect of

Table 8. Comparison of fully relativistic and lowest order ($1/c^2$) perturbation theory estimates of the direct and indirect effects (cm^{-1}).

		Lu		Au	Tl	
		$5\bar{d}$	5d	6s	$6\bar{p}$	6p
Direct effect	pert$^{(a)}$	-5160	-2211	-9640	-7996	-2843
	rel core	-4084	-1624	-14165	-9729	-2604
	NR core	-5583	-2219	-15797	-12467	-3157
Indirect effect	pert$^{(b)}$	17094	15927	-1007	6043	6622
	rel core	17002	15024	237	8309	6638
	NR core	15502	14428	-1395	5571	6085

(a) = $\langle v^{NR}|\hat{F}_{TC}(1,Dir)|v^{NR}\rangle$
(b) = $\langle v^{NR}|\hat{F}_{TC}(1,Ind)|v^{NR}\rangle$

changing the core potential without changing the valence electron dynamics. Although these two ways of calculating the direct and indirect effects do not yield identical results and there is no a priori reason for preferring either of the two methods, it is clear from Table 8 that the results provide quite tight bounds on the magnitudes of the two effects.

The results of Table 8 show not only that perturbation theory can reproduce the indirect effect quite well for all the valence orbitals but also that it gives a good account of the direct effect for p, $\bar{d}$ and d orbitals. However, it seriously underestimates the direct relativistic effect both for the Gold 6s and the Thallium $6\bar{p}$ orbital. The 36% error for Gold, relative to the average of the two fully relativistic estimates, explains the overall 32% error in the perturbation theory prediction $\varepsilon_V^{(1)}$ of the total relativistic correction because the indirect effect is small here. The larger error shown by perturbation theory in the case of the Tl $6\bar{p}$ orbital compared with the Au 6s one is also explained because the total correction in Tl arises from the partial cancellation of the larger direct and indirect effects. The perturbation approach describes the indirect effect quite well but only partially recovers the direct one thus predicting their sum badly.

Further insight into the large direct effects shown by s and $\bar{p}$ orbitals is provided by the ratio Q(r)/P(r) which gives a measure of the local electron velocity as v(r)/c. For the Gold 6s orbital these ratios are approximately 1:3, 1:7, 1:10, 1:25, 1:70 and 1:300 for distances between respectively the nucleus and the first node of P, the second and third nodes of P, the third and fourth nodes, etc. These ratios show that the electron moves very relativistically in the inner spatial regions but non-relativistically in the valence region. This reconciles the large direct relativistic effect for the Gold 6s orbital with the result |43| that the interaction energy of two Hg atoms calculated from undistorted relativistic atomic densities using the free electron gas model of Gordon and Kim |44| is unaffected by substitution of the Schrodinger kinetic energy for the Dirac one. It should be pointed out that the much smaller errors in the perturbation predictions of the larger relativistic corrections in Lutetium compared to those of Au and the Tl $6\bar{p}$ case show that the perturbation approach does not break down for the latter systems solely because the corrections are large. Perturbation theory fails for valence s and $\bar{p}$ orbitals in heavy atoms because it is incapable of describing the large direct relativistic energy stabilization arising from the highly relativistic valence electron motion in inner spatial regions.

It should be pointed out that the failure of lowest order perturbation theory cannot be circumvented by working with a zeroth order equation in which the valence electron moves non-relativistically in the field of the exact relativistic core. This point is fully discussed in sections 4.3 and 5 of |28|. It only remains to point out that the consistency of this approach can be queried because it does not treat the two contributions to the relativistic connection, namely the direct and indirect effects on an equal footing. It treats the non-exchange contribution to the indirect effect to all orders in $1/c^2$ whilst treating the direct effect only to order $1/c^2$.

(b) Divergencies at higher orders in perturbation theory.

The obvious method of overcoming the deficiencies of the lowest order ($1/c^2$) perturbation theory for heavy atoms elucidates above is to extend the perturbation theory to higher orders. At a first glance it would appear that this could be achieved by simply calculating further terms in the Foldy-Wouthuysen transformation defined by (36) and then performing further Foldy-Wouthuysen transformations to remove the remaining higher order operators linking the large and small components. This procedure is described in |26|.

However, it is well known that many of the expectation values of higher order than $1/c^2$ diverge in the point nucleus approximation. Since the theory developed here also diverges at higher order even though it is based on extended nuclear charge distributions, these difficulties will be briefly examined in order to remove the doubt which would otherwise be case on the validity of the lower order

† A point nucleus was used in |30|. The calculations were repeated with the nuclear potential due to a uniform nucleus.

non-divergent expressions. In the first step of the reduction to the non-relativistic limit the Dirac-Fock hamiltonian is transformed according to (36) with $\hat{S} = -i\beta e_1 \underline{\Sigma}\cdot\hat{\underline{p}}/(2c)$. For the mass $(c^2(\beta - 1))$ and Dirac kinetic energy $(c\underline{\alpha}\cdot\hat{\underline{p}})$ terms it is not necessary to resort to the usual expansion

$$\hat{\mathcal{H}}' = \exp(i\hat{S})\hat{\mathcal{H}}\exp(-i\hat{S}) = \hat{\mathcal{H}} + i[\hat{S},\hat{\mathcal{H}}] + \frac{i^2}{2!}[\hat{S},[\hat{S},\hat{\mathcal{H}}]] + \ldots \qquad (70)$$

because the results

$$[\Sigma,\beta] = [\Sigma,e] = \beta e_1 + e_1\beta = 0 \qquad (71)$$

with the 4 x 4 matrix e_1 defined as

$$e_1 = \begin{pmatrix} 0 & I \\ I & 0 \end{pmatrix} \qquad (72)$$

where I is the 2 x 2 identity matrix can be used to show that

$$\begin{aligned}\hat{\mathcal{H}}_{Fr} &= \exp(\beta e_1\underline{\Sigma}.\hat{\underline{p}}/(2c))(c^2(\beta - 1) + ce_1\underline{\Sigma}\cdot\hat{\underline{p}})\exp(-\beta e_1\underline{\Sigma}\cdot\hat{\underline{p}}/2c) \\ &= \beta\{c^2\cos(\tfrac{\hat{p}}{c})+c\hat{p}\,\sin(\tfrac{\hat{p}}{c})\}-c^2+\{c\underline{\Sigma}.\hat{\underline{p}}\,\cos(\tfrac{\hat{p}}{c})-c^2\frac{\underline{\Sigma}.\hat{\underline{p}}}{p}\,\sin(\tfrac{\hat{p}}{c})\}e_1\end{aligned} \qquad (73)$$

Here $\hat{\mathcal{H}}_{Fr}$ is $c\underline{\alpha}\cdot\hat{\underline{p}} + c^2(\beta - 1)$. Although further Foldy-Wouthuysen transformations are applied, the final expression for the energy contains the expectation value of the even part of (73) over the transformed wavefunction $|a'\rangle$ which using (42) can be written as a sum of the expectation value over the non-relativistic wavefunction plus correction terms. It proves instructive to consider the hydrogen atom in the point nucleus approximation for which the non-relativistic 1s wavefunction can be written in terms of the momentum representation as |21|.

$$\begin{aligned}|1s^{NR}(\underline{r})\rangle &= (2\pi)^{-3/2}\int|1s^{NR}(\underline{p})\rangle\exp(i\underline{p}.\underline{r})d\underline{p} \\ |1s^{NR}(\underline{p})\rangle &= \pi^{-1}2\sqrt{2}(p^2 + 1)^{-2}\end{aligned} \qquad (74)$$

Considering only the large components, the expectation value of the even part of (73) over $|1s^{NR}\rangle$ is readily calculated using the momentum representation (74) and found to be

$$\langle 1s^{NR}(\underline{r})|c^2\cos(\tfrac{\hat{p}}{c}) + c\hat{p}\,\sin(\tfrac{\hat{p}}{c}) - c^2|1s^{NR}(\underline{r})\rangle =$$

$$\frac{32}{\pi}\{c^2\int_0^\infty\frac{p^2\cos(p/c)dp}{(p^2+1)^4} + c\int_0^\infty\frac{p^3\sin(p/c)dp}{(p^2+1)^4}\} - c^2 = (c^2+c+1+\frac{2}{3c}-\frac{1}{3c^2})e^{-1/c} - c^2 \qquad (75)$$

Although these expressions are finite and the final function of c involving exp(1/c) can be expanded as a power series in 1/c, it is clear that if the series obtained by expanding cos(p/c) and sin(p/c) are integrated term by term all terms of higher order than $1/c^2$ will diverge, although alternating in sign, because the numerators contain p to powers greater than or equal to eight. However not only are the finite integrals equal to the corresponding terms in the series expansion of the final result but they also equal the corresponding expectation values of the operators of orders upto $1/c^2$ in (37). This shows that the sum of the integrals having coefficients of orders upto and including $1/c^2$ provides an asymptotic representation of (75) because the relation 4.6.3 of |45| is satisfied. Furthermore the divergent integrals are precisely those arising if the higher order terms in the Foldy-Wouthuysen expansion are evaluated. Indeed it is easily seen that the procedure of expanding the integrals entering (75) is exactly equivalent to using the operator expansion (70). In the position representation the divergencies at high momenta are replaced by divergencies at the nucleus which arise from repeated applications of $-i\underline{\nabla}$ on the cusp at the nucleus producing derivatives of delta functions which are highly singular. These results strongly suggest that all the divergencies arise from the term by term integration procedure which is used in the more complicated situation where potential terms are present and hence that all the finite low order terms provide an asymptotic representation |45| of the full series which is therefore both trustworthy and accurate for systems which are not too relativistic.

In this work the point nucleus approximation is avoided because the divergence of Dirac s and $\bar{p}$ orbitals and the cusp in non-relativistic s orbitals which arise at nuclei are sources of divergencies. Any divergence arising in the point nucleus approximation solely because the radial part of the orbital varies as $r^{\gamma-1}$ ($\gamma = (\kappa^2 - (Z/c)^2)^{1/2}$ |32|) at distances (r) close to the nucleus which is absent in extended nuclear treatments because the $r^{\gamma-1}$ dependence is replaced by an r^{ℓ} one is clearly of no fundamental significance. The finite results obtained in the latter case will depend only extremely weakly on the details of the assumed nuclear charge distribution. However any operators whose expectation values are crucially sensitive to this charge distribution and whose convergence depends directly on the actual nuclear size employed have almost certainly arisen solely because the expansion in 1/c has broken down and cannot therefore be used in any calculation of relativistic connections. It is shown elsewhere |26| that the expectation values both of the operator $\hat{p}^6$, which is one of the divergent ones in the point nucleus case discussed above, and of the operator $\hat{p}^2$ SON depend on the nuclear radius r_N as r_N^{-1} and log r_N respectively. Since both these operators occur when the Foldy-Wouthuysen transformation is extended to order $1/c^4$, it is clear that divergencies arise which can not be ignored.

It should be pointed out that for the hydrogen atom with a point nucleus where analytic wavefunctions are available, it has been shown |46| that if all the corrections of order $1/c^4$ to both the operators and the wavefunctions are calculated, then all the divergencies cancell to yield the correct finite result for the energy. However it has not so far proved possible to extend this result to roder $1/c^6$. More importantly it does not appear to be readily possible to carry out a similar calculation fot the Hartree-Fock case because here analytic expressions for the orbitals are not available. This would make it appear that attempts to use higher order perturbation descriptions of relativistic effects in non-pseudo-potential theories are unpromising.

C. Theoretical Basis of Pseudo-potential Methods

1. Lowest order ($1/c^2$) pseudo-potential theories.

The object of this section (C) is to examine the possibility of placing on a sound theoretical basis the relativistic pseudo-potential methods which use the Schrodinger kinetic energy operator. This sub-section (1) examines the lowest order ($1/c^2$) theory as a necessary preliminary to the discussion of the higher order theory presented in the mext section (2). Only closed shell systems are examined here because these illustrate the main principles, the extension to systems containing one electron in an open shell is described in |26| and |27|. The viewpoint taken here is that any pseudo-potential theory using the Schrodinger kinetic energy operator is only soundly based if it can be derived from an equivalent fully relativistic theory based on the Dirac equation.

In the Phillips-Kleinman pseudo-potential theory |47| each valence pseudo-orbital $|\chi_v\rangle$ is constructed as a linear combination of the valence orbital $|v\rangle$ and core orbitals $|c\rangle$. Thus

$$|\chi_v\rangle = c_v|v\rangle + \sum_c c_c|c\rangle \ , \quad \langle\chi_v|\chi_v\rangle = 1 \tag{76}$$

The coefficients c_v and c_c, which are arbitrary, are usually determined by demanding that the kinetic energy of $|\chi_v\rangle$ be minimized thereby generating a smooth nodeless pseudo-orbital having a low amplitude in spatial regions close to the nucleus. The pseudo-orbitals $|\chi_v\rangle$ satisfy

$$\hat{F}_{PK}|\chi_v\rangle = (\hat{F} - \hat{P}_T)|\chi_v\rangle = \varepsilon_v|\chi_v\rangle \tag{77}$$

with

$$\hat{P}_T = \sum_c(\varepsilon_c - \varepsilon_v)|c\rangle\langle c| \tag{78}$$

because all the orbitals $|a\rangle$ are eigenkets of $\hat{F}$ (equation (28)). The relativistic corrections to Phillips-Kleinman pseudo-potential theory are derived by applying the Foldy-Wouthuysen reduction method described in section B.1.b.i to (77) and comparing the result with its non-relativistic equivalent. Application of the transformation defined by (36) to (77) converts this equation to

$$[c^2(\beta - 1) + \frac{\beta\hat{p}^2}{2} + \hat{V}_{loc} + \hat{\mathcal{H}}^{DN} + \tfrac{1}{2}\hat{\mathcal{H}}^{DEd} + \hat{\mathcal{H}}^{SON} + \hat{\mathcal{H}}^{SOEd1}$$
$$+ \hat{\mathcal{H}}^{M} - \exp(i\hat{S})\hat{K}_T\exp(-i\hat{S}) - \sum_c(\varepsilon_c - \varepsilon_v)|c'\rangle\langle c'|]|\chi_v'\rangle = \varepsilon_v|\chi_v'\rangle \tag{79}$$

where $|c'\rangle$ is defined by (38) and (40). The pseudo-potential term is related to the non-relativistic result by using both (42) and $\varepsilon_a = \varepsilon_a^{NR} + \varepsilon_a^{(1)}$ so that this becomes to order $(1/c^2)$

$$\sum_c(\varepsilon_c - \varepsilon_v)|c'\rangle\langle c'| = \hat{P}_T^{NR} + \hat{P}_T^{(1\varepsilon)} - \hat{P}_T^{(1)} \tag{80}$$

where

$$\hat{P}_T^{NR} = \sum_c(\varepsilon_c^{NR} - \varepsilon_v^{NR})|c^{NR}\rangle\langle c^{NR}| \tag{81}$$

$$\hat{P}_T^{(1\varepsilon)} = \sum_c(\varepsilon_c^{(1)} - \varepsilon_v^{(1)})|c^{NR}\rangle\langle c^{NR}| \tag{82}$$

$$\hat{P}_T^{(1)} = \sum_c(\varepsilon_c^{NR} - \varepsilon_v^{NR})(|c^{NR}\rangle\langle c^{(1)}| + |c^{(1)}\rangle\langle c^{NR}|) \tag{83}$$

After substituting (80) into (79), expressing the direct and exchange potentials in (79) according to (39)-(45) as described in section B.1.b.i. and noting that the non-relativistic valence pseudo-orbital satisfies

$$(\hat{F}^{NR} - \hat{P}_T^{NR})|\chi_v^{NR}\rangle = \varepsilon_v^{NR}|\chi_v^{NR}\rangle \tag{84}$$

equation (79) becomes

$$(\hat{F}^{NR} - \hat{P}_T^{NR} + \hat{F}^{(1)} + \hat{P}_T^{(1)} - \hat{P}_T^{(1\varepsilon)})(|\chi_v^{NR}\rangle - |\chi_v^{(1)}\rangle)$$
$$= (\varepsilon_v^{NR} + \varepsilon_v^{(1)})(|\chi_v^{NR}\rangle - |\chi_v^{(1)}\rangle) \tag{85}$$

Here $\hat{F}^{(1)}$ is the sum of the operators describing the direct and indirect effects which arises in the non-pseudo-orbital case. Left multiplication of (85) by $\langle\chi_v^{NR}|$ then shows through (84) that the relativistic correction to the valence orbital eigenvalue is given by

$$\varepsilon_v^{(1)} = \langle\chi_v^{NR}|\hat{F}^{(1)} + \hat{P}_T^{(1)} - \hat{P}_T^{(1\varepsilon)}|\chi_v^{NR}\rangle \tag{86}$$

This result shows that in the pseudo-orbital case the eigenvalue correction contains a contribution $(\hat{P}_T^{(1)} - \hat{P}_T^{(1\varepsilon)})$ from the correction to the pseudo-potential.

The significance of the two further operators $\hat{P}_T^{(1)}$ and $-\hat{P}_T^{(1\varepsilon)}$ which enter the Phillips-Kleinman case is shown by considering a valence orbital $|v^{NR}\rangle$ for which the direct relativistic effect dominates the indirect one. Further operators in addition to $\hat{F}^{(1)}$ must arise because the major contributions to the expectation values of the mass-velocity, spin-orbit coupling and Darwin operators for a noded valence orbital which orthogonal to the core come from spatial regions close to the nucleus where momentum and r^{-3} are large whilst a nodeless pseudo-orbital has a very low amplitude in these inner spatial regions. Hence $\langle\chi_v^{NR}|\hat{F}^{(1)}|\chi_v^{NR}\rangle$ will be very much smaller in magnitude than $\langle v^{NR}|\hat{F}^{(1)}|v^{NR}\rangle$ is often negligible compared with the pseudo-orbital term $\langle\chi_v^{NR}|\hat{P}_T^{(1)} - \hat{P}_T(1\varepsilon)|\chi_v^{NR}\rangle$ which is therefore very important. Furthermore the term $(-\langle\chi_v^{NR}|\hat{P}_T(1\varepsilon)|\chi_v^{NR}\rangle)$ will be positive because core orbitals are dominated by the direct effect so that $\langle\chi_v^{NR}|\hat{P}_T^{(1)}|\chi_v^{NR}\rangle$ must be the largest term for a valence orbital whose behaviour is dominated by this effect. These results show how at least some of the relativistic effects are transferred to the pseudopotential terms. Further insight is gained by noting that a pseudo-orbital is obtained by adding arbitrary core components (76) to the valence orbital $|v'\rangle$ and then considering the expectation value

$$\langle\chi_v^{L'}|\hat{F}^{NR} + \hat{F}^{(1)}|\chi_v^{L'}\rangle = |c_v|^2(\varepsilon_v^{NR} + \varepsilon_v^{(1)}) + \sum_c |c_c|^2(\varepsilon_c^{NR} + \varepsilon_c^{(1)}) \qquad (87)$$

All the off-diagonal elements $\langle v^{L'}|\hat{F}^{NR} + \hat{F}^{(1)}|c^{L'}\rangle$ and $\langle c^{L'}|\hat{F}^{NR} + \hat{F}^{(1)}|d^{L'}\rangle$ where $|d\rangle$ is a core orbital vanish to order $1/c^2$ by vitrue of (46) thus showing that the energy of $|\chi_v^{L'}\rangle$ is lowered relative to that of $|v^{L'}\rangle$ by the core admixture. Hence if the energy of $|\chi_v^{L'}\rangle$ is to remain ε_v the additional terms $-\sum_c(\varepsilon_c^{NR}+\varepsilon_c^{(1)}-\varepsilon_v^{NR}-\varepsilon_v^{(1)})|c_c|^2$ must appear in (87) which can be generated by adding the operator $-\sum_c(\varepsilon_c^{NR} + \varepsilon_c^{(1)} - \varepsilon_v^{NR} - \varepsilon_v^{(1)})|c^{L'}\rangle\langle c^{L'}|$ to the Fock hamiltonian because (76) shows that

$$c_c = \langle c^{L'}|\chi_v^{L'}\rangle \qquad (88)$$

However the non-relativistic orbital $|\chi_v^{NR}\rangle$ is not even approximately an eigenket of $\hat{F}^{NR} + \hat{F}^{(1)}$ so that

$$\begin{aligned}\langle\chi_v^{NR}|\hat{F}^{NR} + \hat{F}^{(1)}|\chi_v^{NR}\rangle &= |c_v|^2(\varepsilon_v^{NR} + \varepsilon_v^{(1)}) + \sum_c |c_c|^2(\varepsilon_c^{NR} + \varepsilon_c^{(1)}) \\ &+ \sum_c (c_v^* c_c\langle v^{NR}|\hat{F}^{(1)}|c^{NR}\rangle + c_c^* c_v\langle c^{NR}|\hat{F}^{(1)}|v^{NR}\rangle) \\ &+ \sum_{c\neq d} c_c^* c_d\langle c^{NR}|\hat{F}^{(1)}|d^{NR}\rangle \end{aligned} \qquad (89)$$

If this energy is to equal $\varepsilon_v^{NR} + \varepsilon_v^{(1)}$ one must not only add the Phillips-Kleinman type terms $-\sum_c(\varepsilon_c^{NR} + \varepsilon_c^{(1)} - \varepsilon_v^{NR} - \varepsilon_v^{(1)})|c^{NR}\rangle\langle c^{NR}|$ to replace $\varepsilon_c^{NR} + \varepsilon_c^{(1)}$ by $\varepsilon_v^{NR} + \varepsilon_v^{(1)}$ but also further terms which exactly cancell all the off-diagonal elements of $\hat{F}^{(1)}$. Left multiplication of (46) by $\langle b^{NR}|$ shows through (42) that to order $1/c^2$

$$\langle b^{NR}|\hat{F}^{(1)}|a^{NR}\rangle = (\varepsilon_b^{NR} - \varepsilon_a^{NR})\langle b^{NR}|a^{(1)}\rangle \quad (b\neq a) \qquad (90)$$

and hence using (63) that

$$\sum_c (c_v^* c_c \langle v^{NR}|\hat{F}^{(1)}|c^{NR}\rangle + c_c^* c_v \langle c^{NR}|\hat{F}^{(1)}|v^{NR}\rangle) + \sum_{c \neq d} c_c^* c_d \langle c^{NR}|F^{(1)}|d^{NR}\rangle$$
$$= \sum_c (\varepsilon_v^{NR} - \varepsilon_c^{NR})(c_v^* c_c \langle v^{NR}|c^{(1)}\rangle + c_c^* c_v \langle c^{(1)}|v^{NR}\rangle)$$
$$+ \sum_{c \neq d} c_c^* c_d (\varepsilon_c^{NR} \langle c^{NR}|d^{(1)}\rangle + \varepsilon_d^{NR} \langle c^{(1)}|d^{NR}\rangle) \quad (91)$$

However, after noting that from $\langle c^{NR}|d^{(1)}\rangle = -\langle c^{(1)}|d^{NR}\rangle$ (cf (63))

$$\sum_{c \neq d} \varepsilon_v^{NR} (\langle c^{(1)}|d^{NR}\rangle + \langle c^{NR}|d^{(1)}\rangle) = 0$$

(91) is seen to be nothing but $-\langle \chi_v^{NR}|\hat{P}_T^{(1)}|\chi_v^{NR}\rangle$. Thus the two corrections to the pseudopotential have been shown to be rather different because $-\hat{P}_T^{(1\varepsilon)}$ is a Phillips-Kleinman type operator which projects the relativistic corrections to the core orbital energies to the valence relativistic corrections whilst $\hat{P}_T^{(1)}$ cancells all the off-diagonal elements of $\hat{F}^{(1)}$ entering $\langle \chi_v^{NR}|\hat{F}^{(1)}|\chi_v^{NR}\rangle$.

The data reported in Table 9 compare for pseudo-orbitals the perturbation theory predictions ($\varepsilon_v^{(1)}$) of the eigenvalue correction with the exact results ($\varepsilon_v - \varepsilon_v^{NR}$). The coefficients (76) defining the pseudo-orbital were determined by the prescription |48| that the pseudo-orbital kinetic energy be a minimum thereby generating a smoot nodeless pseudo-orbital. The systems examined are open shell ones fo which the closed shell theory described above is incomplete. Since the core and valence orbitals are not eigenfunctions of the same hamiltonian, generalized Phillips-Kleinman theory |49| must be used. This yields in addition to the two pseudo-potential corrections $\hat{P}_T^{(1)}$ and $-\hat{P}_T^{(1\varepsilon)}$ described above a further operator denoted $\hat{P}_T^{(1G)}$,

Table 9. Lowest order ($1/c^2$) perturbation theory predictions and components of relativistic corrections to valence pseudo-orbital eigenvalues (cm^{-1}).

	Al(3$\bar{p}$)	Al(3p)	Au(6s)	Tl(6$\bar{p}$)	Tl(6p)
Direct	-2.64	-0.52	-13	-4	-1
Indirect	149.98	105.82	3338	7827	5300
$\langle \chi_v^{NR}|\hat{P}_T^{(1)}|\chi_v^{NR}\rangle$	-192.60	56.71	-48078	-18385	-1814
$\langle \chi_v^{NR}|-\hat{P}_T^{(1\varepsilon)}|\chi_v^{NR}\rangle$	67.89	-46.59	37959	11129	112
$\varepsilon_v^{(1)}$	2.04	119.83	-12765	-2879	3471
$\varepsilon_v - \varepsilon_v^{NR}$	2.07	119.70	-15560	-4158	3481

somewhat analogous to the off-diagonal lagrange multiplier correction for Hartree-Fock orbitals, which is fully described elsewhere |26|. Furthermore the results presented in Table 9 include the expectation value of a further operator $\hat{P}_T^{(2)}$ fully described elsewhere |27| which arises by avoiding the approximation of replacing $\varepsilon_c - \varepsilon_v$ by $\varepsilon_c^{NR} - \varepsilon_v^{NR}$ in the expansion of the pseudo-potential in (79) according to (80). The almost perfect agreement between the perturbation theory predictions and the true corrections for the light atoms confirms that the theory is correct. The results for all the atoms show that the direct relativistic corrections $\langle\chi_v^{NR}|\hat{F}^{(1,Dir)}|\chi_v^{NR}\rangle$ are very small and that it is the pseudo-potential corrections which stabilize s and $\bar{p}$ valence orbitals as predicted. Although this lowest order $(1/c^2)$ theory fails for s and $\bar{p}$ valence pseudo-orbitals in heavy atoms, as it did for Hartree-Fock orbitals, the result that the direct relativistic corrections are almost entirely transferred to the pseudo-potential corrections constitutes one vital step in the development of a more accurate higher order theory. It should be remembered that it was the direct relativistic effect in s and $\bar{p}$ Hartree-Fock orbitals in heavy atoms which was not well described by perturbation theory.

2. Higher order pseudo-potential theories.

It was shown in the last section that the lowest order $(1/c^2)$ perturbation theory failed for valence s and $\bar{p}$ pseudo-orbitals in heavy atoms. The successful development of a more accurate higher order pseudo-potential theory by extending the Foldy-Wouthuysen transformation of (77) to include terms of order $1/c^4$ depends on three key observations. These are

(i) A lowest order $(1/c^2)$ treatment of the direct effect is more than adequate because this effect is small if smooth nodeless pseudo-orbitals are used.

(ii) The pseudo-potential correction can be calculated to order $(1/c^4)$ without introducing any diverging or crucially nuclear size dependent terms (A term is said to be crucially nuclear size dependent if it diverges as the nuclear size tends to zero).

(iii) The freedom of choice in defining the pseudo-orbital can be exploited to eliminate from s and $\bar{p}$ pseudo-orbitals terms varying for small distances r from the nucleus as r^0 and r respectively. This choice is useful because it eliminatec crucially nuclear size dependent terms from expectation values such as $\langle\chi^L|\hat{p}^4|\chi^L\rangle$ which arise not only in the direct corrections (i) but also in the kinetic energy term $\langle\chi_v^{L''}|\frac{p^2}{2}|\chi_v^{L''}\rangle$ where $|\chi_v^{L''}\rangle$ constitutes the transformed pseudo-orbital large components.

Although the second observation need not be shown in detail here, the first step (79) in the derivation |26| indicates how this comes about It is not necessary in (79) to expand the term $\sum_c(\varepsilon_c - \varepsilon_v)|c'\rangle\langle c'|$

according to (80), thereby avoiding the first order $1/c^2$ approximati The utility of the third observation (iii above) is that it enables to avoid introducing the corresponding non-relativistic wavefunction which are somewhat disimiliar to the relativistic orbitals in a heav atom, and work directly with the transformed pseudo-orbital equation This step which is really only fully justifiable with pseudo-orbital eliminates one large source of error.

The results presented in Table 10 compare the valence orbital eigenvalues predicted by this approximation with the true values. T perturbation predictions ($\varepsilon_v{}^P$) were calculated as the expectation va (92) over the large components $|\chi_v{}^{L''}>$ of the Foldy-Wouthuysen trans formed pseudo-potential defined by (101) of |26|.

$$\varepsilon_v{}^p = <\chi_v{}^{L''}|\hat{F}^{(0)} + \hat{P}_T{}^t + \hat{F}^{(1s)} + \hat{F}^{(Hs)}|\chi_v{}^{L''}> \quad (92)$$

$$\hat{F}^{(0)} = \tfrac{1}{2}\hat{p}^2 + \hat{V}_{loc} - \hat{K}_T{}^{LL} - \hat{P}_T{}^{LL} \quad (93)$$

Here $\hat{K}_T{}^{LL}$ and $\hat{P}_T{}^{LL}$ are those portions of the exchange and Phillips-Kleinman pseudo-potentials built solely from the large components, $\hat{P}_T{}^t$ (100b) of |26a| constitutes the remainder of the pseudo-potentia $\hat{F}^{(1s)}$ (100c) of |26a| is the direct relativistic correction whilst $\hat{F}^{(Hs)}$ (100d) of |26a| is a higher order cross term between $\hat{F}^{(1s)}$ and $\hat{P}_T{}^t$. These results require knowledge of the exact relativistic larg components because $|\chi_v{}^{L''}>$ has to be computed from these. Neverthe-less the excellent agreement between the perturbation eigenvalues an the true results does show that the pseudo-potential is soundly base at least as far as incorporating the relativistic effect is concerne

Table 10. Comparison between pseudo-valence orbital energies predict by higher order pseudo-potential theory with exact values (cm^{-1}).

Atom (orbital)	Exact Dirac-Fock	Higher order pseudo-potential numerical εv^P equation (92)	Slater basis expansio 4 Slaters	7 Slater
Hg(6s)	-71990.7	-71990.8	-71988.6	-71990.4
E112(7s)	-98990.8	-98990.9	-98986.7	-98990.0
Tl(6$\bar{p}$)	-46386.7	-46386.8	-46384.0	-46386.3
E113(7$\bar{p}$)	-58251.0	-58250.9	-58247.8	-58250.5

However in any practical application the valence pseudo-orbital will not be known before carrying out the calculation. It is therefore necessary to show that this can be regarded as an unknown function which can be determined by solving some pseudo-potential equation. The results presented in the last two columns of Table 10 were calculated by expanding the radial part of the valence pseudo-orbital in a Slater basis set and determining the expansion coefficients by solving the standard mztrix eigenvalue equation $\underline{F}^{(0)}\ \underline{c} = \underline{s}.\underline{c}.\underline{\varepsilon}$ where $\underline{F}^{(0)}$ is the matrix representation of $F^{(0)}$ (93) in the basis. The remaining terms in the hamiltonian were then added as first order perturbations. The good agreement with the exact results provides very strong evidence that a viable pseudo-potential theory using the Schrodinger kinetic energy operator can be constructed which is soundly based theoretically at least as far as describing the effects of relativity is concerned.

FULLY AB-INITIO FULLY RELATIVISTIC CALCULATIONS BASED ON THE DIRAC EQUATION

A. Introduction

It was shown in the first section of this talk that relativistic effects in even the valence orbitals of heavy atoms are large and that they cannot be adequately described by presently known perturbation theory methods. Although the results presented in the last lecture seem to show that pseudo-potential calculations for molecules containing heavy atoms can be put on a sound theoretical basis as far as incorporating the relativistic effects is concerned, other uncertainties already present in non-relativistic pseudo-potential calculations remain. Although these problems are clearly worthy of further study, it is also worthwhile to develop fully ab-initio methods not using pseudo-potentials which are also fully relativistic in that they are based directly on the Dirac equation. The viewpoint taken is that no such method is worthwhile unless it can be applied to molecules containing very heavy and even super-heavy elements. After all, these are the systems for which relativity is important. After outlining below one such method developed during the past 3½ years applications to ionic solids are described in section 3 whilst preliminary results of a molecular calculation are presented in section D.

B. Programme Outline

The details of the purely numerical methods used in the newly developed program for performing relativistic ab-initio calculations for diatomic molecules will be fully described elsewhere. Only the theoretical methods will be described here.

The wavefunction for the diatomic molecule containing a total of N electrons is written as

$$|\Psi(\underline{r}_1 \ldots \underline{r}_N)\rangle = S\hat{\mathcal{A}}' \left[|\Phi_{ac}(\underline{r}_1 \ldots \underline{r}_{nca})\rangle |\Phi_{bc}(\underline{r}_{nca+1} \ldots \underline{r}_{nca+ncb})\rangle \right.$$
$$\left. \times \; |\Phi_v(\underline{r}_{nca+ncb+1} \ldots \underline{r}_N)\rangle \right] \qquad (94)$$

where S is a normilization constant. Here $|\Phi_{ac}(\underline{r}_1 \ldots \underline{r}_{nca})\rangle$ and $|\Phi_{bc}(\underline{r}_{nca+1} \ldots \underline{r}_{ncb})\rangle$ are wavefunctions for the cores, containing nca and ncb electrons, of the atoms a and b. These core wavefunctions are taken to be single determinants of the Dirac-Fock atomic orbitals of the isolated atoms.

$$|\Phi_{ac}(\underline{r}_1 \ldots \underline{r}_{nca})\rangle = \hat{\mathcal{A}}\left(\prod_{i=1}^{nca} |\phi_{i_a}(\underline{r}_{a,i})\rangle\right) \qquad (95a)$$

$$|\Phi_{bc}(\underline{r}_{nca+1} \ldots \underline{r}_{nca+ncb})\rangle = \hat{\mathcal{A}}\left(\prod_{i=1}^{ncb} |\phi_{i_b}(\underline{r}_{b,i+nca})\rangle\right) \qquad (95b)$$

The quantity $\underline{r}_{\mu,i}$ is the vector describing the position of electron i with respect to nucleus μ, $|\phi_{i_\mu}(\underline{r})\rangle$ is the i th Dirac-Fock atomic orbital for atom μ, $\hat{\mathcal{A}}$ is the anti-symmetrizer whilst $\hat{\mathcal{A}}'$ appearing in (94) is the partial anti-symmetrizer which contains, besides the identity, only permutations interchanging co-ordinates belonging to one of the sets $(\underline{r}_1 \ldots \underline{r}_{nca})$, $(\underline{r}_{nca+1} \ldots \underline{r}_{nca+ncb})$ and $(\underline{r}_{nca+ncb+1} \ldots \underline{r}_N)$ with those belonging to a different set. The function $|\Phi_v(\underline{r}_{nca+ncb+1} \ldots \underline{r}_N)\rangle$ appearing in (94) is an anti-symmetric wavefunction for the valence electrons which is built from those Dirac-Fock atomic orbitals of the isolated atoms not appearing in the core functions (95). The valence function can be taken to be a single anti-symmetrized Hartree product of molecular orbitals which are expanded in the Dirac-Fock atomic orbitals, both occupied and unoccupied, of the isolated atoms, In this case (94) becomes a molecular relativistic SCF wavefunction calculated within a frozen core approximation. This wavefunction can be generalized either by introducing into the valence wavefunction excited molecular orbital configurations or by taking the valence wavefunction to have a relativistic valence bond or kappa valence |50,51| form. The choice for the orbitals included in the cores is input data to the program, all the remaining electrons being included in the valence function $|\Phi_v(\underline{r}_{nca+ncb+1} \ldots \underline{r}_N)\rangle$. Computational expense is the only factor limiting the number of electrons included in the valence function.

The construction of the molecular wavefunction (94) from Dirac-Fock atomic orbitals ensures that all the orbitals lie within some suitably defined positive energy (electron-like) subspace |16-18| and hence that (94) is a valid approximation to an exact eigenfunction of the Brown hamiltonian (3).

The evaluation of the energy of the wavefunction (94) as the expectation value over the relativistic hamiltonian $\hat{\mathcal{H}}_T$ (4) is complicated by the overlap of the atomic orbitals belonging to one atom with those belonging to the other. However since the wavefunction

(94) is invariant with respect to a linear transformation of the core orbitals $|\phi_{i_a}(\underline{r})>$ and $|\phi_{i_b}(\underline{r})>$ among themselves, these orbitals can be replaced by core orbitals $|\psi_{ci}(\underline{r})>$, constructed as linear combinations of the $|\phi_{i_a}(\underline{r})>$ and $|\phi_{i_b}(\underline{r})>$, which are chosen to be orthonormal. Furthermore the wavefunction (94) also remains unchanged if any linear combination of the core orbitals $|\psi_{ci}(\underline{r})>$ is added to the orbitals used to construct the valence wavefunction. Consequently each orbital used to form the valence wavefunction can be replaced by a linear combination of that valence orbital with the core orbitals $|\psi_{ci}(\underline{r})>$ constructed such that the new valence orbital $|\psi_{vj}(\underline{r})>$ is orthogonal to all the core orbitals. With these transformations of the orbitals the wavefunction (94) becomes

$$|\Psi(\underline{r}_1 \ldots \underline{r}_N)> = S'\hat{\mathcal{A}}\,|(\prod_{i=1}^{nca+ncb} |\psi_{ci}(\underline{r}_i)>)\,|\Psi_v(\underline{r}_{nca+ncb+1} \ldots \underline{r}_N)>| \tag{96}$$

where the new valence function $|\Psi_v>$ is constructed from the orbitals $|\psi_{vj}(\underline{r})>$ and is therefore strongly orthogonal |52| to the core. Hence it follows that the total energy (E) of the wavefunction (96), which is equal to that of (94), is given by

$$E = E_c + E_{Ve} \tag{97}$$

with the core energy E_c given by the standard expression

$$\begin{aligned} E_c = &\sum_{i=1}^{nca+ncb} <\psi_{ci}|\hat{\mathcal{H}}_{KE} + \hat{V}_{nuc}(\underline{r})|\psi_{ci}> \\ &+ \sum_{i<j} (<\psi_{ci}\psi_{cj}|r_{12}^{-1}|\psi_{ci}\psi_{cj}> - <\psi_{ci}\psi_{cj}|r_{12}^{-1}|\psi_{cj}\psi_{ci}>) \end{aligned} \tag{98}$$

The valence energy E_{Ve} consists of an effective one-electron energy plus the electrostatic repulsion energy between the valence electrons which can be expressed as a sum of integrals of the type $<\psi_{vi}\psi_{vj}|r_{12}^{-1}|\psi_{vk}\psi_{vl}>$. The effect of the core on the valence orbitals, which is included in the valence effective one-electron energy, is taken into account by simply adding to the nuclear potential operator $\hat{V}_{nuc}(\underline{r})$ the direct and exchange potentials generated to the core orbitals. Thus the valence effective one-electron energy consists of a sum of integrals of the type $<\psi_{vi}|\hat{F}_1|\psi_{vj}>$ defined by

$$\begin{aligned} <\psi_{vi}|\hat{F}_1|\psi_{vj}> = &<\psi_{vi}|\hat{\mathcal{H}}_{KE} + \hat{V}_{nuc}(\underline{r})|\psi_{vj}> \\ &+ \sum_{k\varepsilon c} (<\psi_{vi}\psi_{ck}|r_{12}^{-1}|\psi_{vj}\psi_{ck}> - <\psi_{vi}\psi_{ck}|r_{12}^{-1}|\psi_{ck}\psi_{vj}>) \end{aligned} \tag{99}$$

The Dirac-Fock atomic orbitals used to construct the wavefunction (94), which are purely numerical in form, were computed using the Oxford Dirac-Fock program |20|. All the integrals entering the core energy (98), the effective one-electron energy integrals (99) and the valence-valence repulsion integrals were computed exactly with the new molecular program using numerical methods to be described more fully elsewhere. The details of the choice of the orthogonalized core orbitals will also be reported elsewhere.

C. Lattice Energy Calculations

1. Method of calculation

The lattice energy U(R) of a given crystal structure composed of cations C and anions A for a value R of the cation-anion (AC) nearest neighbour inter-ionic separation is given by

$$U(R) = \frac{M}{R} - n_{AC}V_{AC}(R) - \tfrac{1}{2}n_{AA}V_{AA}(x_{AA}R) - \tfrac{1}{2}n_{CC}V_{CC}(x_{CC}R) - U_{disp}(R) \tag{100}$$

if all the interactions excepting those between the nearest AC, AA and CC pairs are assumed to be composed only of point coulombic plus dispersion energy terms. Here M is the Madelung constant, n_{XY} is the number of nearest Y ions around each X ion and $V_{XY}(R)$ is the interaction potential of the ions X and Y after subtraction of both the dispersion energy and the point coulomb expression q_Xq_Y/R where q_X is the charge on the ion X. The quantities x_{AA} and x_{CC} are purely geometrically factors which relate the AA and CC separations to the AC separation R. The quantity $U_{disp}(R)$ is the total energy arising from the dispersion interactions between the ions.

The lattice energy (100) is dominated by the Madelung term and the nearest neighbour cation-anion potential $V_{AC}(R)$. In this approximation equation (100) can be expressed

$$U(R) = - n_{AC}V^{o}_{AC}(R) - (n_{AC}q_Aq_C - M)/R \tag{101a}$$

$$V^{o}_{AC}(R) = V_{AC}(R) + q_Aq_C/R \tag{101b}$$

The quantity $V^{o}_{AC}(R)$, which is negative for separations R close to the equilibrium nearest-neighbour cation-anion distance in the crystal, is neglecting the dispersion term, the difference between the energy of one cation-anion pair separated by a distance R and that of the infinitely separated ion pair. It should be pointed out that equation (101b) does not imply that any decomposition of $V^{o}_{AC}(R)$ into a purely coulombic q_Aq_C/R and a short range contribution $V_{AC}(R)$ has any significance beyond providing one, possibly not very good, way of defining a short range contribution to the interaction energy. The major problem in the calculation of the lattice energy of a halide of a heavy metal lies in the computation of the non-correlation contribution to $V^{o}_{AC}(R)$. However, this problem is solved because the programme described in the last section can be used to calculate the energy of the wavefunction $\mathcal{A}'(\psi_{cation}\psi_{anion})$ where ψ_{cation} and ψ_{anion} are Dirac-Fock cation and anion wavefunctions of the type (95).

The lattice energy calculation can be refined by retaining the short range anion-anion interaction $V_{AA}(x_{AA}R)$ in equation (100). The potential $V_{AA}(R)$ is calculated as $-q_A^2/R$ plus the interaction energy, neglecting dispersion between two anions separated by a distance R. The interaction between two fluoride or two chloride ions can be calculated by any standard non-relativistic ab-initio method or by the electron gas density functional method |44,53| which has been shown |54| to describe reliably two systems having the same inert gas electronic structures. The contributions to the short range potentials which arise from electron correlation can be calculated by the electron-gas density functional method because this term only makes a minor contribution. Methods for calculating $U_{disp}(R)$ to the lattice energy arising from dispersion in a fully ab-initio fashion are not currently available. However the dispersion term is far from being the major contributer to the lattice energy and can be calculated not unreliably by standard methods described elsewhere |55|.

2. Results and discussion

The lattice energies and nearest neighbour cation-anion separations in four fluoride crystals predicted by calculations including both the short range $F^- - F^-$ interaction and the electron gas correlation energy but excluding the dispersion energy are presented in Table 11.

For both AgF and PbF_2 the computed lattice energies are slightly but significantly smaller than the experimental ones derived from the Born-Haber cycle. The predicted ionic radius of 1.21 Å for Ag^+ is slightly larger than the experimental value of 1.13A whilst the Pb^+

Table 11. Lattice energies and interionic distances predicted by ab-initio calculations without the dispersion energy[(a)]

	Lattice energy kJ/mol		Equilibrium A-C separation (a.u.)	
	calc	expt	calc	expt
AgF	874	953	4.81	4.65
PbF_2	2262	2460	5.12	4.84
E113F	721		5.63	
E116F_2	2097		5.43	

(a) For predictions of slightly different calculations and sources of experimental data see |55|.

radius is more significantly overestimated (1.38 Å compared with the experimental value of 1.21 Å |56|). These results by themselves which have the advantage of being computed using an entirely ab-initio method, show either that AgF and PbF_2 are not fully ionic or, as verified below, that the dispersion energy plays a small but significant role in the cohesion of both these crystals. It should be pointed out that the semi-empirical calculation, neglecting the dispersion energy, which predicts the lattice energy of PbF_2 to be 2433 kJ/mol (|56| Table 2.5) contains an element of circularity in that both the radius of Pb^{++} and the parameter used to describe the short range potential $V_{AC}(R)$ are obtained from experimental data assuming that various lead compounds are fully ionic. Consequently the effects of the neglected dispersion terms will be absorbed by these parameters.

Although the results for AgF and PbF_2 show that the cation radii may be overestimated particularly if the ion is polarizable, the relative smallness of the discrepancies between the computed and the experimental lattice energies shows that the results for E113F and $E116F_2$ can be used as a basis for the prediction of the chemistry of these elements. Furthermore the AgF and PbF_2 results show that for E113F and $E116F_2$ the predicted lattice energies of 721 kJ/mol and 2097 kJ/mol and cation radii of 1.65 Å and 1.54 Å are almost certainly lower and upper bounds respectively to the true lattice energies and radii. Thus the differences between the ab-initio lattice energies and those (748 kJ/mol and 2341 kJ/mol) previously calculated using the Kapustinskii equation and taking the radii of $E113^+$ and $E116^{++}$ to be 1.49 Å and 1.35 Å are sufficiently small that the predictions that E113F will be exothermic and that $E116F_2$ will be strongly exothermic compounds remain unchanged. The predicted heats of formation of E113F and $E116F_2$ are reduced from -125 kJ/mol |57| and -837 kJ/mol |58| to -98 kJ/mol and -595 kJ/mol. The conformation of these qualitative predictions of exothermicity by an ab-initio method which will under - rather than overestimate the lattice energies provides strong evidence that they are correct.

The predictions obtained by adding the dispersion energy to the terms considered in the previous calculation are presented in Table 12. The predicted $E113^+$ and $E116^{++}$ radii of 1.43 Å and 1.36 Å are close to the values of 1.49 Å and 1.35 Å obtained previously |57,58| by using the empirical relation between the ionic radius and that of the outermost occupied Dirac-Fock atomic orbital. Since these predications require the inclusion of the dispersion energy which cannot be introduced in a fully ab-initio fashion, there is not sufficient evidence to distinguish between the present 1.43 Å prediction for the radius of $E113^+$, that of 1.49 Å |57| and that 1.48 Å obtained |59| by extrapolating down Group IIIB. Furthermore the predicted lattice energy of $E116F_2$ of 2279 kJ/mol is little changed from that obtained previously |58| by more empirical methods although the present 813 kJ/mol prediction for E113F is somewhat greater than the previous

Table 12. Lattice energies, interionic distances and cation radii predicted including semi-empiracle calculation of dispersion.

	Lattice energy kJ/mol		Equilibrium A-C separation (a.u.)		Cation radius (Å)	
	calc	expt	calc	expt	calc	expt
AgF	939	953	4.61	4.65	1.11	1.13
PbF_2	2403	2460	4.85	4.84	1.24	1.21
E113F	813		5.21		1.43	
$E116F_2$	2279		5.08		1.36	

estimate |57|. Use of these lattice energies predicts the heats of formation of E113F and $E116F_2$ to be -190 kJ/mol and -777 kJ/mol, the latter being not dissimilar to the semi-empirical result |58|. The close agreement between theory and experiment for AgF and PbF_2 shows that the results for both the ionic radii of $E113^+$ and $E116^{++}$ and the lattice energies of the flourides are probably the most reliable predictions of these quantities currently available, even though the exact magnitude of the dispersion contribution may be a little uncertain.

D. A Preliminary Calculation of $E113_2$

The object of this section is to present a few preliminary results of a fully relativistic, fully ab-initio calculation for the dimer $E113_2$ of eka-Tl. In all the calculations reported here the core was taken to consist of all the occupied Dirac-Fock atomic orbitals on both atoms excepting those belonging to the $7\bar{p}$ subshells. The valence wavefunctions $|\Phi_v>$ are built from both the $7\bar{p}$ and the 7p orbitals, the latter being unoccupied in the ground electronic configuration of an isolated E113 atom. All the calculations were performed for an intermolecular separation of 4.5 a.u. which is about 7% of twice the radius (= 3.040 |60|) of the $7\bar{p}$ atomic orbital. This ensures that the internuclear separation bears approximately the same relation to the radius of the valence orbital as the equilibrium internuclear distance of 5.05 a.u. |61| in Li_2 bears to the radius (3.87 a.u. |39|) of the 2s Hartree-Fock atomic orbital. Furthermore the overlap between the σ symmetry portion of the large components of the $7\bar{p}$ Dirac-Fock atomic orbitals of -0.307 has approximately the same magnitude as that (0.313) between the π symmetry portion of these large components thereby ensuring that the results of the present calculations can be compared with the previous discussion |50,62| of bonding between two $\bar{p}$ orbitals in which the magnitudes of these overlaps were taken to be equal.

The following normalized wavefunctions, which are shown elsewhere |63| to have the correct relativistic symmetry were examined:

$$|\psi_{kv}\bar{p}\text{-}\bar{p}\rangle = S_a\hat{\mathcal{A}}[|core\rangle \frac{1}{\sqrt{2}}(|\bar{p}_A\tfrac{1}{2}\rangle|\bar{p}_B{-}\tfrac{1}{2}\rangle - |\bar{p}_A{-}\tfrac{1}{2}\rangle|\bar{p}_B\tfrac{1}{2}\rangle)] \quad (102a)$$

$$|\psi_{kv}\bar{p}\text{-}p\rangle = S_b\hat{\mathcal{A}}[|core\rangle\tfrac{1}{2}(|\bar{p}_A\tfrac{1}{2}\rangle|p_B{-}\tfrac{1}{2}\rangle + |\bar{p}_A{-}\tfrac{1}{2}\rangle|p_B\tfrac{1}{2}\rangle + |\bar{p}_B\tfrac{1}{2}\rangle|p_A{-}\tfrac{1}{2}\rangle + |\bar{p}_B{-}\tfrac{1}{2}\rangle|p_A\tfrac{1}{2}\rangle)] \quad (102b)$$

$$|\psi_{kv}p\text{-}p\rangle = S_c\hat{\mathcal{A}}[|core\rangle \frac{1}{\sqrt{2}}(|p_A\tfrac{1}{2}\rangle|p_B{-}\tfrac{1}{2}\rangle - |p_A{-}\tfrac{1}{2}\rangle|p_B\tfrac{1}{2}\rangle)] \quad (102c)$$

$$|\psi_{ion}\bar{p}^2\rangle = S_d\hat{\mathcal{A}}[|core\rangle \frac{1}{\sqrt{2}}(|\bar{p}_A\tfrac{1}{2}\rangle|\bar{p}_A{-}\tfrac{1}{2}\rangle + |\bar{p}_B\tfrac{1}{2}\rangle|\bar{p}_B{-}\tfrac{1}{2}\rangle)] \quad (102d)$$

$$|\psi_{ion}p^2\rangle = S_e\hat{\mathcal{A}}[|core\rangle \frac{1}{\sqrt{2}}(|p_A\tfrac{1}{2}\rangle|p_A{-}\tfrac{1}{2}\rangle + |p_B\tfrac{1}{2}\rangle|p_B{-}\tfrac{1}{2}\rangle)] \quad (102e)$$

$$|\psi_{ion}\bar{p}p\rangle = S_f\hat{\mathcal{A}}[|core\rangle\tfrac{1}{2}(|\bar{p}_A\tfrac{1}{2}\rangle|p_A{-}\tfrac{1}{2}\rangle + |\bar{p}_A{-}\tfrac{1}{2}\rangle|p_A\tfrac{1}{2}\rangle + |\bar{p}_B\tfrac{1}{2}\rangle|p_B{-}\tfrac{1}{2}\rangle + |\bar{p}_B{-}\tfrac{1}{2}\rangle|p_B\tfrac{1}{2}\rangle)] \quad (102f)$$

$$|\psi_{kv\text{-}i}\bar{p}\rangle = c_{1g}|\psi_{kv}\bar{p}\text{-}\bar{p}\rangle + c_{2g}|\psi_{ion}\bar{p}^2\rangle \quad (102g)$$

$$|\psi_{cov2}\rangle = c_{1h}|\psi_{kv}\bar{p}\text{-}\bar{p}\rangle + c_{2h}|\psi_{kv}p\text{-}p\rangle \quad (102h)$$

$$|\psi_{cov}\rangle = c_{1i}|\psi_{kv}\bar{p}\text{-}\bar{p}\rangle + c_{2i}|\psi_{kv}\bar{p}\text{-}p\rangle + c_{3i}|\psi_{kv}p\text{-}p\rangle \quad (102i)$$

$$|\psi_{C.I.}\rangle = c_{1j}|\psi_{kv}\bar{p}\text{-}\bar{p}\rangle + c_{2j}|\psi_{kv}\bar{p}\text{-}p\rangle + c_{3j}|\psi_{kv}p\text{-}p\rangle + c_{4j}|\psi_{ion}\bar{p}^2\rangle + c_{5j}|\psi_{ion}p^2\rangle + c_{6j}|\psi_{ion}\bar{p}p\rangle \quad (102j)$$

$$|\psi_{RVB}\rangle = S_K\,[|core\rangle\frac{1}{\sqrt{2}}(|z_A\tfrac{1}{2}\rangle|z_B{-}\tfrac{1}{2}\rangle - |z_A{-}\tfrac{1}{2}\rangle|z_B\tfrac{1}{2}\rangle)] \quad (102k)$$

$$|\psi_{MO}\bar{p}b\rangle = \hat{\mathcal{A}}[|core\rangle S_l\frac{1}{\sqrt{2}}(|\bar{p}_A\tfrac{1}{2}\rangle + |\bar{p}_B\tfrac{1}{2}\rangle)S_l\frac{1}{\sqrt{2}}(|\bar{p}_A{-}\tfrac{1}{2}\rangle + |\bar{p}_B{-}\tfrac{1}{2}\rangle)] \quad (102l)$$

$$|\psi_{SCF}(-i)\rangle = \hat{\mathcal{A}}[|core\rangle|\phi\tfrac{1}{2}{-}1{-}i\rangle|\phi{-}\tfrac{1}{2}{-}1i\rangle] \quad (102m)$$

$$|\psi_{MO}pb\rangle = \hat{\mathcal{A}}[|core\rangle S_n\frac{1}{\sqrt{2}}(|p_A\tfrac{1}{2}\rangle - |p_B\tfrac{1}{2}\rangle)S_n\frac{1}{\sqrt{2}}(|p_A{-}\tfrac{1}{2}\rangle - |p_B{-}\tfrac{1}{2}\rangle)] \quad (102n)$$

$$|\psi_{SCF}(i)\rangle = \hat{\mathcal{A}}\,[|core\rangle|\phi\tfrac{1}{2}1i\rangle|\phi{-}\tfrac{1}{2}1{-}i\rangle] \quad (102o)$$

The energies of the functions (102) were computed exactly by using the orthogonalization procedures described in Section 2 (see Table 13).

The kappa-valence function (102a), relative to which the energies of the other functions (10) are measured, describes |50,63| the formation of a covalent bond between the ground $7\bar{p}$ configurations of the two atoms. Both the smallness of the energy lowering upon mixing with just the ionic configuration $|\psi_{ion}\bar{p}^2\rangle$ involving no promotion of electrons into 7p orbitals and the smallness of the mixing coefficient c_{2g} (Table 14) suggest that the bond has very little ionic character. This result also explains why the molecular orbital function (10l) constructed solely from $7\bar{p}$ atomic orbitals yields a significantly poorer description judged on purely energetic grounds. The function (102b) describes |50,63| the formation of a kappa-valence bond between a $7\bar{p}$ orbital on one atom with a 7p one on the other whilst the function (102c) describes |50,63| the formation of a kappa-valence bond between two 7p orbitals with $|m_j| = \frac{1}{2}$. The energies of these functions relative to the $\bar{p}$-$\bar{p}$ covalent function (102a) are explained by the result that the effective one-electron energy $\langle p_{\frac{1}{2}}' | \hat{F}_1 | p_{\frac{1}{2}}' \rangle$ of the $7p'$ orbital produced by orthogonalizing the 7p orbital against the core is 4.39 e.v. greater than that of the $7\bar{p}'$ orbital produced from the $7\bar{p}$ orbital by a similar orthogonalization provess. The origin of this increase from the 3.18 e.v. $\bar{p}\rightarrow p$ excitation energy |60| in an isolated E113 atom, which may be due to the greater overlap with the core of the other atom by the more extended 7p orbital ($\langle r\rangle = 4.51$ a.u. |60|) compared with that of the $7\bar{p}$ orbital ($\langle r\rangle = 3.04$ a.u. |60|), need not be investigated here. Although the unpromoted kappa-valence function (102a) is the largest contributor (Table 14) to the best purely covalent function which can be constructed from $7\bar{p}$ and 7p atomic orbitals having $|m_j| = \frac{1}{2}$, the admisture of the remaining functions not insignificantly lowers the energy by 0.32 e.v. (Table 13).

Table 13. Energies of $E113_2$ wavefunctions (102) relative to that of the kappa valence functions (102a) built from the ground electronic configuration (e.v.)

$\lvert\psi_{kv}\bar{p}\text{-}\bar{p}\rangle$	0.0	$\lvert\psi_{ion}\bar{p}p\rangle$	3.88	$\lvert\psi_{MO}\bar{p}b\rangle$	1.24
$\lvert\psi_{kv}\bar{p}\text{-}p\rangle$	2.73	$\lvert\psi_{cov2}\rangle$	-0.26	$\lvert\psi_{SCF}(-i)\rangle$	0.39
$\lvert\psi_{kv}p\text{-}p\rangle$	7.47	$\lvert\psi_{cov}\rangle$	-0.32	$\lvert\psi_{MO}pb\rangle$	8.77
$\lvert\psi_{ion}\bar{p}^2\rangle$	2.92	$\lvert\psi_{C.I.}\rangle$	-0.42	$\lvert\psi_{SCF}(i)\rangle$	1.59
$\lvert\psi_{ion}p^2\rangle$	9.14	$\lvert\psi_{RVB}\rangle$	6.44	$\lvert\psi_{kv\text{-}i}\bar{p}\rangle$	-10^{-5}

Table 14. Coefficients $c_{i\mu}$ (i = 1,6) defining the optional wavefunctions (102)(a)

	$c_{1\mu}$	$c_{2\mu}$	$c_{3\mu}$	$c_{4\mu}$	$c_{5\mu}$	$c_{6\mu}$
$\|\psi_{kv-i}>$	1.000	0.001				
$\|\psi_{cov2}>$	0.901	-0.264				
$\|\psi_{cov}>$	0.887	-0.246	-0.091			
$\|\psi_{C.I.}>$	0.819	-0.202	-0.066	0.027	-0.041	-0.141
$\|\psi_{SCF}(-i)>$	0.833	0.289				
$\|\psi_{SCF}(i)>$	0.954	-0.150				

(a) For full definition of coefficients in $|\psi_{SCF}(-i)>$ and $|\psi_{SCF}(i)>$ see |63|.

This energy lowering is likely to be comparable with the binding energy because the dissociation energy of the analogous Tl_2 system has been measured |64| to be only 0.6 e.v. Comparison of the energy of the function (102c) with that of (102h) shows that 79% of the energy lowering comes from the covalent $\bar{p}$-p function (102b). The further energy lowering upon admixture of the ionic functions (102d), (102e) and (102f) to produce the best function that can be constructed from the configurations considered in this paper is sufficiently small that it can be concluded that the bonding is essentially covalent.

The composition of the best covalent function (102i) is of interest because this is the simplest relativistic wavefunction which will reduce in the limit of small relativistic effects to the relativistic valence bond wavefunction (102k) which describes the formation of a purely covalent Heitler-London singlet bond between two pure p_z orbitals. The large components of the orbitals $|z_\mu \pm \frac{1}{2}>$ defined by (10) of |50|, become respectively pure $|p_o\alpha>$ and $|p_o\beta>$, where the subscript denotes the ℓ_z eigenvalue, in the limit that the radial functions in the large components of the $7\bar{p}$ and 7p orbitals become identical. The failure of this function to describe even approximately the bonding and its energy relative to that of $|\psi_{kv}\bar{p}-\bar{p}>$ are explained by realizing that the energy needed to promote one E113 atom from the $7\bar{p}$ state to the $(-|\bar{p}\frac{1}{2}> + \sqrt{2}|p\frac{1}{2}>)/\sqrt{3}$ valence state in the molecular environment is approximately $\frac{2}{3}$ 4.39 = 2.93 e.v.

The energy lowering on passing from the simplest relativistic molecular orbital function (102l) to the SCF function (102m) obtained by solving the usual matrix eigenvalue equation $\underline{F}.\underline{C} = \varepsilon\underline{S}.\underline{C}$ for the molecular orbital $|\phi\tfrac{1}{2}\text{-}1\text{-}i\rangle$ defined through

$$|\phi\tfrac{1}{2}\mp 1\mp i\rangle = c_{1\pm}S_{\pm}\frac{1}{\sqrt{2}}(|\bar{p}_A\tfrac{1}{2}\rangle \pm |\bar{p}_B\tfrac{1}{2}\rangle) + c_{2\pm}S_{\pm 2}\frac{1}{\sqrt{2}}(|p_A\tfrac{1}{2}\rangle \pm |p_B\tfrac{1}{2}\rangle)$$

is still insufficient to yield an energy lower than that of the pure unhybridized kappa-valence function (102a).

The results for $E113_2$ presented here show that molecular electronic structure can be investigated using a fully relativistic, and ab-initio computation.

ACKNOWLEDGEMENTS

I wish to thank my two research students Mr P Marketos and Mr C P Wood without whose work, none of the numerical results would have been obtained. Mr Marketos performed the pseudo-potential and perturbation theory calculations, whilst Mr Wood developed the fully relativistic ab-initio programme and then used this to obtain the results for ionic solids and $E113_2$. I also thank NATO and the Royal Society for partial financial support enabling me to attend this Advanced Study Institute.

REFERENCES

|1| Pyper, N.C., and Marketos, P., 1981, J. Phys., B14, 1387.
|2| Desideuo, A.M., and Johnson, W.R., 1971, Phys. Rev., A3, 1267.
|3| Loddi, M.A.K., 1978, Superheavy Elements, (Oxford Pergamon Press).
|4| Das, G., and Wahl, A.C., 1976, J. Chem. Phys., 64, 4672.
|5| Datta, S.N., Ewig, C.S., and van Wager, J.R., 1978, Chem. Phys. Letts., 57, 83.
|6| Snijders, J.G., and Baerends, E.J., 1978, Molec. Phys., 36, 1789.
|7| Lee Y.S., Ermler, W.C. and Pitzer, K.S., 1977, J. Chem. Phys., 67, 5861.
|8| Ermler, W.C., Lee, Y.S., Pitzer, K.S., and Winter, N.W., 1978, J. Chem. Phys., 69, 976.
|9| Kahn, L.R., Hay, P.J., and Cowan, R.D., 1978, J. Chem. Phys., 68,2386.
|10| Hay, P.J., Wadt, W.R., Kahn, L.R., and Bobrowicz, F.W., 1978, J. Chem. Phys., 69, 984.
|11| Hafner, P., and Schwartz, H.W., 1978, J. Phys., B11, 217.
|12| Phillips, J.C., and Kleinman, L., 1959, Phys. Rev., 116, 287.
|13| Cowan, R.D., and Griffin, D.C., 1976, J. Opt. Soc. Am., 66, 1010.
|14| Pyper, N.C. and Grant, I.P., 1978, J. Chem. Soc. Faraday II, 74, 1885.

|15| Blume, M., and Watson, R.E., 1963, Proc. Roy. Soc., A271, 565.
|16| Brown, G.E., and Ravenhall, D.G., 1951, Proc. Roy. Soc., A208, 552.
|17| Mittleman, M., 1971, Phys. Rev., A4, 893.
|18| Mittleman, M., 1972, Phys. Rev., A6, 2395.
|19| Dirac, P.A.M., 1958, The Principles of Quantum Mechanics, (Oxford University Press).
|20| Grant, I.P., Mayers, D.F., and Pyper, N.C., 1976, J. Phys., B9, 2777.
|21| Bethe, H.A., and Salpeter, E.E., 1957, Quantum Mechanics of One and Two Electrons Atoms (Berlin: Springer Verlag).
|22| Mann, J.B., and Johnson, W.R., 1971, Phys. Rev., A4, 41.
|23| Foldy, L.L., and Wouthuysen, W.A., 1950, Phys. Rev., 78, 29.
|24| Pyper, N.C., 1980, Molec. Phys., 39, 1327.
|25| Itoh, T., 1965, Rev. Mod. Phys., 37, 159.
|26| Pyper, N.C., 1980, Molec. Phys., 39, 1327 (erratum 26a ibid.
|27| Pyper, N.C., 1981, Molec. Phys., 42, 1059. 49, 949.)
|28| Pyper, N.C., and Marketos, P., 1981, Molec. Phys., 42, 1073.
|29| Meyers, D.F., 1957, Proc. Roy. Soc., A241, 93.
|30| Rose, S.J., Grant, I.P., and Pyper, N.C., 1978, J. Phys., B11, 1171.
|31| Nesbet, R.K., 1955, Proc. Roy. Soc., A230, 312.
|32| Grant, I.P., 1970, Adv. Phys., 19, 747.
|33| Roothaan, C.C.J., 1960, Rev. Mod. Phys., 32, 179.
|34| Rose, S.J., Pyper, N.C., and Grant, I.P., 1978, J. Phys., B11, 755.
|35| Kim, Y.K., 1967, Phys. Rev., 154, 17.
|36| Grant, I.P., and Pyper, N.C., 1976, J. Phys., B9, 761.
|37| Blume, M., and Watson, R.E., 1962, Proc. Roy. Soc., A270, 127.
|38| Froese-Fischer, C., 1967, Can. J. Phys., 45, 1501.
|39| Froese-Fischer, C., 1972, At. Data. Tables, 4, 301.
|40| Walker, T.E.H., and Richards, W.G., 1970, J. Chem. Phys., 52, 1311.
|41| Trivedi, H.P., and Richards, W.G., 1980, J. Chem. Phys., 72, 3438.
|42| Pyper, N.C., and Marketos, P., 1981, J. Phys., B14, 4469.
|43| Pyper, N.C., Grant, I.P., and Gerber, R.B., 1977, Chem. Phys. Letts., 49, 479.
|44| Gordon, R.G., and Kim, Y.S., 1972, J. Chem. Phys., 56, 3122.
|45| Morse, P.M., and Feshbach, H., 1953, Methods of Theoretical Physics (McGraw Hill).
|46| Morrison, J.D., and Moss, R.E., 1980, Molec. Phys., 41, 491.
|47| Phillips, J.C., and Kleinman, L., 1959, Phys. Rev., 116, 287.
|48| Cohen, M.H., and Heine, V., 1961, Phys. Rev., 122, 1821.
|49| Weeks, J.D., Haizi, A., and Rice, S.A., 1969, Adv. Chem. Phys., 16, 283.
|50| Pyper, N.C., 1980, Chem. Phys. Letts., 74, 554.
|51| Pyper, N.C., 1980, Chem. Phys. Letts., 73, 385.

|52| McWeeney, R., and Sutcliffe, B.T., 1969, Methods of Molecular Quantum Mechanics (Academic Press, New York).
|53| Clugston, M.J., 1978, Adv. Phys., 27, 893.
|54| Wood, C.P., and Pyper, N.C., 1981, Molec. Phys., 43, 1371.
|55| Wood, C.P., and Pyper, N.C., 1981, Chem. Phys. Letts., 81, 395.
|56| Johnson, D.A., 1968, Some Thermodynamic Aspects of Inorganic Chemistry (Cambridge University Press).
|57| Pyper, N.C., unpublished work.
|58| Grant, I.P., and Pyper, N.C., 1977, Nature, 265, 715.
|59| Keller, D.L., Burnett, J.L., Carlson, T.A., and Nestor, C.W., 1970, J. Phys. Chem., 74, 1127.
|60| Pyper, N.C., and Grant, I.P., 1981, Proc. Roy. Soc., A339, 525.
|61| Huber, K.P., and Herzberg, G., 1979, Constants of Diatomic Molecules, Van Nostrand.
|62| Pyper, N.C., 1982, Phil. Trans. Roy. Soc., in press.
|63| Wood, C.P., and Pyper, N.C., 1981, Chem. Phys. Letts., 84, 614.
|64| Balducci, G., and Piacente, V., 1980, Chem. Comm., 1287.

THE GEOMETRY OF SPACE TIME AND THE DIRAC EQUATION

Jaime Keller

D.E.Pg., Fac. de Química
Universidad Nacional Autónoma de México
04510, México, D.F.

ABSTRACT

The main object of this paper is the derivation of relativistic quantum mechanics from the vector form of relativity. The basic geometry of Minkowski space time can be expressed in terms of the Clifford algebra C_{16} generated from the 4 basis vectors of L_4. From the four momentum conservation equation

$$p^0\gamma_0 + p^1\gamma_1 + p^2\gamma_2 + p^3\gamma_3 = p'^0\gamma'_0 + p'^1\gamma'_1 + p'^2\gamma'_2 + p'^3\gamma'_3 ,$$

where the γ_μ are vectors of observer S, γ'_μ of reference system S', and the multivector form of the Lorentz transformations

$$\gamma'_\alpha = R\gamma_\alpha\tilde{R}$$

(here R is given in terms of multivectors of S only), we obtain

$$p^\alpha\gamma_\alpha R = p'^\alpha R\gamma_\alpha$$

where the explicit dependence on the vectors of S' has disappeared. We may then make a further Lorentz transformation with the even multivector (I is a bivector, and Q is in fact a gauge transformation)

$$Q = e^{-I(\underline{p}\cdot\underline{x} + \gamma_5\underline{p}'\cdot\underline{x}')/\hbar} \; ; \; I^2 = -1 \; ; (\partial_\alpha = \partial/\partial x_\alpha \; ; \partial'_\alpha = \partial/\partial x'_\alpha)$$

and define $\psi = RQ$ to obtain the eigenvalue equation

$$\gamma_\alpha g_{\alpha\alpha}\partial_\alpha\psi I = -g_{\alpha\alpha}\partial'_\alpha\psi\gamma_5 I\gamma_\alpha$$

we will call the "Full Dirac equation" in multivector form.

In this paper we present the derivation of this last equation, its reduction to the standard Dirac equation, its solution for the free particle, and symmetries. As the energy momentum conservation relation holds for a system of several particles the method is the same for the case of two or more particle system. Finally the vectors and multivectors can be represented by four matrices and new forms of the many electron problems can be obtained.

INTRODUCTION

An entirely new procedure of deriving the Dirac's wave equation is presented here. It allows the generalization of the spin 1/2 field equations in a way more suitable to understand the physical meaning of the different terms in the (multivector) wave function and the behaviour under the basic symmetries of space time. In previous papers we developed this equation in matrix form[1].

The new wave equation, which we will call full Dirac equation (fD), can be studied, in particular, under the operation of duality rotation. Duality rotations mix space time vectors with space time three-vectors as they are dual to each other. The duality rotation D^μ_α can be performed on each of the four space time vectors independently, the wave function can be single valued, multivalued or invariant upon each of the D^μ_α. This will be on the origin of symmetry restrictions to the possible solutions to the full Dirac equation. Because the gauged Dirac equation, with the electromagnetic field as a gauge field, requires the duality rotation to be global (on all four space time vectors simultaneously), only those symmetry constrained Dirac fields which have the same valuedness upon all four D^μ_α can interact with the electromagnetic field according to standard electrodynamics.

The set of elementary matter fields described by the full Dirac equations are given the generic name: symmetry constrained Dirac fields[1b], or diracons in short.

First, the full Dirac equation is derived from the energy-momentum space-time vector, then duality rotation symmetries of the fD equation are studied and the resulting set of diracons are classified. The basic collection of gauge fields and their implications are presented. Finally, we comment on the implications of this work in the analysis of quantum mechanics.

LORENTZ TRANSFORMATIONS IN MULTIVECTOR FORM

Consider the tangent space $\mathcal{D}(x)$ at point x of Minkowski space time consisting of the vectors $\{\gamma_\mu; \mu = 0,1,2,3\}$. This orthonormal vectors

$$\gamma_\mu = \Box\, x^\mu \,;\; \gamma_0^2 = 1, \gamma_1^2 = \gamma_2^2 = \gamma_3^2 = -1 \tag{1}$$

have their metric given by the symmetric product (· product)

$$g_{\mu\nu} = \gamma_\mu \cdot \gamma_\nu \equiv \frac{1}{2}\left(\gamma_\mu\gamma_\nu + \gamma_\nu\gamma_\mu\right), \tag{2}$$

and can be used to define multivectors from their antisymmetric product (Λ product, well known but not widely used in physics[2-18])

$$\gamma_{\mu\nu\ldots\lambda} \equiv \gamma_\mu \Lambda \gamma_\nu \Lambda \ldots \Lambda \gamma_\lambda \tag{3}$$

defined by recursion, if b is an n-vector and a is a 1-vector

$$a \Lambda b \equiv \frac{1}{2}\left(ab + (-1)^n ba\right) = a \Lambda b_1 \Lambda b_2 \Lambda \ldots \Lambda b_n \tag{4}$$

The dot product of multivectors is also defined by recursion. Following Hestenes[12] we define the total or geometrical product of multivectors as the sum of the symmetric and antisymmetric parts

$$ab = a \cdot b + a \Lambda b \tag{5}$$

The observer S_1 at x has at his disposal the 16 different basic multivectors or d-numbers (d = Dirac), including the scalar 1 and the pseudoscalar $\gamma_5 = \gamma_0\gamma_1\gamma_2\gamma_3$, which constitute the Clifford algebra C_{16} at $\mathcal{D}(x)$. The pseudoscalar γ_5 can be used to define the duality operation $a_D = *a = \gamma_5 a$ which transforms a scalar into a pseudoscalar, a vector into a trivector and a bivector into another bivector. The 16 basic multivectors are then grouped into two sets of 8 elements, each set dual of the other. As $\gamma_5^2 = -1$ then $** = -1$ and is not identical with Hodge's star, differing from it by $i = \sqrt{-1}$. This vector Clifford algebra has been used to study the Lorentz transformation and its application for classical fields,

$$\mathcal{D}(x) \to \mathcal{D}'(x) = R(x)\mathcal{D}(x)R^{-1}(x) \tag{6}$$

with R being the sum of a scalar, a bivector and a pseudoscalar such that

$$RR^{-1} = R^{-1}R = 1 \tag{7}$$

and

$$R^{-1} = \pm \tilde{R} \tag{8}$$

($\tilde{R}$ is a d-number in which the order of all products has been reversed) can be written in terms of generators

$$R = \pm e^{-\frac{1}{2} B\theta} \tag{9}$$

the B being the unit bivectors required to satisfy (8), they are in the hyperplane of the Lorentz rotation by an angle θ. The vectors associated with a second observer S_2:

$$\alpha_\mu = R\gamma_\mu R^{-1} \tag{10}$$

will again be orthogonal and have the same norm, but not necessarily constant. R operates on all members of C_{16}.

QUANTUM MECHANICS DERIVED FROM RELATIVISTIC MECHANICS

Consider now the energy momentum vector relation $\underline{p} = \underline{p}'$ for a "particle" moving with respect to observer S_1

$$p^0\gamma_0 + p^1\gamma_1 + p^2\gamma_2 + p^3\gamma_3 + p'^0\gamma'_0 + p'^1\gamma'_1 + p'^2\gamma'_2 + p'^3\gamma'_3 \tag{11}$$

the prime components are those computed by S_1 when the particle's movement is referred to a S'_1 inertial system.

Here

$$\gamma'_\alpha = R\gamma_\alpha\tilde{R} \tag{12}$$

and

$$\gamma_\alpha = \tilde{R}\gamma'_\alpha R \tag{13}$$

then we can write (11) as

$$p^\alpha\gamma_\alpha = p'^\alpha R\gamma_\alpha\tilde{R} \tag{14}$$

where the explicit dependence on the vectors of S'_1 has disappeared, and obtain, on multiplying by R on the right, the relation

$$p^\alpha\gamma_\alpha R = p'^\alpha R\gamma_\alpha \tag{15}$$

and the transpose of it

$$\tilde{R}p^\alpha\gamma_\alpha = p'^\alpha\gamma_\alpha\tilde{R}\,. \tag{16}$$

The dot product of (16) and (15) give, of course, the standard scalar invariant

$$(p^{\beta}\gamma_{\beta}R)\cdot(\tilde{R}p^{\alpha}\gamma_{\alpha}) = (p'^{\beta}R\gamma_{\beta})\cdot(p'^{\alpha}\gamma_{\gamma}\tilde{R}) \tag{17}$$

such that

$$p^{\beta}p^{\alpha}g_{\beta\alpha} = p'^{\beta}p'^{\alpha}g_{\beta\alpha} \ . \tag{18}$$

We may then make a further Lorentz transformation with the even multivector

$$Q = e^{-I(\underline{p}\cdot\underline{x} + \gamma_5\underline{p}'\cdot\underline{x}')/\hbar} \ ; \quad I^2 = -1 \tag{19}$$

with the particular choice $I\cdot B = 0$ in such a way that it represents a space rotation around the direction of motion by the angle $\underline{p}\cdot\underline{x}$. On the other hand $I\gamma_5\cdot B = 1$ and it represents a boost from S' to a S''_1. The operator Q is otherwise a phase factor which will be used to define

$$RQ = \psi \quad \text{and} \quad \tilde{\psi} = \tilde{Q}\tilde{R} \ , \tag{20}$$

to obtain the eigenvalue equation

$$\gamma_{\alpha}g_{\alpha\alpha}\partial_{\alpha}\psi I = - g_{\alpha\alpha}\partial'_{\alpha}\psi\gamma_5 I\,\gamma_{\alpha} \tag{21}$$

which we will call the full Dirac equation in multivector form.

The unit bivector I can be changed into any other unit bivector because this only amounts to a redefinition of the system S_1 and S'_1, then we do not lose any generality by the particular choice made here,where the gauge transformation character of Q is evident as a freedom to rotate in a plane perpendicular to the motion, moreover the phase factor Q can always be brought into the form (19).

The $\gamma_5\gamma_{\mu} = \gamma_{\mu}^{D}$ are the dual of the γ_{μ} and linearly independent, ψ is now an even d-number depending on $\{x^{\mu}\}$ and $\{x'^{\mu}\}$ and (21) may be properly called a quantum-mechanical like particle's wave equation. The particular choice $I = \gamma_1\gamma_2$ and $B = \gamma_3\gamma_0$ can be made in (20) to reflect the fact that two conserved currents can be defined which will be considered,

$$J_0 = \tilde{\psi}\gamma_0\psi \quad \text{and} \quad J_3 = \tilde{\psi}\gamma_3\psi \tag{22}$$

The γ_5 in the exponent of Q is needed to have both the bivector I and its dual to keep the two terms in the exponent of (19) linearly independent phase factors of ψ. The $g_{\mu\mu}$ appear in (21) in order to have the correct signs for the p^{μ} and p'^{μ} because time- and spatial- like terms in $\underline{p}\cdot\underline{x}$ and $\underline{p}'\cdot\underline{x}'$ have opposite signs.

With the particular choice of S'_1 such that

$$\underline{p}' = m_0\gamma_0' \, , \tag{23}$$

(21) reduces to

$$\gamma_\alpha g_{\alpha\alpha}\partial_\alpha\psi I = m_0\psi\gamma_0 \tag{24}$$

which is now the Dirac equation in multivector form. It has been derived here without the use of the standard postulates of quantum mechanics, our analysis constitutes then a new way of obtaining the basic operators of quantum theory. Moreover it justifies equation (24) which has been discussed in length by Hestenes[12] who obtained it in a different way. It has the same form as the one derived by Gürsey[19], with ψ being a four spinor, in his study of the nucleons.

SYMMETRIES AND SYMMETRY CONSTRAINED DIRAC PARTICLES

As we have mentioned, the choice of a particular rotation plane, for example $I = I_3 = \gamma_5\sigma_3 = \gamma_1\gamma_2$ is arbitrary and the orthogonal bivectors $I = I_2 = \gamma_5\sigma_2$ or $I = I_1 = \gamma_5\sigma_1$ could as well be considered. The wave equations (21) and (24) are then SU(2) symmetric and, because $I = -\tilde{I}$, they are also U(1) symmetric. Then it is SU(2) × U(1) symmetric as basic gauge transformations symmetries.

An examination of the fD shows that a physically equivalent theory could have been obtained considering the substitution of $\underline{p}$ by its dual $\underline{p}^D$ which is a trivector, this concept embodies the well known duality principle, where representation space and momentum space are considered to be dual to each other in the physical sense. The interesting consequences of this idea will be considered elsewhere.

The gauge transformations admit other symmetries which leave (21) invariant set of a four duality rotation transformation:

$$\left.\begin{aligned} \gamma_\mu \to \gamma_\mu^K &= \cos\theta_\mu\gamma_\mu + \sin\theta_\mu\gamma_\mu^D \\ \gamma_\mu^D \to \gamma_\mu^{DK} &= -\sin\theta_\mu\gamma_\mu + \cos\theta_\mu\gamma_\mu^D \end{aligned}\right\} \tag{25}$$

or in multivector notation

$$\gamma_\mu \to \gamma_\mu^K = e^{\gamma_5\theta^\mu t/2}\gamma_\mu e^{-\gamma_5\theta^\mu t/2} = e^{\gamma_5\theta^\mu t}\gamma_\mu = \gamma_\mu e^{-\gamma_5\theta^\mu t} \tag{26}$$

The second and third equalities because γ_5 anticonmutes with all γ_μ and γ_μ^D. The duality rotation factors can be incorporated directly in the ψ, with

$$\psi^K = e^{\gamma_5 t_\mu \theta^\mu} \psi . \tag{27}$$

The dependence on duality rotations should be considered for the wave function and the quantities, like the field's stress-energy tensor, constructed from it. For this reason, charged particles should be described by a wave function single or double valued with respect to the 4 duality rotations θ^μ going from 0 to 2π. For the same reason eq. (24) can only be written down for charged particles or for systems of diracon fields with $\Sigma t_1 = \Sigma t_2 = \Sigma t_3 = \Sigma t_0$ in order to keep (24) Lorentz invariant. Fields with $t_\alpha \neq t_\beta$, $\alpha \neq \beta$ will then be rest-mass-less fields and uncharged.

There are 3 space-like gauge duality rotations, equivalent among themselves $t_i/2 = a_i$ and 1 time-like $t_0/2 = {}_1b$. This could be made more explicit by doing polar transformations[1].

The duality rotation constrained Dirac fields should present a relation between the a_i, the b and the particles of the different matter fields; the three relative a_1, a_2 and a_3 are like the three "colors" and the series of b can be related to strangeness and charm in the form: b = 1/2, first family of leptons and building block of barions, b = 1 to be the second family or the strange-charmed family and b = n/2 in general the nth family of particles. I suggest the different fields, which will be Dirac fields constrained by symmetry should take the generic name of diracons as they all obey (a generalized, multivector) Dirac equation. Table I presents the first two families according to the previous ideas.

The dynamics of the matter fields, obtained by gauging (24), are:

1) Symmetric in a_1, a_2, a_3 and b, with no change in their values.
2) a_1 and a_2 and a_3 change simultaneously (for example by 1/2).
3) A set of a_i changing into a different value.
4) b changing into a new value.

Field dynamics 1) may correspond to the normal electromagnetic interaction. Field dynamics 2), to the weak action of diracon fields, whereas 3) is just the gluon type of interaction changing one color particle into another. Field dynamics 4) will change particles from one family to another.

The theory of diracons is more general than SU(5) because it contains different families (different a_i and b values), and contains the basis for a SU(3) and a SU(2) × U(1) theory of elementary

Table I. Simplest Combinations of Quantum Numbers t/2 Corresponding to Duality Rotation of the Wave Function ψ of a Symmetry Constrained Dirac particle (Diracon)

a_1	a_2	a_3	b	Σa	Possible identification of the corresponding matter field
0	0	0	$\pm\frac{1}{2}$	0	neutrino
$\pm\frac{1}{2}$	0	0	$\pm\frac{1}{2}$	$\pm\frac{1}{2}$	lepton (b = $\frac{1}{2}$) three types,
0	$\pm\frac{1}{2}$	0	$\pm\frac{1}{2}$	$\pm\frac{1}{2}$	probable component of mesons
0	0	$\pm\frac{1}{2}$	$\pm\frac{1}{2}$	$\pm\frac{1}{2}$	and barions.
$\pm\frac{1}{2}$	$\pm\frac{1}{2}$	0	$\pm\frac{1}{2}$	$\pm\frac{2}{2}$	lepton (b = $\frac{1}{2}$) three types,
0	$\pm\frac{1}{2}$	$\pm\frac{1}{2}$	$\pm\frac{1}{2}$	$\pm\frac{2}{2}$	probable component of mesons
$\pm\frac{1}{2}$	0	$\pm\frac{1}{2}$	$\pm\frac{1}{2}$	$\pm\frac{2}{2}$	and barions.
$\pm\frac{1}{2}$	$\pm\frac{1}{2}$	$\pm\frac{1}{2}$	$\pm\frac{1}{2}$	$\pm\frac{3}{2}$	electron
$\pm\frac{1}{2}$	$\pm\frac{1}{2}$	$\pm\frac{1}{2}$	± 1	$\pm\frac{3}{2}$	neutrino (b = 1)
± 1	$\pm\frac{1}{2}$	$\pm\frac{1}{2}$	± 1	± 2	lepton (b = 1) three types,
$\pm\frac{1}{2}$	± 1	$\pm\frac{1}{2}$	± 1	± 2	probable component of heavy
$\pm\frac{1}{2}$	$\pm\frac{1}{2}$	± 1	± 1	± 2	mesons and barions
± 1	± 1	$\pm\frac{1}{2}$	± 1	$\pm 2\frac{1}{2}$	lepton (b = 1) three types,
± 1	$\pm\frac{1}{2}$	± 1	± 1	$\pm 2\frac{1}{2}$	probable component of heavy
$\pm\frac{1}{2}$	± 1	± 1	± 1	$\pm 2\frac{1}{2}$	mesons and barions
± 1	± 1	± 1	± 1	± 3	muon

particles. From the corresponding gauge transformations the symmetry properties of the different force mediating boson fields and the conditions to observe diracons with electromagnetic interactions may be derived. The possibility of diracons interacting with the electromagnetic and other gauge field appears as a symmetry condition . If the duality rotations phase factors can be factorized, the particles will interact electromagnetically, if the duality rotations phase factor cannot be factorized then the field's particles cannot obey standard electrodynamics. In the case of a composite particle the total wave function will be a product of the constituent fields ψ, their product may have the required combination of duality rotation phase factors and the composite particle obey standard electrodynamics. Symmetry constrained electrodynamics is not to be ruled out for other diracon fields or combination of fields.

The opposite procedure can be considered, the symmetries of the interaction carrying fields being fixed and, as a consequence, the symmetries of the diracon fields (in particular under duality rotations) be obtained.

A composite particle will obey (12) as a whole, then the analysis can be carried through for it, or internal and external coordinates can be used. A particularly interesting case is a system of rest-mass-less diracons, each one should move with a directed velocity $|u_n| = c$ equal to the velocity of light but the composite has $v < c$. If we further require the internal angular momentum to be quantized a scheme similar to Mandelstam's[21] series for hadrons arises. This elementary particle "chemistry" will be reported elsewhere.

THE FULL DIRAC EQUATION AND QUANTUM MECHANICS

The multivector equation (24) factorices into four independent equations, each one corresponding to the standard Dirac equation, once a matrix representation of the vectors γ_μ is selected (we prefer the choice γ_0 and γ_{21} diagonal, Table II) and a Dirac four spinor $\Psi = \psi u$ is defined. The matrix u is a column matrix with only the one element being different from zero, according to which column of ψ is to be selected. The i in the Dirac equation are the diagonal terms of $\gamma_2\gamma_1$.

For a composite n-particle system the size of the matrices can be enlarged by a factor n, this allows the possibility of keeping the particles independent or not.

The procedure described in equations (12)-(22) and the analysis of equations (23)-(26) are basic for the interpretation

Table II. Matrix Representation of the Space-Time Dirac-Clifford Algebra

$$\gamma_1 = \begin{pmatrix} 0 & 0 & 0 & -1 \\ 0 & 0 & -1 & 0 \\ 0 & 1 & 0 & 0 \\ 1 & 0 & 0 & 0 \end{pmatrix} \quad \gamma_2 = \begin{pmatrix} 0 & 0 & 0 & i \\ 0 & 0 & -i & 0 \\ 0 & -i & 0 & 0 \\ i & 0 & 0 & 0 \end{pmatrix} \quad \gamma_3 = \begin{pmatrix} 0 & 0 & -1 & 0 \\ 0 & 0 & 0 & 1 \\ 1 & 0 & 0 & 0 \\ 0 & -1 & 0 & 0 \end{pmatrix}$$

$$\gamma_3\gamma_2 = \begin{pmatrix} 0 & i & 0 & 0 \\ i & 0 & 0 & 0 \\ 0 & 0 & 0 & i \\ 0 & 0 & i & 0 \end{pmatrix} \quad \gamma_3\gamma_1 = \begin{pmatrix} 0 & -1 & 0 & 0 \\ 1 & 0 & 0 & 0 \\ 0 & 0 & 0 & -1 \\ 0 & 0 & 1 & 0 \end{pmatrix} \quad \gamma_2\gamma_1 = \begin{pmatrix} i & 0 & 0 & 0 \\ 0 & -i & 0 & 0 \\ 0 & 0 & i & 0 \\ 0 & 0 & 0 & -i \end{pmatrix}$$

$$\gamma_0 = \begin{pmatrix} 1 & 0 & 0 & 0 \\ 0 & 1 & 0 & 0 \\ 0 & 0 & -1 & 0 \\ 0 & 0 & 0 & -1 \end{pmatrix} \qquad \gamma_5 = \gamma_0\gamma_1\gamma_2\gamma_3 = \begin{pmatrix} 0 & 0 & i & 0 \\ 0 & 0 & 0 & i \\ i & 0 & 0 & 0 \\ 0 & i & 0 & 0 \end{pmatrix}$$

The other 8 terms can be obtained by multiplication with γ_5 on this matrices, and a change in sign when needed. A multiplication of the 16 matrices by i could be used to complexify the representation, the Dirac algebra being 4×4 complex matrices can give a total of 32 elements, this is the number of elements of a 5-dimentional space or of a complex 4-dimentional space.

of the multivector wave function. It is first clear that the Dirac theory is a way of relating the energy-momentum vector from two different reference frames using the basic set γ_μ of only one observer. The wave function contains the information of the second (particle at rest or proper) frame and of the action associated with the particle. Knowing this, the relations between the observer's frame and the particle's rest frame can be accounted for and the action phase factor found.

Equally interesting is the case of the gauged Dirac equation to account for interaction carrying fields. In this case the energy momentum vector depends on time and position, as well as the particle proper time-like vector relative to the observer

$$\alpha_0^2 = (\alpha_0\gamma_0)(\gamma_0\alpha_0) = \frac{1}{1 - v^2} \tag{28}$$

and the square of the wave function gets a time dilatation factor. A new type of question may now be asked: what is the relative probability of finding a particle in the neighbourhood of two space time points $\underline{x}_1$ and $\underline{x}_2$? The ratios of the absolute values of the time-like vectors (usually called $\rho(\underline{x}_2)/\rho(\underline{x}_1)$) obtained from (10) contain this, probabilistic information, which is the origin of the probabilistic interpretation of quantum mechanics. In equation (21) the notion of trayectory has been lost.

REFERENCES

1. J. Keller, Rev. Soc. Quim. Mex., 25, 28, January 1981.
   J. Keller, Int.J.Theor.Phys., Dirac's birthday issue (1982).
   J. Keller, in the Proceedings of the Int. Conf. on "Mathematics of the Physical Space Time", Mexico, 1981, J.Keller Ed., U.N.A.M., pag. 117 (1982).
2. F. Sauter, Z. Physik 63, 803 (1930), ibid 64, 295 (1930).
   A. S. Eddington, Relativity Theory of Protons and Electrons (MacMillan, New York, 1936) and references therein.
3. A. Sommerfeld, Atombau und Spektrallinien (Braunschweig 1939) Vol. II, 217.
   P. K. Rasevskii, Trans. Amer. Math. Soc. 6, 1 (1957).
4. M. Riesz, Comptes Rendus du Dixieme Congres des Mathematiques des Pays Scandinaves (Copenhagen 1946) 123.
5. M. Riesz, Comptes Rendus du Douzieme Congres des Mathematiques des Pays Scandinaves (Lund 1953), 241.
6. M. Riesz, Lecture Series 38 (University of Maryland), 1958.
7. S. Teitler, Nuovo Cimento Suppl. 3, 1 (1965).
8. S. Teitler, Nuovo Cimento Suppl. 3, 15 (1965).
9. S. Teitler, J. Math. Phys. 6, 1976 (1965).
10. S. Teitler, J. Math. Phys. 7, 1730 (1966).
11. S. Teitler, J. Math. Phys. 7, 1739 (1966).
12. D. Hestenes, Space-Time Algebra, Gordon and Breach, New York (1966).
    D. Hestenes, J. Math. Phys., 16, 556 (1975) and references therein.
13. N. Salingaros and M. Dresden, Phys. Rev. Lett. 43, 1, (1979).
    N. Salingaros, J. Math. Phys. 22, 226, (1981).
14. T. K. Greider, Phys. Rev. Lett. 44, 1718, (1980).
15. G. Casanova, C.R. Acad. Sc. Paris, 270 Serie A, 1202 (1970).
16. R. Boudet, Comptes Rendus 272, Serie A, 767, (1971).
17. P. Quilichini, Comptes Rendus 273, Serie B, 829 (1971).
18. R. Boudet, Comptes Rendus Acad.Sc.Paris 278, Serie A, 1063 (1974).
19. F. Gürsey, Nuovo Cimento 7, 411 (1958).
20. S. Mandelstam, Phys.Rep. 13C, 259 (1974); ibid 67C, 109 (1980).

Perturbation Theory of a Relativistic Particle in Central Fields

C. K. Au

Department of Physics and Astronomy
University of South Carolina
Columbia, South Carolina 29208 USA

We present summary results of a bound-state perturbation theory for a relativistic spinless (Klein-Gordon) and a relativistic spin-half (Dirac) particle in central fields due to scalar or fourth-component vector-type interactions for an arbitrary bound state. The reduction of the wave equations to Ricatti form enables a decoupling between the pair of coupled first order differential equations on the large and small component Dirac wave functions or a decoupling of the second order differential equation in the Klein-Gordon case. All corrections to the energies and wave functions, including corrections to the positions of the nodes in excited states, are expressed in quadratures in a hierarchical scheme, without the use of either the Green function or the sum over intermediate states.

ON THE PERSISTANCE OF NON-RELATIVISTIC CORRELATION EFFECTS IN RELATIVISTIC ATOMS*

Donald R. Beck

Physics Department
Michigan Technological University
Houghton, MI 49931

ABSTRACT

It is shown[1] that the large discrepancies between single and multi-configurational Dirac-Fock calculations[2] and experiment recently obtained for certain Binding and Auger Energies of Ar, Kr and Xe can be removed by including easily computed Symmetric Exchange of Symmetry (SEOS) correlations.

The effect of explicitly projecting out lower states (at the Dirac-Fock level) was seen to shift the 1s hypersatellite line[3] in Hg by about 2 eV. It was noted that chemical shifts for all levels involved in this problem may not be the same, due to their differing charges.

1. Details concerning this material will appear in Phys. Rev. A (in press).
2. N. Beatham, I. P. Grant, B. J. McKenzie and S. J. Rose, Phys. Scr. 21, 423 (1980).
3. e.g. K. Schreckenbach, H. G. Börner, and J. P. Desclaux, Phys. Letts. 63A, 330 (1977).

*Partial support from Michigan Technological University is gratefully acknowledged.

PARITY NON-CONSERVING EQUATIONS OF MOTION METHOD

C.P. Botham, A-M. Martensson and P.G.H. Sandars

Clarendon Laboratory, University of Oxford
Parks Road
Oxford

We suggest that a relativistic version of the equations of motion method may prove a useful procedure for making a priori calculations of parity non-conservation in heavy atoms. After setting out the basic features of the method we discuss its application to closed and open shells. We then consider the additional features caused by the parity non-conserving term in the Hamiltonian. We outline the procedure used to solve the resulting coupled radial equations and carry out a complete calculation on the $1s^2\ {}^1S_0 \rightarrow 1s2s\ {}^3S_1$ transition in He as an initial test for our method. Our value $A = 1.4036 \times 10^{-4}\ s^{-1}$ for the parity conserving M1 transition rate is in good agreement with previous calculations. For the parity non-conserving E1 amplitude we obtain excellent agreement between length and velocity gauges with a value

$$R = \text{Imag.}\frac{E1}{M1} = 3.450 \times 10^{-10}$$

A preliminary calculation on the $6p_{\frac{1}{2}}^2\ J{=}0 \rightarrow 6p_{\frac{1}{2}}\ 6p_{\frac{3}{2}}\ J{=}1$ transition in Pb has also been carried out giving $R = -15 \times 10^{-8}$.

PSUEDO-RELATIVISTIC CALCULATIONS ON THE ELECTRON STRUCTURE AND SPECTRUM OF $PtCl_4^{2-}$

Edward A. Boudreaux

Department of Chemistry
University of New Orleans
New Orleans, Louisiana, 70148, U.S.A.

The Self-Consistent Modified Extended Hückel (SC-MEH) molecular orbital method, a non-parameterized "atoms-in-molecules" computational technique, which may incorporate atomic input data either semi-empirically or via free atom HF-SCF calculations, has been shown to be capable of accurately calculating electronic properties of first row transition metal halide complexes and the Xenon Flourides[1-3]. This report describes an SC-MEH calculation on $PtCl_4^{2-}$ in both non-relativistic and relativistic formalisms.

Two types of calculations were made, one in which the relativistic splitting on the Pt-5s,5p,5d,6s,6p, levels was treated as a perturbation on the initial M.O. eigenvalues, and a second whereby complete diagonalization of the H and S matrices was carried out using relativistic basis functions. Although the results obtained with these two approaches are similar, in that the filled M.O.'s are numbered 55 thru 32, there are quantitative differences in the net orbital energies and A-O characteristics. It was further observed that a well defined minimum in the curve for the potential surface at the observed inter nuclear distance is not particularly sensitive to the specific details of the relativistic influences, so long as they are included in at least some average way. However, only a 5s,5p,5d,6s,6p,-Pt basis produces good results, as those obtaining only 5d,6s,5p,-Pt basis were quantitatively poor.

REFERENCES

1. A. Dutta-Ahmed & E.A. Boudreaux, Inorg. Chem. 12, 1597 (1973).
2. L.E. Harris & E.A. Boudreaux, Inorg. Chim. Acta. 9, 245 (1974).
3. E.A. Boudreaux, E.S. Elder & L.E. Harris, Proc. 11th Int. Coord. Chem. Conf. Heifa, Jerusalem, 1968, p. 536.

ATOMIC STRUCTURE CALCULATIONS USING THE RELATIVISTIC RANDOM-PHASE APPROXIMATION

K. T. Cheng

Argonne National Laboratory*
Argonne, Illinois 60439

The relativistic random-phase approximation (RRPA) is an approximate relativistic many-body theory designed to treat atomic transitions in atoms and highly stripped ions where both relativity and correlation are important. The governing equations can be derived from the time-dependent Hartree-Fock theory. The RRPA has been applied to the studies of discrete excitation and photoionization processes. The multichannel RRPA treatment of photoionization is described here, and selected examples of the RRPA calculations are given on total cross sections, branching ratios, angular distributions, spin polarizations and autoionization resonances.

*Work supported by the U. S. Department of Energy, Office of Basic Energy Sciences under Contract W-31-109-Eng-38.

THE SPINNING ELECTRON

Jens Peder Dahl

Department of Chemical Physics
Technical University of Denmark
DTH 301, DK-2800 Lyngby, Denmark

The properties of the 3-dimensional quantum mechanical rotor are reviewed, and the conditions studied under which it is possible to construct a local relativistic dynamics in the so-called instant form.[1] Let the position of the rotor in space and time be given by the coordinates (x_1, x_2, x_3, ict), and the orientation of the internal orthogonal axis system $(\underline{e}_1, \underline{e}_2, \underline{e}_3)$ by the Euler angles (α, β, γ). The spin operators (s_1, s_2, s_3) which generate rotations of the internal system, will then have the form

$$s_1 = i\hbar(\sin\alpha \frac{\partial}{\partial\beta} + \cot\beta\cos\alpha\frac{\partial}{\partial\alpha} - \frac{\cos\alpha}{\sin\beta}\frac{\partial}{\partial\gamma})$$

$$s_2 = i\hbar(-\cos\alpha \frac{\partial}{\partial\beta} + \cot\beta\sin\alpha \frac{\partial}{\partial\alpha} - \frac{\sin\alpha}{\sin\beta}\frac{\partial}{\partial\gamma})$$

$$s_3 = -i\hbar \frac{\partial}{\partial\alpha}$$

Together with the operators $\zeta_i = \underline{s}\cdot\underline{e}_i$ $(i=1,2,3)$, with which they commute, they form an O(4) algebra. The common eigenfunctions of $s^2 = s_1^2+s_2^2+s_3^2 = \zeta_1^2+\zeta_2^2+\zeta_3^2$, of s_3 and of ζ_3 are $D^s_{mn}(\alpha,\beta,\gamma)$, with $s = 0,\frac{1}{2},1,\ldots$ We prove that relativity requires that the probability amplitude $\Psi(x_1,x_2,x_3,\alpha,\beta,\gamma,t)$ for the rotor be built over the $s = \frac{1}{2}$ spin space, and that Ψ must satisfy an equation which is equivalent to the Dirac equation. The properties of this equation and its solutions are then reconsidered in the new perspective, with special emphasis on the symmetry of the problem. We conclude that a Dirac particle is neither more nor less than a particle for which it is possible to talk about an orientation in space.[2]

REFERENCES

1. P.A.M. Dirac, Rev.Mod.Phys. 21: 392 (1949).
2. J.P. Dahl, Mat.Fys.Medd.Dan.Vid.Selsk. 39, no. 12 (1977).

RELATIVISTIC EFFECTS ON EXCITATION

ENERGY OF MOLECULES

Ernest R. Davidson

Department of Chemistry BG-10
University of Washington
Seattle, Washington 98195

Fock-Dirac calculations with discrete basis sets tend to collapse into negative energy states. Coupled Hartree-Fock and variational calculations with the Breit-Pauli hamiltonian are also unstable. Hence, it is desirable to use first-order perturbation theory. The accuracy of first-order theory for atoms has recently been demonstrated by E.R. Davidson et al.[1] For atoms lighter than neon, the error is only 0.2% of the total relativistic correction, while for argon the error is 1%.

Relativistic corrections can have an important chemical affect on promotion energies. The s → p excitation energy of carbon has a 100 cm^{-1}. This causes a 15 cm^{-1} correction to the $^3B_1$ - $^1A_1$ energy gap of CH_2[2] and a 28 cm^{-1} correction to the excitation energy of CH_2O[3].

It is expected that the d → s excitation energies of transition metals could be affected by several tenths of an eV.

REFERENCES:

1. E.R. Davidson, Y. Ishikawa, and G. Malli, Chem. Phys. Lett. in press.
2. E.R. Davidson, D. Feller, and P. Phillips, Chem. Phys. Lett. 76, 416 (1980).
3. P. Phillips and E.R. Davidson, Chem. Phys. Lett. 78, 230 (1981).

SATELLITE STRUCTURE ACCOMPANYING PHOTOIONIZATION OF TRANSITION METAL COMPOUNDS AND ADSORBATES

Hans-Joachim Freund

Lehrstuhl für Theoretische Chemie
Universität zu Köln
Köln, West-Germany

ABSTRACT

The ionization of CO complexes and CO adsorbates is known to be accompanied by strong satellite structure due to intensity borrowing by excited states,i.e. two-particle-hole states derived from the hole state under consideration. This process is driven by electron-relaxation of the remaining electrons in the system leading to a screened hole state. For core ionizations the redistribution of electrons upon ionization is particularly significant due to the localized character of the hole. The electron distribution in the ion is calculated self-consistantly and compared with the neutral molecular ground state distribution. This comparison allows a classification of the electron redistribution as "metal-to-CO","CO-to-metal" if electrons are transferred accordingly or as "local-CO" and "local-metal" if only part of the molecular moity is involved. The energy of those states, involving the transfer of electrons between the metal and the CO ligand is sensitive to the metal-CO bond strength in the neutral molecular ground state. The analogies between transition metal compounds and transition metal adsorbates in this respect is discussed in detail with emphasis on the question of how many metal atoms are sufficient to describe the ionization of an adsorbate.

Ref.: H.J. Freund,E.W. Plummer Phys.Rev. B23,4859(1981)
H.J. Freund,E.W.Plummer,W.R. Salaneck,R.W.Bigelow J.Chem.Phys. 75,4275 (1981)

RELATIVISTIC CORRECTION TO THE ENERGY CURVE FROM THE NUCLEUS-NUCLEUS INTERACTION IN DIATOMIC MOLECULES

Hans H. Grelland

Department of Chemistry
University of Oslo
P.O. Box 1033-Blindern, Oslo 3, Norway

We have obtained a formula for the relativistic contribution to the energy-versus-distance due to the nucleus-nucleus interaction in diatomic Born-Oppenheimer molecules.[1] It leads to a bond shortening of 0.007 A and an increase in the lowest vibration frequency from 165.2 cm^{-1} to 165.4 cm^{-1} for the Au_2 molecule.

The derivation of such a formula should be based on a proper Einsteinian treatment of the two-body problem. Several proposals on how to do this have been published.[2] The theory put forward by T. Aaberge[3-5] has certain advantages in our context. It is built constructively from very general assumptions, with a minimum of arbitrary hypotheses. For instance, it does not introduce the Lorenz-invariant time parameter as an additional postulate, but derives its existence from an analysis of the relativistic transformations.[3] Moreover, Aaberge has presented explicitly an exact solution to the relativistic quantum Coulomb problem[4] and to the classical Kepler problem[5] thus giving a basis for the present application.

REFERENCES

1. H.H. Grelland, J.Phys.B:Atom.Molec.Phys. 13(1980)L389.
2. L.P. Horwitz and F. Rohrlich, Phys.Rev.D. 24,6(1981)1528, and references therein.
3. T. Aaberge, Helv.Phys.Acta 48(1975)163;50(1977)917.
4. T. Aaberge, "The system of two electrically charged Einstein relativistic particles of spin 0". To be published.
5. T. Aaberge, Gen.Rel.Grav. 10,11(1979)897.

RELATIVISTIC DENSITY FUNCTIONAL THEORY

E.K.U. Gross and R.M. Dreizler

Institut für Theoretische Physik der Universität
D-6000 Frankfurt, Federal Republic of Germany

The total energy of a relativistic many Fermion system can be represented as a functional of the one-particle density alone. We construct this functional in a systematic fashion by means of a gradient expansion technique due to Kirzhnits[1] which has been applied so far only to nonrelativistic[1,2] or weakly relativistic[3] systems.

Applications[4,5] of these functionals are envisaged for atomic and molecular structure as well as heavy ion scattering problems.

REFERENCES

1. D.A. Kirzhnits, Field Theoretical Methods in Many Body Systems, Pergamon Press (Oxford 1967).
2. E.K.U. Gross and R.M. Dreizler, Z.Phys. A302, 103 (1981).
3. E.K.U. Gross and R.M. Dreizler, Phys.Lett. 81A, 447 (1981).
4. E.K.U. Gross and R.M. Dreizler, Phys.Rev. A20, 1798 (1979).
5. A. Toepfer, E.K.U. Gross and R.M. Dreizler, Phys.Rev. A20, 1808 (1979); Z.Phys. A298, 167 (1980).

QUASI-RELATIVISTIC MODEL POTENTIAL METHOD

M. Klobukowski and S. Huzinaga

Department of Chemistry, University of Alberta
Edmonton, Alberta, CANADA

The model potential method of Bonifacic and Huzinaga has been improved by careful determination of the model potential parameters and the orbital basis functions. Calculations on a series of molecules indicate that the model potential method is capable of giving results which agree very well with the results of all-electron calculations.

There have been strong indications that inclusion of some of the relativistic effects is necessary in studying the properties of molecules containing heavy atoms.

In the present work an approximate method of including some of the relativistic effects into the model potential method is proposed. The relativistic effects are introduced on the quasi-relativistic level based on the atomic relativistic Hartree-Fock method of Cowin and Griffin. The method has been successfully applied to Ag_2 and AgH. The molecular calculations were done with the program MOLECULE modified for the model potential calculations by O. Gropen and S. Huzinaga. Once the quasi-relativistic model potential is established the rest of the calculation is done in completely non-relativistic fashion. The spin-orbit effects have not been included here, but they may be treated later using the perturbational method.

REX: A RELATIVISTICALLY PARAMETERIZED MOLECULAR ORBITAL METHOD

Lawrence L. Lohr

Department of Chemistry
University of Michigan
Ann Arbor, MI 48109, USA

ABSTRACT

We have outlined[1] a relativistically parameterized version of extended Hückel theory (EHT) called REX. This method incorporates relativistic effects by its use of atomic orbital basis sets with an $|\ell sjm\rangle$ quantization and by its systematic parameterizations based on Desclaux' atomic relativistic Dirac-Fock (DF) and non-relativistic Hartree-Fock (HF) calculations. No spin-orbit hamiltonian need be specified, as the diagonal hamiltonian matrix elements in the $|\ell sjm\rangle$ basis are taken as the DF atomic orbital energies, and the off-diagonal elements are taken to be proportional to the product of the overlap matrix element in the $|\ell sjm\rangle$ basis and the arithmetic mean of the corresponding diagonal elements. In its simplist form REX employs single Slater atomic basis functions, with exponents ζ obtained by a fit to DF electron mean radii. The alternative use of ζ's, fit to HF mean radii, and of HF orbital energies provides the non-relativistic reference calculation. We have also developed a multi-ζ version of REX and presented[2] the results of its use in a study of orbital energies of Group IV tetrahalides and tetramethyls. In addition, we have made the REX FORTRAN program available for general use.[3] We have applied[4] the method to explore the effects of relativity upon molecular orbital energies and compositions in uranium compounds including UO_2^{2+}, $UO_2Cl_4^{2-}$, UF_6, UCl_6, UCl_4, $U(BH_4)_4$, and $U(C_8H_8)_2$. Reasonable estimates of molecular spin-orbit splittings are obtained. Results are also discussed[5] for the homonuclear clusters Ge_9^{4-}, Ge_9^{2-}, Sn_4^{2-}, Sn_5^{2-}, Sn_9^{4-}, Pb_5^{2-}, Pb_9^{4-}, Bi_9^{5+}, and the heteronuclear clusters $PbSn_4^{2-}$, $SnGe_8^{4-}$, $GeSn_8^{4-}$, $PbSn_8^{4-}$, and $TlSn_8^{5-}$. The results include binding energies per atom, rearrangement energies, and charge distributions. Significant differences are found for the heavier element clusters between the results obtained using the relativistic parameterization and those obtained using the non-relativistic parameterization.

REFERENCES

1. L. L. Lohr and P. Pyykkö, Relativistically parameterized extended Hückel theory, _Chem. Phys. Lett._ 62:333 (1979).
2. L. L. Lohr, M. Hotokka, and P. Pyykkö, Relativistically parameterized extended Hückel calculations. 2. Orbital energies of Group-IV tetrahalides and tetramethyls, _Int. J. Quantum Chem._ 18:347 (1980).
3. L. L. Lohr, M. Hotokka, and P. Pyykkö, REX: Relativistically parameterized extended Hückel program, _QCPE_ 12:387 (1980).
4. P. Pyykkö and L. L. Lohr, Relativistically parameterized extended Hückel calculations. 3. Structure and bonding for some compounds of uranium and other heavy elements, _Inorg. Chem._ 20:1950 (1981).
5. L. L. Lohr, Relativistically parameterized extended Hückel calculations. 5. Charged polyhedral clusters of germanium, tin, lead, and bismuth atoms, _Inorg. Chem._ 20:4229 (1981).

RELATIVISTIC EFFECTS IN THE INNER SHELL IONIZATION PROCESS

F. Bary Malik and J. Tsekrisme Ndefru

Department of Physics and Astronomy
Southern Illinois University
Carbondale, Illinois 62901

The relativistic effects for processes involving inner-shell electrons already become pronounced for elements around $Z \cong 12$. This is obvious because $\langle V^2 \rangle = V_H^2 Z^2 = 2.2 \times 10^6$ (m/s)Z^2 and for $Z \cong 12$, V is about 9% of the velocity of light. It is clear also from the fact that the expectation value of the Breit operator taken with respect to Löwdin type of wave functions is $\cong -(\alpha^2/4)Z^4 (1 - 1/(2Z))$. For the inner-shell ionization process we expect therefore, the relativistic effect to be significant already for the K-shell ionization of Si or ionization of ions having an effective charge around $Z \cong 14$. This is indeed found in the detailed calculation of the K and L shell ionization elements by Scofield, Ndefru, and Malik in the plane wave Born approximation. Not only is it important to use the Dirac equation to describe the motion of the involved electrons, it is necessary to use the Møller operator to describe the interaction between them. Because of the relativistic effects both the energy and the Z dependence of the cross section deviates sharply from the non-relativistic case used widely in many calculations such as stopping power and in rate equations to determine temperature of plasmas or stars.

RELATIVISTIC HARTREE-FOCK CALCULATIONS

WITHIN SCALAR BASIS SETS

Franz Mark

Max-Planck-Institut für Strahlenchemie
Stiftstr. 34-36
D-4330 Mülheim a.d. Ruhr

In relativistic calculations on molecules problems arise when employing the linear expansion technique. These problems such as basis sensitivity, the need for large basis sets, appearance of non-physical solutions and numerical instability of the SCF process are due to the basis set approximation of the relativistic kinetic energy operator $\vec{\sigma}\vec{p}$. The first two problems are substantially diminished by introducing quasi-nonrelativistic eigenvalue equations. These are obtained from the Dirac-Fock equations within the context of the elimination method by the limiting process $c \to \infty$ (F. Mark and F. Rosicky). All above mentioned problems are resolved by using a new representation of $\vec{\sigma}\vec{p}$ (F. Mark and W.H.E. Schwarz). This new representation

$$[\vec{\sigma}\vec{p}]_{mod}=[\vec{\sigma}\vec{p}]\{[\vec{\sigma}\vec{p}]^{-1}[\vec{p}^2][\vec{\sigma}\vec{p}]^{-1}\}^{1/2}$$

is obtained by imposing two conditions. First, the representation of the nonrelativistic kinetic energy operator be reproduced:

$$[\vec{\sigma}\vec{p}]_{mod}[\vec{\sigma}\vec{p}]^{+}_{mod}=[\vec{p}^2].$$

Second, the norm

$$||[\vec{\sigma}\vec{p}]^{-1}\{[\vec{\sigma}\vec{p}]_{mod}-[\vec{\sigma}\vec{p}]\}||$$

be minimized. Due to the first condition, for any basis for which $[\vec{\sigma}\vec{p}]$ is nonsingular the relativistic energy converges to the nonrelativistic one for $c \to \infty$. The results of calculations using Gaussian lobe functions for the He-like systems of nuclear charge Z = 2,50 and 90 and for the molecules H_2, Li_2, LiH and F_2 demonstrate the power and scope of the approach.

ENERGY GRADIENT WITH THE EFFECTIVE

CORE POTENTIAL APPROXIMATION

Keiji Morokuma

Institute for Molecular Science
Myodaiji, Okazaki 444, Japan

The efficiency of the energy gradient technique in the ab initio MO method has been established by now for calculating equilibrium and transition state geometries, the reaction coordinate and vibrational frequencies of polyatomic molecules. The direct application of the gradient method, however, to systems containing heavy atoms has been very limited, because the method still requires too much computer time. The use of the effective core potential approximation (ECPA) will enable one to circumvent this practical difficulty and to open a way of applying the energy gradient method to systems such as transition metal complexes. The relativistic ECPA can be easily implemented.

For the energy gradient with the model hamiltonian, one needs the matrix elements of the derivative of the ECP with respect to the nuclear coordinates, in addition to the matrix elements required for usual gradient calculations. The derivative of the ECP can be complicated, but its matrix elements can be easily computed, without calculating the derivative of the ECP, by using the translational invariance of the matrix element

$$\langle\chi_\alpha|\partial V_\gamma^{ECP}/\partial R_\gamma|\chi_\beta\rangle = -(\langle\partial\chi_\alpha/\partial R_\alpha|V_\gamma^{ECP}|\chi_\beta\rangle+\langle\chi_\alpha|V_\gamma^{ECP}|\partial\chi_\beta/\partial R_\beta\rangle).$$

We have completed the program for this purpose[1] and applied it to the transition state of oxidative addition reaction[2]:

$$Pt(PH_3)_2 + H_2 \rightarrow Pt(H)_2(PH_3)_2$$

REFERENCES

1. K. Kitaura, S. Obara, and K. Morokuma, Chem. Phys. Lett. 77, 452 (1981).
2. K. Kitaura, S. Obara and K. Morokuma, J. Am. Chem. Soc. 103, 289 (1981).

ON INTERELECTRONIC MAGNETIC AND RETARDATION EFFECTS WITHIN RELATIVISTIC HARTREE-FOCK THEORY

Franz Rosicky

Theoretische Festkörperphysik, FB 1o, Universität
Duisburg GH, D-41oo Duisburg, Germany

Two different forms and uses of interelectronic beyond-Coulomb interactions are discussed[1] within a version of the molecular analytical relativistic Hartree-Fock formalism[2,3].

In the first approach the Breit-Hamiltonian B at the whole is taken into account by the use of first order perturbation theory. The magnetic part of B is expressed in terms of the Coulomb- and exchange integrals well known from the non-relativistic LCAO-method. Thus only the retardation part of B leads to new integrals which, using Gaussian functions for the basis set, are evaluated explicitly.

The second approach is to include the magnetic interaction only into the SCF-procedure[4] which results in new matrix Hartree-Fock equations. Finally, following the lines of our formalism[2,3], these equations are recast into super-matrix form.

REFERENCES

1. F. Rosicky, Chem. Phys. Letters 85: 195 (1982)
2. F. Mark, H. Lischka and F. Rosicky, Chem. Phys. Letters 71: 5o7 (198o)
3. F. Mark and F. Rosicky, Chem. Phys. Letters 74: 562 (198o)
4. W. Buchmüller, Phys. Rev. A 18: 1784 (1978)

ON PROBLEMS IN RELATIVISTIC ALL-ELECTRON AND PSEUDOPOTENTIAL LCAO-MO APPROACHES

W.H. Eugen Schwarz

Theoretical Chemistry
The University
D-5900 Siegen 21, Germany

ABSTRACT

The problems connected with ab initio relativistic LCAO-MO-SCF calculations are discussed. They are mainly due to mixing in of low kinetic energy states and not of negative energy states[1]. Different strategies for improvement are compared. The method of the 'improved representation of the momentum operator' is recommended, and numerical results are presented[2].

Relativistic pseudopotentials may be derived by applying the Foldy-Wouthuysen transformation before or after the Philips-Kleinman transformation of the Hamiltonian. The problems connected with both strategies are pointed out. It is noted that the Kahn-Pitzer[3] method which is free from most problems can be put on a sound theoretical basis by applying the FW transformation to the relativistic atomic orbitals[4].

REFERENCES

1. W.H.E. Schwarz and H. Wallmeier, Mol.Phys. in press 1982; W.H.E. Schwarz and E. Wechsel-Trakowski, Chem. Phys. Let. 85 (1982) 94
2. F. Mark and W.H.E. Schwarz, Phys. Rev. Let. submitted 1981; J. Chem. Phys. to be submitted 1982; F. Mark, this volume
3. K.S. Pitzer, this volume
4. W.H.E. Schwarz, Mol. Phys. to be submitted 1982

RELATIVISTIC CALCULATIONS OF SUPER HEAVY QUASI-MOLECULES

W.-D. Sepp and B. Fricke

Physics Department, University of Kassel
D-35oo Kassel, Federal Republic of Germany

Correlation diagrams play an important role in the interpretation of electronic inner shell excitation processes in near adiabatic heavy ion collisions. Two main conditions must be met in the calculation of super heavy quasi-molecules: (i) Since inner shell electrons of heavy atomic systems are involved in the excitation process the problem has to be solved fully relativisticly; (ii) Since the electronic clouds of the two colliding ions (or atoms) strongly overlap all electrons have to be included in order to get the correct electronic screening.

The Dirac-Fock-Slater (DFS) molecular electronic wavefunctions are expanded in (numerically calculated) atomic DFS-wavefunctions of the colliding partners. The advantages of this approach are threefold: (i) The basis states are physically adapted to the quasi-molecular problem; (ii) The basis states are practically orthogonal to the molecular negative energy solutions so we have no 'Brown-Ravenhall' type of disease; (iii) Only a limited number of basis states (typically 5o AO's for a heavy system like Pb-Pb) are needed. Thus the dimensions of the Overlap and Fock matrices are relatively small. The calculation of the Overlap and Fock matrices is done with the Discrete Variational Method (DVM)[1].

Various heavy systems have been calculated[2].

REFERENCES

1. See for the general method: A. Rosén and D.E. Ellis, 1975, J.Chem.Phys. 62, 3o39
2. See e.g.: T. Morović, W.-D. Sepp, and B. Fricke, 1982, Z.Phys. A3o4, 79

RELATIVISTIC EFFECTS IN RANDOMLY DISORDERED BINARY ALLOYS

Julie Staunton*, B.L. Gyorffy*, and P. Weinberger**

* H.H. Wills Physics Lab., University of Bristol,Bristol
**Institut für Technische Elektrochemie, Technische
Wien, Austria

There are many disordered alloys with interesting electronic properties which vary smoothly with concentration, c. For a randomly, disordered binary alloy, A_cB_{1-c}, a single electron is considered to move through a lattice with sites randomly occupied by effective A- and B-type potentials, which include many-body effects as well as the ionic potentials. The averaged, electronic properties are calculated from the configurationally averaged, single particle Green Function, $\langle G_\gamma \rangle$. The Coherent Potential Approximation (CPA) is the best, single site approximation to estimate $\langle G_\gamma \rangle$. By modelling the random alloy by a lattice of randomly arranged, spherically symmetric, muffin-tin, effective A and B potentials and, thus, using multiple scattering theory, the KKR-CPA method (Korringa-Kohn-Rostoker method within the CPA) has been successful, over the past few years, in describing a wide range of averaged electronic properties of several transition metal random alloys in a "first principles" fashion. When an alloy contains atomic constituents of high atomic number (Z > 50), the averaged, electronic structure of the alloy must include relativistic effects. The relativistic extension to the KKR-CPA method is developed (1) and then the method is demonstrated by describing RKKR-CPA calculations for several Gold-Platinum and Nickel-Platinum alloys. Interesting, new, relativistic alloying effects emerge in both of these systems, particularly when the averaged, electronic structures are analysed in wave-vector and energy-dependent detail.

REFERENCES

1 J. Staunton, B.L. Gyorffy, and P.Weinberger, J. Phys. F (Metal Physics) 10:2665 (1980).

ATOMIC LEVEL STRUCTURE AND TRANSITION PROBABILITIES

IN SUPERSTRONG MAGNETIC FIELDS *)

Günter Wunner

Institut für Theoretische Physik
Universität Erlangen-Nürnberg
D-8520 Erlangen, West Germany

In recent years the study of the properties of matter in superstrong magnetic fields ($B \gtrsim 10^4 T$) has greatly been stimulated by the discovery of such fields in the atmospheres of white dwarfs ($B \sim 10^3 - 10^5 T$) and in the surroundings of neutron stars ($B \sim 10^6 - 10^9 T$). In these fields, the magnetic forces are comparable with, or larger than, the Coulomb binding forces. Consequently all atomic structure is drastically changed -atoms become needle-shaped and cross sections highly anisotropic-, and all relevant atomic quantities have to be calculated anew. In this context, much work has been spent on the nonrelativistic treatment of energy levels[1] and electromagnetic transitions[2] of the hydrogenic sequence. Although relativistic effects had been expected to become important[3] when the characteristic (Larmor) length of the motion perpendicular to B in quantized Landau orbits becomes smaller than the Compton wave length, i.e. $a_L = (2\hbar/eB)^{1/2} \leq \lambda\!\!\!^{-}_C$, or $B \geq B_{cr} = 4.41 \cdot 10^9 T$, the solution of the Dirac equation for the H atom in strong magnetic fields[4] proves corrections to the energy levels to be negligible up to $B \sim 10^{10} - 10^{11} T$, the reason being that, although the motion perpendicular to the magnetic field indeed becomes relativistic, the motion parallel to B, which essentially gives rise to the Coulomb binding, still turns out to be fairly nonrelativistic even up to such fields.

REFERENCES

1. H.C. Praddaude, Phys.Rev.A, 6, 1321 (1972)
   J. Simola and J. Virtamo, J.Phys.B, 11, 3309 (1978)
2. G. Wunner and H. Ruder, Astrophys.J., 242, 828 (1980)
   G. Wunner, H. Ruder, and H. Herold, Astrophy.J., 247, 374 (1981)
3. M.L. Glasser and J.I. Kaplan, Phys.Lett.A, 53, 373 (1975)
4. K.A.U. Lindgren and J. Virtamo, J.Phys.B., 14, 3465 (1979)

*) Work supported in part by the Deutsche Forschungsgemeinschaft.

PARTICIPANTS

DIRECTOR:

PROFESSOR G.L. MALLI, Department of Chemistry and Theoretical Sciences Institute, Simon Fraser University, Burnaby, B.C. V5A 1S6, CANADA.

EXECUTIVE COMMITTEE:

DR. J.P. DESCLAUX, Centre D'Etudes Nucleaires de Grenoble, DRF/ Laboratoire d'Interactions Hyperfines, 85X-38041 Grenoble Cedex, FRANCE.
DR. I.P. GRANT, Theoretical Chemistry Department, University of Oxford, 1 South Parks Road, Oxford OX1 3TG, UNITED KINGDOM.
PROFESSOR J. LADIK, Institut fur Physikalische und Theoretische Chemie, Universitat Erlangen Nurnberg, D-8520 Erlangen, Egerlandstrase 3, FEDERAL REPUBLIC OF GERMANY.
PROFESSOR G.L. MALLI, Department of Chemistry and Theoretical Sciences Institute, Simon Fraser University, Burnaby, B.C. V5A 1S6, CANADA.
PROFESSOR K.S. PITZER, Department of Chemistry, University of California, Berkeley, California 94720, U.S.A.
PROFESSOR C.C.J. ROOTHAAN, Departments of Physics and Chemistry, University of Chicago, Chicago, Illinois 60637, U.S.A.

LECTURERS:

DR. J.P. DESCLAUX, Centre D'Etudes Nucleaires de Grenoble, DRF/ Laboratoire d'Interactions Hyperfines, 85X-38041 Grenoble Cedex, FRANCE.
DR. I.P. GRANT, Theoretical Chemistry Department, University of Oxford, 1 South Parks Road, Oxford OX1 3TG, UNITED KINGDOM.
DR. P.J. HAY, University of California, Los Alamos Scientific Laboratory, P.O. Box 1663, Los Alamos, New Mexico 87545, U.S.A.
DR. D.D. KOELLING, Argonne National Laboratory, 9700 S. Cass Avenue, Argonne, Illinois 60439, U.S.A.
PROFESSOR J. LADIK, Institut fur Physikalische und Theoretische Chemie, Universitat Erlangen Nurnberg, D-8520 Erlangen, Egerlandstrase 3, FEDERAL REPUBLIC OF GERMANY.
DR. A.H. MacDONALD, National Research Council of Canada, Ottawa K1A 0R6, CANADA.
PROFESSOR G.L. MALLI, Department of Chemistry and Theoretical Sciences Institute, Simon Fraser University, Burnaby, B.C. V5A 1S6, CANADA.

PROFESSOR K.S. PITZER, Department of Chemistry, University of California, Berkeley, California 94720, U.S.A.
DR. N.C. PYPER, Department of Theoretical Chemistry, Cambridge University, Cambridge, England, UNITED KINGDOM.
PROFESSOR C.C.J. ROOTHAAN, Departments of Physics and Chemistry, University of Chicago, Chicago, Illinois 60637, U.S.A.
PROFESSOR J. SUCHER, Department of Physics and Astronomy, University of Maryland, College Park, Maryland 20742, U.S.A.
DR. C.Y. YANG, Surface Analytic Research Inc., 465 A Fairchild Drive, Suite 128, Mountain View, California 94043, U.S.A.
DR. T. ZIEGLER, Department of Chemistry, University of Calgary, 2500 University Dr. N.W., Calgary, Alberta, CANADA.

STUDENTS:
P.J.C. AERTS, University of Groningen, Chemisch Laboratorium, Nyenborgh 16, 9747A Groningen, THE NETHERLANDS.
G. ASPROMALLIS, Theoretical and Physical Chemistry Institute, E.I.E. 48, Vas. Constantinou Avenue, Athens 501/1, GREECE.
C.K. AU, Department of Physics and Astronomy, University of South Carolina, Columbia, South Carolina 29208, U.S.A.
E. BAERENDS, Sheikundig Laboratorium, Vrije Universiteit, 1081 HV Amsterdam, THE NETHERLANDS.
K. BALASUBRAMANIAN, Department of Chemistry, University of California, Berkeley, California 94720, U.S.A.
S. BANNA, Vanderbilt University, Department of Chemistry, Nashville, Tennessee 37235, U.S.A.
N.A. BAYKARA, The Scientific and Technical Research Council of Turkey, Marmara Research Institute, P.O. Box 141, Kadikoy, Instanbul, TURKEY.
D.R. BECK, Department of Physics, Michigan Technical University, Houghton, Michigan 49931, U.S.A.
B. BERGERSEN, Department of Physics, University of British Columbia, Vancouver, British Columbia, CANADA.
R. BINNING, Hiram College, Box 1926, Hiram, Ohio 44234, U.S.A.
E.A. BOUDREAUX, Department of Chemistry, University of New Orleans, Lake Front, New Orleans, Louisiana 70122, U.S.A.
P. CEYZERIAT, Labo. S.I.M. Faculte des Sciences, 43 Bd du 11 Nov. 1918, 69622 Villeurbanne, FRANCE.
M. CHEN, Department of Physics, University of Oregon, Eugene, Oregon 97403, U.S.A.
K.T. CHENG, Argonne National Laboratory, 9700 South Cass Avenue, Argonne, Illinois 60439, U.S.A.
H. CHERMETTE, Institut de Phys. Nucleaire, Universite C. Bernard Layon I, 43 Bd. 11/11/1918-F69622, Villeurbanne, FRANCE.
E. CLEMENTI, I.B.M., P.O. Box 390, Bldg. 703-1, Poughkeepsie, New York 12602, U.S.A.
I. CSIZMADIA, Department of Chemistry, University of Toronto, Toronto, Ontario M5S 1A1, CANADA
J.P. DAHL, Chemistry Department B, Chemical Physics, The Technical University of Denmark, DTH 301, DK-2800, Lyngby, DENMARK.

E.R. DAVIDSON, Department of Chemistry, University of Washington, Seattle, Washington 98195, U.S.A.
T.W. DINGLE, Department of Chemistry, University of Victoria, Box 1700, Victoria, British Columbia V8W 2Y2, CANADA.
H. ERNST-CHR, Department of Chemistry and Chemical Engineering, University of Saskatchewan, Saskatoon, Saskatchewan S7N OWO, CANADA.
H. FREUND, Lehrstuhl f. Theor. Chemie, Greinstr. 4, 5000 Koln 41, FEDERAL REPUBLIC OF GERMANY.
P.H.N. FRIES, D.R.F./L.I.H., Centre d'Etudes Nucleaires de Grenoble, 85X, F38041, Grenoble, Cedex, FRANCE.
H.H. GRELLAND, Department of Chemistry, University of Oslo, Blindern, Oslo 3, NORWAY.
E.K.U. GROSS, Institut für Theoretische, Physik der Universitaet, D-6000 Frankfurt, Robert-Mayer-Str.8, FEDERAL REPUBLIC OF GERMANY.
K. HELFRICH, I.N. Stranski-Institut, Technische Universitat Berlin, Ernst Reuter-Platz 7, Telefunken-Haus 15 O.G., D-1000 Berlin 10, FEDERAL REPUBLIC OF GERMANY.
P.S. HERMAN, Department of Chemistry, University of Iowa, Iowa City, Iowa 52242, U.S.A.
S. HUZINAGA, Department of Chemistry, University of Alberta, Edmonton, Alberta, CANADA.
Y. ISHIKAWA, Department of Chemistry, Simon Fraser University, Burnaby, British Columbia V5A 1S6, CANADA
H.W. JONES, Department of Physics, Florida A & M University, College of Science and Technology, Florida 32307, U.S.A.
K.R. KARIM, Department of Physics, College of Arts and Sciences, 122 Science 1, University of Oregon, Eugene, Oregon 97403, U.S.A.
J. KELLER, Fac. de Quimica, Universidad Nacional Autonoma de Mexico, Apartado 70-528, Ciudad Universitaria, 04510 MEXICO, D.F.
L. LAAKSONEN, Abo Akademi, Department of Physical Chemistry, SF-20 500 Abo 50, Port Hansgatan 3-5, FINLAND.
W.G. LAIDLAW, Department of Chemistry, University of Calgary, Calgary, Alberta T2N 1N4, CANADA
W.A. LESTER, JR., Lawrence Berkeley Laboratory, NRCC, Bldg. 50D, Berkeley, California 94720, U.S.A.
L.L. LOHR, Department of Chemistry, University of Michigan, Ann Arbor, Michigan 48109, U.S.A.
F.B. MALIK, Department of Physics, Southern Illinois University, Carbondale, Illinois 62901, U.S.A.
G. MARCONI, Laboratorio Frae-CNR, Via Castagnoli 1, 40126 Bologna, ITALY.
F. MARK, Institut fur Strahlenchemie, Im Max-Planck-Institut fur Kohlenforschung, Striftstrasse 34-36, D-4330 Mulheim a.d. Ruhr 1, FEDERAL REPUBLIC OF GERMANY.
J. MARUANI, C.M.O.A. due C.N.R.S., 23 Rue du Maroc, 75019 Paris, FRANCE.
P. MOHR, Gibbs Laboratory, Physics Department, Yale University, New Haven, Connecticut 06520, U.S.A.
K. MOROKUMA, Institute for Molecular Science, Myodaiji, Okazaki 444, JAPAN.

C. NICOLAIDES, Theoretical and Physical Chem. Inst., E.I.E. 48, Vas. Constantinou Avenue, Athens 501/1, GREECE.
P. PHILLIPS, Department of Chemistry, University of Washington, Seattle, Washington 98185, U.S.A.
M. PLISCHKE, Physics Department, Simon Fraser University, Burnaby, British Columbia V5A 1S6, CANADA.
A. RAUK, Department of Chemistry, University of Calgary, Calgary, Alberta T2N 1N4, CANADA.
F. ROSICKY, FB10 Theoretische Physik, Universitat Duisburg, D-4100 Duisburg 1, FEDERAL REPUBLIC OF GERMANY.
D. SALAHUB, Departement de Chimie, Universite de Montreal, C.P. 6210, Succ. A. Montreal, Quebec H3C 3V1, CANADA.
P.G.H. SANDARS, Clarendon Laboratory, Oxford, England, UNITED KINGDOM.
D. SCHUCH, Institut fur Physikalische und Theoretische Chemie, J.W. Goethe Universitat, Frankfurt, Robert-Mayer Str. 11, D-6000 Frankfurt Main, FEDERAL REPUBLIC OF GERMANY.
W.H.E. SCHWARZ, Theoretical Chemistry, University of Siegen, POB 210209, D-5900 Siegen 21, FEDERAL REPUBLIC OF GERMANY.
W.D. SEPP, Gesamthochschule Kassel, Fachbereich Physik, Heinrich-Plett-Str. 40, 3500 Kassel, FEDERAL REPUBLIC OF GERMANY.
D.G. SHANKLAND, AFIT/ENP, Wright-Patterson Air Force Base, Ohio 45433, U.S.A.
F. SOLLIEC, Maitre-Assistant, U.E.R. de Physique, Universite Paris 7, 75230 Paris, Cedex 05, FRANCE.
J. STAUNTON, H.H. Wills Physics Lab, University of Bristol, Royal Fort, Tyndall Avenue, Bristol BS8 1T1, UNITED KINGDOM.
O. STEINBORN, Institut fur Chemie, Universitat Regenburg Universitat-sstrase, 31, D-8400 Regensburg, FEDERAL REPUBLIC OF GERMANY.
J.S. TSE, Chemistry Division, National Research Council of Canada, MRL, M-12, Rm. 113, Ottawa, Ontario K1A 0R9, CANADA.
J.A. TUSZYNSKI, Physics Department, University of Calgary, Calgary, Alberta T2N 1N4, CANADA
W. VON NIESSEN, Institut fur Physikalische Chemie, Technische Universitat Braunschweig, D-3300 Braunschweig, FEDERAL REPUBLIC OF GERMANY.
S.H. VOSKO, Department of Physics, University of Toronto, Toronto, Ontario M5S 1A7, CANADA.
H. WALLMEIER, Ruhr-Universitat Bochum, Atbeilung Chemie, Lehrstuhl fur Theoretische Chemie, 4630 Bochum-Querenburg, FEDERAL REPUBLIC OF GERMANY.
L. WILK, Department of Physics, University of Toronto, Toronto, Ontario M5S 1A7, CANADA.
G. WUNNER, Lehrstuhl fur Theoretische Physik, Universitat Erlangen-Nurnberg, Gluckstrasse 6, 8520 Erlangen, FEDERAL REPUBLIC OF GERMANY.
B. ZYGELMAN, City College of New York, Physics Department, Convent Avenue, 137 Street, New York City, New York 10025, U.S.A.

INDEX